Springer
Tokyo
Berlin
Heidelberg
New York
Barcelona
Hong Kong
London
Milan
Paris
Singapore

M. Kasahara (Ed.)

Major Histocompatibility Complex

Evolution, Structure, and Function

With 148 Figures, Including 5 in Color

 Springer

Masanori Kasahara, M.D., Ph.D.
Professor
Department of Biosystems Science, School of Advanced Sciences
The Graduate University for Advanced Studies
Hayama 240-0193, Japan

QR
184
.315
.M35x
2000

Cover art credit: The following DNA sequences, obtained from the National Center for Biotechnology Information (NCBI) GenBank database (Nucleic Acids Res 27: 12-17, 1999), were used in the cover design: GI 1644236, 176820, 22891, 2864659, 435020, 164004, 415004, 1644445, 992571, 4185618, 214593, 213822, 5231102, 2598201. The crystal structure of the mouse H2 molecule (MMDB ID 4627) was obtained from the MMDB databank (Trends Biochem Sci 22: 314-316, 1999) and visualized using the Cn3D program (Nucleic Acids Res 27: 240-243, 1999). The GenBank and MMDB databases and the Cn3D program are provided by NCBI at the NCBI website (http://www.ncbi.nlm.nih.gov/).

Cover design by Makoto Yawata

ISBN 4-431-70276-8 Springer-Verlag Tokyo Berlin Heidelberg New York

Printed on acid-free paper

Typesetting: Camera-ready by authors/editors
Printing and binding: Best-set Typesetter Ltd., Hong Kong
SPIN: 10728600

Preface

Every biological system is the outcome of evolution and has a history all its own. This history dictates how the system works and why it has certain properties and not others. This is why we need to study not only the structure and function, but also the history of the system. This argument undoubtedly applies to the study of the immune system and also to the study of the major histocompatibility complex (MHC).

Since 1989, researchers of various scientific disciplines who share a deep interest in MHC evolution have held a meeting every two years to discuss their latest research developments, exchange ideas, and foster friendship. Together with my colleagues Drs. Naoyuki Takahata and Yoko Satta, I organized the Sixth International Workshop on MHC Evolution in Hayama, Japan, May 25–29, 1999. This volume is the proceedings of that conference. It covers diverse topics pertinent to MHC evolution, including the origin of the adaptive immune system, the organization of the MHC in humans and other model vertebrates, MHC–parasite co-evolution, and the nature and origin of MHC polymorphism. I hope that this book will be of interest not only for MHC researchers and immunologists, but also for other specialists who are interested in the evolution of biological systems in general.

I would like to express my sincere gratitude to all my colleagues in the Department of Biosystems Science for their enthusiastic support in organizing the Workshop. I am particularly indebted to Dr. Makoto Yawata, Mr. Takashi Suzuki, and Ms. Atsuko Usami, whose assistance enabled the Workshop to proceed smoothly. Dr. Yawata and Mr. Suzuki also spent a great deal of effort in helping me edit this book. Without their dedicated help, it would have been impossible to publish this book in a timely manner. I also thank The Ministry of Education, Science, Sports and Culture (Monbusho) of Japan and The Graduate University for Advanced Studies for their generous financial support. Last but not least, I thank the staff of Springer-Verlag Tokyo for their assistance and cooperation.

Masanori Kasahara
Hayama, Japan
Autumn, 1999

Contents

1. Origin of the MHC

2. Genome organization of the MHC

3. Function and evolutionary dynamics of MHC genes

4. Natural killer gene complex

5. MHC-pathogen coevolution

6. MHC polymorphism

Contributors

Laurent Abi Rached, *Unité INSERM 119, Cancérologie Expérimentale, 27 bd Leï Roure, 13009 Marseille, France*

Marielle Afanassieff, *Department of Molecular Biology, Beckman Research Institute, City of Hope National Medical Center, Duarte, CA 91010, USA; INRA, Station de Pathologie Aviaire et Parasitologie, Nouzilly 37380, France*

Miguel Alvarez, *Department of Immunology and Molecular Biology, Hospital Universitario "12 de Octubre", Universidad Complutense de Madrid, Carretera de Andalucia s/n. 28041 Madrid, Spain*

Asako Ando, *Department of Genetic Information, Division of Molecular Life Science, Tokai University School of Medicine, Bohseidai, Isehara, Kanagawa-ken 259-1193, Japan*

Hitoshi Ando, *Department of Genetic Information, Division of Molecular Life Science, Tokai University School of Medicine, Bohseidai, Isehara, Kanagawa-ken 259-1193, Japan*

Antonio Arnaiz-Villena, *Department of Immunology and Molecular Biology, Hospital Universitario "12 de Octubre", Universidad Complutense de Madrid, Carretera de Andalucia s/n. 28041 Madrid, Spain*

Charles Auffray, *CNRS, UPR 420, Genetique Moleculaire et Biologie du developpement, Villejuif, France*

Keith T. Ballingall, *International Livestock Research Institute (ILRI), Box 30709, Nairobi, Kenya*

Dagmar Bauer, *Max von Pettenkofer-Institut, Lehrstuhl Virologie, Ludwig-Maximilians-Universität München, Pettenkoferstr. 9a, 80336 München, Germany*

Hans-Peter Beck, *Department of Medical Parasitology, Swiss Tropical Institute, Basel, Switzerland*

Stephan Beck, *The Sanger Centre, Wellcome Trust Genome Campus, Hinxton, Cambridge CB10 1SA, UK*

Ann B. Begovich, *Department of Human Genetics, Roche Molecular Systems, Alameda, CA, USA*

Tomas F. Bergström, *University of Washington Genome Center, University of Washington, Seattle, WA, USA*

Jemma Berry, *Centre for Molecular Immunology and Instrumentation, University of Western Australia, Nedlands, Western Australia 6907, Australia*

Rainer Blasczyk, *Department of Transfusion Medicine, Hannover Medical School, Carl-Neuberg-Str. 1, 30625 Hannover, Germany*

Jan Böhme, *Center of Biotechnology, NOVUM, University College of South Stockholm/Karolinska Institute, S-141 57, Huddinge, Sweden*

Barry Boettcher, *Department of Biological Sciences, The University of Newcastle, Newcastle, Australia*

W. Elwood Briles, *Department of Biological Sciences, Northern Illinois University, DeKalb, IL 60115, USA*

Michael G. Brown, *Washington University School of Medicine, St. Louis, MO 63110, USA*

Teodorica L. Bugawan, *Department of Human Genetics, Roche Molecular Systems, Alameda, CA, USA*

Patrick Carnegie, *Centre for Molecular Immunology and Instrumentation, University of Western Australia, Nedlands, Western Australia 6907, Australia*

Maria Jose Castro, *Department of Immunology and Molecular Biology, Hospital Universitario "12 de Octubre", Universidad Complutense de Madrid, Carretera de Andalucia s/n. 28041 Madrid, Spain*

Sonia Cattley, *Centre for Molecular Immunology and Instrumentation, University of Western Australia, Nedlands, Western Australia 6907, Australia*

Gareth Chelvanayagam, *Human Genetics Group, John Curtin School of Medical Research, The Australian National University, Canberra, Australia*

L. William Clem, *Department of Microbiology and Immunology, University of Mississippi Medical Center, Jackson, MS 39216, USA*

Françoise Coudert, *INRA, Station de Pathologie Aviaire et Parasitologie, Nouzilly 37380, France*

Bart Currie, *Menzies School of Health Research, Darwin, Australia*

Eric H. Davidson, *Division of Biology 156-29, California Institute of Technology, Pasadena, CA 91125, USA*

Roger L. Dawkins, *Centre for Molecular Immunology and Instrumentation, University of Western Australia, Nedlands, 6008, Western Australia, Australia*

Louis Du Pasquier, *Basel Institute for Immunology, Grenzacherstrasse 487, CH-4005 Basel, Switzerland*

Anke Ehlers, *Institut für Immungenetik, Universitätsklinikum Charité, Humboldt-Universität zu Berlin, Spandauer Damm 130, 14050 Berlin, Germany*

Shirley A. Ellis, *Institute for Animal Health, Compton, Newbury, RG20 7NN, UK*

Henry A. Erlich, *Department of Human Genetics, Roche Molecular Systems, Alameda, CA, USA; Children's Hospital Oakland Research Institute, Oakland, CA, USA*

David T. Evans, *New England Regional Primate Research Center, One Pine Hill Drive, Southborough, MA 01772-9102, USA*

Felipe Figueroa, *Max-Planck-Institut für Biologie, Abteilung Immungenetik, D-72076 Tübingen, Germany*

Martin F. Flajnik, *Department of Microbiology and Immunology, University of Maryland at Baltimore, Room 13-009, 655 West Baltimore Street, Baltimore, MD 21201, USA*

Michael Flores, *Department of Biology, N108 Howell Science Complex, East Carolina University, Greenville, NC 27858, USA*

Simon Forbes, *Department of Pathology, Division of Immunology, University of Cambridge, Tennis Court Road, Cambridge CB2 1QP, UK*

Tatsuo Fukagawa, *Division of Evolutionary Genetics, Department of Population Genetics, National Institute of Genetics, Mishima, Shizuoka-ken 411-8540, Japan*

Xiaojiang Gao, *IRSP, SAIC Frederick, NCI-FCRD, Frederick, MD, USA; Human Genetics Group, John Curtin School of Medical Research, The Australian National University, Canberra, Australia*

Silvana Gaudieri, *Centre for Molecular Immunology and Instrumentation, University of Western Australia, Nedlands, Western Australia 6907, Australia; Center for Information Biology, National Institute of Genetics, Mishima, Shizuoka-ken 411-8540, Japan*

Daniel E. Geraghty, *Human Immunogenetics Program, Fred Hutchinson Cancer Research Center, Seattle, USA*

Ulla B. Godwin, *Department of Biology, N108 Howell Science Complex, East Carolina University, Greenville, NC 27858, USA*

Takashi Gojobori, *Center for Information Biology, National Institute of Genetics, Mishima, Shizuoka-ken 411-8540, Japan*

Eduardo Gomez-Casado, *Department of Immunology and Molecular Biology, Hospital Universitario "12 de Octubre", Universidad Complutense de Madrid, Carretera de Andalucia s/n. 28041 Madrid, Spain*

Ronald M. Goto, *Department of Molecular Biology, Beckman Research Institute, City of Hope National Medical Center, Duarte, CA 91010, USA*

Sarah Grams, *Department of Human Genetics, Roche Molecular Systems, Alameda, CA, USA*

B. Rosemary Grant, *Department of Ecology and Evolutionary Biology, Princeton University, Princeton, NJ 08544-1003, USA*

Peter R. Grant, *Department of Ecology and Evolutionary Biology, Princeton University, Princeton, NJ 08544-1003, USA*

Eberhard Günther, *Division of Immunogenetics, University of Göttingen, Heinrich-Düker-Weg 12, D-37073 Göttingen, Germany*

Ulf Gyllensten, *Department of Genetics and Pathology, Rudbeck Laboratory, University of Uppsala, Uppsala, Sweden*

Jennifer Ha, *Department of Molecular Biology, Beckman Research Institute, City of Hope National Medical Center, Duarte, CA 91010, USA*

Kaori Habara, *Center for Information Biology, National Institute of Genetics, Mishima, Shizuoka-ken 411-8540, Japan*

Tina M. Hambuch, *Museum of Vertebrate Zoology and Departmen of Integrative Biology, University of California, Berkeley, CA 94720, USA*

Katsuko Hashiba, *Saitama Prefectural University, 820 Sannomiya, Koshigaya, Saitama 343-8540, Japan*

Hartmut Hengel, *Max von Pettenkofer-Institut, Lehrstuhl Virologie, Ludwig-Maximilians-Universität München, Pettenkoferstr. 9a, 80336 München, Germany*

Adrian V.S. Hill, *Institute of Molecular Medicine, University of Oxford, Oxford, UK*

Kari Högstrand, *Department of Immunology, Stockholm University, S-106 91 Stockholm, Sweden*

Jill Hollenbach, *Department of Integrative Biology, University of California, Berkeley, CA, USA*

Edward C. Holmes, *The Wellcome Trust Centre for the Epidemiology of Infectious Disease, Department of Zoology, University of Oxford, South Parks Road, Oxford OX1 3PS, UK*

Jennie Hui, *Centre for Molecular Immunology and Instrumentation, University of Western Australia, Nedlands, Western Australia 6907, Australia*

Toshimichi Ikemura, *Division of Evolutionary Genetics, Department of Population Genetics, National Institute of Genetics, Mishima, Shizuoka-ken 411-8540, Japan*

Hidetoshi Inoko, *Department of Genetic Information, Division of Molecular Life Science, Tokai University School of Medicine, Bohseidai, Isehara, Kanagawa-ken 259-1193, Japan*

Sofia Ioannidu, *Division of Immunogenetics, University of Göttingen, Heinrich-Düker-Weg 12, D-37073 Göttingen, Germany*

Yoshihide Ishikawa, *Japanese Red Cross Central Blood Center, Tokyo, Japan*

Takeo Juji, *Japanese Red Cross Central Blood Center, Tokyo, Japan*

Günther Jung, *Institut für Organische Chemie, Eberhard-Karls-Universität Tübingen, Auf der Morgenstelle 18, D-72076 Tübingen, Germany*

Shigehiko Kanaya, *Department of Electrical and Information Engineering, Faculty of Engineering, Yamagata University, Yonezawa, Yamagata-ken 992-8510, Japan; Division of Physiological Genetics, Department of Ontogenetics, National Institute of Genetics, Mishima, Shizuoka-ken 411-8540, Japan*

Masanori Kasahara, *Department of Biosystems Science, School of Advanced Sciences, Graduate University for Advanced Studies, Hayama 240-0193, Japan; CREST, Japan Science and Technology Corporation, Hayama 240-0193, Japan*

Kouichi Kashiwase, *Japanese Red Cross Central Blood Center, Tokyo, Japan*

Carol M. Kiekhaefer, *Wisconsin Regional Primate Research Center, 1220 Capitol Court, Madison, WI 53715-1299, USA*

Miyuki Kinebuchi, *Department of Pathology, School of Medicine, Sapporo Medical University South-1, West-17, Chuo-ku, Sapporo 060-8556, Japan*

Jan Klein, *Max-Planck-Institut für Biologie, Abteilung Immungenetik, Corrensstrasse 42, D-72076 Tübingen, Germany*

William Klitz, *School of Public Health, University of California, Berkeley, CA, USA*

Miki Komatsu-Wakui, *Department of Human Genetics, Graduate School of Medicine, University of Tokyo, Tokyo, Japan; Japanese Red Cross Central Blood Center, Tokyo, Japan*

Vladimir I. Konenkov, *Institute of Clinical Immunology, Novosibirsk, Russia*

Katja Kotsch, *Department of Transfusion Medicine, Hannover Medical School, Carl-Neuberg-Str. 1, 30625 Hannover, Germany*

Karin Kriener, *Max-Planck-Institut für Biologie, Abteilung Immungenetik, Corrensstrasse 42, D-72076 Tübingen, Germany*

Yoshihiro Kudo, *Department of Electrical and Information Engineering, Faculty of Engineering, Yamagata University, Yonezawa, Yamagata-ken 992-8510, Japan*

Jerzy K. Kulski, *Centre for Molecular Immunology and Instrumentation, University of Western Australia, Nedlands, 6008, Western Australia, Australia*

Eileen A. Lacey, *Museum of Vertebrate Zoology and Department of Integrative Biology, University of California, Berkeley, CA 94720, USA*

Sue Lester, *Red Cross Blood Transfusion Service, Adelaide, Australia*

Javier Longas, *Department of Immunology and Molecular Biology, Hospital Universitario "12 de Octubre", Universidad Complutense de Madrid, Carretera de Andalucia s/n. 28041 Madrid, Spain*

Natalie Longman, *Centre for Molecular Immunology and Instrumentation, University of Western Australia, Nedlands, Western Australia 6907, Australia*

Leslie Louie, *School of Public Health, University of California, Berkeley, CA, USA*

Ernesto Lowy, *Department of Immunology and Molecular Biology, Hospital Universitario "12 de Octubre", Universidad Complutense de Madrid, Carretera de Andalucia s/n. 28041 Madrid, Spain*

Anthony Luyai, *International Livestock Research Institute (ILRI), Box 30709, Nairobi, Kenya*

Steven J. Mack, *Children's Hospital Oakland Research Institute, Oakland, CA, USA; Department of Human Genetics, Roche Molecular Systems, Alameda, CA, USA*

Bernard Marasa, *International Livestock Research Institute (ILRI), Box 30709, Nairobi, Kenya*

Patricia Martinez, *Centre for Molecular Immunology and Instrumentation, University of Western Australia, Nedlands, Western Australia 6907, Australia*

Jorge Martinez-Laso, *Department of Immunology and Molecular Biology, Hospital Universitario "12 de Octubre", Universidad Complutense de Madrid, Carretera de Andalucia s/n. 28041 Madrid, Spain*

Akihiro Matsuura, *Department of Pathology, School of Medicine, Sapporo Medical University South-1, West-17, Chuo-ku, Sapporo 060-8556, Japan*

Werner E. Mayer, *Max-Planck-Institut für Biologie, Abteilung Immungenetik, Corrensstrasse 42, D-72076 Tübingen, Germany*

James McCluskey, *Red Cross Blood Transfusion Service, Adelaide, Australia*

Thomas J. McConnell, *Department of Biology, N108 Howell Science Complex, East Carolina University, Greenville, NC 27858, USA*

Declan J. McKeever, *International Livestock Research Institute (ILRI), Box 30709, Nairobi, Kenya*

Diogo Meyer, *Department of Integrative Biology, University of California, Berkeley, CA 94720-3140, USA*

Marcia M. Miller, *Department of Molecular Biology, Beckman Research Institute, City of Hope National Medical Center, Duarte, CA 91010, USA*

Norman W. Miller, *Department of Microbiology and Immunology, University of Mississippi Medical Center, Jackson, MS 39216, USA*

Shigeki Mitsunaga, *Japanese Red Cross Central Blood Center, 4-1-31 Hiroo, Shibuya-ku, Tokyo 150-0012, Japan*

Frank Momburg, *Abteilung für Molekulare Immunologie, Deutsches Krebsforschungszentrum, Im Neuenheimer Feld 280, 69120 Heidelberg, Germany*

Pablo Morales, *Department of Immunology and Molecular Biology, Hospital Universitario "12 de Octubre", Universidad Complutense de Madrid, Carretera de Andalucia s/n. 28041 Madrid, Spain*

W. Ivan Morrison, *Institute for Animal Health, Compton, Newbury, RG20 7NN, UK*

Kiyoshi Naruse, *Department of Biological Sciences, Graduate School of Science, University of Tokyo, Bunkyo-ku, Tokyo 113-0033, Japan*

Masaru Nonaka, *Department of Biological Sciences, Graduate School of Science, University of Tokyo, Bunkyo-ku, Tokyo 113-0033, Japan*

David H. O'Connor, *Wisconsin Regional Primate Research Center, 1220 Capitol Court, Madison, WI 53715-1299, USA*

Jun Ohashi, *Department of Human Genetics, Graduate School of Medicine, University of Tokyo, Tokyo, Japan*

Yuko Ohta, *Department of Microbiology and Immunology, University of Maryland at Baltimore, Room 13-009, 655 West Baltimore Street, Baltimore, MD 21201, USA*

Colm O'hUigin, *Max-Planck-Institut für Biologie, Abteilung Immungenetik, Corrensstrasse 42, D-72076 Tübingen, Germany*

Paola Oliveri, *Division of Biology 156-29, California Institute of Technology, Pasadena, CA 91125, USA*

Pierre Pontarotti, *Unité INSERM 119, Cancérologie Expérimentale, 27 bd Leï Roure, 13009 Marseille, France*

Sylvie Quiniou, *Department of Microbiology and Immunology, University of Mississippi Medical Center, Jackson, MS 39216, USA*

Jonathan P. Rast, *Division of Biology 156-29, California Institute of Technology, Pasadena, CA 91125, USA*

Olga Rickards, *Dipartimento di Biologia, Universita di Roma "Tor Vergata", Rome, Italy*

Ricardo Rojo, *Department of Immunology and Molecular Biology, Hospital Universitario "12 de Octubre", Universidad Complutense de Madrid, Carretera de Andalucia s/n. 28041 Madrid, Spain*

Isabel Rubio, *Department of Immunology and Molecular Biology, Hospital Universitario "12 de Octubre", Universidad Complutense de Madrid, Carretera de Andalucia s/n. 28041 Madrid, Spain*

Muhammad Saha, *Department of Paediatrics, National University of Singapore, Singapore*

Marina L. Sartakova, *Institute of Clinical Immunology, Novosibirsk, Russia*

Takehiko Sasazuki, *Department of Genetics, Medical Institute of Bioregulation, Kyushu University, Fukuoka, Japan*

Akie Sato, *Max-Planck-Institut für Biologie, Abteilung Immungenetik, Corrensstrasse 42, D-72076 Tübingen, Germany*

Yoko Satta, *Department of Biosystems Science, School of Advanced Sciences, Graduate University for Advanced Studies, Hayama, Kanagawa 240-0193, Japan*

Anthony A. Scalzo, *University of Western Australia, Nedlands, Western Australia 6907, Australia*

Jung Won Seo, *Division of Immunogenetics, University of Göttingen, Heinrich-Düker-Weg 12, D-37073 Göttingen, Germany*

Takashi Shiina, *Department of Genetic Information, Division of Molecular Life Science, Tokai University School of Medicine, Kanagawa, Japan*

Akihiro Shima, *Department of Biological Sciences, Graduate School of Science, University of Tokyo, Bunkyo-ku, Tokyo 113-0033, Japan; Department of Integrated Biosciences, Graduate School of Frontier Science, University of Tokyo, Bunkyo-ku, Tokyo 113-0033, Japan*

Naoki Shimbara, *Department of Biomedical R&D, Sumitomo Electric Industries, 1 Taya-cho, Sakae-ku, Yokohama 244, Japan*

Karen A. Staines, *Institute for Animal Health, Compton, Newbury, RG20 7NN, UK*

Lori L. Steiner, *Department of Human Genetics, Roche Molecular Systems, Alameda, CA, USA*

Mark Stoneking, *Max-Planck-Institute for Evolutionary Anthropology, Leipzig, Germany; Department of Anthropology, Pennsylvania State University, University Park, PA, USA*

Vina Suraj-Baker, *Department of Human Genetics, Roche Molecular Systems, Alameda, CA, USA; Department of Medicine, Tulane University School of Medicine, New Orleans, LA, USA*

Takashi Suzuki, *Department of Biosystems Science, School of Advanced Sciences, Graduate University for Advanced Studies, Hayama 240-0193, Japan*

Naoyuki Takahata, *Department of Biosystems Science, School of Advanced Sciences, Graduate University for Advanced Studies, Hayama, Kanagawa 240-0193, Japan*

Nobuyuki Tanahashi, *The Tokyo Metropolitan Institute of Medical Science, and CREST, Japan Science and Technology Corporation (JST), 18-22 Honkomagome, Bunkyo-ku 3-chome, Tokyo 113-8613, Japan*

Keiji Tanaka, *The Tokyo Metropolitan Institute of Medical Science, and CREST, Japan Science and Technology Corporation (JST), 18-22 Honkomagome, Bunkyo-ku 3-chome, Tokyo 113-8613, Japan*

Vincent P.K. Titanji, *University of Buea, Cameroon*

Katsushi Tokunaga, *Department of Human Genetics, School of International Health, Graduate School of Medicine, University of Tokyo, Hongo 7-3-1, Bunkyo-ku, Tokyo 113-0033, Japan*

Elizabeth Trachtenberg, *Children's Hospital Oakland Research Institute, Oakland, CA, USA*

John Trowsdale, *Department of Pathology, Division of Immunology, University of Cambridge, Tennis Court Road, Cambridge CB2 1QP, UK*

Barbara Uchanska-Ziegler, *Institut für Immungenetik, Universitätsklinikum Charité, Humboldt-Universität zu Berlin, Spandauer Damm 130, 14050 Berlin, Germany*

Keiko Udaka, *Department of Biophysics, Graduate School of Science, Kyoto University, Sakyo-ku, Kyoto 606-8502, Japan*

Anthony Veale, *National Centre for Epidemiology and Population Health, The Australian National University, Canberra, Australia*

Armin Volz, *Institut für Immungenetik, Universitätsklinikum Charité, Humboldt-Universität zu Berlin, Spandauer Damm 130, 14050 Berlin, Germany*

Lutz Walter, *Division of Immunogenetics, University of Göttingen, Heinrich-Düker-Weg 12, D-37073 Göttingen, Germany*

David I. Watkins, *Wisconsin Regional Primate Research Center, 1220 Capitol Court, Madison, WI 53715-1299, USA*

Karl-Heinz Wiesmüller, *Evotec BioSystems GmbH, Grandweg 64, D-22529 Hamburg, Germany*

Melanie R. Wilson, *Department of Microbiology and Immunology, University of Mississippi Medical Center, Jackson, MS 39216, USA*

Makoto Yawata, *Department of Biosystems Science, School of Advanced Sciences, Graduate University for Advanced Studies, Hayama 240-0193, Japan; First Department of Internal Medicine, Yokohama City University School of Medicine, Yokohama 236-0004, Japan*

Wayne M. Yokoyama, *Washington University School of Medicine, St. Louis, MO 63110, USA*

Ruth Younger, *The Sanger Centre, Wellcome Trust Genome Campus, Hinxton, Cambridge CB10 1SA, UK*

Andreas Ziegler, *Institut für Immungenetik, Universitätsklinikum Charité, Humboldt-Universität zu Berlin, Spandauer Damm 130, 14050 Berlin, Germany*

Peter A. Zimmerman, *Laboratory of Parasitic Diseases, National Institutes of Health, Bethesda, MD, USA*

Rima Zoorob, *CNRS, UPR 420, Genetique Moleculaire et Biologie du developpement, Villejuif, France*

1. Origin of the MHC

Jaws and AIS

Jan Klein, Akie Sato, and Werner E. Mayer

Max-Planck-Institut für Biologie, Abteilung Immungenetik, Corrensstrasse 42, D-72076 Tübingen, Germany

Summary. In considering the origin of the anticipatory (adaptive) immune system (AIS), we have developed several theses. We propose that the AIS emerged not in a big bang, but gradually, in a piecemeal fashion. Different components of the AIS emerged at different times. For example, rudimental antigen processing systems existed before the protostome-deuterostome split; lymphocytes may have appeared before the divergence of gnathostomes from the agnathans; the RAG-based diversification mechanism emerged close to the origin of gnathostomes; and so on. Whether living agnathans possess *Mhc, Tcr*, and *Ig* genes in some form remains an open, but answerable question. Non-vertebrates have not evolved the AIS because this option is not included in their developmental program. (Jawed) vertebrates evolved the AIS not because they are plagued by parasites more than non-vertebrates, but because their developmental program and their body plan allowed it. There is no single factor (such as the appearance of jaws) responsible for the emergence of the AIS. The system is the result of a set of particular circumstances and of a particular pattern of embryonic development. The vertebrate developmental pattern is contingent on specific changes in the genome, both in the regulatory and structural genes. The trigger for the changes may have been a wholesale genomic amplification leading to the formation of new regulatory circuits. The amplification provided genes (e.g., those of the *Mhc* proper), cells (e.g., clonally-selected lymphocytes), tissues (e.g., neural crest), and organs (e.g., the thymus) necessary for the full realization of the AIS. Although the impetus for the emergence of the AIS was endogenous, the evolution of the AIS was driven by selection exerted by parasites. The selection-driving factors included increase in body and tissue complexity, increase in the efficiency of the circulatory systems, lengthening of lifespan, and reduction in fecundity. None of these factors are unique to vertebrates, but together they make vertebrates more vulnerable to parasites. The time interval required for the agnathan-gnathostome transition remains undetermined; it may have been as long as >160 million years.

Key words. Adaptive immune system, Major histocompatibility complex, Agnatha, Gnathostomata, Lampreys

AIS versus NAIS

The various mechanisms by which organisms resist assaults from parasites can be classified into two broad categories — those effected by the anticipatory (adaptive) immune system (AIS) and those comprising the nonanticipatory (nonadaptive) immune system (NAIS; see Klein and Horejsí 1997). The former rely on the interplay of three receptor sets which together anticipate all ligands (antigenic determinants) that parasites can present to the host, except those present in the host itself (i.e., the system discriminates between self and nonself). The latter involve a variety of unrelated receptors, each specific for a particular ligand usually shared by many different parasites (Klein 1999). The three receptor sets of the AIS are the major histocompatibility complex (Mhc) molecules, the T-cell receptors (Tcr), and the B-cell receptors (Bcr) or immunoglobulins (Ig). Each of the Mhc receptors is broadly specific for a set of peptides which have certain physico-chemical characteristics in common. The specificity of these receptors is germ line-determined and it varies from one allele to another among individuals of a population. The specificity of the Tcr and Bcr is determined in part in the germ line by large sets of separate gene segments and in part by somatic mechanisms involving shuffling of the segments and mutations. The Tcr are specific for the Mhc receptors and the peptides bound to them, whereas the Bcr recognize antigens independently of the Mhc molecules. Both the Tcr and Bcr are expressed clonally, each T or B lymphocyte expressing a receptor of a single specificity. An encounter with the matching ligand leads to the activation of the lymphocyte, expansion of the clone, and the generation of memory cells which can be rapidly expanded upon a re-encounter with the same antigen.

There is now a general agreement that invertebrates (nonchordates) lack the AIS and rely exclusively on the NAIS for protection against parasites. This conclusion is based primarily on the failure to find any evidence for the presence of AIS receptors (Mhc, Tcr, Bcr) in any of the invertebrate taxa examined. Indeed, in the one nonchordate animal whose genome has been sequenced in its entirety, the nematode *Caenorhabditis elegans,* there is no sign of the three sets of AIS receptor genes (The *C. elegans* Sequencing Consortium 1998). Less certain is the time of AIS emergence within the chordates (Fig. 1). Some recent developments in the molecular biology of the AIS have been widely interpreted as favoring the notion that the AIS may be restricted to the jawed vertebrates, the Gnathostomata, and hence that it may have arisen during the transition from the jawless vertebrates, the Agnatha, to Gnathostomata (henceforth the A-G transition). This view, too, is in the main based on the failure of several laboratories to clone agnathan *Mhc, Tcr,* and *Ig* genes. Here we take a critical look at the evidence for the absence of the AIS in agnathans, examine what it may take to pinpoint the origin of the AIS more precisely, and speculate on the evolutionary incentives that may have been behind the emergence of the AIS.

The search for a needle in a haystack

Mhc class I and class II genes have been identified in all vertebrate classes except the Agnatha (Kasahara et al. 1995). No homologs of *Mhc* class I and class II genes have been identified in any of the invertebrate taxa (for classification of invertebrates see Brusca and Brusca 1990). Similarly, there is no evidence for the presence of genes homologous to those coding for the Tcrs or Igs outside of the group of jawed vertebrates, (Klein 1989, 1997). Below, we consider briefly the approaches that have been used to reach this conclusion.

In our laboratory, we used two approaches aimed at identifying the *Mhc* genes in two representatives of agnathans, the Sea Lamprey (*Petromyzon marinus*) and the Freshwater Lamprey (*Lampetra fluviatilis*; W. E. Mayer, unpublished data). Originally, we availed ourselves of the standard method based on polymerase chain reaction (PCR) amplification using degenerate primers corresponding to conserved segments of known *Mhc* sequences. This approach yielded different genomic

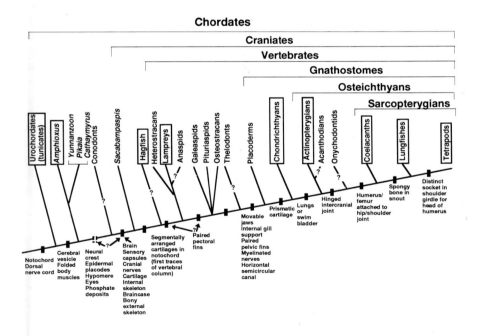

Fig. 1. Cladogram of the chordate phylum. Closed rectangles indicate the presumed appearance of selected shared, derived characters. Dotted rectangles and question marks indicate an alternative time of character appearance and questionable relationships, respectively. Extant taxa are enclosed in rectangles; all other vertically listed taxa are extinct.

sequences accidentally amplified by the primers, but no sequence even distantly related to known *Mhc* genes. Several other laboratories have, apparently, had a similar experience in the search for not only the *Mhc* but also the *Tcr* and the *Ig* genes. For the second approach, we prepared cDNA libraries by reverse transcription of mRNA isolated from lamprey organs that have been claimed to contain lymphoid tissues. Clones were picked up randomly from the libraries and the inserts were sequenced, but again no candidates for *Mhc* homologs could be found among the numerous identifiable and unidentifed sequences (W. E. Mayer, unpublished data; see Table 1).

The AIS also involves, in addition to the immunological Holy Trinity *(Mhc, Tcr, Ig)*, other gene sets without which the system, as it is known in gnathostomes, could not function. The failure to identify some of these sets in extant agnathans has been interpreted as further evidence for the restriction of AIS to gnathostomes (e.g., Kasahara 1998). Two of these gene systems in particular have been considered informative in this regard — the recombination activating genes *(RAGs)* and the genes coding for the immunoproteasome β subunits *(PSMBs)*. The former are essential components of the process that joins selected *V* segments to selected *D*, and *J* segments and so leads to somatic diversification of the *Tcrs* and *Igs* (Schatz et al. 1992). The latter respond to interferon-γ during an immune response to produce the PSMB8 (=LMP7), PSMB9 (=LMP2), and PSMB10 (=MECL1) subunits which replace the constitutively expressed PSMB5 (=X), PSMB6 (= Y), and PSMB 7 (= Z) subunits, respectively (Tanaka and Kasahara 1998). The resulting immunoproteasome then digests the proteins into peptides tailored for loading onto newly synthesized class I molecules (Tanaka et al. 1997; Pamer and Cresswell 1998). Both gene sets are well conserved among various gnathostomes and therefore should not have been too difficult to find in agnathans if these possessed them. But, as in the case of other AIS-specific genes, the search for them proved to be futile (Bernstein et al. 1996; Greenhalgh and Steiner 1995; Kandil et al. 1996; Nonaka et al. 1997). While one could argue that the agnathan *Mhc, Tcr*, and *Ig* genes may have diverged so much from their

Table 1. Search for AIS genes by sequencing clones from cDNA libraries (Werner E. Mayer, unpublished data)

Class and species	No. of clones sequenced	
	Identified	Unidentified
Cephalochordata		
Branchiostoma lanceolatum	65	329
Hyperoartia (lampreys)		
Petromyzon marinus	63	294
Lampetra fluviatilis	126	361

gnathostome counterparts that they have escaped detection by standard methods, this argument is not applicable to the *RAG* and *PSMB* genes.

In reality, however, the putative absence of *RAG* and *PSMB8, 9,* and *10* orthologs in agnathans is not evidence that this vertebrate group lacks the AIS. Regarding the former, one can well imagine that in its initial stages of evolution the AIS operated without somatic diversification of *Tcr* and *Ig* genes and that the repertoire of the two receptors was wholly encoded in the germ line. The repertoire might have been more limited than that established later when the *VDJ* recombination mechanism began to operate when recombinase genes of a transposon were inserted into the vertebrate genome (Agrawal et al. 1998; Hiom et al. 1998), but selection ensured that these few genes fulfilled the basic needs of the organism in terms of defense against infectious agents. In fact, this is in essence the situation that still exists in cartilaginous fishes in which the *V(D)JC* units are repeated many time over in the genome (Rast and Litman 1998).

In regard to the PSMB subunits involved in the formation of immunoproteasomes, recent studies indicate that proteasomes isolated from invertebrate cells process longer peptides into shorter ones of the type found bound to Mhc class I molecules in vertebrate cells (Niedermann et al. 1997). Furthermore, insect cells transfected with *Mhc* class I, *TAP*, and *TAPBP* (tapasin) genes generate Mhc-bound peptides recognizable by activated T cells (Schoenhals et al. 1999).

We conclude, therefore, that there is no definitive evidence for the absence of the AIS in agnathans. There is circumstantial evidence in support of this contention but not of the kind that would, in court, suffice to convict a person of a capital crime. The question then arises: can the absence of the AIS in agnathans be proven definitely at all? As in all situations requiring proof of something that may not exist, definitive proof of nonexistence of the AIS may seem all but impossible. Yet, in fact, one can think of several ways of resolving the issue. The most obvious one is that by which the AIS was discovered in the first place — by immunization of a living agnathan and demonstrating the presence of antibodies in the serum. Unfortunately, this approach suffers from a certain credibility gap as a result of a number of publications in the past offering exactly this kind of proof (e.g., Kobayashi et al. 1985; Hanley et al. 1990; Varner et al. 1991), only to be shown later to constitute false alarms: the "antibody" turned out to be the third component of the complement cascade (C3; Fujii et al. 1992; Hanley et al. 1992; Ishiguro et al. 1992). Nevertheless, if antibodies are produced in response to an antigenic challenge of agnathans, it should be possible to isolate, identify, and characterize them. Considering the history of this effort, however, it is not surprising that immunologists are not exactly racing toward this goal.

At the genomic level, there are several options open for any serious attempt at resolving the issue of the existence of *Mhc, Tcr*, and *Ig* genes in agnathans. One is to sequence the entire genome (or an entire cDNA complement) of a living jawless vertebrate. Although the DNA content of living agnathans is on the low side, the posited problem is probably not a sufficient incentive for any laboratory to embark on a lamprey or hagfish genome sequencing project any time soon. The next best alternative to sequencing the entire genome is to sequence the chromosomal region in which any of the three systems could be expected to reside if they existed in agnathans. Here the prospects are best for the *Mhc* region which has been shown by Trachtulec et al. (1997) to be part of an ancient, syntenic group whose beginnings extend all the way back to the period before the divergence of proto- and deuterostomes. It should be possible to identify this pre-*Mhc* region in lamprey or hagfish by the presence of, for example, the *PSMB, RING3,* and *RXRB* genes, and then sequence it to find out whether it contains the class I and class II genes.

Interrogation of the supporting cast

The *Mhc, Tcr,* and *Ig* are the preeminent, but not the only players in the AIS extravaganza. Their act is dependent on a large cast of supporting characters, who all had to be in place when the three leading characters made an entrance onto the scene. It is therefore more than an idle curiosity to inquire into the origin of the other dramatis personae. Here we discuss only three of these additional groups of performers in the AIS play.

The origin of lymphocytes

The performances of the three prima donnas — *Mhc, Tcr,* and *Ig* — are intimately tied to the existence of specialized cells, the lymphocytes. The relationship of the *Tcr* and *Ig* with the lymphocytes is so close that the expression of these two genes can be taken as these cells' defining feature (Klein 1988). Lymphocytes, of course, do much more in the AIS play than provide a setting for the expression of *Tcr, Ig,* and *Mhc* genes. They initiate an immune response, secrete cytokines, kill other cells, perpetuate the memory of an encounter with the antigen, and carry out many other specialized functions. The performance of these functions requires the possession of special capabilities on the part of the performer. These include the abilities to vary the life span from hours to years, to oscillate between a resting state and proliferation, to circulate through the body or settle down in a tissue, and to use different genetic programs depending on the circumstances. These abilities were probably acquired by the cells gradually, either before or concurrently with the evolution of the AIS.

Little is known about the evolutionary origins of lymphocytes, however. All gnathostomes presumably possess lymphocytes which display all or most of the characteristics of mammalian lymphocytes, including the presence of Tcrs and Igs in their plasma membranes. Immunological literature is also replete with descriptions of lymphocytes in agnathans, but these reports are colored by taking the presence of the AIS in this vertebrate group for granted. Any small, round cell with scanty cytoplasm and a large, round, slightly indented nucleus containing coarse masses of chromatin was, in the past, considered to be a lymphocyte, in both vertebrates and invertebrates (Ratcliffe and Rowley 1981). In reality, all the reports on the presence of lymphocytes in animals other than gnathostomes must now be viewed with skepsis. Even if it could be demonstrated that some of these cells are related ontogenetically to gnathostome lymphocytes, it would be incorrect to parade them under this label, if one were to use the ability to rearrange *Tcr* or *Ig* gene segments as the defining characteristic of a lymphocyte (Klein 1988).

The issue whether agnathans possess lymphocyte-like cells is not going to be resolved by morphological but by molecular investigations. In gnathostome ontogeny all blood cells, including lymphocytes, originate from a common, self-renewing, pluripotential hemopoietic stem cell of mesodermal origin (Orkin 1996). Although the mode of hemopoietic stem cell derivation from the mesoderm may differ among various gnathostome classes (Smith and Albano 1993) and the site of hemopoiesis may change during the development of an individual (Robb 1997), generally the differentiation pathways appear to be conserved (Mathei-Prevot and Perrimon 1998). The pathways are determined by sequential activation of genes via sets of transcription factors interacting with the regulatory regions of their target genes (Sieweke and Graf 1998). After the initial commitment to either the erythroid, myeloid, or lymphoid lineages, further commitments follow in each lineage. In the mammalian lymphoid lineage, some 20 different transcription factors have thus far been shown to participate in the various stages of differentiation (Georgopolous 1997; Sieweke and Graf 1998; Reya and Grosschedl 1998), of which two sets are of particular interest in the present context — the Ikaros and the PU.1 families.

The Ikaros family consists of at least three related zinc finger-type transcription factors —Ikaros, Aiolos, and Helios — which apparently interact with one another to form multimeric complexes participating in the activation of different target genes (Georgopolous et al. 1997). They bind to DNA sites with a core sequence of GGGAA/T in the regulatory regions of several target genes involved in lymphopoiesis (Tdt, RAG1, IL2R, λ5, VpreB, lck). Ikaros is expressed in hemopoietic stem cells and mature lymphocytes; Aiolos, specifically in the lymphocytic lineage; and Helios, in the T-cell lineage only. The PU.1 family consists of the Ets-type transcription factors with the ability to bind to the purine-rich (PU) box of their target genes (Fisher and Scott 1998). One of the factors, Spi-B, is a cellular counterpart of the oncogene Spi-1 produced by the Friend/Rauscher spleen focus-forming retrovirus (Ray et al. 1992). The expression of the Spi-B gene is restricted to the lymphoid lineage and largely to its B-cell sublineage. Mammalian Spi-B protein is essential for antigen-dependent expansion of B cells, T-dependent immune respones, and maturation of germinal centers (Su et al. 1997).

Although restricted in their expression to the lymphoid lineage in gnathostomes, the *Ikaros*- and *PU.1*-family genes have been found to be expressed in an agnathan, the lamprey. The Ikaros family is represented in the lamprey by at least two transcription factors, probably Aiolos and Helios, although the exact identity of the encoding genes has not, as yet, been established definitely (W. E. Mayer, C. O'hUigin, and J. Klein, in preparation). The PU.1 family is represented by at least one member, tentatively identified as the *Spi-B* homolog (S. Shintani, C. O'hUigin, and J. Klein, in preparation). These findings could mean one of two things. Either the ancestors of the Ikaros and PU.1 families of transcription factors had a different function in agnathans than they later assumed in gnathostomes, or their function has remained the same and the lamprey does indeed possess lymphocyte-like cells which either do or do not express *Tcr* and *Ig* genes. At any rate, it is conceivable that major hemopoietic lineages were established before the emergence of the AIS and that the lymphocyte-like cells originally had mainly secretory or some other function in the NAIS. Whatever the case might have been, the presence of Ikaros- and PU.1-family transcription factors in agnathans suggests that the AIS did not appear out of the blue in one large evolutionary leap. Rather, it may have arisen gradually through a combination of two processes: first, the procurement of systems that evolved originally to serve other functions and second, the generation of new systems with new functions. Hemopoiesis-controlling genes might be an example of the former, and the *Mhc, Tcr,* and *Ig* genes examples of the latter.

The origin of the thymus

Most lymphoid organs (e.g., lymph nodes and the bursa of Fabricius) appeared relatively late in vertebrate evolution and are therefore present only in some gnathostomes. The one organ one might expect to have emerged early in the evolution of the AIS is the thymus, the site of the differentiation of *Tcr*-expressing lymphocytes. In all gnathostomes, the thymus originates within the branchial (gill, visceral) arches (Brachet 1921) from four embryonic sources : the pharyngeal endoderm, the head mesoderm, the lateral plate mesoderm, and the neural crest (Le Douarin and Jotereau 1975; Turpen et al. 1997). The pharyngeal endoderm contributes the thymic epithelium, which in the thymic primordium forms a mass of closely apposed cells, but in the maturing thymus develops a loose, mesh-like appearance. The head mesoderm gives rise to the endothelium of the blood vessels that invade the thymus. The lateral plate mesoderm is the original source of hemopoietic stem cells and thus ultimately of the lymphocyte progenitors, the thymocytes, that populate the thymus infiltrating the epithelial meshwork. The neural crest cells give rise to the mesenchyme which develops into the connective tissue of the septa separating individual thymic lobules, the capsule enveloping the thymus, and the connective tissue of the invading blood vessels. The neural crest, however, provides much more than that, since if its cells are prevented from immigrating into the primordium, the organ fails to develop. It is apparently the interaction between the endodermal and neural crest cells that induces the formation of the thymus.

Because the neural crest cells are essential for the formation of the thymus, it is of interest to know how these cells behave in the agnathans. Jawless vertebrates, of course, have a neural crest, as indeed all vertebrates do, neural crest being one of this subphylum's distinguishing characteristics. In a recent study, Horigome and coworkers (1999) investigated the behavior of neural crest cells in the pharyngule stage of agnathan embryonic development. They discovered that in the lamprey, as in gnathostomes, the neural crest cells of the pharyngule's head region migrate ventrally beneath the surface of the ectoderm to colonize the phyryngeal arches. In both the lamprey and the gnathostomes, three independent populations of cephalic neural crest cells originate in three distinct hindbrain bulges, the rhombomeres, and stream to different pairs of arches. One population originates in rhombomere 1 (r1) and the region of the neural rod rostral to r1, and migrates to the first pharyngeal arch, which in gnathostomes transforms into the upper and lower jaws, as well as the masticatory muscles. The second population arises in r4 and heads for the second pharyngeal arch, which in gnathostomes forms the hyoid. The third population issues from r6 and regions caudal to it, and distributes itself among the remaining pharyngeal arches (3 through 7), which in jawed fishes support the gills and also give rise to the thymus. In the lamprey, the first and the second pharyngeal arches develop into the muscles and cartilage of the upper lip and the velum, an apparatus for suction and pumping water. The remaining arches form the cartilage of the branchial basket. It appears, therefore, that despite the large difference in head construction between agnathans and gnathostomes, the genetic programs specifying the distribution of the neural crest cells among the pharyngeal arches and thus laying the foundation for the early morphogenesis of the head region are shared between the two vertebrate groups. The programs must have been established before the divergence of the jawless and jawed vertebrates. The morphological differences in the agnathan and gnathostome head construction must therefore be controlled by genes that act at later stages of development.

Whether lampreys possess a thymus remains an open question. The reports on thymus presence rely on the observation that the pharyngeal region of the lamprey is marked by the accumulation of "lymphocyte-like" cells (see, for example, Good et al. 1972; Ardavin and Zapata 1988). They therefore hinge on the notion that cells that *look* like lymphocytes really *are* lymphocytes. Since there has not been, until recently, any other evidence to support this notion, the issue of the presence or absence of the thymus could not be resolved. Now that the analysis of molecular markers characterizing the lymphocytic lineages is being extended to the agnathans, the situation is changing. Whichever way the controversy will ultimately be resolved, however, the embryological studies indicate that the lamprey possesses the infrastructure necessary for thymus development. The endodermal and mesodermal elements are all in place and the neural crest cells are capable of undertaking the pilgrimage to the sites at which gnathostomes sprout their thymi. If it turns out that lampreys do *not* develop thymi, they are probably only a few developmental steps away from doing so.

The origin of the antigen processing systems

The anticipatory immune response begins with the presentation of Mhc-bound peptides to the Tcr (Klein and Horejsí 1997). A prerequisite for this "antigen presentation" is a sequence of "antigen processing" events: the right peptides must be generated, delivered to and loaded onto the Mhc molecules. Each of these steps involves a number of components encoded in separate genes (van Endert 1999). The available evidence indicates that the basic components necessary for antigen processing are not only present in agnathans, but that they were in fact established long before the emergence of vertebrates.

In the case of class I molecules, peptides destined for loading onto the Mhc receptors are generated by two systems, the proteasomes (Baumeister et al. 1998) and cytosolic enzymes such as the leucine aminopeptidase (Beninga et al. 1998). The proteasome system was established as far back in evolution as the Archaea and has been degrading proteins into peptides ever since. Moreover, recent evidence indicates that peptides generated by proteasomes of organisms such as the yeast and the fruit fly, which lack the *Mhc*, are of the correct size and composition for loading onto mammalian Mhc molecules (Niedermann et al. 1997; Schoenhals et al. 1999). Similarly, endopeptidases are an old family of enzymes which were functioning long before the *Mhc* appeared on the scene (e.g., Guenet et al. 1992).

Peptides generated by proteasomes or soluble cytosolic proteases are delivered to the site of Mhc class I chain biosynthesis in the endoplasmic reticulum by the transporters associated with antigen-processing, the TAPs (Elliott 1997). The transporters are members of the ATP-binding cassette (ABC) protein family which arose before the divergence of Archaea, Bacteria, and Eukarya (Hughes 1994). In phylogenetic trees, the mammalian TAPs cluster with eukaryotic Pgp and related molecules of purple bacteria. This observation has led Hughes (1994) to suggest that the ancestor of *TAP* and *Pgp* genes was a mitochondrial gene translocated to the nuclear genome. Following the creation of the *Mhc* genes, some members of the family apparently specialized in translocating, preferentially fitting peptides into the jaws of the class I molecules.

Finally, the loading of the peptides concurrently with the assembly of the class I molecules is accomplished with the help of several molecules including calnexin, calreticulin, ERp57, and TAPBP (van Endert 1999), each of which has a long evolutionary history. Similarly, the generation of peptides loaded onto the class II molecules relies on mechanisms established long before the emergence of the AIS (Geuze 1998). Taken together, all these observations indicate that the individual steps in the processing sequence were already in place before the emergence of the *Mhc*. When the *Mhc* finally appeared on the scene, a few new steps had to be added, but mostly the new system could fall back on what was already available and adapt the steps to specialized tasks.

Why AIS?

The jaw hypothesis

Whether the AIS emerged with the gnathostomes or whether agnathans possess some simple form of the AIS remains, therefore, an open question. Whatever the case may be, however, it is now generally agreed that nonvertebrates lack the AIS. And since vertebrates represent only a tiny fraction of the Metazoa, it is entertaining to speculate on the reasons that may have prompted the evolution of the AIS.

If the AIS did, indeed, first appear in jawed vertebrates, it would be tempting to associate its emergence with the development of the jaws. This view has been championed by Matsunaga and his colleagues (Andersson and Matsunaga 1996; Matsunaga 1998) who believe, along with some evolutionary biologists (Jorgensen 1966; Mallatt 1984), that the early vertebrates were microphagous, feeding on minute particles (phytoplankton and fine detritus) suspended in the water. The development of jaws potentiated an ecological shift from suspension feeding to predation by providing the early gnathostomes with the means of grasping their prey and cutting it into pieces. Alas, according to Matsunaga, the hard parts of the swallowed prey (bones and fish scales, crustacean carapaces, mollusc shells) often injured the digestive tract and the infected lacerations then provided a major selection pressure for the evolution of the AIS, particularly in the tissues surrounding the digestive tract (the gut-associated lymphoid tissue or GALT).

The jaw hypothesis of the AIS origin can, however, be assailed on several grounds. To begin with, not everybody agrees that the earliest vertebrates and all the agnathans were microphagous. Several evolutionary biologists (e.g., Jollie 1982; Gans and Northcutt 1983) have argued that the first vertebrates were, in fact, predators. Even if they were not, however, some jawless vertebrates, such as conodonts, apparently *were* predators. Although the position of conodonts in the phylogenetic system of chordates remains unresolved, recent fossil data (reviewed by Aldridge and Purnell 1996) strongly support the assignment of these tiny worm-like creatures to the vertebrate subphylum. Nobody has ever proposed that conodonts had jaws, but their possession of a biomineralized feeding apparatus in the mouth, with toothlike structures, blades, and ramiform bars, is undisputed. In fact, for a long time, fossilized remains of the feeding apparatus were the only evidence for these animals' existence. The term "conodont" was originally used exclusively in reference to these artful structures and only later, when impressions of the whole body were found, was it applied to a group of organisms. The recently reported evidence of microwear detectable on the bones of the apparatus (Purnell 1995) strongly supports the view that conodonts, despite their small size and their otherwise soft bodies, were predators. The microwear also indicates that conodonts should have suffered abrasions to their digestive tract the same way as Matsunaga postulates for the early gnathostomes. While we have no way of ever learning whether conodonts possessed a simple AIS, we can rest assured that according to the paleontological data they did not possess jaws. So, either the AIS evolved without jaws, or predation and the injuries associated with it did not necessarily lead to the appearance of the AIS.

Moreover, vertebrates are not the only predators among the Metazoa. The living cephalopods among the molluscs, for example, are rapacious predators, as were their extinct ancestors, the ammonites. Cephalopods such as the *Nautilus* even possess "jaws" of a sort, a pair of powerful, highly mobile, beak-like horny structures which they use effectively to cut and tear prey such as fish, shrimp, crabs, and other molluscs. Their mode of life should therefore expose them to a similar danger of digestive tract laceration as Matsunaga postulates for predatory gnathostomes. Yet, cephalopods have not evolved an AIS.

There is no question that the tissues of the inner surface of the digestive tract are heavily populated by lymphoid cells constituting, either in a diffuse or aggregated form, the GALT. The digestive tract may have indeed been the original site at which the AIS began to evolve, as was suggested 30 years ago by Fichtelius (1970). This observation does not, however, necessarily support the jaw hypothesis. Since the densest concentration of lymphoid cells is in the part of the digestive tract that receives processed and hence uninjurious food stuff (the intestine), whereas it is much lighter in the parts (mouth, pharynx) in which the potential for injury from unaccomodating prey is greatest, it is difficult to see how the existence of the GALT supports the jaw hypothesis. Furthermore, there are plenty of jawed vertebrates that do not lead a predatory existence and have not done so for many millions of years and yet they all possess strongly developed GALT. It seems therefore more reasonable to assume that GALT owes its existence to needs such as the control of the natural flora in the intestine and the capture of antigenic substances at the surfaces at which they are most likely to enter the body. Hence we find it improbable that the acquisition of jaws by gnathostomes was responsible for the development of the AIS. If it contributed at all, it was only one of many factors in the establishment of the AIS.

If not jaws, what then?

If jaws were not the trigger for the appearance of the AIS, was it one of the other transformations characterizing the A-G transition? The list of the transformations is too long to be dealt with exhaustively in the space available (see Fig. 1). Instead, we merely catalog here some of the major changes to illustrate the nature of the transition and to describe the context in which the AIS may have arisen (Kent and Miller 1997; Walker and Liem 1994; Janvier 1996).

The appearance of jaws was accompanied by several other modifications of the digestive system. The jaws sprouted teeth, probably by transfiguration from special types of scales that covered the bodies of the extinct jawless fishes (Lumsden 1981). The placoid scales, part of the dermal armor, first changed to dentin- and amelogenin-containing denticles on the surface of the head and these then extended into the oropharyngeal cavity and evolved into teeth. The straight tube of the agnathan digestive tract extending from the pharynx to the anus began to compartmentalize and diversify into a muscular chamber, the stomach, and the intestine with a well-developed spiral valve increasing the epithelial area available for absorption. The stomach became connected to the liver by special blood vessels. The digestive enzyme-secreting cells, which in jawless fish were scattered in the submucosa of the digestive tract, organized themselves in jawed vertebrates into a discrete organ, the pancreas, in which they were joined ultimately by hormone-synthesizing cell islets. Similarly, the hemopoietic cells, which in jawless fish are found diffusely distributed in the walls of their digestive tracts, tend to form a separate organ, the spleen, in gnathostomes.

Dramatic changes also affected the fin system during the A-G transition. The hagfishes possess a single unpaired fin in the caudal region; the lampreys have unpaired caudal and dorsal fins. In both groups, the fins are comprised of two skin surfaces apposed back-to-back and strengthened by cartilaginous rays or radials. In lampreys, each ray is associated with a single radial muscle which enables the fin web to undulate. Jawed fishes have both unpaired and paired fins. The unpaired dorsal, caudal, and anal fins come in a variety of sizes and shapes; they were all lost during the transition from fishes to tetrapods. The paired fins are of two types, pectoral and pelvic, which in tetrapods have been transformed into forelimbs and hind limbs, respectively. The gnathostome fin rays are anchored in cartilaginous or bony base elements which, in the case of unpaired fins, rest on the vertebral column. The paired fins articulate with cartilaginous or bony girdles comprised of both dermal and endoskeletal elements.

The postcranial (axial) skeleton of living agnathans consists of the notochord and the fin skeleton. In the lamprey, in addition, short, paired, cartilaginous processi sprout out at regular intervals along the line of apposition between the notochord and the overlying spinal cord - the arcualia or lateral neural cartilages, resembling the neural arches of vertebrae. Whether they represent a commencement of vertebrae evolution or vestiges of vertebrae the ancestors of these animals may have once possessed remains unresolved. In gnathostomes, the spinal cord is surrounded, along its length, by a protective cartilaginous or bony sheath, the vertebral column, segmented into individual vertebrae. The flexible column is morphologically differentiated into abdominal, caudal, and tail regions. It is flanked by a segmented axial musculature with the trunk muscles divided by a horizontal septum, in addition to the myosepta which separates one muscle segment from the next. Modern agnathans possess only the myosepta; they lack the horizontal septum.

Other major changes associated with the A-G transition include the addition of a horizontal semicircular canal to the labyrinth of the inner ear (hagfishes have only the posterior and lampreys the anterior and posterior vertical canals) enabling both lateral and vertical orientation; the wrapping of the nerve fibers into a myelin sheath for improved insulation and message carrying ability; the expansion of the pressure-sensitive lateral line (enclosed or supported by specialized scales) over the entire body length; the development of separate male and female reproductive ducts from different parts of kidney tissue, allowing sperm to pass through the excretory system; and the development of two nostrils (instead of one in agnathans), one on each side of the head, and two distinct olfactory tracts leading to separate olfactory bulbs.

There is nothing on this list of changes that could be interpreted as leading to an increased parasitic exposure and thus predicating the development of the AIS. To us, this observation indicates simply that the search for a direct link between a morphological innovation and immunity to parasites is misplaced and futile. Although the innovations may enable the evolving organisms to enter new niches and thus become exposed to new parasites, such a transition is not peculiar to any particular group of organisms. It is something every evolving group has to cope with and since, before vertebrates, all invertebrates managed to rise to the occasion without developing the AIS, there is no reason why vertebrates, or more specifically gnathostomes could not have done the same.

The most remarkable feature of the A-G transition may seem to be that so many seemingly disjointed changes occurred all at once. The truth of the matter may be, however, that the changes were neither as disjointed as they appear to be, nor did they occur simultaneously. As for the former, changes may seem autonomous when observed at the level of an adult organism, but are often found to be interconnected in their embryonic development. Even more interdependence is now being discovered at the level of the development-controlling genes. The *Hox* genes (Holland and Garcia Fernandez 1996) are perhaps the best known but not the only example of a gene cluster exerting a control over a wide variety of characters, ranging from the determination of the anterior-posterior body axis, over the metamerism of the hindbrain, to the specification of the paired appendages in gnathostomes. Here, a limited change in the number and composition of the genes in the cluster can have a profound effect on different body parts.

As for the suddenness of the changes associated with the A-G transition, here, too, the appearance is deceitful. For one thing, possession by extant gnathostomes of an array of characters lacking in extant agnathans does not necessarily signify that all these characters arose simultaneously, without transitory forms. Although the fossil record is too fragmentary to document the A-G transition fully, it provides nevertheless clear evidence for the nonsimultaneity of the appearance of a number of different characters. Ostracoderm fossils have been found, for example, with pectoral fins but without pelvic fins (Janvier 1996). This observation indicates that jaws (ostracoderms being extinct jawless fishes), pectoral fins, and pelvic fins evolved at different times. Similar evidence is also available for other combinations of characters.

The other thing to point out is the uncertainty about the length of the A-G transition interval. One often encounters statements expressing astonishment over the shortness of the interval and the number of changes that occurred in it. The figure of 20 million years (my) is frequently mentioned without a proper documentation. Notwithstanding the fact that 20 my is not that short a time, even on the evolutionary scale on which the shape of the fish heads can change within less than 12,000 (Greenwood 1981; Johnson et al. 1996), and the shape of a bird's beak in only a few generations (Grant and Grant 1993), it is not clear to us where the confidence in such an estimate comes from. As the issue is of relevance to the topic under discussion, it is perhaps worth devoting a few lines to the time estimates before we use the information on the A-G transition to draw some conclusions about the emergence of the AIS. How much time for the A-G transition was there really?

The length of the A-G transition interval

When did the gnathostomes diverge from the agnathans? Theoretically, it should be possible to answer this question by two approaches — one based on paleontology, the other on molecular biology. As regards the former, the question can be rephrased: from which agnathan group did the first gnathostomes diverge? Relevant to this question are four extinct groups — Osteostraci, Thelodonti, Acanthodii, and Placodermi, as well as two extant groups — Chondrichthyes (cartilaginous fishes) and Osteichthyes (bony fishes). The first two of the four extinct groups are jawless, the last two are jawed vertebrates. The four groups are generally believed to have been involved in the transition from the jawless to the jawed phenotype, but unfortunately the sequence of appearance of the groups remains unresolved (Carroll 1988; Forey and Janvier 1993; Janvier 1996; Maisey 1996).

Osteostracans were distinguished by a large semicircular shield encasing a "skull" made of perichondrial bone, upturned tail, pectoral fins, and a specialized sensory field on the sides of the head shield. They lived during the Silurian and Devonian periods. The possession of perichondrial (cellular) bone and pectoral fins, two characters otherwise present only in gnathostomes, suggests that osteostracans may have been on the evolutionary lineage leading to jawed vertebrates.

Thelodonts come in two quite diverse types — flat-bodied and deep-bodied — which may be only distantly related to each other. They have a pipette-like mouth, forked tail, pointed scales, and special pharyngeal denticles. They may have possessed a stomach which is otherwise absent in all other agnathans. It is primarily this character that has led some paleontologists to the conclusion that thelodonts may have been closely related to gnathostomes; in fact, the possibility has not been excluded conclusively (well preserved fossils being rare) that thelodonts possessed jaws. Like osteostracans, thelodonts are known from the Silurian and Devonian periods. Either the osteostracans or the thelodonts could be the closest relatives of gnathostomes; paleontologists are divided in their opinions in this regard.

The two characteristic features of the acanthodians are, first, the presence of long spines in front of all fins except the caudal fin and second, the diamond-shaped, onion-structured scale crowns in which layers of orthodentine alternate with layers of mesodentine. Because of their vague resemblance to elasmobranchs, acanthodians are often referred to as "spiny sharks" and regarded by some paleontologists to be related to osteichthyans. The shape of the acanthodian head with large, anteriorly placed eyes reminds other paleontologists of early actinopterygians (i.e., ray-finned bony fishes). Acanthodians are also united with actinopterygians by two other characteristics — similar gill structure (alas, now documented only by a drawing based on a specimen that has since been lost) and the presence of three pairs of otoliths (ear stones) in the labyrinth region. Acanthodians thrived in the Silurian and became extinct in the Early Permian period, but scraps of acanthodian-like fossils have recently been reported from the Late Ordovician period.

Placoderms (literally "plated skins") owe their name to two sets of plated armor, one covering the head and the other the anterior part of the trunk. In most placoderms the two sets are connected by a flexible joint. The plates are made of dermal bone composed of semidentine, a special type of hard tissue in which the unipolar odontocytes were entirely enclosed so that they left behind drop-shaped spaces. Placoderms were extremely successful in the Devonian seas and fresh waters all over the world. They emerged in the Early Silurian and became extinct suddenly at the end of the Permian period.

The acanthodians and the placoderms are undisputed gnathostomes, but their relationship to each other and to the two major living groups of jawed vertebrates, the chondrichthyans and the osteichthyans, has been debated for over a century and no consensus has emerged from the dispute as yet. The four main possibilities that have been proposed are these (Fig. 2): first, the placoderms are the sister group of all other gnathostomes; second, the placoderms and chondrichthyans are one sister group, while the acanthodians and the osteichthyans are another sister group, the two groups sharing a common ancestor; third, the acanthodians are the sister group of the chondrichthyans, while the placoderms are the sister group of the osteichthyans, and the two groups again derive from a common ancestor; and fourth, acanthodians are the sister group of all the other gnathostomes. For our purpose, one would like to know which of the four groups emerged first, but the answer obviously depends on which of the four schemes one favors. Formally, the oldest should be the acanthodians because theirs are the earliest gnathostome fossils found thus far (Late Ordovician Harding Stone in the United States, dating to 450 my ago). Paradoxically, however, many paleontologists believe that acanthodians are more closely related to bony than to cartilaginous fishes and therefore that they emerged after the placoderms and chondrichthyans, whose fossil record goes back only to Early Silurian some 420 my ago. Hence, if the presumed acanthodian-actinopterygian sister group relationship is real, the present fossil record must be incomplete and the earliest jawed vertebrates either remain to be discovered or left no record. At present, therefore, paleontologists can neither identify nor date the agnathan group most closely related to the gnathostomes, nor

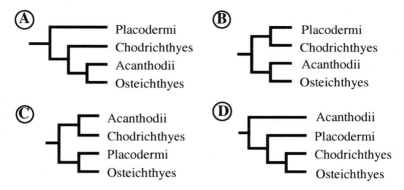

Fig. 2. Four competing hypothesis of the relationships among extinct (Placodermi, Acanthodii) and extant (Chondrichthyes, Osteichthyes) Gnathostomata.

can they pinpoint the oldest gnathostomes. The whole issue of the agnathan-gnathostome split thus remains unresolved and the question as to how long it took for the early vertebrates to develop jaws cannot, at present, be answered by the paleontological method.

The molecular clock method is not an independent way of estimating divergence time, since ultimately the clock must be calibrated by dates provided by the fossil record. The calibration point can, however, be chosen to lie in a time interval for which the record is reasonably reliable for a particular vertebrate group. In a recent study, Kumar and Hedges (1998) calibrated the vertebrate molecular clock by the divergence of birds and mammals. The lizard-like ancestors of the early birds, the diapsids, and of the early mammals, the synapsids, first appear in the fossil record in the Carboniferous period, approximately 310 my ago. The divergence of diapsids and synapsids therefore presumably occurred, at the latest, at about this time. The 310 my might be an underestimate, but other considerations suggest that it may not be too far off from the true date. By applying the calibrated clock to 658 genes of 207 vertebrate species, Kumar and Hedges estimated the divergence times of the various vertebrate taxa. For the emergence of the agnathans and chondrichthyans, they arrived at values of 564 ± 74.6 my and 528 ± 56.4 my, respectively, which gives some 36 my as the maximum time available for the transition from jawless to jawed vertebrates. However, even if the calibration point were accurate, the range of the estimate is tremendous because of the large standard errors of the divergence time — from 0 to 166 my. In this sense, the molecular method does not help much either and the problem remains unresolved. We simply don't know how much time there was for the development of the jaws and of all the other features associated with the emergence of jawed vertebrates, including the appearance of the AIS.

An interpretation: our favorite just so story

Against this background of the A-G transition, we project the following rendering of events that might have led to the emergence of the AIS. We propose that the AIS appeared not in a "big bang", a scenario favored for example by Marchalonis and colleagues (1998), not suddenly within a short period of time, but gradually, over an extended interval. The various elements of the AIS were added to the emerging system in a piecemeal fashion at different times and have therefore different origins and histories. The beginning of some of them (e.g., the elements of antigen processing) go back to times before the divergence of protostomes and deuterostomes, of others (e.g., the development of lymphocytes) to the inception of vertebrates, and of others still to the A-G transition.

This transition itself was anything but sudden. It took place step by step, each step adding a suite of distinguishing characters to the existing ones in the evolutionary line leading to the gnathostomes. We consider the issues of the presence of Mhc, Tcr, and Ig molecules in the living agnathans as unresolved, but resolvable. Although it is reasonable to assume that the living agnathans lack the immunological Holy Trinity as it exists in gnathostomes, it has not been excluded that they possess primigenial forms of one or more members of the Trinity. Further, it is also reasonable to postulate that RAG-mediated recombinational diversification of receptors is absent in living agnathans, but whether this absence is the result of a loss or a representation of a state that existed before the RAG transposition remains an open question. The existence of a simple, entirely germ line-encoded anticipatory system without a recombinational diversification mechanism and with or without mutational somatic diversification is a viable possibility. Unresolved remains also the question whether the *Mhc, Tcr,* and *Ig* clusters and the different classes of genes within each cluster emerged more or less simultaneously or in succession. If the ancestors of the three receptors had different functions before they were recruited into the AIS, the issue may turn out to be difficult to resolve. Also, when discussing the origin of the three molecules, it is important to specify which part of each molecule one is referring to, since the various parts may have different histories. The primigenial *Mhc* gene was almost certainly assembled from parts derived from at least two different sources, one providing the peptide-binding domain and the other the immunoglobulin-like domain (Klein and O'hUigin 1993). Similarly, although the V domains of the Tcr and Ig are related to their respective C domains, it remains to be demonstrated that in each case the V and the C domains come from the same ancestral molecule. It is therefore misleading to claim that Mhc, Tcr, and Ig molecules (genes) are of common origin: only parts of them are. Finally, although lampreys and hagfishes are the closest living jawless relatives of gnathostomes, there were almost certainly forms among the extinct agnathans that were closer to jawed fishes than the ones that survived ultimately (Janvier 1996). We will, of course, never be able to examine the immune system of these closest relatives of gnathostomes, but their existence should caution us from relying too heavily on living agnathans for information on the primordial AIS. Not only because of their taxonomic status, but also because of their lifestyles, which may not be characteristic of the entire agnathan group, they may be misinforming us in this regard. Also, we must keep in mind that lampreys and hagfishes are at least as far apart from each other phylogenetically as mammals and reptiles are. Nevertheless, if our deductions are correct, it can be expected that extant agnathans will be found to possess several elements of the AIS.

Although invertebrates have evolved some of the components of the AIS, they have not acquired an actual AIS because the various body plans they produced have not allowed them to do so. The appearance of an AIS with clonally distributed receptors was contingent upon opting for certain developmental pathways when the choices for the various body plans were made during evolution. By choosing certain plans, the invertebrates forfeited the possibility of developing the structures (molecules, cells, tissues, and organs) necessary for the emergence of the AIS. Invertebrates, just like vertebrates, would have undoubtedly benefited from possessing the AIS and they would have probably evolved it if they could have. The fact that they haven't suggests strongly that their developmental programs preclude this option. Instead, they have been forced to explore alternatives and in this regard they have done quite well. There is thus nothing uncommon about the restriction of the AIS to the (jawed) vertebrates; in truth, innovations in general are contingent on the generation of the right conditions for their appearance. Take cartilage and bone, for example. Many an invertebrate could have put them to good use had it been able to evolve them. This it could not do, however, and so the occurrence of cartilage and bone remains limited to the only metazoan group in which the developmental program includes this option, the same group that evolved the AIS — the vertebrates.

Not enough is known about the genetic control of vertebrate development to specify the various parts of the program needed for the evolution of the AIS. One of them, however, must have been the part responsible for the appearance of the neural crest, that remarkable embryonic tissue involved in a whole array of evolutionary innovations. Neural crest cells are distinguished not only by their amazing ability to migrate to various corners of the body, but also by their truly protean differentiation potentials (Maderson 1987). Depending on the embryonic environment they encounter, they can give rise to neurons, adrenal medulla, epidermal pigment cells, skull cartilage and bones, corneal endothelium and stroma, tooth papillae, connective tissue of salivary glands, and many other cells and tissues. As mentioned earlier, they are also necessary for the development of the thymus, which in turn is needed for the differentiation of one class of lymphocytes with clonally expressed receptors. Besides the neural crest, there must be several other (jawed) vertebrate-specific parts of the developmental program on which the development of the AIS depends.

At a more basic level, the development of new cells, tissues, and organs in (jawed) vertebrates is contingent on the appearance of new genes, on the assumption of new functions by existing genes, and on the placement of existing genes under new controls by regulatory elements. The key to the understanding of how the AIS arose is probably in the comprehension of the new regulatory pathways ushered in during the invertebrate-vertebrate and the A-G transitions. It appears that both these transitions have been associated with dramatic expansions of parts of the genome which may have provided the necessary material basis for the alterations in embryonic patterning and the appearance of a new body plan (Holland 1998). The main evidence for the wholesale genome amplifications is the existence of the paralogous groups (Kasahara 1999), but the alternative explanation, the congress of selected genes into specific chromosomal regions (Hughes 1998; Skrabanek and Wolfe 1998), remains a possibility. Ultimately, suitable statistical tests will have to be applied to distinguish between these two alternatives. The function of the AIS relies on the participation a large number of genes (Klein 1997) and these were apparently produced by gene duplications, by whatever mechanism they occurred.

According to the proposed interpretation, there was no single reason, no single factor responsible for the emergence of the AIS. Rather, the AIS arose when the pattern of embryonic development allowed it to arise. When that happened, further evolution of the AIS was driven by the selection pressure provided by the parasites. Because every ecological niche abounds with parasites of all kinds, it is unlikely that it was the quantity or "quality" of parasites that provided the selection pressure for the evolution of the AIS. Instead, as suggested in an earlier publication (Klein 1989), the increased pressure on the emerging vertebrates came from a greater vulnerability of their bodies to a parasitic load. The vulnerability increased because of a conspiration of factors, including increased tissue and body complexity, development of a very efficient circulatory system, tendency toward longer life spans, and tendency toward a decreasing number of progeny. It was the combination of some such factors that might have conferred selective advantage to vertebrates with the emerging AIS. The effect of the individual factors may not be restricted to vertebrates, but only vertebrates found themselves in a position to respond to the challenge by developing the AIS. Had they not been in that position, they would have either died out, adapted the NAIS to the challenge, or changed the direction of their evolution.

Acknowledgments

We thank Ms. Jane Kraushaar for editorial assistance. J.K. thanks Springer-Verlag for sponsoring the *Immunogenetics* lecture on which this manuscript is based.

References

Agrawal A, Eastman QM, Schatz DG (1998) Transposition mediated by RAG1 and RAG2 and its implications for the evolution of the immune system. Nature 394:744-751

Aldridge RJ, Purnell MA (1996) The conodont controversies. Trends Ecol Evol 11:463-648

Andersson E, Matsunaga T (1996) Jaw, adaptive immunity and phylogeny of vertebrate antibody VH gene family. Res Immunol 147:233-240

Ardavin CF, Zapata A (1988) The pharyngeal lymphoid tissue of lampreys. A morpho-functional equivalent of the vertebrate thymus? Thymus 11:59-65

Baumeister W, Walz J, Zühl F, Seemüller E (1998) The proteasome: Paradigm of a self-compartmentalizing protease. Cell 92:367-380

Beninga J, Rock KL, Godberg AL (1998) Interferon-gamma can stimulate post-proteasomal trimming of the N terminus of an antigenic peptide by inducing leucine aminopeptidase. J Biol Chem 273:18734-18742

Bernstein RM, Schluter SF, Bernstein H, Marchalonis JJ (1996) Primordial emergence of the recombination activating gene 1 (*RAG1*): Sequence of the complete shark gene indicates homology to microbial integrases. Proc Natl Acad Sci USA 93:9454-9459

Brachet A (1921) Traité d'Embryologie des Vertébrés. Masson, Paris

Brusca RC, Brusca GJ (1990) Invertebrates. Sinauer Associates, Sunderland, MA

Carroll RL (1988) Vertebrate Paleontology and Evolution. W.H. Freeman, New York

Elliot T (1997) How does TAP associate with MHC class I molecules? Immunol Today 14:375-379

Fichtelius KE (1970) Cellular aspects on phylogeny of immunity. Lymphology 1:50-59

Fisher RC, Scott EW (1998) Role of PU.1 in hematopoiesis. Stem Cells 16:25-37

Forey P, Janvier P (1993) Agnathans and the origin of jawed vertebrates. Nature 361:129-134

Fujii T, Nakamura T, Sekzawa A, Tomonaga S (1992) Isolation and characterization of a protein from hagfish serum that is homologous to the third component of the mammalian complement system. J Immunol 148:117-123

Gans C, Northcutt RG (1983) Neural crest and the origin of vertebrates: a new head. Science 220:268-274

Georgopolous K (1997) Transcription factors required for lymphoid lineage commitment. Curr Opin Immunol 9:222-227

Georgopolous K, Winandy S, Avitahl N (1997) The role of the Ikaros gene in lymphocyte development and homeostasis. Annu Rev Immunol 15:155-176

Geuze HJ (1998) The role of endosomes and lysosomes in MHC class II functioning. Immunol Today 19:282-287

Good RA, Finstad J, Litman J (1972) Immunology. In: Hardisty MW, Potter IC (Eds) The Biology of Lampreys. Academic Press, London, pp 405

Grant BR, Grant PR (1993) Evolution of Darwin's Finches caused by a rare climatic event. Proc R Soc Lond B Biol Sci 251:111-117

Greenhalgh P, Steiner LA (1995) Recombination activating gene 1 (*RAG1*) in zebrafish and shark. Immunogenetics 41:54-55

Greenwood PH (1981) The Haplochromine Fishes of the East African Lakes. Cornell University Press, Ithaca, New York

Guenet C, Lepage P, Harris BA (1992) Isolation of the leucine aminopeptidease gene from *Aeromonas proteolytica*. Evidence for an enzyme precursor. J Biol Chem 267:8390-8395

Hanley PJ, Seppelt IM, Gooley AA, Hook JW, Raison RL (1990) Distinct Ig H chains in a primitive vertebrate, *Eptatretus stouti*. J Immunol 145:3823-3828

Hanley PJ, Hook JW, Raftos DA, Gooley AA (1992) Hagfish humoral defense protein exhibits structural and functional homology with mammalian complement components. Proc Natl Acad Sci USA 89:7910-7914

Hiom K, Melek M, Gellert M (1998) DNA transposition by the RAG1 and RAG2 proteins: a possible source of oncogenic translocations. Cell 94:463-470

Holland PWH (1998) Major transitions in animal evolution: a developmental genetic perpective. Amer Zool 38:829-842

Holland PWH, Garcia Fernandez J (1996) *Hox* genes and chordate evolution. Dev Biol 173:382-395

Horigome N, Myojin M, Ueki T, Hirano S, Aizawa S, Kuratani S (1999) Development of cephalic neural crest cells in embryos of *Lampetra japonica*, with special reference to the evolution of the jaw. Dev Biol 207:287-308

Hughes AL (1994) Evolution of the ATP-binding-cassette trans-membrane transporters of vertebrates. Mol Biol Evol 11:899-910

Hughes AL (1998) Phylogenetic tests of the hypothesis of block duplication of homologous genes on human chromosomes 6, 9 and 1. Mol Biol Evol 15:854-870

Ishiguro H, Kobayashi K, Suzuki M, Titani K, Tomonaga S, Kurosawa Y (1992) Isolation of a hagfish gene that encodes a complement component. EMBO J 11:829-837

Janvier P (1996) Early Vertebrates. Oxford University Press, Oxford

Johnson TC, Scholz CA, Talbot MR, Kelts K, Ricketts RD, Ngobi G, Beuning K, Ssemmanda I, McGill JW (1996) Late Pleistocene dessication of Lake Victoria and rapid evolution of cichlid fishes. Science 273:1090-1093

Jollie M (1982) What are the 'Calcichordata'? and the larger question of the origin of the chordates. Zool J Linnean Soc 75:167-188

Jorgensen CB (1966) Biology of Suspension Feeding. International Series of Monographs in Pure and Applied Biology /Zoology Division, 27. Pergamnion Press, Oxford

Kandil E, Namikawa C, Nonaka M, Greenberg A, Flajnik MF, Ishibashi T, Kasahara M (1996) Isolation of low molecular mass polypeptide complementary DNA clones from primitive vertebrates: implications for the origin of MHC class I-restricted antigen presentation. J Immunol 156:4245-4253

Kasahara M (1998) What do the paralogous regions in the genome tell us about the origin of the adaptive immune system? Immunol Rev 166:159-175

Kasahara M (1999) Genome paralogy: a new perspective on the organization and origin of the major histocompatibility complex. Curr Topics Microbiol Immunol 248:in press

Kasahara M, Flajnik MF, Ishibashi T, Natori T (1995) Evolution of the major histocompatibility complex: A current overview. Transpl Immunol 3:1-20

Kent GC, Miller L (1997) Comparative Anatomy of the Vertebrates. 8th ed. WC Brown, Dubuque, IA

Klein J (1988) The evolution, ontogeny, and physiological function of lymphocytes. In: Bray MA, Morley J (eds) The Pharmacology of Lymphocytes. Springer-Verlag, Berlin, pp 11-36

Klein J (1989) Are invertebrates capable of anticipatory immune response? Scand J Immunol 29:499-505

Klein J (1997) Homology between immune response in vertebrates and invertebrates: Does it exist? Scand J Immunol 46:558-564

Klein J (1999) Self-nonself discrimination, histocompatibility, and the concept of immunology. Immunogenetics, in press

Klein J, Horejsi V (1997) Immunology. 2nd ed. Blackwell Scientific Publishers, Oxford

Klein J, O'hUigin C (1993) Composite origin of major histocompatibility complex genes. Curr Opin Genet Dev 3:923-930

Kobayashi K, Tomonaga S, Hagiwara K (1985) Isolation and characterization of immunoglobulin of hagfish, *Eptatretus burgeri*, a primitive vertebrate. Mol Immunol 22:1091-1097

Kumar S, Hedges SB (1998) A molecular timescale for vertebrate evolution. Nature 392:917-920

Le Douarin NM, Jotereau FV (1975) Tracing of cells of the avian thymus through embryonic life in interspecific chimeras. J Exp Med 142:17-40

Lumsden AGS (1981) Evolution and adaptation of vertebrate mouth. In: Osborn JW (ed) Dental Anatomy and Embryology. Blackwell Scientific Publications, Oxford, pp 88-154

Maderson FPA (ed) (1987) Developmental and Evolutionary Aspects of the Neural Crest. John Wiley, New York

Maisey JG (1996) Discovering Fossil Fishes. Henry Holt and Co., New York

Mallatt J (1984) Feeding ecology of the earliest vertebrates. Zool J Linnean Soc 82:261-272

Marchalonis JJ, Schluter SF, Bernstein RM, Hohman VS (1998) Antibodies of sharks: revolution and evolution. Immunol Rev 166:103-122

Mathei-Prevot B, Perrimon N (1998) Mammalian and Drosophila Blood: JAK of all trades? Cell 92:697-700

Matsunaga T (1998) Did the first adaptive immunity evolve in the gut of ancient jawed fish? Cytogenet Cell Genet 80:138-141

Niedermann G, Grimm R, Geier E, Maurer M, Realini C, Gartmann C, Soll J, Omura S, Rechsteiner MC, Baumeister W, Eichmann K (1997) Potential immunocompetence of proteolytic fragments produced by proteasomes before evolution of the vertebrate immune system. J Exp Med 186:209-220

Nonaka M, Namikawa-Yamada C, Sasaki M, Salter-Cid L, Flajnik MF (1997) Evolution of proteasome subunits _ and LMP2: complementary DNA cloning and linkage analysis with MHC in lower vertebrates. J Immunol 159:734-740

Orkin SH (1996) Development of the hematopoietic system. Curr Opin Genet Dev 6:597-602

Pamer E, Cresswell P (1998) Mechanisms of MHC class I-restricted antigen processing. Annu Rev Immunol 16:323-358

Purnell MA (1995) Microwear on conodont elements and macrophagy in the first vertebrates. Nature 374:798-800

Rast JP, Litman GW (1998) Towards understanding the evolutionary origins and early diversification of rearranging antigen receptors. Immunol Rev 166:79-86

Ratcliffe NA, Rowley AF (1981) Invertebrate Blood Cells. Vol. 1 and 2. Academic Press, London

Ray D, Bosselut R, Ghysdael J, Mattei M-G, Tavitian A, Moreau-Gachelin F (1992) Characterization of Spi-B, a transcription factor related to the putative oncoprotein Spi-1/PU.1. Mol Cell Biol 12:4297-4304

Reya T, Grosschedl R (1998) Transcriptional regulation of B-cell differentiation. Curr Opin Genet Dev 10:158-165

Robb L (1997) Origin pinned down at last? Curr Biol 7:R10-R12

Schatz DG, Oettinger MA, Schlissel MS (1992) V(D)J recombination: molecular biology and regulation. Annu Rev Immunol 10:359-383

Schoenhals GJ, Krishna RM, Grandea AG, Spies T, Peterson PA, Yang Y, Fruh K (1999) Retention of empty MHC class I molecules by tapasin is essential to reconstitute antigen presentation in invertebrate cells. EMBO J 18:743-753

Sieweke MH, Graf T (1998) A transcripton factor party during blood cell differentiation. Curr Opin Genet Dev 8:545-551

Skrabanek L, Wolfe KH (1998) Eukaryote genome duplication - where's the evidence? Curr Opin Genet Dev 8:694-700

Smith JC, Albano RM (1993) Mesoderm induction and eythroid differentiation in early vertebrate development. Sem Devel Biol 4:315-324

Su GH, Chen H-M, Muthusamy N, Garrett-Sinha LA, Baunoch D, Tenen DG, Simon MC (1997) Defective B cell receptor-mediated responses in mice lacking the Ets protein, Spi-B. EMBO J 16:7118-7129

Tanaka K, Kasahara M (1998) The MHC class I ligand-generating system: roles of immunoproteasomes and the interferon-g-inducible proteasome activator PA28. Immunol Rev 163:161-176

Tanaka K, Tanahashi N, Tsurumi C, Yokota KY, Shimbara N (1997) Proteasomes and antigen processing. Adv Immunol 64:1-38

The *C. elegans* Sequencing Consortium (1998) Genome and sequence of the nematode *C. elegans*: a platform for investigating biology. Science 282:2012-2018

Trachtulec Z, Hamvas RMJ, Forejt J, Lehrach HJ, Vincek V, Klein J (1997) Linkage of TATA-binding protein and proteasome subunit C5 genes in mice and humans reveals synteny conserved between mammals and invertebrates. Genomics 44:1-7

Turpen JB, Kelley CM, Mead PE, Zon LI (1997) Bipotential primitive-definitive hematopoietic progenitors in the vertebrate embryo. Immunity 7:325-334

van Endert PM (1999) Genes regulating MHC class I processing of antigen. Curr Opin Immunol 11:82-88

Varner J, Neame P, Litman GW (1991) A serum heterodimer from hagfish *(Eptatretus stouti)* exhibits structural similarity and partial sequence identity with immunoglobulin. Proc Natl Acad Sci USA 88:1746-1750

Walker WF, Liem KF (1994) Functional Anatomy of the Vertebrates. An Evolutionary Perspective. 2nd ed. Saunders College Publishing, Fort Worth, TX

The MHC paralogous group: listing of members and a brief overview

Masanori Kasahara[1,2], Makoto Yawata[1,3], and Takashi Suzuki[1]

[1]Department of Biosystems Science, School of Advanced Sciences, The Graduate University for Advanced Studies, Shonan Village, Hayama 240-0193, Japan
[2]CREST (Core Research for Evolutional Science and Technology), Japan Science and Technology Corporation, Hayama 240-0193, Japan
[3]First Department of Internal Medicine, Yokohama City University School of Medicine, Yokohama 236-0004, Japan

Summary. Human chromosomes 1, 9, and 19 contain the regions paralogous to the major histocompatibility complex (MHC). Since our initial description, the number of gene families with copies in the MHC and these paralogous regions has been increasing steadily and now counts 37. There are at least 50 gene families that do not have copies in the MHC but share paralogous copies among the paralogous regions on chromosomes 1, 9, and 19, or between two of them. Thus, the MHC paralogous group is made up of more than 80 gene families with diverse structures, functions, and patterns of expression. Here we present the updated listing of gene families constituting this paralogous group. Systematic identification of the members of the MHC paralogous group offers a unique opportunity to deduce the organization of the primordial MHC, and to study the fate of duplicated genes.

Key words. Major histocompatibility complex, *HLA*, Genome paralogy, Duplication, Paralogous region

Introduction

The two terms, which are commonly used to define the evolutionary relationship of genes, are orthologous and paralogous. Orthologous genes are those that diverged by speciation events. Thus, the human *NOTCH1* gene is orthologous to the mouse *Notch1* gene. On the other hand, genes within a single species that arose by duplication are termed paralogous genes. For example, the human genome contains four structurally related *NOTCH* genes designated *NOTCH1* through *NOTCH4*. These *NOTCH* genes are paralogous to one another. Occasionally, closely linked sets of paralogous genes are found on more than two

27

chromosomal segments. This phenomenon is referred to as genome paralogy, and such chromosomal segments are called paralogous regions.

Thanks to the rapid advances in gene mapping and genome analysis, numerous examples of putative genome paralogy have been described recently (Nadeau and Kosowsky 1991; Lundin 1993; Ruddle et al. 1994; Ruvinsky and Silver 1997; Spring 1997; Hallböök et al. 1998; Kasahara 1998; Ollendorff et al. 1998; Patton et al. 1998; Pébusque et al. 1998; Postlethwait et al. 1998). One example of genome paralogy, which has evoked much interest both in terms of the origin of adaptive immunity and genome evolution, concerns the major histocompatibility complex (MHC) (Kasahara 1997, 1999a; Kasahara et al. 1997; Abi Rached et al. 1999; see also the chapter by L. Abi Rached and P. Pontarotti, this volume). Despite its relatively recent discovery (Kasahara et al. 1996; Katsanis et al. 1996), the MHC paralogous group already surpasses any other known paralogous groups in the number of constituting gene families and the amount of available phylogenetic information (Kasahara 1999b, c). Therefore, this paralogous group is quickly establishing itself as a prototype of genome paralogy.

Here, we present the updated, comprehensive listing of gene families that constitute the human MHC paralogous group. Unlike the previous one (Kasahara 1999b), this listing contains the members that do not have copies within the MHC — the category of genes that have been only insufficiently analyzed thus far. Systematic identification of the members of the MHC paralogous group offers a unique opportunity to deduce the organization of the primordial MHC, and to study the fate of duplicated genes.

Listing of members of the MHC paralogous group

HLA-encoded genes with paralogous copies on chromosomes 1, 9, and/or 19

The human MHC, known as the *HLA* complex, occupies ~4 megabases of the chromosomal segment on 6p21.3. The regions established to be paralogous to *HLA* are located on chromosomes 1, 9, and 19, more specifically on 1q21-q25/1p11-p32, 9q33-q34, and 19p13.1-p13.3. There is compelling evidence that human chromosome 1 underwent a pericentric inversion after the divergence of the human and chimpanzee lineages (Yunis 1982; Maresco et al. 1996). This inversion most likely accounts for the occurrence of the paralogous regions on both arms of chromosome 1.

Recently, the region of the *HLA* complex, spanning from *KNSL2* (*HSET*) to *HLA-F*, was completely sequenced (http://www.sanger.ac.uk/HGP/Chr6/MHC). Taking advantage of this sequence information, we examined whether the genes mapping to this interval have paralogous copies and if so, to which regions of the human genome such copies map. Counting genes with multiple copies within the MHC, such as class I and class II genes, as single entities, a total of 97 genes were subjected to this analysis. We included in this analysis only those genes whose functionality is well established (for the list of the genes, see Kasahara 1999b) [1]. The class II region contains at least 24 functional genes when class II genes are counted as a single entity. Among them, 12 were found to have paralogous copies in at least one of the paralogous regions on chromosomes 1, 9, and 19 (Table 1). The remaining 12 genes did not have any paralogous copies or had paralogous copies only within the MHC. An example of the latter is the MHC class II genes that most likely increased their copies by tandem duplication. Of the 51 genes mapping to the class III region, 16 had paralogous copies on chromosomes 1, 9, and/or 19 (Table 1). Of the 22 genes located between *MICB* and *HLA-F* in the class I region, five (class I, *POU5F1* , *DDR* , *TUBB*, and *KIAA0170*) had paralogous copies in at least one of the paralogous regions on chromosomes 1, 9, and 19, counting class I genes as a single entity (Table 1).

Feder et al. (1996) discovered a non-classical class I gene, designated *HFE* (the gene, the mutation of which causes hemochromatosis), ~4 megabases telomeric from the *HLA-F* gene. *HLA-F* is located at the telomeric border of the traditionally defined class I region. Thus, if we define the MHC as a contiguous stretch of DNA containing class I and class II genes, the size of the human MHC increases almost twofold from ~4 to ~8 megabases. At least four gene families that map telomeric from *HLA-F* (*GABBR* , *OLFR*, the histone gene cluster, and the HMG17 gene family coding for high-mobility group proteins) have paralogous copies on chromosome 1, 9, or 19 (Table 1). Thus, in total, the extended class I region contains at least nine members of the MHC paralogous group.

Members of the MHC paralogous group that do not have copies within the MHC

Some gene families such as *NOTCH* and *RING3* have copies on all four paralogous regions (Table 1). However, such gene families are rare; most gene families listed in Table 1 have only two or three paralogous copies. This suggests that some gene families should share paralogous copies only among 9q33-q34, 1q21-q25/1p11-p32, and 19p13.1-p13.3 or between two of them. To identify such members, we performed a computer search of the NCBI UniGene database (http://www.ncbi.nlm.nih.gov/Unigene/). This search identified at least 50 gene families that qualified as members of the MHC paralogous group (Table 2). Thus, in total, the MHC paralogous group is comprised of more than 80 members at present. Compared with the MHC, the paralogous segments on chromosomes 1, 9, and 19 are less well characterized. Therefore, many more members with no copies in the MHC are likely to be discovered as the Human Genome Project progresses.

---

1. As discussed below, the chromosomal duplication that formed the MHC paralogous group appears to have taken place close to the origin of vertebrates. We, therefore, searched only for those paralogous copies presumed to have emerged in the vertebrate or chordate lineage.

Table 1. Members of the MHC paralogous group with copies in the *HLA* complex [a]

MHC	Chromosome 9	Chromosome 1	Chromosome 19
Gene symbol	Gene symbol (Location)	Gene symbol (Location)	Gene symbol (Location)
CLASS II			
SYNGAP		*NGAP* (1q25)	
RPL12L	*RPL12* (D9S1821-D9S159)		
BING1	*ZNF-X* (9q34)		
RGL2 (*RALGDS2*)	*RALGDS* (9q34)		
B3GALT4 [b]		*B3GALT2* (1q31)	*B3GALT-like* (19p13.1)
RPS18 (*KE3*)		*RPS18-like* (D1S196-D1S210)	
RING1 (*RNF1*)		*BAP-1* (D1S2640-D1S461)	
RXRB	*RXRA* (9q34.3)	*RXRG* (1q22-q23)	
COL11A2	*COL5A1* (9q34.2-q34.3)	*COL11A1* (1p21)	
RING3 (*RNF3*)	*RING3L* (9q34)	*BRDT* (1p21-p22)	*HUNKI* (19p13.1)
LMP2, LMP7	*PSMB7* (9q34.11-q34.12)		
TAP1, TAP2 [c]	*ABC2* (9q34)		
CLASS III			
NOTCH4	*NOTCH1* (9q34.3)	*NOTCH2* (1p13-p11)	*NOTCH3* (19p13.2-p13.1)
PBX2	*PBX3* (9q33-q34)	*PBX1* (1q23)	
LPAATA (*G15*)	*LPAATB* (9q34.3)		
PPT2 (*G14*)		*PPT1* (1p32)	
CREBL1 (*CREB-RP, G13*)		*ATF6* (1q21)	

Table 1. (continued)

MHC	Chromosome 9	Chromosome 1	Chromosome 19
Gene symbol	Gene symbol (Location)	Gene symbol (Location)	Gene symbol (Location)
TNTX, TNXB2	HXB (9q33)	TNR (1q24)	
C4B C4A	C5 (9q34.1)		C3 (19p13.3-p13.2)
G9A		KIAA0067 (D1S514-D1S2635)	
NG22			clone 24686 [AF070636] (D19S413-D19S221)
HSPA1B HSPA1A HSPA1L	HSPA5 (9q34)	HSPA6 (D1S2635-D1S2844) HSPA7 (1q21.3)	
CLIC1 [d] (NCC27)	CLIC3 (9q34)	p64H1 [AF109196.1] (D1S2843-D1S417)	
NG30		NG30-like [AB001915] (D1S203-D1S2865)	
BAT2	KIAA0515 (D9S159-qTEL)	Clone 598F2 [AL021579] (1q23.1-q24.3)	
AIF1		AIF1-like [AL031283] (D1S434-D1S2843)	
TNFSF3 (LTB) [e] TNFSF2 (TNF) TNFSF1 (LTA)	TNFSF8 (CD30L) (9q33)	TNFSF4 (OX40L) (1q25) TNFSF6 (FasL) (1q23)	TNFSF7 (CD70) (19p13) TNFSF9 (4-1BBL) (19p13.3)
BAT1			BAT1-like [U90426] (D19S221-D19S226)

CLASS I

Class I genes		CD1 (1q21-q23)	
POU5F1 (OTF3) [f]		POU3F1 (Oct-6) (1p34.1)	
DDR (NTRK4) [g]	NTRK2 (9q22.1)	NTRK1 (1q21-q22)	
TUBB (TUBB1)	TUBB2 [X02344] (9q34)		
KIAA0170		MSA (1q25)	

Table 1. (continued)

MHC	Chromosome 9	Chromosome 1	Chromosome 19
Gene symbol	Gene symbol (Location)	Gene symbol (Location)	Gene symbol (Location)
GABBR1	GABBR2 (D9S287-D9S176)		
OLFR2 [h]	OLFR3 (9q21-q22, 9q34)		OLFR (19p13.1)
H1, H2A, H2B, H3, H4		H2A, H2B, H3, H4 (1q21)	
HMG17-like (NHC)		HMG17 (1p36.5-p35)	

a, Genes are arranged from centromere (top) to telomere (bottom). When no officially approved gene symbols are available, the names that appear in the UniGene database are used. Alternative names are in parentheses.

For some genes, GenBank accession numbers are shown in square brackets.

b, The β 1, 3 galactosyl transferase family has at least two more paralogous copies : *B3GALT3* on 3q25 and a *B3GALT-like* gene on 21q22.3. Whether the latter is a functional gene is not known.

c, The human genome contains more than 50 ABC transporter genes. Thus, it is questionable whether *TAP* and *ABC2* diverged by chromosomal duplication.

d, *CLIC2* is located on Xq28.

e, The TNF ligand superfamily has at least six additional paralogous copies : *TNFSF10* (*TRAIL*) on 3q26, *TNFSF5* (*CD40 ligand*) on Xq26, *TNFSF11* (*RANKL*) on 13q14, *TNFSF12* (*TWEAK*) on 17p13, *TNFSF13* (*April*) on chromosome 17, and *TNFSF14* (*HVEM-L*) on chromosome 16. It is not known when the divergence of these members took place.

f, The *POU3F4* gene, which maps to Xq21.2, might have diverged from *POU5F1* early in vertebrate or chordate evolution.

g, The neurotrophic tyrosine kinase receptor family has one more paralogous copy, designated *NTRK3*, that maps to 15q25.

h, Olfactory receptor genes have many copies all over the genome.

Fate of duplicated genes

The MHC and its three paralogous regions appear to have emerged as a result of chromosomal or block duplication (see below). If we assume that they arose by two rounds of chromosomal duplication, the maximum number of paralogous copies per gene family is four. Therefore, the observation that most gene families listed in Tables 1 and 2 have only two or three paralogous copies suggests that a substantial number of paralogous copies were inactivated and then lost from the genome.

The *HLA* class II region contains ~24 functional genes when class II genes are

Table 2. Members of the MHC paralogous group that do not have copies within the *HLA* complex

	Chromosome 1		Chromosome 9		Chromosome 19	
	Gene symbol [a]	Location	Gene symbol	Location	Gene symbol	Location
v-abl Abelson murine leukemia viral oncogene-homologue	ABL2 (NM_005158)	1q24-q25	ABL1 (X16416)	9q34.1		
Cargo selection protein			ADFP (X97324)	D9S1808-D9S162	TIP47 (AF057140)	pTEL-D19S413
Adenylate kinase [b]	AK2 (U84371)	1p34	AK1 (J04809)	9q34.1		
Transcription factor	B1F2 (AF124247)	1q31-q32.1	GCNF (X99975)	9q33-q34.1		
Calcium channel, voltage-dependent, L type, alpha subunit [c]	CACNA1E (L29384)	1q25-q31	CACNA1B (M94173)	9q34	CACNA1A (NM_000068)	19p13.1
Cyclin-dependent kinase inhibitor 2	CDKN2C (AF041248)	1p32	CDKN2A (L27211) CDKN2B (L36844)	9p21 9p21	CDKN2D (U40343)	19p13
Clone 677 and clone 510D11	Clone 510D11 (Z98044)	D1S214-D1S244			Clone 677 (AF091091)	19p13
Clone 774I24 and KIAA0634	Clone 774I24 (AL031290)	1q24.1-q24.3	KIAA0634 (AB0145534)	D9S177-D9S1864		
Calponin [d]	CNN3 (S80562)	1p22-p21			CNN1 (D17408)	19p13.2-p13.1
Cathepsin [e]	CTSS (M90696) CTSK (NM_000396)	1q21 1q21	CTSL (X91755)	9q21-q22		
DNA-damage-inducible transcript	DDIT1 (M60974)	1p34-p12			MYD118 (AF090950)	qTEL-D19S413

Table 2. (continued)

	Chromosome 1		Chromosome 9		Chromosome 19	
	Gene symbol	Location	Gene symbol	Location	Gene symbol	Location
Dynamin			DNM1 (NM_004408)	9q34	DNM2 (NM_004945)	D19S221-D19S226
Lysophosphatidic acid G-protein-coupled receptor [f]	EDG1 (M31210)	D1S2865-D1S418	EDG2 (U80811)	D9S176-D9S279	EDG4 (AF011466)	19p12
			EDG3 (AF022139)	9q22.1-q22.2	EDG6 (NM_003775)	19p13.3
Ephrin-A [g]	EFNA3 (U14187)	1q21-q22			EFNA2 (NM_001405.1)	19p13.3
	EFNA1 (NM_004428.1)	1q21-q22				
	EFNA4 (NM_005227.1)	1q21-q22				
ELAV (embryonic lethal, abnormal vision)-like	ELAVL4 (M62843)	1p34	ELAVL2 (U12431)	9p21	ELAVL1 (U38175)	19p13.2
					ELAVL3 (L26405)	19p13.2
Rab6 GTPase activating protein	IDN4-GGTR14 (AB019492)	1q24-q25	DKFZp586D2123 (AK050195)	D9S258-D9S1821		
	IDN4-GGTR9 (AB019493)	1q24-q25				
	IDN4-GGTR6 (AB019489)	1q24-q25				
	IDN4-GGTR7 (AB019490)	1q24-q25				
	IDN4-GGTR8 (AB019491)	1q24-q25				
Class I cytokine receptor	IL12RB2 (U64198)	1p31.3-p31.2			WSX-1 (AF053004)	D19S221-D19S226

Table 2. (continued)

	Chromosome 1		Chromosome 9		Chromosome 19	
	Gene symbol	Location	Gene symbol	Location	Gene symbol	Location
Interleukin 6 receptor	IL6R (X58298)	1q21	IL11RA (U32324)	9p13	CRLF1 (NM_004750.1)	19p12
Insulin receptor	INSRR (J05046)	Chr 1			INSR (X02160)	19p13.3-p13.2
Protein tyrosine kinase	JAK1 (M64174)	1p31.3	JAK2 (AF005216)	9p24	TYK2 (X54637)	19p13.2
					JAK3 (U09607)	19p13.1
v-jun avian sarcoma virus 17 oncogene homolog	JUN (J04111)	1p32-p31			JUNB (M29039)	19p13.2
					JUND (X51346)	19p13.2
KIAA0677 and KIAA0876	KIAA0677 (AB014577)	D1S2843-D1S417			KIAA0876 (AB020683)	pTEL-D19S413
Laminin, gamma	LAMC1 (NM_002293.1)	1q31	LAMC3 (NM_006059.1)	9q31-q34		
Lamin	LMNA (M13452)	1q21.2-q21.3			LMNB2 (M94363)	19p13.3
MADS box transcription enhancer factor 2 [h]	MEF2D (L16794)	1q12-q23			MEF2B (X68502)	19p12
Neuronal olfactomedin related ER localized protein	MYOC (AF049793)	1q24	Clone 23876 (AF035301)	D9S159-qTEL		
CCAAT box-binding transcription factor	NFIA (U07809)	1p31.3-p31.2	NFIB (U85193)	9p24.1	NFIC (X12492)	19p13.3
					NFIX (U18759)	19p13.3

Table 2. (continued)

	Chromosome 1		Chromosome 9		Chromosome 19	
	Gene symbol	Location	Gene symbol	Location	Gene symbol	Location
Natriuretic peptide receptor A/guanylate cyclase	NPR1 (X15357)	1q21-q22	NPR2 (NM_003995)	9p21-p12		
SH2-containing protein NSP [i]	NSP2 (AF124250.1)	D1S2865-D1S418			NSP1 (AF124249)	pTEL-D19S413
IB3089A	PAC 1026E2 (AL035289)	1q24.1-q25.3	IB3089A (AF027734)	9q32-q33		
NY-CO-3	PAC 262D12 (Z99297)	1q23.3-q24.3	NY-CO-3 (AF039688)	D9S159-qTEL		
Angiopoietin-like factor	PAGA (L19184.1)	1p34.1			CTD6 (Z22548)	D19S221-S226
PTPL1-associated RhoGAP1	PARG1 (NM_004815)	D1S2865-D1S418			KIAA0223 (D86976)	19p13.3
Phosphodiesterase 4B, cAMP-specific [j]	PDE4B (L12686)	1q31			PDE4A (L20965) PDE4C (U66346)	19p13.2 D19S899-D19S407
Peptidyl-prolyl isomerase	PIN1L (U82382)	1p31			PIN1 (U49070)	19p13
Phosphatidylinositol-4-phosphate 5-kinase, type I	PIP5K1A (U78575)	D1S514-D1S2635	PIP5K1B (X92493)	9q13-q21.1	PIP5K1G (AB011161)	19p13.3
Paired homeobox protein	PMX1 (Z97200)	1q24	PRX2 (U81600)	D9S159-qTEL		
Phosphatidic acid phosphatase type 2 [k]	PPAP2B (AB000889)	D1S417-D1S476			PPAP2C (AF047760)	pTEL-D19S413

Table 2. (continued)

	Chromosome 1		Chromosome 9		Chromosome 19	
	Gene symbol	Location	Gene symbol	Location	Gene symbol	Location
Protein kinase C-like	PRKCL2 (U33052)	D1S2865-D1S418			PRKCL1 (D26181)	19p13.1-p12
Prostaglandin E receptor [1]	PTGER3	1p31.2			PTGER1	19p13.1
Prostaglandin-endoperoxide synthase	PTGS2 (M90100)	1q25.2-q25.3	PTGS1 (M59979)	9q32-q33.3		
RAS oncogene family member	RAB3B (M28214)	1p32-p31			RAB3A (M28210)	19p13.2
RAD23 (S. cerevisiae) homolog [m]	RAD23B-like (Hs.75563)	D1S203-D1S2865	RAD23B-like (Hs.75563)	D9S176-D9S279	RAD23A (D21235)	19p13.2
Regulator of G-protein signalling [n]	RGS1 (NM_002922)	1q31	RGS3 (U27655)	9q31-q33		
	RGS2 (NM_002923)	1q31				
	RGS4 (NM_005613)	D1S2635-D1S2844				
	RGS5 (NM_003617)	1q23				
	?RGS7 (U32439)	D1S459-D1S304				
	?RGS13 (AF030107)	D1S461-D1S2622				
	RGS16 (NM_002928)	1q25-q31				
SWI/SNF related, matrix associated, actin-dependent regulator of chromatin			SMARCA2 (D26155)	9p24-p23	SMARCA4 (D26156)	pTEL-D19S413
Spectrin, alpha, non-erythrocytic	SPTA1 (NM_003126)	1q21	SPTAN1 (J05243)	9q33-q34		

Table 2. (continued)

	Chromosome 1		Chromosome 9		Chromosome 19	
	Gene symbol	Location	Gene symbol	Location	Gene symbol	Location
Syntaxin-binding protein 1			STXBP1 (D63851)	9q34.1	Hunc18b2 (AB002559)	pTEL-D19S143
Helix-loop-helix protein	TAL1 (M29038)	1p32	TAL2 (NM_005421)	9q31	LYL1 (M22637)	19p13.2
Transducin-like enhancer of split [o]			TLE4 (M99439)	D9S153-D9S264	TLE1 (M99435)	19p13.3
					TLE2 (M99436)	19p13.3
Vav oncogene	VAV3 (AF067817)	D1S2865-D1S418	VAV2 (S76992)	9q34.1	VAV1 (X16316)	19p13.2

a, Accession numbers or UniGene Hs. numbers are given in parentheses. Officially approved gene symbols are available for only a fraction of genes. When such symbols are not available, clone names or the names that appear in the UniGene database are shown. The genes are arranged in an alphabetical order.

b, *AK3* on 9q24-p13.

c, *CACNA1C*, *CACNA1D*, and *CACNA1F* appear to have diverged from the other *CACNA1* genes before the emergence of chordates.

d, *CNN2* maps to chromosome 1, 9, 10, 12, or 19. Its exact location is not known.

e, The gene for cathepsin O maps to 4q31-q32. This gene may have diverged from *CTSS*/*CTSK*/*CTSL* close to the origin of vertebrates.

f, The location of *EDG5* is not known.

g, *EFNA5* maps to D5S495-D5S492.

h, *MEF2A* and *MEF2C* map to 15q26 and 5q14, respectively.

i, The location of *NSP3* is not known.

j, *PDE4D* maps to 5q12.

k, *PPAP2A* maps to D5S474-D5S491.

l, *PTGER2* and *PTGER4* map to 5p13.1.

m, *RAD23B* maps to 3p25.1.

n, *RGS6* and *RGS12* map to 14q24.3 and 4p16.3, respectively.

o, *TLE3* maps to chromosome 15.

counted as a single entity. If we assume that the ancestors of all of these genes were present in a pre-duplicated chromosomal segment, two rounds of chromosomal duplication should have produced a total of ~96 (~24 x 4) genes. The number of paralogous genes actually retained is 41 (24 genes in the MHC class II region plus 17 genes on chromosomes 1, 9, and 19). Thus, ~43% of duplicated genes appear to have survived and acquired new functions. This type of analysis is likely to give a minimal estimate for two reasons. First, not all genes currently residing in the class II region may have existed in a pre-duplicated chromosomal segment. Second, the number of paralogous copies may increase as genome analysis progresses. Similar analysis can be performed for the genes residing in *HLA* class III region. Here, at least 79 paralogous copies (51 genes in the MHC class III region plus 28 genes on chromosomes 1, 9, and 19) out of a total of ~204 (~51 x 4) have survived two rounds of chromosomal duplication. Thus, the survival rate is ~39%. Therefore, combined analysis of the genes mapping to the class II and class III regions suggests that ~40% of genes have survived two rounds of duplication and achieved functional divergence. (We did not perform similar analysis for the genes mapping to the *HLA* class I region because the accurate number of functional genes in this region is not known.)

Conservation of gene order

Information currently available is not sufficient to assess fully how well the gene orders are conserved among the four paralogous regions. However, preliminary analysis indicates that they are poorly conserved (Kasahara 1999a). Thus, if the four paralogous regions have descended from a common ancestral region, each of them must have experienced numerous inversions and/or intrachromosomal translocations after its inception.

Origin of the MHC paralogous group

Block duplication versus functional clustering

Two explanations have been proposed as to how the paralogous regions emerged. The first is that they are descended from a common ancestral region and emerged as a result of large-scale block or chromosomal duplications. The second is that they represent assemblages of genes brought into proximity by selective forces. The gene families constituting the MHC paralogous group are quite diverse in terms of structure, function, gene size, and expression pattern (Tables 1 and 2). Furthermore, the number of gene families constituting the MHC paralogous group is quite large. These observations suggest strongly that the major mechanism involved in the formation of the MHC paralogous group was large-scale block duplications.

Relationship of the four paralogous regions

At least two rounds of duplication are required to create four paralogous regions. Phylogenetic analysis of the genes coding for tenascins (*TNX, TNR, HXB*), PBX, type V/XI collagens (*COL*), retinoid X receptors (*RXR*), and NOTCH indicates that the copies located on chromosomes 1 and 9 are more closely related to each other than they are to the corresponding copies on the MHC (Katsanis et al. 1996; Endo et al. 1997; Kasahara 1997, 1999a; Hughes 1998). Thus, it appears that the paralogous regions on chromosomes 1 and 9 shared an immediate common ancestral region. The relationship of 19p13.1-p13.3 to the other paralogous regions is less clear. If 19p13.1-p13.3 and 6p21.3 were to share an immediate common ancestral region as suggested by some of the gene families, the existence of the four paralogous regions would be accounted for by postulating two rounds of duplication. If this is not the case, one must postulate a more complex scenario: at least three rounds of duplication and cluster loss events.

Role of duplications in the emergence of the adaptive immune system

Accumulated evidence suggests that the prototypes of the four paralogous regions including the MHC emerged for the first time in a common ancestor of jawed vertebrates after its separation from jawless vertebrates (Kasahara 1999a, b). Thus, the genome of jawless vertebrates is unlikely to have a full-fledged MHC region. Instead, it is likely to contain one or two primitive MHC-like regions, which should be called the primordial MHC, proto-MHC, or Ur-MHC. This prediction is consistent with the observation that the adaptive immune system characterized by the MHC and antigen-specific rearranging receptors has been identified only in jawed vertebrates (Kasahara et al. 1995; Klein and Sato 1998; Du Pasquier and Flajnik 1999; Litman et al. 1999). The MHC paralogous group encodes several accessory molecules of the adaptive immune system, such as LMP (low molecular mass polypeptide), VAV1, NOTCH1, C4, and various members of the tumor necrosis factor ligand superfamily including the Fas ligand (Tables 1 and 2). These molecules appear to have emerged as a result of the chromosomal duplications that formed the MHC paralogous group. These observations have lead to the proposal that the formation of the four MHC paralogous regions played an important role in the emergence of the adaptive immune system (Kasahara et al. 1997).

Genome organization of the primordial MHC

Systematic identification of the members of the MHC paralogous group enables one to infer the assemblage of genes that existed in a pre-duplicated region, and hence, offers a unique opportunity to deduce the organization of the primordial MHC.

Experimental evidence accumulated thus far is largely consistent with the assumption that the primordial MHC was made up of the precursors of the members of the MHC paralogous group. For example, *RING3*, *PSMB* (proteasome, ß-type subunit), and *RXR* are members of the MHC paralogous group. Thus, we can expect that the MHCs of jawed vertebrates contain the genes coding for RING3, LMP, and RXRB. This is indeed the case with the MHC of the bony fish (Takami et al. 1997). The former two genes, for which the information is currently available, are also MHC-encoded in *Xenopus laevis* (Flajnik et al. 1999). More direct evidence for the existence of the pre-duplicated region was provided by Trachtulec et al. (1997), who found that the precursor genes of at least four members of the MHC paralogous group (the genes coding for the RXR-like, PBX-like, NOTCH-like, and tenascin-like molecules) are located within a distance of 5.8 megabases on chromosome III of the nematode *Caenorhabditis elegans*. More recent work indicates that *NOTCH-* and *PBX*-like genes are also tightly linked in the genome of the sea urchin, *Strongylocentrotus purpuratus* (see the chapter by J. Rast et al., this volume). Thus, there is now convincing evidence that a primitive form of the pre-duplicated region exists in invertebrates.

Some members of the MHC paralogous group must have lost all but one copy. If the only remaining copy maps to human chromosome 1, 9, or 19, it is impossible to realize that its precursor was originally a member of the primordial MHC. However, some gene loss events must have occurred independently in each vertebrate class, order, or family. Thus, as previously pointed out (Kasahara 1999a), a gene that maps to human chromosome 19 can have its corresponding copy in the MHC of *Fugu rubripes*. Therefore, it is vital to collect and synthesize relevant information from many vertebrate and invertebrate species if we wish to reconstruct the complete organization of the primordial MHC.

Concluding remarks

Nearly 30 years ago, Ohno (1970) proposed that genome-wide duplications took place at least once at the stage of fish or amphibians. Recently, this hypothesis was refined taking into account rapidly accumulating genome mapping data. Sharman and Holland (1996) proposed that widespread gene duplication took place close to vertebrate origins and again in a common ancestor of jawed vertebrates after its separation from jawless fishes; they suggested that the latter duplication presumably involved tetraploidization of the genome. Sidow (1996) proposed two rounds of genome-wide duplication, one in a common ancestor of all jawed and jawless vertebrates, and the other in a common ancestor of jawed vertebrates after its separation from jawless fishes. Paralogous regions of various lengths appear to be scattered all over the human genome. Furthermore, in most cases, they appear to have arisen close to the origin of vertebrates. These observations favor the idea that paralogous regions represent the remnants of genome-wide duplication events. This most likely applies also to the MHC paralogous group. An abundance of members and phylogenetic information confers upon the MHC paralogous group a distinct advantage in scrutinizing the origin of paralogous groups in the vertebrate genome. Studies are in progress in our laboratory to determine how many paralogous regions are present in the genome of jawless fishes. Detailed structural analysis of the paralogous region(s) in this class of vertebrates should also provide important insights into the origin of adaptive immunity.

Acknowledgments

We thank Tatsuya Ota for his helpful comments. This work was supported in part by Grants-in-Aid for Scientific Research from The Ministry of Education, Science, Sports and Culture of Japan, and a grant provided by The Naito Foundation. This article is contribution No. 14 from the Department of Biosystems Science, School of Advanced Sciences, The Graduate University for Advanced Studies, Hayama 240-0193, Japan.

References

Abi Rached L, McDermott MF, Pontarotti P (1999) The MHC big bang. Immunol Rev 167:33-44

Du Pasquier L, Flajnik MF (1998) Origin and evolution of the vertebrate immune system. In: Paul WE (ed) Fundamental Immunology. 4th ed. Lippincott-Raven, Philadelphia-New York, pp 605-650

Endo T, Imanishi T, Gojobori T, Inoko H (1997) Evolutionary significance of intra-genome duplications on human chromosomes. Gene 205:19-27

Feder JN, Gnirke A, Thomas W, Tsuchihashi Z, Ruddy DA, Basava A, Dormishian F, Domingo RJ, Ellis MC, Fullan A, Hinton LM, Jones NL, Kimmel BE, Kronmal GS, Lauer P, Lee VK, Loeb DB, Mapa FA, McClelland E, Meyer NC, Mintier GA, Moeller N, Moore T, Morikang E, Prass CE, Quintana L, Starnes SM, Schatzman RC, Brunke KJ, Drayna DT, Risch NJ, Bacon BR, Wolff RK (1996) A novel MHC class I-like gene is mutated in patients with hereditary haemochromatosis. Nat Genet 13:399-408

Flajnik MF, Ohta Y, Namikawa-Yamada C, Nonaka M (1999) Insights into the primordial MHC from studies in ectothermic vertebrates. Immunol Rev 167:59-67

Hallböök F, Lundin L-G, Kullander K (1998) *Lampetra fluviatilis* neurotrophin homolog, descendant of a neurotrophin ancestor, discloses the early molecular evolution of neurotrophins in the vertebrate subphylum. J Neurosci 18:8700-8711

Hughes AL (1998) Phylogenetic tests of the hypothesis of block duplication of homologous genes on human chromosomes 6, 9, and 1. Mol Biol Evol 15:854-870

Kasahara M (1997) New insights into the genomic organization and origin of the major histocompatibility complex: Role of chromosomal (genome) duplication in the emergence of the adaptive immune system. Hereditas 127:59-65

Kasahara M (1998) What do the paralogous regions in the genome tell us about the origin of the adaptive immune system? Immunol Rev 166:159-175

Kasahara M (1999a) The chromosomal duplication model of the major histocompatibility complex. Immunol Rev 167:17-32

Kasahara M (1999b) Genome dynamics of the major histocompatibility complex: insights from genome paralogy. Immunogenetics 49:in press

Kasahara M (1999c) Genome paralogy: A new perspective on the organization and origin of the major histocompatibility complex. Curr Top Microbiol Immunol 248:in press

Kasahara M, Flajnik MF, Ishibashi T, Natori T (1995) Evolution of the major histocompatibility complex: a current overview. Transplant Immunol 3:1-20

Kasahara M, Hayashi M, Tanaka K, Inoko H, Sugaya K, Ikemura T, Ishibashi T (1996) Chromosomal localization of the proteasome Z subunit gene reveals an ancient chromosomal duplication involving the major histocompatibility complex. Proc Natl Acad Sci USA 93:9096-9101

Kasahara M, Nakaya J, Satta Y, Takahata N (1997) Chromosomal duplication and the emergence of the adaptive immune system. Trends Genet 13:90-92

Katsanis N, Fitzgibbon J, Fischer EMC (1996) Paralogy mapping: Identification of a region in the human MHC triplicated onto human chromosomes 1 and 9 allows the prediction and isolation of novel *PBX* and *NOTCH* loci. Genomics 35:101-108

Klein J, Sato A (1998) Birth of the major histocompatibility complex. Scand J Immunol 47:199-209

Litman GW, Anderson MK, Rast JP (1999) Evolution of antigen binding receptors. Annu Rev Immunol 17:109-147

Lundin LG (1993) Evolution of the vertebrate genome as reflected in paralogous chromosomal regions in man and the house mouse. Genomics 16:1-19

Maresco DL, Chang E, Theil KS, Francke U, Anderson CL (1996) The three genes of the human FCGR1 gene family encoding $Fc\gamma RI$ flank the centromere of chromosome 1 at 1p12 and 1q21. Cytogenet Cell Genet 73:157-163

Nadeau JH, Kosowsky M (1991) Mouse map of paralogous genes. Mamm Genome 1:S433-S460

Ohno S (1970) Evolution by Gene Duplication. Springer-Verlag, New York

Ollendorff V, Mattei M-G, Fournier E, Adelaide J, Lopez M, Rosnet O, Birnbaum D (1998) A third human CBL gene is on chromosome 19. Int J Oncol 13:1159-1161

Patton SJ, Luke GN, Holland PWH (1998) Complex history of a chromosomal paralogy region: Insights from amphioxus aromatic amino acid hydroxylase genes and insulin-related genes. Mol Biol Evol 15:1373-1380

Pébusque M-J, Coulier F, Birnbaum D, Pontarotti P (1998) Ancient large-scale genome duplications: phylogenetic and linkage analyses shed light on chordate genome evolution. Mol Biol Evol 15:1145-1159

Postlethwait JH, Yan Y-L, Gates MA, Horne S, Amores A, Brownlie A, Donovan A, Egan ES, Force A, Gong Z, Goutel C, Fritz A, Kelsh R, Knapik E, Liao E, Paw B, Ransom D, Singer A, Thomson M, Abduljabbar TS, Yelick P, Beier D, Joly J-S, Larhammar D, Rosa F, Westerfield M, Zon LI, Johnson SL, Talbot WS (1998) Vertebrate genome evolution and the zebrafish gene map. Nat Genet 18:345-349

Ruddle FH, Bentley KL, Murtha MT, Risch N (1994) Gene loss and gain in the evolution of the vertebrates. Development 1994 Suppl:155-161

Ruvinsky I, Silver LM (1997) Newly identified paralogous groups on mouse chromosomes 5 and 11 reveal the age of a T-box cluster duplication. Genomics 40:262-266

Sharman AC, Holland PWH (1996) Conservation, duplication, and divergence of developmental genes during chordate evolution. Netherlands J Zool 46:47-67

Sidow A (1996) Gen(om)e duplications in the evolution of early vertebrates. Curr Opin Genet Dev 6:715-722

Spring J (1997) Vertebrate evolution by interspecific hybridization - are we polyploid? FEBS Lett 400:2-8

Takami K, Zaleska-Rutczynska Z, Figueroa F, Klein J (1997) Linkage of *LMP*, *TAP*, and *RING3* with *Mhc* class I rather than class II genes in the zebrafish. J Immunol 159:6052-6060

Trachtulec Z, Hamvas RMJ, Forejt J, Lehrach HR, Vincek V, Klein J (1997) Linkage of TATA-binding protein and proteasome subunit C5 genes in mice and humans reveals synteny conserved between mammals and invertebrates. Genomics 44:1-7

Yunis JJ (1982) The origin of man: a chromosomal pictorial legacy. Science 215:1525-1530

The MHC « Big-Bang » : duplication and exon shuffling during chordate evolution. A hypothetico-deductive approach

Laurent Abi Rached and Pierre Pontarotti

Unité INSERM 119, Cancérologie Expérimentale, 27 bd Leï Roure, 13009 Marseille, France

Summary. The major histocompatibility complex (MHC) is a genomic region that has for many years interested immunologists, population geneticists, and evolutionary biologists (Klein 1986). Here we present a hypothesis regarding the evolution of this chromosomal region and a strategy that we have developed to test this hypothesis.

Key words. MHC, Evolution, Duplication

The MHC « Big-Bang »

We have shown (Abi Rached et al. 1999; Henry et al. 1999) that most of the genes located in the major histocompatibility complex (MHC) of vertebrates result from cis duplication and exon shuffling, and constitute several multigenic families. The exon shuffling occurred close to and before the radiation of jawed vertebrates (Fig. 1). This conclusion is based on the fact that the genes issued from such rearrangements (in particular MHC class I and class II genes) and their corresponding function are only found in jawed vertebrates (Klein and O'hUigin 1993).

In the vicinity of these genes forming multigenic families are found other genes called anchor genes, which are single copy in this portion of genome. It should be noted that the anchor genes are stable in evolution, and have evolved more slowly than genes forming multigenic families. Indeed unlike genes forming multigenic families, anchor genes are found in non-vertebrate species. Some of the MHC anchor genes are found in three other genomic regions representing paralogous regions (Kasahara et al. 1997). These could be the remains of two large scale duplications that occurred after the separation of cephalochordate ancestors and those of jawed vertebrates (Holland et al. 1994) (Fig. 2). Our hypothesis is that the genetic information contained in these regions became

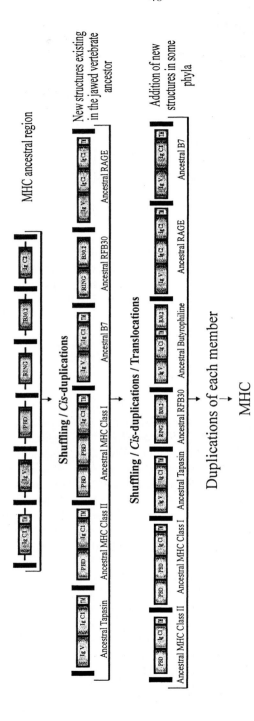

Fig. 1. Apparition of the MHC multigenic family precursors from few ancestral domains, and by a quick round of gene cis-duplication and exon shuffling. The ancestral domains are : Ig (Immunoglobulin) C1, C2 or V; PBD (peptide binding domain) ; RING (zinc finger domain) and B30.2 (a domain of about 160 amino acids whose function is unknown (Henry et al. 1998)). Beside these domains are black boxes representing anchor gene-encoded domains.

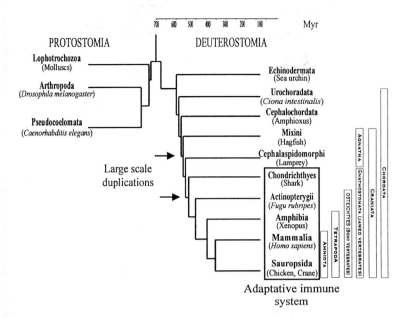

Fig. 2. Schematic phylogenetic tree of the triploblastic metazoan. Adapted from the « Tree of Life Project » (http://phylogeny.arizona.edu/tree/phylogeny.html). Dating is from Doolittle et al. (1996) and Kumar and Hedges (1998).

redundant in those species whose ancestors' genomes underwent two rounds of duplication. Genomic rearrangement (duplication, exon shuffling, etc.) was permitted in this region, even if important genes may have been deleted during this process, with no deleterious effect (Fig. 3). As a corollary, such chromosomal rearrangements must have been absent in species where large scale duplications did not occur, since gene deletions would have been deleterious.

At this point in the article, we would like to introduce two neologisms which can be useful for the comprehension of the following parts. Structural plesiogenes are genes which are structurally close to their ancestor genes (encoding proteins with the same domain organization). Structural apogenes are genes in which the structure is rearranged compared to the ancestral structures (encoding proteins with different domain organization). In a more general manner, plesiogenes are genes which are functionally close to their ancestral genes, while apogenes are genes whose function is derived from the ancestral one (in this case the structure can be rearranged or not). This idiolect is based on definitions used in comparative anatomy i.e. plesiomorphy which designates an ancestral character and apomorphy which designates a derived character.

A resume of our hypothesis follows: During the evolution history of jawed vertebrates, duplications of the whole ancestral MHC region have occurred. This ancestral region contained 1) anchor genes and 2) ancestral genes which gave rise to the apogenes present in the actual MHC. Each structural apogene then duplicated and gave rise to different multigenic families. In species where rounds of large scale duplications did not occur, structural plesiogenes should be found, and furthermore such structural plesiogenes should still be linked to anchor genes. The analysis of the MHC-like regions in such species is a first step toward the demonstration of our hypothesis (Fig. 4).

Test of the hypothesis

At this stage of the discussion we have to define the criteria for the choice of the model species. In order to facilitate the analysis, it is necessary to work with a representative of the phylogenetically closest phylum to the jawed vertebrates in which no MHC is present and whose ancestor genome has not undergone two rounds of duplication.

The hagfish (Hyperotreti) and lamprey (Hyperoartia) are representatives of the two closest phyla of jawed vertebrates. They do not possess an MHC. However, their genomes seem to have undergone one round of large scale duplication (Holland et al. 1994) (Fig. 2). Therefore, according to our hypothesis, it is possible that rearrangements occurred in one or the other duplicate MHC-like copies (or indeed in both of them).

The next closest phylum is the cephalochordate phylum, one of the actual representative of which is the amphioxus. It does not possess an MHC and its genome does not seem to have undergone any large scale duplication via tetraploidization. We think therefore that the MHC-like region of the amphioxus contains anchor genes in synteny with structural plesiogenes.

The anchor gene lineage evolved more slowly than the structural apogene lineage (Abi Rached et al. 1999). It will be therefore easier to evidence anchor genes in such species than structural plesiogenes (we do not know how the structural plesiogenes evolved, but as we only have access to the sequence of apogenes, the corresponding plesiogenes will be difficult to show). The evidence for structural plesiogenes can be sought by sequencing the regions which are in the vicinity of the anchor genes, followed by bioinformatic analysis.

Our first results show that large scale duplications of the region occurred after the amphioxus/vertebrate separation, which is the first step toward proving our hypothesis (this analysis has been done by the cloning of anchor genes and phylogenetic analysis (Fig. 5)). The structural plesiogenes related to class I and class II genes are currently searched for by sequencing cosmids containing the anchor genes.

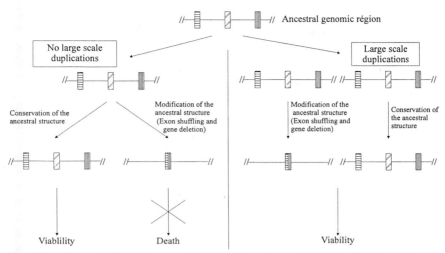

Fig. 3. Possible evolution schemes of a genomic region and their implications. Schemes without and with large scale duplications are shown on the left and right, respectively.

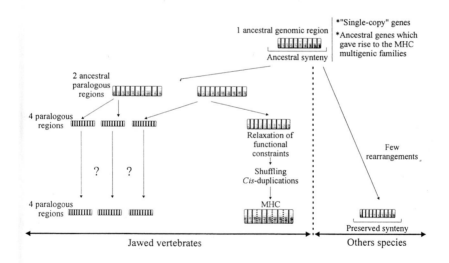

Fig. 4. A model explaining the evolution of an MHC ancestral region. Two distinct cases are considered: the first one in species where two rounds of large scale duplication occured and the second one in species where these events did not occur. In the first case, genomic rearrangement was permitted due to the redundancy of the genetic information, and led to the apparition of the MHC multigenic families, whereas in the second case, these rearrangements were not permitted because of deleterious effects.

The origin of the "Big-Bang"

We have hypothesized that the big rearrangements of this region occurred because of the lack of functional constraint due to genetic redundancy (Abi Rached et al. 1999). However this explanation, even if necessary, is not sufficient. Indeed, how can one explain such a high level of duplication and exon shuffling?

In principle, the probability for a duplication being fixed in a given population is proportional to the level of duplications for a given gene (Kimura 1983). The rate of fixation for a duplication of a given gene is not known, but in the case of a pseudogene, which should be selectively neutral, cis duplications are very rarely described. One way to explain these big rearrangements is that the ancestors of MHC class I and class II genes were in a very unstable region.

Another explanation for this phenomenon is divergent selection, meaning that duplication events were selected for in populations. The selection can be direct, in that case MHC class I and class II gene ancestors were under positive selective pressure in the beginning. This selection can be indirect, for example by a hitchhiking effect, whereby the ancestors of MHC class I and class II genes were perhaps physically close to genes that were under selective pressure (duplications do not concern only the gene which is under selective pressure but also surrounding sequences). Thus genes which surround a gene under selective

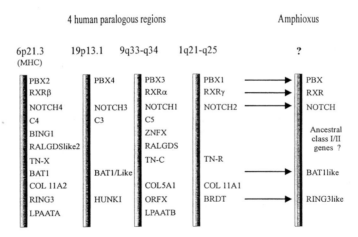

Fig. 5. Representation of the four human paralogous regions and the presumed region in the amphioxus. Arrows indicate orthologous relationships between the human genes and the amphioxus gene.

pressure can become fixed. Concerning this point, the presence of olfactory receptors (OR) within the mammalian MHC has to be emphasized (Gallinaro et al. 1997). It is possible that the MHC ancestor genes and the genes coding for the OR had been linked in the common ancestor of jawed vertebrates before the apparition of the MHC, and as proposed in the succeeding paragraph, OR may have driven the evolution of the MHC.

OR belong to the seven transmembrane domain (7TM) family which has been found in different vertebrate classes and invertebrates such as *Drosophila melanogaster* and *Caenorhabditis elegans*. It is thus likely that 7TM OR genes were present in their common ancestor. By contrast MHC functions are only found in the jawed vertebrate phylum. Furthermore, it is probable that the OR family has been under positive selective pressure from the metazoan origin since they are under selective pressure both in vertebrates and in *Caenorhabditis elegans* (Hughes and Hughes 1993; Robertson 1998). MHC-like genes and OR could therefore have been linked in the ancestors of jawed vertebrates, and we propose that positive selection acted on OR present in the ancestral MHC region. This positive selection could have acted indirectly on those genes present around the ancestral olfactory cluster by a hitchhiking effect, leading to fixation of new MHC-like genes. In parallel, OR could have been under balancing selection or any form of selection that keeps two or more alternative variants in a population longer than expected under genetic drift alone. Such selection leads to an accumulation of variants at linked neutral sites, and thus to an 'excess of variation' around the selected site (Aquadro 1992), as in the case of MHC-like genes. Therefore directional and balancing selections on OR may have allowed fixation of duplicate MHC-like genes as well as polymorphism induction, thereby driving the earliest development of the adaptive immune system. It should be noted that coduplications are apparent in the present-day genome, and that olfaction is still a major driving force in evolution. Other examples of coduplications implicating OR are also found in our present-day genome (Ayer-LeLievre et al. 1999).

The presence of OR in the MHC region of the amphioxus could support the "Big-Bang" driven olfaction hypothesis.

References

Abi Rached L, McDermott MF, Pontarotti P (1999) The MHC big bang. Immunol Rev 167:33-44

Aquadro CF (1992) Why is the genome variable? Insights from Drosophila. Trends Genet 8:355-362

Ayer-LeLievre C, Henry J, Gallinaro H, Avoustin P, Ribouchon MT, Bouissou C, Pontarotti P (1999) The olfactory receptor gene cluster in the major histocompatibility complex (MHC) of man and mouse: new insights into the evolution of the vertebrate olfactory receptor gene family. in press

Doolittle RF, Feng DF, Tsang S, Cho G, Little E (1996) Determining divergence times of the major kingdoms of living organisms with a protein clock. Science 271:470-477

Gallinaro H, Amadou C, Avoustin P, Ribouchon MT, Bouissou C, Lapointe F, Pontarotti P, Ayer-LeLièvre C (1997) Identification of new olfactory receptor genes located close to the human major histocompatibility complex. In: Charron D (ed) Genetic Diversity of HLA: Functional and Medical Implication. EDK, Vol. 2, pp 233-236

Henry J, Mather IH, McDermott MF, Pontarotti P (1998) B30.2-like domain proteins: update and new insights into a rapidly expanding family of proteins. Mol Biol Evol 15:1696-1705

Henry J, Miller MM, Pontarotti P (1999) Structure and evolution of the extended B7 family. Immunol Today 20:285-288

Holland PW, Garcia-Fernandez J, Williams NA, Sidow A (1994) Gene duplications and the origins of vertebrate development. Development (Suppl):125-133

Hughes AL, Hughes MK (1993) Adaptive evolution in the rat olfactory receptor gene family. J Mol Evol 36:249-254

Kasahara M, Nakaya J, Satta Y, Takahata N (1997) Chromosomal duplication and the emergence of the adaptive immune system. Trends Genet 13:90-92

Kimura M (1983) The Neutral Theory of Molecular Evolution. Cambridge University Press, Cambridge, UK

Klein J (1986) Natural History of the Major Histocompatibility Complex. Wiley-Interscience Publication, New York-Chichester-Brisbane-Toronto-Singapore

Klein J, O'hUigin C (1993) Composite origin of major histocompatibility complex genes. Curr Opin Genet Dev 3:923-930

Kumar S, Hedges SB (1998) A molecular timescale for vertebrate evolution. Nature 392:917-920

Robertson HM (1998) Two large families of chemoreceptor genes in the nematodes *Caenorhabditis elegans* and *Caenorhabditis briggsae* reveal extensive gene duplication, diversification, movement, and intron loss. Genome Res 8:449-463

Relationships among the genes encoding MHC molecules and the specific antigen receptors

Louis Du Pasquier

Basel Institute for Immunology, Grenzacherstrasse 487, CH-4005 Basel, Switzerland

Summary. The three essential elements of the adaptive immune system of vertebrates: MHC class I-II molecules, T-cell receptor and immunoglobulin (Ig) share one relatively rare Ig superfamily (Igsf) domain type with each other, the C1 domain (Du Pasquier and Chrétien 1996). C1 domains are characterized by their strand/loop composition (Williams and Barclay 1988) and so far have not been found outside gnathostome vertebrates, the only ones that somatically rearrange the genes of their antigen specific receptors (Du Pasquier 1999). This could be taken as an indication of close evolutionary relationships among the genes encoding such molecules and could give clues towards the understanding of the genesis of the immune system of vertebrates. I attempted to investigate these relationships by looking at more C1-containing genes or pseudogenes discovered in the databases to see whether patterns of homology and linkage emerged. The main result has been the identification of many genes encoding C1 domains, which are regularly associated with V domain genes that do not undergo somatic rearrangement. With such an architecture they could resemble what was the ultimate ancestor of the somatically rearranging receptor genes. Moreover, some of these genes are found on conserved syntenic groups, landmarks of the duplications that, through whole genome duplication or other mechanisms, have characterized the evolution of vertebrates. Several of them map to the MHC.

Key words. C_1 domain, Immunoglobulin superfamily

Identification of C1 containing genes in the MHC

C1 domains are now found in several molecules where they had not been identified earlier (Du Pasquier 1999). Figure 1 shows a summary of the architecture of molecules with C1 domains, Figure 2 shows sequence comparisons, and Figure 3 presents a tentative phylogenetic tree of the C1 domains. With a C2 domain from a human Igsf member and CTH as an outgroup (Chrétien et al. 1998) the tree suggests that the C1 domains of

54

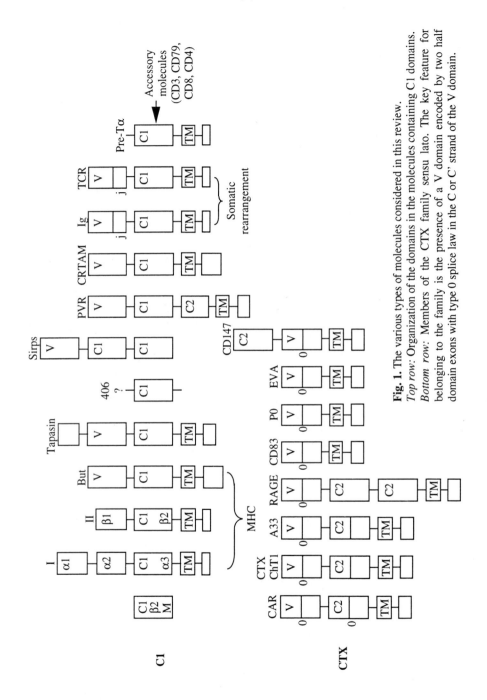

Fig. 1. The various types of molecules considered in this review. *Top row:* Organization of the domains in the molecules containing C1 domains. *Bottom row:* Members of the CTX family sensu lato. The key feature for belonging to the family is the presence of a V domain encoded by two half domain exons with type 0 splice law in the C or C' strand of the V domain.

```
C$HLAA2    ------------DAPKTHMTHH--AVSDH--EATLRCWALSFYP-AEITL    X60764
C$MR1      ------------EPPLVRVNRK-ETFPG---VTALFCKAHGFYP-PEIYM    AF073485
C$ZAG      ------------DPPSVVVTSH--QAPGE--KKKLKCLAYDFYP-GKIDV    P25311
C$FCRN     ----------KEPPSMRLKAR-PSSPG---FSVLTCSAFSFYP-PELQL    NP004098
C$CD1      ------------VKPKAWLSRG--PSPGP-GRLLLVCHVSGFYP-KPVWV    M14664
C$DR1B     ------------VHPKVTVYPS-KTQPLQ-HHNLLVCSVSGFYP-GSIEV    U83583
BETA2M     ----------IQRTPKIQVYSRHPAENGK--SNFLNCYVSGFHP-SDIEV    Y00567
C$SIRP1    ------------RAKPSAPVVSGPAARATPQHTVSFTCESHGFSP-RDITL    D86043
C$SIRP2    ------------VPPTLEVTQQ--PVRAE-NQVNVTCQVRKFYP-QRLQL    D86043
C$PVR      ------------EGTQAVLRAK-KGQDDK--VLVATCTSANGKP--PSVV     JC4024
CTAPAS$HUM ------------VSLMPATLAR--AAPGE-APPELLCLVSHFYPSGGLEV    NP003181
CTAP$CHR12 -------VILFLILSPASPKVRLSLANEAL--LPTLICDIAGYYP-LDVVV    AC005840
C$MUHUM1   ---------DVYLLPPAREQLN--LRES----ATITCLVTGFSP-ADVFV    X57086
C$TCRB     ------EDLKNVFPPEVAVFEPSEAEISHTQKATLVCLATGFYP-DHVEL    M12886
C$TCRG     ------------VSPKPTIFLPSIAETKLQKAGTYLCLLEKFFP-DVIKI    M16768
CCRTAM     ------------FKPILEASVIR-KQNGE-EHVVLMCSTMRSKP--PPQI    AF001622
BUTY.SEQ   ------------LGSDPHISMQ-VQENGE---ICLECTSVGWYP-EPQVQ    NP001723
C1$POLIO   ------------KPKNQAEAQKVTFSQD-PTTVALCISKEGRP--PARI     X80038
C$AC406    -------FFSLTASPASRLLLDQVGMKEN--EDKYMCESSGFYP-EAINI    AC000406
           1.......10........20........30........40........50

C$HLAA2    TWQ-RDGE------DQTQDTEL----VETR-PAGDGTF----QKWAAVVV    X60764
C$MR1      TWM-KNGE------EIVQEIDY----GDIL-PSGDGTY----QAWASIEL    AF073485
C$ZAG      HWT-RAGE------VQEPELR---GDVL-HNGNGTY----QSWVVVAV     P25311
C$FCRN     RFL-RNG--------LAAGTGQ----GDFG-PNSDGSF----HASSSLTV    NP004098
C$CD1      KWM-RGE-------QEQQGTQP----GDIL-PNADETW----YLRATLDV    M14664
C$DR1B     RWF-RNGQ--------EEKTGVVS---TGLI-HNGDWTF----QTLVMLET    U83583
BETA2M     DLL-KNG-------ERIEKVEH----SDLS-FSKDWSF----YLLYYTEF    Y00567
C$SIRP1    KWF-KNV-------NELSDFQT-----NVDPVGESVSYS--IHSTAKVVL    D86043
C$SIRP2    TWL-ENGN-------VSRTETA---STVT-ENKDGTYNW--MSWLLVNV    D86043
C$PVR      SWET-RLK-------GEARVPGD---SGTP-MAPVTVI---SRYRLVPS     JC4024
CTAPAS$HUM EWELRGGPGGR-----SQKAEGQRWLSALR-HHSDGSVSLSGHLQPPPVT    NP003181
CTAP$CHR12 TWT-REELGGSP--AQVSGASF-----SSLR-QSVAGTY----SISSSLTA    AC005840
C$MUHUM1   QWM-QRGQP------LSPEKYVTS---APMPEPQAPGRY----FAHSILTV    X57086
C$TCRB     SWWVNGKEVHS---GVSTDPQP----LKEQPALNDSRY----CLSSRLRV    M12886
C$TCRG     HWQEKKS-------NTILGSQ-----EGNT-MKTNDTY----MKFSWLTV    M16768
CCRTAM     TWLLGNS-------MEVSGGTL----HEFETDGKKCN-----TTSTLIIH    AF001622
BUTY.SEQ   EWT-SKG-------EKFPSTS------ESRNPDEEGLF----TVAASVII    NP001723
C1$POLIO   SWLSSLD-------WEAKETQV-----SGTLAGTVTVT----SRFTLVPS    X80038
C$AC406    TWE-KQTQKFPHPIEISEDVIT----GPTI-KNMDGTFNV--TSCLKLNS    AC000406
           ........60........70........80........90.......100

C$HLAA2    PSG----QEQRYTCHVQHE----GLP-KPLT--LRW-------    92     X60764
C$MR1      DPQ----SSNLYSCHVEHC----GVH-MVLQ--VPQE------    93     AF073485
C$ZAG      PPQ----DTAPYSCHVQHS----SLA-QPLV--VPWEAS----    94     P25311
C$FCRN     KSG----DEHHYCCIVQHA----GLA-QPL-------------    87     NP004098
C$CD1      VAG----EAAGLSCRVKHS----SLE-GQDIV-LYW-------    93     M14664
C$DR1B     VPR----SGEVYTCQVEHP----SVT-SPLLT-VEW-------    95     U83583
BETA2M     TPT----EKDEYACRVNHV----TLS-QPKI--VKWDRDM---    99     Y00567
C$SIRP1    TRED---VHSQVICEVAHV----TLQGDPLR--GTANLSETIR    107    D86043
C$SIRP2    SAHR---DDVKLTCQVTHI----GQP-AVSKS-HDLKVSA---    100    D86043
C$PVR      REA----HQQSLACIVNYHMD--RFK-ESLTLNVQYEPEV---    100    JC4024
CTAPAS$HUM TEQ----HGARYACRIHHP----SLP-ASGRS-AEVTL-----    107    NP003181
CTAP$CHR12 EPGS---AGATYTCQVTHI----SLE-EPLGA-STQVV-----    108    AC005840
C$MUHUM1   SEEEW-NTGETYTCVVAHE----ALP-NRVT---ERTV-----    100    X57086
C$TCRB     SATFWQNPRNHFRCQVQFY----GLS-ENDEW-TQDRAK----    115    M12886
C$TCRG     PEKS---LDKEHRCIVRHENNKNGVD-QEIIF-PPIKTD----    104    M16768
CCRTAM     TYG----KNSTVDCIIRHR----GLQGRKLV--APFRFEDLV-    100    AF001622
BUTY.SEQ   RDT----STKNVSCYIQNL----LLG-QEKK--V--------    88     NP001723
C1$POLIO   GRA----DGVTVTCKVEHE----SFEEPALIP-VTLSVRYP--    100    X80038
C$AC406    SQED---PGTVYQCVVRHA----SLH-TPLR--SNFTLTA---    112    AC000406
           .......110.......120.......130.......140...
```

Fig. 2. Alignment of C1 domains from the molecular types depicted in Fig. 1. Grayed amino acids correspond to the residues either shared, or equivalent from one sequence to another as defined by the Clustal X program.

butyrophilin, poliovirus receptor, and CRTAM are the oldest. The tree amply portrays the concept that HLA, CD1 class I genes diverged from each other long ago and belong to ancient paralogous groups presumably created by whole genome duplication (Kasahara et al. 1996; Kasahara 1999). Indeed the linkage group of these molecules might help their classification especially if syntenies and paralogous groups were conserved. Figure 5 gives the positions of these genes on various human chromosomes. All these positions have been determined or confirmed with the help of "Gene Map 99". Some of them are found in the MHC. Given the number of Ig domains in the animal kingdom it is not surprising that some Igsf genes are found in or near the MHC. However, with C1 domains one deals with a much more restricted population of molecules. Their linkage to MHC or paralogous regions is therefore much more likely to be significant.

C1 genes can be found in the MHC and paralogous regions

Genes of the tapasin family

Tapasin itself can be found in the human MHC (Herberg et al. 1998). A related gene, AC005840, has recently been identified on 12p13 close to the alpha2 macroglobulin gene on a syntenic group identified recently as a paralogue to part of the chromosome 6 MHC region (Kasahara 1999; Du Pasquier 1999). Several ests (eg AA099467, R25738) correspond to this gene but so far they are all incomplete. Similarly, the N-terminal region of the poliovirus receptor (PVR) (V-C1-C2-TM—Cy) is made of one V and one C1 domain and it is homologous to the V-C region of tapasin. PVR is found on 19q13. This region with its class I gene FCRGT (Kandil et al. 1995) could perhaps also be a paralogue of MHC. One would have to admit a pericentric inversion to explain the presence of the rest of the paralogue on 19p13. Indeed the phylogenetic analysis of FCRN suggests that, unlike ZAG (Ueyama et al. 1991) and MR1 (Hashimoto et al. 1995), it diverged from the other class I genes long ago.

Butyrophilin

Butyrophilin is a family of genes whose members are located in the MHC (Henry et al. 1997). The V and C1 domains are followed, after the transmembrane region, by a B30 domain. All phylogenies show the butyrophilin family as the progeny of a very ancient lineage whether for the V or C domain (Rached et al. 1999).

In all, 4 different families of C1 containing genes are represented: tapasin, butyrophilin, MHC I and MHC II. What are these genes and could such links between the MHC be non-fortuitous?

These MHC-linked C1 genes are associated with a V gene that does not rearrange somatically

Tapasin and butyrophilin are characterized by the presence of a V domain encoded by a single exon. Their product is generated without somatic rearrangement. These genes could therefore resemble those of ancestral antigen specific receptors. In this speculation I am going to assume that what occurred in the genesis of the immune system was very Darwinian: variation preceded selection. In other words, generation of the somatic rearrangement of receptors came first allowing for the production of a large variety of ligands. I further assume that the genes of antigen presenting molecules of the MHC were assembled and used later for the purpose of selection. Their assembly would imply the joining of a peptide binding region gene to a C1 domain, as was originally suggested by Flajnik et al. (1991), borrowed from the collection of

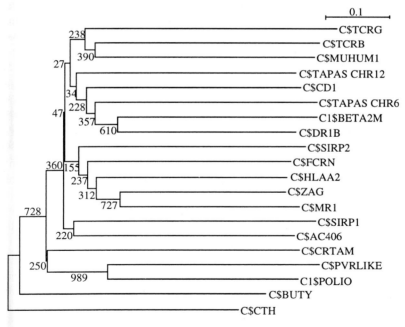

Fig. 3. Phylogenetic tree obtained using the neighbour joining method (clustal X) with the exclusion of positions with gaps for the C1 domain of human. The outgroup is a C2 domain from a CTX family member. Note the early branching out of the butyrophilin.

receptor ancestors present in the class III region. I have no satisfactory explanation as to why the somatically rearranging receptor genes are not in the MHC linkage group. I believe, however, that the original copy of these genes was located in the MHC linkage group giving an easier explanation for the sharing of C1 domains. Most of the time it is argued that the MHC class I and II genes entered the ancestral class III region late in evolution for no special reason other than perhaps the linkages provided some selective advantage later on (Kasahara 1999). I would propose that this was the case for the sharing of C1 domains. These ancestral receptor genes, or rather their copies, transferred to other chromosomes later taking with them their C1 domains. They were then modified by the introduction of the rearranging machinery with the split between F and G strands in V genes. Why were C1 domains preferred and differentiated only in those species achieving somatic rearrangement? Perhaps the major distinctive trait of C1 domains is to belong to molecules that interact with co-receptors (CD3 for TCR, CD79 for Ig, CD4 and CD8 for MHC) via regions that precisely differentiate C1 from C2 domains. It would be worthwhile to find out whether molecules like CRTAM, tapasin, butyrophilin, or related molecules interact with the equivalent of co-receptors.

Linkages with CTX family members

C1 domain genes, MHC and CTX homologues in mammals

In *Xenopus*, CTX is a gene encoding an Igsf member (V-C2 transmembrane-cytoplasmic segment) characterized by a V domain made of 2 half domain exons (splice law type 0) that do not rearrange somatically. Loosely linked to the MHC, it withstands gene duplication following whole genome duplication without deletions within the series of polyploid *Xenopus* (Chrétien et al. 1996; Du Pasquier et al. 1999). Duplicated forms of CTX homologues also seem to exist in zebrafish (Fig. 4). Because of its stability it could be an interesting "tag" for the duplications that occurred during evolution. Perhaps its mammalian homologues would show the same behaviour. Indeed many CTX homologues have been found in the human and mouse genome (Chrétien et al. 1998). Their chromosomal location is indicated in Figure 5.

The architecture of CTX, as well as its dimerization apparently linked to the presence of a diglycine bulge in the G strand of some of the isotypes, makes it another putative relative of the antigen specific receptor (Du Pasquier and Chrétien 1996). The exon-intron structure of the V domain suggests that it diverged from a non-split V domain gene independently from the process leading to TCR where the intron has been introduced between strands F and G instead of within C' in CTX. The phylogenies of the V and C domains are in favour of this hypothesis (Du Pasquier 1999). This implies the existence of an ancestral non-

```
                                       1         *  :.   * *
     JAM  MGTEGKAGRKLLFLFTSMILGSLVQGKGSVYTAQSDVQVPENESIKLTCTY  U89915
   ZFCTX2  ----------------------------------------VKENEGVDLQCSY  A1626248
   ZFCTX1  ----------------------------VLLKSTNSKPWVNEFESIELSCMI  A1588817
      CTX  --------MSFLLFITLGLSLTALSHCVQVTIQNPIINVTSGQNATLYCTY  U43330
  JAM$XEN1  ----------------------ALAGVTAPDPTITVKEGDSPDLRCSY  A10311296
            1.......10........20........30........40........50.


                   .  ::*:  .          .  :.   :        *.
     JAM  SG----FSSPRVEWKFVQG----STTALVCYNSQ-ITAP-YADRVTFSSSG  U89915
   ZFCTX2  TSDF--GATPRVEWKFKDLK---GSQTLVYFDGK-PTGQ-YTGRVTMYDKG  A1626248
   ZFCTX1  ESIT--TTKPRIEWKKIKN----GDPSYVYFDNQ-ISGD-LERRAKIREPA  A1588817
      CTX  ILNNQNKNNLVIQWNIFQAKSQNQETVFFYQNGQSLSGPSYKNRVTAAMSP  U43330
  JAM$XEN1  TSDY--I-KPRVEWKFVNKD---QETSFVFYDGG-LTTS-YKDRA------  A10311296
            .......60........70........80.........90........100..


                                          V | C
                                          ←--→
     JAM  ----ITFSSVTRKDNGEYTCMVSEEGGQ-NYGEVSIHLTVLVPPS  132  U89915
   ZFCTX2  ----LRFNKVTRADTGDY--------------------------  71   A1626248
   ZFCTX1  T---LVILNATRSDSADYRCEVTAPNDQKSFDEILISLTVRVKPV  109  A1588817
      CTX  GNATITISNMQSQDTGIYTCEVLNLPES--SGQGKILLTVLVPPS  137  U43330
  JAM$XEN1  -------------------------------------------  63   A10311296
            .....110.......120.......130.......140.......
```

Fig. 4. New CTX family members in vertebrates ZF CTX zebrafish molecules Jam$Xen1 Jam homologue of *Xenopus* J402n21, a human MHC linked related molecule. Grayed amino acids correspond to residues either shared, or equivalent from one sequence to another as defined by the Clustal X program.

split V gene giving rise to the two types of V domains in a way similar to that already proposed after examination of the CD4 V domain gene (Littman and Gettner 1987).

In *Xenopus*, the CTX gene is loosely linked to the MHC, and this is the aspect that is currently of interest to us. In the human genome, some of CTX's close (A33) or very distant (J402n21 see below) homologues can be found in the MHC or in MHC paralogous regions (Fig. 5). Of the other genes sharing with CTX only the split V domain with splice law type 0, RAGE and CD83 can be found in the MHC near the class II region or in the area telomeric to the human MHC region in the 6p21.3-p23 region. The paralogous region 1q21-25, with V domains of the P0 family (splice law type 0) would also correspond to this category.

Newly identified human CTX related molecules

After finding the homologue CTXhumx (provisory nomenclature) on chromosome X (Du Pasquier 1999) I found two other genes related to CTX. One, 7023 (estAC0007023) is only represented so far by its C domain characterized by the two extra cysteins. However, as of yet it has not been

	6p21-3	9q33-q34	12p12-13	19p13.1 p13.3	19q13	1q21-q25	11p13	11q22-q24	21q21.2 q11	20p13	(X)
Comp	C4	C5	α2mac	C3							
V no int.	Tapasin 1C7* Butyro	Tapas-like V2320	Tapas-like AC 5840		PVR V1335			PVR2 CRTAM		SIRP	(H5159q14) Buty-like CTH-like
C1	Tapasin Butyro Class 1 + II Pre-Tα		Tapasin AC 5840		PVR FCRN(I)	CD1 MR1	AC406	PVR2 CRTAM		SIRP	
V0 (CTX)	RAGE CD 83 1C7* J402N21			CD 147		P0 A33?		EVA CTH1 V1235	CTH2 HCAR		CTXHumx q22
Others	MHC paral	MHC paral	MHC paral	MHC paral	KIRs ILTs / Ets / MAG CD33 CD22 / DYRK 1B	MHC paral / DYRK 3 (1q32)	RAGs / Ets (p11.12)	CD3γδε / Ets / NCAM	Erg Down Ets / DYRK 1	LJ simil. Sialoadh.	ILTsf1 q25 / XLP q25 NCAM.1 (xq28)

Fig. 5. Linkage groups of C1 containing molecules and of CTX family members. All sequences from est or HTG are still incomplete. Comp: Complement. V no int.: Variable domain without intron. Vo: Variable domain encoded by 2 half domain exons with a splice law type 0. Paral: Paralogous. More details about these molecules in Du Pasquier 1999.

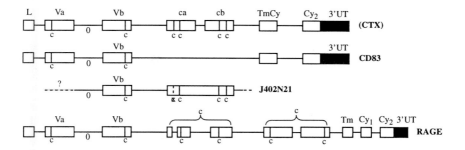

Fig. 6. The MHC linked CTX related (V domain encoded by 2 exons) members. Exon/intron organization of the various genes described in the text. Intron size not drawn to scale.

mapped. The other, j402n21 (Fig. 6) on chromosome 6p21, was found in a human HTG recently released (3 June 1999) with the Vb exon and its acceptor site indicating a splice law of type 0. Its sequence is as follows (Vb+C2) and Blast searches give the best score with the CTX family member CTH (Chrétien et al. 1998):

WFKNGKPARMSKRLLVTRNDPELPAVTSSLELIDLHFSDYG
TYLCMASFPGAPVPDLSVEVVPPEVEPSSQDVRQALGRPVL
LRCSLLRGSPQRIASAVWRFKGQLLPPPPVVPAAAEAPDHA
ELRLDAVTRDSSGSYECSVSNDVGSAACLFQVSGERP.

CTX family C2 domain is characterized in all other homologues by an extra disulfide bridge and by an intron in the middle of the gene. It is encoded here by a single exon and has lost one of the extra cysteins replaced by a serine (bold in the sequence). Some identical ests (e.g. aa481631 from germinal centers, aa922696 from adenocarcinoma) considered similar to NCAM 2 are found in the data base suggesting that the gene can be expressed. Blast searches suggest that the closest relative is a CTX family member. However, at this point the molecule could still be a member of the NCAM family where splicing can also occur within an Ig domain gene. As more and more V domains of this type are found the chances of not forming a single lineage increase and the classification may have to be revised later.

In addition, as far as CTX distantly related molecules and genes are concerned, a molecule with features similar to its V domain has been found in the MHC in the human class III region. Among other interesting features, the gene 1 C7 (Neville and Campbell 1999) shows alternate splicing forms which make it behave like a V gene of CTX since one of its variants uses a type 0 splice site in the C/C' strands like CTX.

In fact the MHC seems to contain representatives of all the members that could have been at the origin of major elements of the immune system before the introduction of the somatic rearrangement. Whether the ancestral receptor genes left the class III region before or after the introduction of rearrangement will probably remain unsolved. The presence of the rearranging receptor on chromosomes 14q11 (α and δ) 7q25 (b) and 7p15 (g) for TCR, 22q11.2 (lambda), 14q32 (heavy chain) and 2q333 (kappa) for Ig suggests that the genes that gave rise to the somatically rearranging receptors were already outside the MHC when the mechanism was introduced. The original genes may have been in a situation similar to the modern CRTAM or AC406 (Du Pasquier 1999) (Fig. 5). Perhaps the presence near the MHC if this gene for the pre T-cell receptor with a 'C1-like domain' is not a simple coincidence (Fehling et al. 1995; Saint-Ruf et al. 1998).

C1-CTX family linkages outside MHC and its paralogues

When not associated to the MHC or its paralogues, C1 genes are often linked to genes of CTX family members wherever they are. Most of the time these linkages are concerned with the immune system functions or molecules. Perhaps the CTX V no intron split is very ancient (Du Pasquier 1999). A new association is the one between genes of the KIR family and those of the ancient receptor lineage, with the discovery that an ILT like molecule is on the same linkage group as the X-linked lymphoproliferative disease on Xq25 not very far from the CTX family member CTXHumx. Chromosomes 21, 19 and 11 already outlined for their Igsf contents contain conserved syntenies with transcription factors of the ets family (Anderson and Rothenberg 1999) and DYRKs. The region 11q23 contains many genes of the family and the translocations in this area often lead to leukaemia.

Other C1 genes

The Zag delipidating agent looks like a class I molecule and contains an Igsf C1 domain but the heavy chain is not associated with beta 2m. Its gene sits on chromosome 7q22 far away from the TCR beta locus (7q35). Another isolated C1 gene, AC406, with strong homology to cold-blooded vertebrate MHC or Ig C1 domains (Du Pasquier 1999) is on chromosome 11p12-13 close to the RAG genes and next to Poliovirus receptor-like genes. The SIRPs molecules have their genes on 20p13 linked to sialoadhesin and a NCAM-L1 like molecule revealing perhaps a synteny conservation with 19q13 (Fig. 5) via the SIRP/PVR homology. Some of the non MHC linked C1 or CTX members are linked to the Ets transcription factor linkage groups (Anderson and Rothenberg 1999) (Fig. 5). Some of these transcription factors are involved in lymphocyte differentiation.

Conclusion

It is beyond the scope of this survey to estimate whether all the new sequences presented here fit into a model of whole genome duplication because the phylogenies have not yet been performed in detail. But this study certainly offers several new regions to be evaluated in this context (e.g. 19q13, 11q22-24, several regions on X). The main hypothesis is that the linkage to the MHC of Ig superfamily members with C1 domains is not fortuitous.

Abbreviations used in the figures

AC 406	C_1 domain isolated from the HTG AC000406
Beta 2M	Beta 2 microglobulin
Buty or But	Butyrophilin
C1 $ polio	PRR_2 gene related to human poliovirus receptor
CAR	Coxsackie virus receptor
CD1	Human CD1
ChT1	Chicken homolog of CTX
CRTAM	MHC class I restricted T cell associated molecule
CTH	Human homolog of CTX
Down	Down syndrome associated protein
DR_1B	MHC class II
DYRK	Dual specificity tyrosine phosphorylation regulated kinase
Ets, Erg	Transcription factors
EVA	Epithelial V-like antigen
FCRN	Fc receptor of the neonate
HLAA2	Human MHC class I A_2 α chain
ILT	Immunoglobulin like transcript
JAM	Junction adhesion molecule
JAM$Xen1	Jam related *Xenopus* molecule
KIR	Killer inhibitory receptor
MAG	Myelin associated glycoprotein
MR1	Class 1-like MHC molecule of chromosome 1
Muhum 1	Immunoglobulin MC_1 domain
Pre-Tα	Pre T-cell receptor
PVR	Polio virus receptor family of chromosome 19
PVR like	Human chromosome 11 poliovirus receptor like molecule
RAGE	Receptor for advanced glycosylation end products
SIRP	Signal regulatory protein
Tapas	Human MHC linked tapasin
Tap chr 12	Human tapasin like gene on chromosome 6
TCR BG	T cell receptor Beta or gamma
XLP	X-linked lymphoproliferative disease

ZAG   Zign alpha-2 glycoprotein (class I like)
ZFCTX   Zebrafish CTX homolog

All the above molecules are referred to in detail in Du Pasquier (1999).

Acknowledgements

I thank Lucy Trippmacher and Allison Dwileski for their help in the preparation of this manuscript. The Basel Institute for Immunology was founded and is supported by F. Hoffmann-La Roche Ltd., Basel, Switzerland.

References

Anderson MK, Rothenberg EV (1999) Transcription factor expression in lymphocyte development: Clues to the evolutionary origins of lymphoid cell lineages. In: Du Pasquier L, Litman GW (eds) Current Topics in Microbiology and Immunology. Sringer Heidelberg, 248:in press

Chrétien I, Courtet M, Marcuz A, Katuevo K, Vainio O, Du Pasquier L (1998) CTX, a *Xenopus* thymocyte receptor, defines a molecular family conserved throughout vertebrates. Eur J Immunol 28:in press

Chrétien I, Robert J, Marcuz A, Garcia-Sanz JA, Courtet M, Du Pasquier L (1996) CTX, a novel molecule specifically expressed on the surface of cortical thymocytes in *Xenopus*. Eur J Immunol 26:780-791

Du Pasquier L (1999) The phylogenetic origin of the antigen specific receptors. In: Du Pasquier L, Litman GW (eds) Current Topics in Microbiology and Immunology. Springer Heidelberg, 248:in press

Du Pasquier L, Chrétien I (1996) CTX, a new lymphocyte receptor in *Xenopus* and the early evolution of Ig domains. Res Immunol 147:218-226

Du Pasquier L, Courtet M, Chrétien I (1999) Duplication and MHC linkage of the CTX family of genes in *Xenopus* and in mammals. Eur J Immunol in press

Fehling HJ, Laplace C, Mattei MG, Saint-Ruf C, von Boehmer H (1995) Genomic structure and chromosomal location of the mouse pre-T-cell receptor alpha gene. Immunogenetics 42:275-281

Flajnik MF, Canel C, Kramer J, Kasahara M (1991) Which came first, MHC class I or class II? Immunogenetics 33:295-300

Hashimoto K, Hirai M, Kurozawa Y (1995) A gene outside the human MHC related to classical HLA class I genes. Science 269:693-695

Henry J, Ribouchon MT, Depetris D, Matteï MG, Offer C, Tazi-Ahnini R, Pontarotti P (1997) Cloning, structural analysis, and mapping of the *B30* and *B7* multigenic families to the major histocompatibility complex (MHC) and other chromosomal regions. Immunogenetics 46:383-395

Herberg JA, Sgouros J, Jones T, Copeman J, Humphray SJ, Sheer D, Cresswell P, Beck S, Trowsdale J (1998) Genomic analysis of the Tapasin gene, located close to the TAP loci in the MHC. Eur J Immunol 28:459-467

Kandil E, Noguchi M, Ishibashi T, Kasahara M (1995) Structural and phylogenetic analysis of the MHC class I-like Fc receptor gene. J Immunol 154:5907-5918

Kasahara M (1999) The chromosomal duplication model of the major histocompatibility complex. Immunol Rev 167:17-32

Kasahara M, Hayashi M, Tanaka K, Inoko H, Sugaya K, Ikemura T, Ishibashi T (1996) Chromosomal localization of the proteasome Z subunit gene reveals an ancient chromosomal duplication involving the major histocompatibility complex. Proc Natl Acad Sci USA 93:9096-9101

Littman DR, Gettner SN (1987) Unusual intron in the immunoglobulin domain of the newly isolated murine CD4 (L3T4) gene. Nature 325:453-455

Neville MJ, Campbell RD (1999) A new member of the Ig superfamily and a V-ATPase G subunit are among the predicted products of novel genes close to the TNF locus in the human MHC. J Immunol 162:4745-4754

Rached LA, McDermott MF, Pontarotti P (1999) The MHC big bang. Immunol Rev 167:33-44

Saint-Ruf C, Lechner O, Feinberg J, von Boehmer H (1990) Genomic structure of the human pre-T cell receptor alpha chain and expression of two mRNA isoforms. Eur J Immunol 11:3824-3831

Ueyama H, Niwa M, Tada T, Sasaki M, Ohkubo I (1991) Cloning and nucleotide sequence of a human Zn-alpha 2-glycoprotein cDNA and chromosomal assignment of its gene. Biochem Biophys Res Commun 177:696-703

Williams AF, Barclay AN (1988) The immunoglobulin superfamily domains for cell surface recognition. Annu Rev Immunol 6:381-405

Conserved linkage among sea urchin homologs of genes encoded in the vertebrate MHC region

Jonathan P. Rast, Paola Oliveri, and Eric H. Davidson

Division of Biology 156-29, California Institute of Technology, Pasadena, CA 91125, USA

Summary. The echinoderms, along with their sistergroup the hemichordates, form an outgroup to the chordates within the deuterostomes. As such, they offer a potentially valuable source of information from which to infer the primitive state of chordate chromosomal organization. As part of an ongoing Sea Urchin Genome Project we have constructed Bacterial Artificial Chromosome (BAC) and P1 Artificial Chromosome (PAC) libraries from the purple sea urchin (*Strongylocentrotus purpuratus*). We have isolated a number of sea urchin genes that have vertebrate homologs encoded in the MHC locus. BAC and PAC clones that contain these genes have been identified and are being characterized. Two complement factor genes, an S. purpuratus C3/4/5 homolog and a Bf/C2 homolog, are both encoded within a single 140 kb BAC clone insert. We have also found convincing evidence for linkage between *S. purpuratus* homologs of Notch and a PBX homeodomain transcription factor, vertebrate homologs of which are encoded in close proximity within the MHC class III region. These observations suggest a remarkable conservation of large scale synteny within the deuterostomes, and may lead to the identification of divergent sea urchin relatives of vertebrate class I or II MHC genes.

Key words. Echinoderm, Complement, Synteny, MHC, Evolution

All evidence now indicates that rearranging immune system genes are confined to the gnathostomes within the vertebrates (reviewed in Litman et al. 1999; Rast and Litman 1998). The immunoglobulin (Ig) superfamily genes which undergo rearrangement, however, clearly arose from a nonrearranging ancestor. Classical and nonclassical major histocompatibility complex (MHC) genes (i.e., genes of the structural form of the antigen presenting MHC genes) are likewise unknown outside of the jawed vertebrates, but are certainly derived, at least in part, from ancestral Ig-like domain genes (Flajnik et al. 1999). Homologs of these genes may be present in the jawless vertebrates and in invertebrate deuterostomes. The isolation of such genes, and characterization of their structure and function may enlighten us as to the nature of the primitive vertebrate immune, recognition system prior to the acquisition of rearranging mechanisms.

As a class, the genes encoding immune system cell surface proteins tend to diverge from one another at extreme rates (Hughes 1997; Murphy 1993), and homologs are likely to be difficult to isolate by direct methods (i.e., library screening or PCR). For example, high divergence rates were responsible for the slow progress of T cell receptor (TCR) homolog isolation among the gnathostomes. The extent of TCR divergence among jawed vertebrates, in which only two stretches of three or four amino acids are conserved, places these genes at the limits of PCR-based isolation methods. Once isolated, however, these sequences are clearly recognizable as homologous to the four mammalian TCR types (Rast et al. 1997). Thus, the failure to isolate nonrearranging Ig-TCR or MHC homologs from invertebrate deuterostomes by these methods, where they would be expected to be even more divergent than among the gnathostomes, can not be taken as evidence of their absence.

Fortunately, other strategies for finding divergent but homologous receptors are becoming accessible. Highly divergent genes are being identified in EST projects. Information from these projects may in the future contribute to an understanding of immune system evolution, by reference to the growing collection of functional information from mammalian systems. Conservation of syntenic gene linkage provides another approach. Genes with conserved sequences or genes already in hand from other sources can be used as markers for chromosomal regions that are likely to contain interesting divergent genes. This strategy is rendered practical with the advent of BAC and PAC cloning, in combination with the mass of chromosomal organization data that is emerging from vertebrate and protostome genome projects. We are using the latter strategy in an effort to identify classical or nonclassical MHC homologs in the purple sea urchin, *Strongylocentrotus purpuratus*. Whether or not true MHC homologs are found, this work will begin to define the extent of chromosomal gene linkage conservation among the deuterostomes, a more general and even more important objective.

The phylogeny of deuterostomes

Sea urchins belong to the phylum Echinodermata, which together with the chordates and hemichordates constitute the deuterostomes. These phyla are all more closely related to one another than any one is to any protostome phylum (e.g., arthropods, nematodes, annelids and mollusks). Though the three deuterostome phyla display very divergent adult morphologies, they share a number of molecular and developmental characters that are absent in other bilaterians. Aspects of immunity which are not found in protostomes also are likely to be shared among deuterostome animals. A current simplified consensus relationship among the deuterostomes is illustrated in Figure 1.

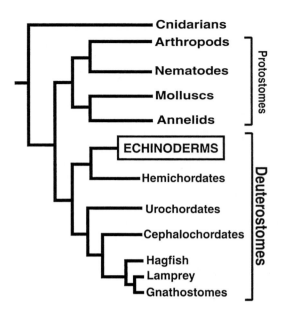

Fig. 1. The interrelationships of selected representative metazoan phyla and chordate subgroups. All deuterostome phyla are shown. The vertebrates (lamprey + hagfish + gnathostomes) are illustrated according to Forey and Janvier (1993). Relations of the protostome phyla are according to Aguinaldo et al. (1997), Balavoine (1997), de Rosa et al. (1999), and Halanych et al. (1995). Hemichordate placement is according to Castresana et al. (1998). The Cnidarians (e.g., jellyfishes) are included to root the bilaterians (protostomes + deuterostomes), but the monophyly of the protostomes is uncertain. Relationships among these groups is reviewed by Adoutte et al. (1999).

Immunity in the sea urchin

Immunity in the sea urchin is mediated by a number of leukocyte like cell types collectively known as coelomocytes (reviewed in Smith and Davidson 1994). A major component of these coelomocytes are macrophage like cells that are able to recognize and engulf a wide range of foreign particles, including bacteria. They also include granular cell types that are capable of releasing various clotting factors and antibacterial molecules. In spite of much effort, no sign of immune memory or any other attribute of vertebrate-type adaptive immunity has been demonstrated in these organisms. Nonetheless they are long-lived animals and they are capable of recovering from apparent states of severe disease. They have no obvious lifestyle characteristic that renders them exempt from the requirement for a sophisticated immune system, and in fact this coelomocyte-based immune system apparently functions very efficiently.

Despite the absence of an adaptive immune system, the coelomocytes of S. *purpuratus* express a number of genes that are homologs of those expressed by vertebrate leukocytes. Effector gene homologs of the vertebrate complement factor genes C3/4/5 (Al-Sharif et al. 1998) and Bf/C2 (Smith et al. 1998) were identified in a coelomocyte EST project (Smith et al. 1996). Phylogenetic analyses of these sea urchin complement factor genes indicate that they are related to the common ancestor of the set of paralogous genes which resulted from a later duplication within a vertebrate ancestor. Transcription factor homologs of vertebrate NFκB, GATA1/2/3 and CBFα (a runt domain protein) genes are also expressed in coelomocytes (Pancer et al. 1999). The extent of homology between this network of control and effector genes and that of the vertebrate immune system is presently unknown. Whether the antigen recognition systems also exhibit homology with those of vertebrates is a question that awaits the isolation of the sea urchin genes responsible for antigen binding.

Conserved linkage of some sea urchin homologs of MHC encoded genes

The isolation of a C3/4/5 complement factor homolog from S. *purpuratus* (Al-Sharif et al. 1998; Smith et al. 1996) initiated a search for linked genes that might indicate conserved chromosomal organization with respect to the mammalian MHC class III region. The rationale is that the S. *purpuratus* gene is homologous to the common ancestor of the three vertebrate complement factor paralogs, C3, C4, and C5. The vertebrate C4 genes lie in the MHC Class III region itself and the C3 and C5 genes lie in chromosomal regions that encode other Class III paralogs (Kasahara 1999a). Thus it is plausible that linkage relationships may be conserved between the vertebrate MHC Class III region and the chromosomal region surrounding the S. *purpuratus* C3/4/5 gene. Furthermore, an S. *purpuratus* Bf/C2 homolog appears to be linked within 140 kb of the C3/4/5 gene. Probes derived from the sea urchin C3/4/5 and Bf/C2 genes cohybridize to the insert of a 140 kb BAC clone. A simplified map of this clone is shown in Figure 2. Evidence for linkage is found also in cohybridization of these probes to restriction fragments separated in pulsed-field gels. The insert from this BAC clone is being sequenced as part of a project to characterize the region around the S. *purpuratus* complement genes.

A single Notch gene was isolated from the sea urchin *Lytechinus variegatus* (Sherwood and McClay 1997). This gene appears to be phylogenetically related to the ancestor of the four vertebrate Notch homologs found in the MHC and in its three paralogous loci (Kasahara 1999a). In vertebrates, Notch genes are closely linked to PBX - type homeobox transcription factor genes in three of

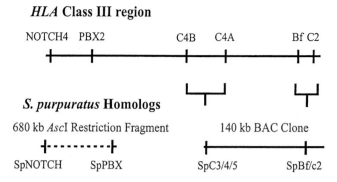

Fig. 2. Genomic organization of two pairs of *S. purpuratus* genes that have homologs encoded in the vertebrate MHC Class III region. Co-hybridization of SpNotch and SpPBX probes to blots of *Asc*I restriction fragments separated by pulsed-field gel electrophoresis (PFGE) suggest that these genes are linked within 680 kb. This interpretation is supported by co-hybridization to larger fragments in similar analyses using other restriction enzymes. Probes complementary to the *S. purpuratus* C3/4/5 and Bf/C2 complement factor homologs hybridize to a single BAC clone containing a 140 kb insert. Probes corresponding to these genes also co-hybridize to restriction fragments on PFGE blots. Branching diagrams between the vertebrate and sea urchin sequences indicate tandem duplications that probably occurred within the vertebrates. The linkage status between the two *S. purpuratus* clusters is being investigated. Note that the diagram is not to scale.

these loci (Katsanis et al. 1996; Kasahara 1999a). Relatively close linkage of homologs of these genes is also evident in two ecdysozoan protostomes, *viz.* in *Drosophila melanogaster* and in *Caenorhabditus elegans* (Trachtulec et al. 1997). We have isolated the *S. purpuratus* homologs of both of these genes. Each appears to be single copy. As expected, phylogenetic analyses suggest that the *S. purpuratus* PBX homolog diverged from the common ancestor of the vertebrate PBX transcription factors prior to their duplication. Linkage between the *S. purpuratus* Notch and PBX genes is evident from DNA blot analyses where probes to both genes hybridize to a 680 kb *Asc*I fragment. This is illustrated in Figure 2. Cohybridization to larger restriction fragments is also evident. We will use the BAC-end sequence resource (discussed below) to determine the linkage status between these genes and the sea urchin complement genes.

In addition to the small linkage groups described above we have both PAC and BAC clones containing other *S. purpuratus* homologs of genes encoded in the mammalian MHC region. These include the *S. purpuratus* homolog of the proteosome subunit LMP/X (Kasahara et al. 1996) and the homolog of the RNA helicase, BAT1(Peelman et al. 1995). These genes also appear to be single copy in the sea urchin by analyses of multiple cDNA sequences and DNA blots. We are currently investigating linkage among the BAT1 and LMP/X genes to those genes discussed above and are attempting to isolate other homologs of genes encoded within the vertebrate MHC and paralogous regions. In addition we are using insert DNA prepared from these BAC clones as probes to screen arrayed cDNA libraries in order to identify adjacent genes.

BAC-end sequencing project

The discovery of conserved syntenic relationships will be greatly facilitated by a BAC-end sequencing project (Siegel et al. 1999; Venter et al. 1996) that has been initiated to characterize the genome of S. *purpuratus*. This work is part of an ongoing Sea Urchin Genome Project in collaboration with G. Mahairas and L. Hood at the High Throughput Sequencing Center, University of Washington. For this purpose a high coverage genomic BAC library was constructed with an average insert size of 130 kb. 80,000 BAC clone ends will be sequenced to establish a set of sequence-tagged-connectors (STC) separated by an average distance of 10 kb (the haploid size of the S. *purpuratus* genome is approximately 8×10^8 bp). A restriction fragment length fingerprint from the inserts of each of these clones will also be characterized. Thus we will be able to efficiently identify overlapping BAC clones and expand coverage surrounding interesting genes. This may not only enable the identification of additional sea urchin homologs of genes encoded within the vertebrate MHC region, but may also lead to the identification of genes that are relevant to an understanding of the origins of jawed vertebrate adaptive immunity.

Conclusions

The presence of conserved chromosomal architecture over vast evolutionary distances may imply an underlying large-scale order that is as yet unappreciated. The intermingling of transcriptional control elements is one factor that may constrain chromosomal organization and be a major driving force behind the conservation of syntenic relationships. S. *purpuratus* offers tremendous advantages as a system for the study of transcriptional regulation especially for genes that are expressed during embryonic development, as for example are the SpPBX and SpNotch genes. The complexity of metazoan transcriptional regulation requires the analysis of spatial and quantitative aspects of transcription from many modified promoter constructs before a mechanistic understanding can be reached (Arnone and Davidson 1997; Yuh et al. 1998). The ease of producing transgenic sea urchin embryos and large preparations of embryonic nuclear proteins facilitates this work. It may be possible to uncover shared transcriptional control elements among closely linked genes and to develop and test hypotheses as to the causes of linkage conservation.

Three chromosomal regions have been identified in the human and mouse that appear to contain a concentration of genes that are paralogous to those encoded within the MHC region (Kasahara 1999a; Kasahara et al. 1996; Katsanis et al. 1996; Pebusque et al. 1998). These loci are likely to have resulted from a series of early vertebrate large block duplications of a prototypic MHC-like region. The list of genes conserved among these paralog groups within mouse and humans is continually growing (Kasahara 1999b). Conservation of these regions within the vertebrates is perhaps the strongest evidence that the chromosomal organization of genes can be exploited as a tool to find homologs in divergent taxa. Gene linkage among these vertebrate paralog groups has probably been conserved over a period of more than 450 million years. This corresponds to the majority of time separating the modern deuterostomes from their common ancestor, lending credence to the idea that gene linkage may be conserved between echinoderms and vertebrates.

As more detailed chromosomal maps become available from vertebrate and protostome species, chromosomal organization will become a more widely applicable means to isolate genes. Like the protochordates, the echinoderms should be a relatively simple system in which to analyze linkage conservation in reference to the vertebrates. Many gene homologs that are dispersed among the four chromosomal regions that are paralogous to the MHC region of mammals may be found "collapsed" into single linkage groups in the invertebrate deuterostomes. Contributions from this project and similar projects described elsewhere in this volume, investigating a range of deuterostome representatives, will over the next several years further elevate the MHC region as a leading model for the study of chromosomal evolution.

Acknowledgments

We would like to thank L. Courtney Smith for sharing prepublication data on the sea urchin complement genes. The Sea Urchin Genome Project is funded by the Stowers Institute for Medical Research.

References

Adoutte A, Balavoine G, Lartillot N, de Rosa R (1999) Animal evolution. The end of the intermediate taxa? Trends Genet 15:104-108

Aguinaldo AM, Turbeville JM, Linford LS, Rivera MC, Garey JR, Raff RA, Lake JA (1997) Evidence for a clade of nematodes, arthropods and other moulting animals. Nature 387:489-493

Al-Sharif WZ, Sunyer JO, Lambris JD, Smith LC (1998) Sea urchin coelomocytes specifically express a homologue of the complement component C3. J Immunol 160:2983-2997

Arnone MI, Davidson EH (1997) The hardwiring of development: organization and function of genomic regulatory systems. Development 124:1851-1864

Balavoine G (1997) The early emergence of platyhelminths is contradicted by the agreement between 18S rRNA and Hox genes data. C R Acad Sci III 320:83-94

Castresana J, Feldmaier-Fuchs G, Yokobori S, Satoh N, Paabo S (1998) The mitochondrial genome of the hemichordate *Balanoglossus carnosus* and the evolution of deuterostome mitochondria. Genetics 150:1115-1123

de Rosa R, Grenier JK, Andreeva T, Cook CE, Adoutte A, Akam M, Carroll SB, Balavoine G (1999) Hox genes in brachiopods and priapulids and protostome evolution. Nature 399:772-776

Flajnik MF, Ohta Y, Namikawa-Yamada C, Nonaka M (1999) Insight into the primordial MHC from studies in ectothermic vertebrates. Immunol Rev 167:59-67

Forey P, Janvier P (1993) Agnathans and the origin of jawed vertebrates. Nature 361:129-134

Halanych KM, Bacheller JD, Aguinaldo AM, Liva SM, Hillis DM, Lake JA (1995) Evidence from 18S ribosomal DNA that the lophophorates are protostome animals. Science 267:1641-1643

Hughes AL (1997) Rapid evolution of immunoglobulin superfamily C2 domains expressed in immune system cells. Mol Biol Evol 14:1-5

Kasahara M (1999a) The chromosomal duplication model of the major histocompatibility complex. Immunol Rev 167:17-32

Kasahara M (1999b) Genome dynamics of the major histocompatibility complex: insights from genome paralogy. Immunogenetics 49: in press.

Kasahara M, Hayashi M, Tanaka K, Inoko H, Sugaya K, Ikemura T, Ishibashi T (1996) Chromosomal localization of the proteasome Z subunit gene reveals an ancient chromosomal duplication involving the major histocompatibility complex. Proc Natl Acad Sci USA 93:9096-9101

Katsanis N, Fitzgibbon J, Fisher EMC (1996) Paralogy mapping: identification of a region in the human MHC triplicated onto human chromosomes 1 and 9 allows the prediction and isolation of novel PBX and NOTCH loci. Genomics 35:101-108

Litman GW, Anderson MK, Rast JR (1999) Evolution of antigen binding receptors. Annu Rev Immunol 17:109-147

Murphy PM (1993) Molecular mimicry and the generation of host defense protein diversity. Cell 72:823-826

Pancer Z, Rast JP, Davidson EH (1999) Origins of immunity: transcription factors and homologues of effector genes of the vertebrate immune system expressed in sea urchin coelomocytes. Immunogenetics 49:773-786

Pebusque MJ, Coulier F, Birnbaum D, Pontarotti P (1998) Ancient large-scale genome duplications: phylogenetic and linkage analyses shed light on chordate genome evolution. Mol Biol Evol 15:1145-1159

Peelman LJ, Chardon P, Nunes M, Renard C, Geffrotin C, Vaiman M, Van Zeveren A, Coppieters W, van de Weghe A, Bouquet Y, et al. (1995) The BAT1 gene in the MHC encodes an evolutionarily conserved putative nuclear RNA helicase of the DEAD family. Genomics 26:210-218

Rast JP, Anderson MK, Strong SJ, Luer C, Litman RT, Litman GW (1997) α, β, γ, and δ T cell antigen receptor genes arose early in vertebrate phylogeny. Immunity 6:1-11

Rast JP, Litman GW (1998) Towards understanding the evolutionary origins and early diversification of rearranging antigen receptors. Immunol Rev 166:79-86

Sherwood DR, McClay DR (1997) Identification and localization of a sea urchin Notch homologue: insights into vegetal plate regionalization and Notch receptor regulation. Development 124:3363-3374

Siegel AF, Trask B, Roach JC, Mahairas GG, Hood L, van den Engh G (1999) Analysis of sequence-tagged-connector strategies for DNA sequencing. Genome Res 9:297-307

Smith LC, Chang L, Britten RJ, Davidson EH (1996) Sea urchin genes expressed in activated coelomocytes are identified by expressed sequence tags. Complement homologues and other putative immune response genes suggest immune system homology within the deuterostomes. J Immunol 156:593-602

Smith LC, Davidson EH (1994) The echinoderm immune system. Characters shared with vertebrate immune systems and characters arising later in deuterostome phylogeny. Ann N Y Acad Sci 712:213-226

Smith LC, Shih CS, Dachenhausen SG (1998) Coelomocytes express SpBf, a homologue of factor B, the second component in the sea urchin complement system. J Immunol 161:6784-6793

Trachtulec Z, Hamvas RM, Forejt J, Lehrach HR, Vincek V, Klein J (1997) Linkage of TATA-binding protein and proteasome subunit C5 genes in mice and humans reveals synteny conserved between mammals and invertebrates. Genomics 44:1-7

Venter JC, Smith HO, Hood L (1996) A new strategy for genome sequencing. Nature 381:364-366

Yuh CH, Bolouri H, Davidson EH (1998) Genomic cis-regulatory logic: experimental and computational analysis of a sea urchin gene. Science 279:1896-1902

2. Genome organization of the MHC

Physical mapping of the class I regions of the rat major histocompatibility complex

Eberhard Günther, Sofia Ioannidu, and Lutz Walter

Division of Immunogenetics, University of Göttingen, Heinrich-Düker-Weg 12, D-37073 Göttingen, Germany

Summary. By using a PAC library of BN rat ($RT1^n$) DNA, contigs have been established that encompass the centromeric class I region, RT1-A, of the rat MHC and a large part of the telomeric class I region, RT1-C/M. The $RT1-A^n$ region is shown to contain three class I genes and is located between the Sacm2l and Ring1 genes. Sequence data indicate that the H2-K and RT1-A regions are localized at orthologous positions. For the $RT1-C/M^n$ region four contigs could be established so far, which do not yet overlap and encompass about 1.8 Mb. At least about 50 class I genes or gene fragments could be identified, among them H2-T and H2-M-like genes, as well as framework genes that are known from the H2 and HLA complex. The order of the framework genes in the RT1-C/M region is the same as in the HLA class I region and the H2-D/Q/T/M region and the class I gene clusters occur at similar positions. Thus the overall organization of the (telomeric) MHC class I region does not differ between these species.

Key words. Rat, Class I genes, Framework genes, PAC cloning

The major histocompatibility complex (MHC) of the laboratory rat (*Rattus norvegicus*), the RT1 complex, maps to the telomeric end of the short arm of chromosome 20 (Helou et al. 1998). The overall organization of the RT1 complex (Günther 1996; Gill et al.1997; Rolstad et al. 1997) is similar to that of the mouse H2 complex. In both species the MHC contains two gene regions that encode class I molecules, the RT1-A region corresponding to H2-K, and the RT1-C/M region (also designated RT1-E/C/M, RT1-E/C-grc) corresponding to H2-D/Q/T/M. The two class I regions are separated by the class II and class III regions (Fig. 1). In human, no equivalent of the RT1-A and H2-K regions is found. It is commonly assumed that RT1-A and H2-K represent insertions due to translocation of class I genes from the telomeric class I region. The chromosomal orientation of the various MHC regions is concordant in human, mouse and rat. The HLA class I, the H2-D/Q/T/M and RT1-C/M regions map telomeric to the class III region, whereas the class II region is centromeric to the latter. The RT1-A and H2-K regions are centromeric to the class II region (or in the centromeric

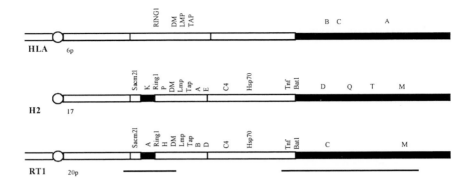

Fig. 1. Schematic representation of the human, mouse and rat MHC. Class I gene regions are shown in black. Only some of the MHC genes are listed for general orientation and localization of the contigs established. The regions described here are marked by bars.

part of the class II region). A hallmark of the genomic structure of the MHC class I regions in mouse and human is the presence of clusters of class I genes between regions of framework genes that are unrelated to class I genes. The class I gene clusters appear to be situated at orthologous positions and with conserved order in both species (Klein et al. 1998; Amadou 1999; Amadou et al. 1999).

In the rat, physical mapping data of the MHC on the basis of genomic libraries are available only to a very limited degree. For the RT1-A (Walter et al. 1996), class II (Diamond et al. 1989; Arimura et al. 1995a,b) and RT1-C/M regions (Jameson et al. 1992; Yuan et al. 1996, 1999; Salgar et al. 1998) cosmid contigs, mostly of small size, have been reported. Furthermore, a YAC contig of the RT1-M region has been published (Lambracht-Washington et al. 1999). We have made use of a rat PAC library that has become available recently (Woon et al. 1998) in order to establish a physical map of the rat MHC. The library represents the MHC of the $RT1^n$ haplotype. A complete physical map will be reported here for the RT1-A region based on a contig of 300 kb and a partial map of the RT1-C/M region based on four contigs encompassing about 1.8 Mb.

Materials and methods

The analysis is based on PAC library RPCI-31 constructed from the genome of the inbred BN strain ($RT1^n$ haplotype) by Woon et al. (1998) and distributed by the Resource Center of the German Human Genome Project in Berlin, Germany (library 712). The library covers the genome tenfold; chimeric clones have not been detected (Woon et al. 1998). Screening of the library was carried out with DNA pools or micro-arrayed filters. Probes of the Sacm2l and Ring1 genes were used for identifying clones of the RT1-A region and probes of the Bat1 and class I genes for RT1-C/M region screening. The class I probes either encompassed exon 1, intron 1, exon 2 and intron 2 ("exon 2" probe) or part of intron 3 and exon 4 ("exon 4" probe) derived from the class Ia gene RT1-Au (Walter et al. 1995). Primers for establishing probes of genes Tnfa, Bat1, Pou5fl, Tcf19, Prg1, Pnuts, RT.BM1, Znf173, Rfb30, RT1-M4 (microsatellite) and Mog were prepared on the basis of published sequences. The Grc probe was prepared according to the pGRC1.4 sequence (Yuan et al. 1996). Probes for markers A224T, I162T, H125C were based on sequences of rat PAC ends. Probes of mouse STS markers 255D16T and 261A5T were prepared according to Amadou et al. (1999). Gna-rs1, Tctex5 and Tu42 probes were obtained from Dr. Pontarotti, Marseille. Southern blot analysis of PAC DNA was performed with various restriction enzymes. The class I-hybridizing fragments were defined on the basis of Bam HI restriction and designated by the respective length. Ordering of PAC clones was primarily based on the overlap of common restriction fragments and genes. Sequencing of PCR amplificates obtained from PAC clones was performed automatically (ABI Prism 310).

Results

Physical map of the RT1-A^n region

The physical map established for the RT1-A region of the RT1^n haplotype is shown in Figure 2. The figure summarizes the analysis of 10 overlapping PAC clones. Seven clones contain the Sacm2l as well as Ring1 genes, two contain Ring1 but not Sacm2l, and one contains Sacm2l but not Ring 1. The contig is anchored in the class II region, since some clones hybridize with the RT1-Hb and one with the RT1-DMb probes (Walter and Günther 1998). At the centromeric side the Syngap, Hset, Daxx and Bing1 genes are found.   They have

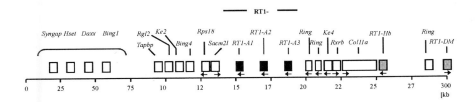

Fig. 2. Physical map of the RT1-A^n and flanking regions. The map is based on 10 overlapping PAC clones. Class I genes are marked by black, class II genes by grey and framework genes by white boxes. Arrows indicate direction of transcription. Genes in brackets have not yet been ordered. For information on the framework genes, see Walter and Günther (1998) and Beck and Trowsdale (1999).

not yet been ordered in the rat. The Tapbp, Rgl2, Bing4, Rps18, Sacm2l genes and the Ring1, Ring2, Ke4, Rxrb, Col11a2 genes show the same order and occur at orthologous positions in the MHC of human (Tubby, EMBL/GenBank database accession number AL031228) and mouse (Rowen et al., accession number AF100956).

The interval that carries the class I genes of the RT1-A region is defined by Sacm2l and Ring1. It encompasses about 60 kb. In the RT1 haplotype analysed here three class I genes can be identified that will be designated A1, A2 and A3. They are about 20 kb apart from each other, the distance between A1 and Sacm2l is about 13 kb and that between A3 and Ring1 is about 10 kb. The sequences of the A1 and A2 genes correspond to those reported as RT1-A1^n and RT1-A2^n by Joly et al. (1998). The sequence of the third gene, RT1-A3, is most similar to the RT1-V gene (Joly et al. 1997).

In the mouse the H2-K and H2-K1 genes are found between Sacm2l and Ring1 (Rowen et al. see above), i.e., at an orthologous position compared to the RT1-A genes of the rat.

We have started to determine the nucleotide sequences centromeric of Ring1 and telomeric of Sacm2l. The sequences were compared with the corresponding regions of human (accession number see above) and mouse (accession number see above) using the DOTTER software (Sonnhammer and Durbin 1995). The sequence downstream of the Ring1 stop codon, i.e. towards the A3 gene, was highly homologous to that downstream of RING1 in human and of Ring1 in the mouse for 400 nucleotides. Then homology between the human sequence on the one hand and the rat/mouse sequences on the other hand disappeared abruptly, whereas for the mouse and rat sequences homology continued. The position of homology loss might represent the demarcation of the RT1-A and H-2K insertion. Sequences obtained downstream of the Sacm2l gene indicate that homology between human and rat/mouse ends about 2870 nucleotides after the stop codon of Sacm2l at a position that is identical in rat and mouse. Again this position could mark the limit of the RT1-A and H2-K insertions. Thus the RT1-A/H-2K regions can be delineated still more precisely inside the Sacm2l/Ring1 interval. Furthermore, with respect to these preliminary sequence data, the insertion of the class I genes could have occurred at the same position in the rat and mouse MHC.

Physical map of the RT1-C^n region

The PAC clones identified so far can be arranged in four contigs that do not yet overlap. Contig 1 (Fig. 3) is anchored in the class III region, because it contains the Tnfa (Kirisits et al. 1994) and Bat1 (Nair et al. 1992) genes. The contig encompasses 12 PAC clones. By Southern blot analysis with class I probes, 20 Bam HI restriction fragments were found that hybridized with the class Ia exon 2 probe, the exon 4 probe or both. The fragments are provisionally designated by the size of the hybridizing Bam HI fragment. On the basis of exon 4-positive fragments, one could extrapolate that about 15 class I genes are present in this contig. Some of them hybridized also with the exon 2 probe and therefore might represent full-size genes. In cases where separate but neighboring exon 2 and exon 4 hybridizing fragments were found, both together might represent a single gene. Partial sequences of the first six "genes" revealed similarity of more than 90 % to rat class Ia genes. The exon 2 sequence of "gene" 4 is identical to that of clone 4 reported by Wang et al. (1996). The telomeric end of contig 1 carries the Pou5f1 (Oct3, Oct4, Otf3; Rosner et al, 1990; Uehara et al. 1991) and Tcf19 (Krishnan et al. 1995) framework genes.

Contig 2 (Fig. 4) encompasses 12 PAC clones. At its one end the framework genes Prg1 (Pietzsch et al. 1998), Pnuts (Fb19, Ppp1r10; Allen et al. 1998; Totaro et al. 1998) and Gna-rs1 (Hsr1, Gnl1; Vernet et al. 1994) are localized. Furthermore, the contig contains 20 fragments that hybridize with the exon 2 probe and 13 fragments that hybridize with the exon 4 probe. In the vicinity of the H125C marker a cluster of 11 to 12 fragments is found, which does not show any exon 4-hybridization. One of these fragments (designated 2) has been subcloned and sequenced. It represents half of exon 2 but no further class I-like sequences are detectable in the flanking parts. Partial sequencing of genes hybridizing with the exon 4 probe revealed that fragment 4.4 is the RT.BM1 (RT1-S3) gene (Parker et al. 1991; Salgar et al. 1998). Southern blot analysis with a grc probe, which was designed according to the pGRC1.4 probe (Yuan et al. 1996), maps the Grc region (Kunz et al. 1980) to an interval indicated in Figure 4. The position of

Contig 1

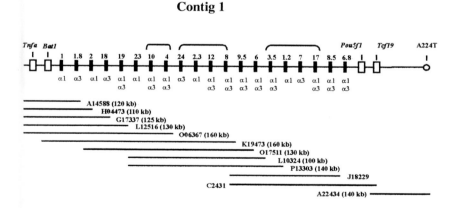

Fig. 3. Scheme of contig 1 representing the beginning of the RT1-C/M region. Class I genes/gene fragments (black boxes) are identified by hybridization with an exon 2 or exon 4 probe, marked by α1 and α3, respectively, and designated by the length (kb) of the hybridizing Bam HI fragment. Framework genes are shown by white boxes. Genes in brackets have not yet been ordered. PAC clones are designated by their official number, the approximate length being added in brackets. Distances between boxes and lengths of PAC clones are not to scale.

STS marker H125C defines a region that shows sequence homology to a mouse MHC sequence deposited in the EMBL/Genbank (Evans et al., accession number AC005807).

Contig 3 (Fig. 5) encompasses 6 PAC clones so far and contains 10 BamHI restriction fragments that hybridize with the exon 4 probe. Since in the mouse the STS marker 255D16T (Amadou et al. 1999) maps to the beginning of the M region, most of the class I genes/gene fragments shown in Figure 5 might represent M homologs of the rat. So far the number of PAC clones of this contig is relatively small compared, e.g. to contig 1, and it is possible that contig 3 has to be extended to the "right".

Order and orientation of contigs 2 and 3 shown in Figures 4 and 5 are tentative and follow the map of corresponding markers in the mouse H2-Q/T/M region. According to preliminary evidence both contigs can be bridged by a single PAC that contains a further class I hybridizing fragment.

Contig 4 (Fig. 6) consists so far of only 6 PAC clones and carries the framework genes Znf173 (Afp; Chu et al. 1995), Rfb30 (Henry et al. 1997) and Tctex5 (Ha et al. 1991) on the one end and Mog (Pham-Dinh et al. 1995) and Tu42 (Amadou et al. 1999) on the other end. The interval defined by these two framework gene groups harbors three class I genes. One of them has been identified as the H2-M4 homolog on the basis of a microsatellite marker described by Lambracht-Washington et al. (1998). Thus this contig represents

Contig 2

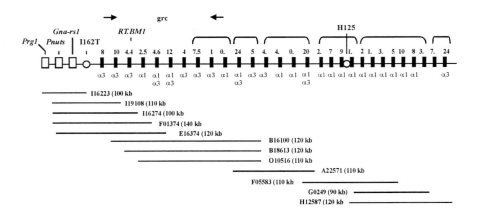

Fig. 4. Scheme of contig 2 of the RT1-C/M region. Arrows mark the grc region as defined by hybridization with a grc probe. For further explanation, see the legend to Figure 3.

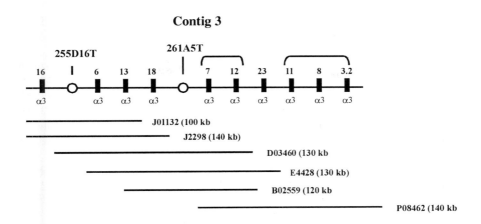

Fig. 5. Scheme of contig 3 of the RT1-C/M region. For explanation, see the legend to Figure 3.

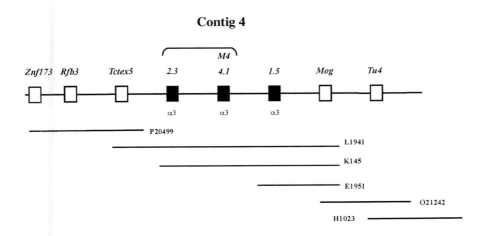

Fig. 6. Scheme of contig 4. For explanation, see the legend to Figure 3.

part of the region that is homologous to the H2-M region. The order of the Znf173, Tctex5, RT1-M4 and Mog genes is identical to that described by Lambracht-Washington et al. (1999) for the RT1^{lvl} haplotype on the basis of a YAC contig. Contig 4 most likely is telomeric to contig 3, because, on the basis of RT1 recombinants, the Mog and RT1-M4 genes map distally to the other RT1-C class I genes (Lambracht et al. 1995).

Several PAC clones that carry the rat homologs of the H2-M2 and H2-M3 genes have been isolated, but not yet linked among each other and to contig 4. Since none of the PAC clones included in contigs 1 to 4 reacted with H2-M2 and H2-M3 probes, the corresponding rat genes map telomeric to Mog.

Considering the lengths of the individual PAC clones the four contigs described here encompass about 1.8 Mb.

Discussion

The physical map of the centromeric class I region, RT1-A, and a preliminary version of a physical map of part of the telomeric class I region, RT1-C/M, of the rat MHC are presented. The map information is based on a PAC library and reflects the RT1^{n} haplotype. This haplotype has not been studied genetically as intensively as other RT1 haplotypes like a, c, l, or u. This is in part due to the relative lack of intra-MHC recombinants (see Günther 1996). On the other hand the BN rat that carries the RT1^{n} haplotype is an extremely useful and widely applied strain in immunology.

We have cloned completely the RT1-A^{n} region, one of the two class I regions of the rat MHC. This region is flanked by the framework genes Sacm2l at the centromeric and Ring1 at the telomeric side and encompasses about 60 kb. So far class I genes that clearly fulfil the criteria of class Ia genes have been found only in the RT1-A region (Günther and Wurst 1994; Rolstad et al. 1997). It has been shown recently that some RT1 haplotypes carry one, others two functional class Ia genes in the RT1-A region (Joly et al. 1996). We have now identified three class I genes in the RT1-A region of BN rats. Two of these genes, A1 and A2, were known by sequence before (Joly et al. 1998), the third, A3, is similar to the RT1-V gene (Joly et al. 1997). Whether expression and function of the A3 gene are class Ia-like is not yet known.

The organization of the RT1-A region and its vicinity described for the RT1^{n} haplotype is very similar to that reported for the RT1^{u} haplotype (Walter and Günther 1998, 1999) except that for the latter haplotype only one or possibly two class I genes are found in the Sacm2l/Ring1 interval.

Partial sequences obtained from both sides of the RT1-A region indicate that, compared to the corresponding human sequence, homology stops at a distinct position downstream of Sacm2l and Ring1, respectively. These positions appear to be identical in rat and mouse. This could indicate that the H2-K and RT1-A regions have the same origin in a common ancestor species. H2-K and RT1-A could then represent the same insertion of class I genes into the centromeric part of what is the class II region in human. It is, however, noteworthy that the RT1-A and H2-K regions are replete of repeats the presence of which could have disturbed the original genomic situation. Furthermore, the repeats could have served as permissive sites for inserting class I genes, an event that also might have occurred repeatedly during mouse/rat evolution. One could also speculate that a H2-K/RT1-A-like region had got lost secondarily in human.

Starting with the telomeric side of the RT1 class III region we have established four so far not yet overlapping clonal contigs, each encompassing several 100 kb. The contigs represent parts of the RT1 complex, because they carry genes already mapped to the rat MHC. Thus contig 1 carries the Tnfa gene (mapped to the rat MHC by Lambracht et al. (1991) and Kirisits et al. (1994)) and Bat1 gene (Helou et al. 1998), contig 2 carries the RT.BM1 gene (Parker et al. 1990), and contig 4 includes the Mog and M4 genes (Lambracht et al. 1995). Only contig 3 cannot yet be formally assigned to the RT1 complex in this manner.

The RT1-C/M contigs together contain 41 Bam HI fragments that hybridize with a class I exon 4 probe, and 15 react also with the exon 2 probe. Therefore, these fragments could represent full-size class I genes. Furthermore 21 Bam HI fragments hybridized with the exon 2 probe only. Since this probe is derived from a class Ia gene the respective exons 2 should be sufficiently homologous to class Ia genes and represent H2D,L, Q, but not H2-T, M homologs. The majority of the exon 2-hybridizing Bam HI fragments presumably constitute full-size class I genes together with the neighboring exon 4-hybridizing fragments. In contig 2 a region containing a cluster of exon 2-hybridizing fragments was detected that did not hybridize with the exon 4 probe. One of the fragments has been sequenced and found to be one half of exon 2. A similar type of fragment has been described in the HLA class II region as HLA-Z1 encompassing 87 bp of exon 4 (Beck et al. 1996). Among the other class I genes/gene fragments, some could be identified by partial sequencing as genes already known, e.g. RT.BM1 (Parker et al. 1990, 1991). It is difficult at present to give a more precise information on the number of class I genes or gene fragments that are present in the RT1-C/M^n region, since some exon 2 and exon 4-hybridizing fragments might represent one gene and longer restriction fragments might carry two class I genes. Furthermore, contig 3 could be more extended, and the class I genes telomeric to Mog like the H2-M2 and H2-M3 homologs, have not yet been analysed in more detail. Nevertheless, the number of class I genes/gene fragments might be in the range postulated to be 61 in the RT1^{av1} haplotype (Jameson et al. 1992) and 61 in the RT1^{r21} haplotype (Yuan et al. 1996).

The region that hybridizes with a pGRC1.4-like probe and therefore defines the Grc region (Yuan et al. 1996) could be delineated in contig 2. The respective PAC clones will be helpful to identify the genes in this region or its vicinity that are responsible for the Grc mutant phenotype (Kunz et al. 1980; Melhem et al. 1993).

In contigs 1, 2 and 4 groups of framework genes could be identified that occur also in the mouse and human MHC. These genes occupy orthologous positions in conserved order in the mouse H2-D/Q/T/M and HLA class I regions. So far the framework genes found in the RT1-C/M region show the same order as in human and mouse (Fig. 7). In addition, several framework genes had not been fine-mapped so far in the MHC and could be physically localized, like Prg1 (synonyms Gly96, Dif2; Pietzsch et al. 1998) or Pnuts (synonyms Ppp1r10, Fb19; Allen et al. 1998; Totaro et al. 1998).

The regions where class I gene clusters are interspersed appear to be at orthologous sites in mouse and human (Amadou et al. 1999). We now show that the class I clusters of the RT1-C/M region are found at orthologous positions compared to mouse H2-D/Q/T/M and HLA class I regions (Fig. 8). This does not imply that the class I genes themselves are orthologous. Orthology between mouse and rat class I genes appears to exist in the case of some H-2T-like genes, as reported earlier (Gill et al. 1997; Günther 1996; Rolstad et al. 1997). An example is RT1.BM1 (contig 2) that appears to be orthologous to H-2T23 (Parker et al. 1991). Furthermore some of the H2-M genes have been shown to have rat orthologs (Lambracht et al. 1995; Wang et al. 1995), like M4 in contig 4. In general the data presented here indicate that contig 1 mainly contains genes that show very high sequence similarity to class Ia genes. Contig 2 might contain H2-T-like genes and contigs 3 and 4 H2-M-like genes. When the sequences of the RT1-C/M class I genes are available, a clearer picture of the evolution of these genes will become apparent.

So far the RT1-C/M contigs encompass about 1.8 Mb, so that this region is at least of the same size as the HLA class I region. The further analysis of the telomeric class region of the rat MHC is in progress. The gaps between the contigs have to be closed and mapping of the region telomeric of Mog will be completed

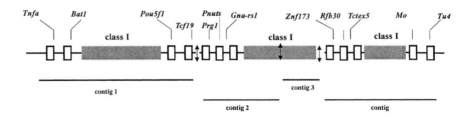

Fig. 7. Summary of the contigs shown in Figures 3 to 6. The class I clusters are shown as grey boxes, framework genes as single white boxes. Gaps between contigs are indicated by vertical double arrows.

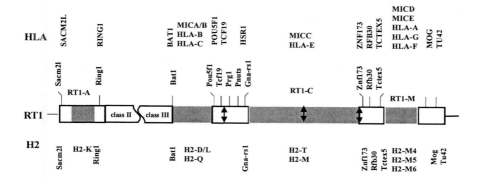

Fig. 8. Synopsis of the mapping data so far obtained for the two RT1 class I regions, of the orthology of framework genes and positions of class I clusters compared to mouse and human MHC. The figure is based on the contigs shown in Figures 2 to 7 and on references Amadou et al. (1999) and Shiina et al. (1999).

(Fig. 8). PAC clones carrying H2-M2, H2-M3 and olfactory receptor genes have already been isolated. As a further step the complete nucleotide sequence of the two class I regions has to be determined. The results obtained so far already demonstrate that the overall genomic organization of the telomeric class I region of the rat MHC is very similar to that found in the HLA and H2 systems.

Acknowledgements

The technical assistance of Ms. Diana Otto is gratefully acknowledged. The authors are grateful to the Resource Center of the German Human Genome Project, Berlin, Germany, for providing filters and clones. The study is supported by EU grant PL96562.

References

Allen PB, Kwon YG, Nairn AC, Greengard P (1998) Isolation and characterization of PNUTS, a putative protein phophatase 1 nuclear targeting subunit. J Biol Chem 273:4089-4095

Amadou C (1999) Evolution of the Mhc class I region: the framework hypothesis. Immunogenetics 49:362-367

Amadou C, Kumánovics A, Jones EP, Lambracht-Washington D, Yoshino M, Fischer Lindahl K (1999) The mouse major histocompatibility complex: some assembly required. Immunol Rev 167:211-221

Arimura Y, Tang WR, Koda T, Kakinuma M (1995) Cloning and analysis of a new rat major histocompatibility complex class II gene, RT1.DOa. Immunogenetics 42:156-158

Arimura Y, Tang WR, Koda T, Kakinuma M (1995) Structure of novel rat major histocompatibility complex class II genes RT1.Ha and Hb. Immunogenetics 41:320-325

Beck S, Abdulla S, Alderton RP, Glynne RJ, Gut IG, Hosking LK, Jackson A, Kelly A, Newell WR, Sanseau P, Radley E, Thorpe KL, Trowsdale J (1996) Evolutionary dynamics of non-coding sequences within the class II region of the human MHC. J Mol Biol 255:1-13

Beck S, Trowsdale J (1999) Sequence organization of the class II region of the human MHC. Immunol Rev 167:201-210

Chu TW, Capossela A, Coleman R, Goei VL, Nallur G, Gruen JR (1995) Cloning of a new "finger" protein gene (ZNF173) within the class I region of the human MHC. Genomics 29:229-239

Diamond AG, Hood LE, Howard JC, Windle M, Winoto A (1989) The class II genes of the rat MHC. J Immunol 142:3268-3274

Gill TJ III, Salgar SK, Yuan X-J, Kunz HW (1997) Current status of the genetic and physical maps of the major histocompatibility complex in the rat. Transplant Proc 29:1657-1659

Günther E (1996) Current status of the molecular genetic analysis of the rat major histocompatibility complex. Folia Biol 42:129-145

Günther E, Wurst W (1984) Cytotoxic T lymphocytes of the rat are predominantly restricted by RT1.A and not RT1.C-determined major histocompatibility complex class I antigens. Immunogenetics 20:1-12

Ha H, Howard CA, Yeom YI, Abe K, Uehara A, Artzt K, Bennett D (1991) Several testis-expressed genes in the mouse t-complex have expression differences between wild-type and t-mutant mice. Dev Genet 12:318-332

Helou K, Walter L, Günther E, Levan G (1998) Cytogenetic orientation of the rat major histocompatibility complex (MHC) on chromosome 20. Immunogenetics 47:166-169

Henry J, Ribouchon M-T, Depetris D, Matteï M-G, Offer C, Tazi-Ahnini R, Pontarotti P (1997) Cloning, structural analysis, and mapping of the B30 and B7 multigenic families to the major histocompatibility complex (MHC) and other chromosomal regions. Immunogenetics 46:383-395

Jameson SC, Tope WD, Tredgett EM, Windle JM, Diamond AG, Howard JC (1992) Cloning and expression of class I major histocompatibility complex genes of the rat. J Exp Med 175:1749-1757

Joly E, Leong L, Coadwell WJ, Clarkson C, Butcher GW (1996) The rat MHC haplotype RT1^c expresses two classical class I molecules. J Immunol 157:1551-1558

Joly E, Graham M, Coadwell J, Deverson E, Leong L, Le Rolle A, Gonzalez A, Butcher GW (1997) Novel expressed MHC class I genes of the rat: RT1-U, RT1-V, RT1-Y and RT1-Z. Rat Genome 3:133-137

Joly E, Le Rolle A-F, Gonzalez AL, Mehling B, Stevens J, Coadwell WJ, Hünig T, Howard JC, Butcher GW (1998) Co-evolution of rat TAP transporters and MHC class I RT1-A molecules. Curr Biol 8:169-172

Klein J, Sato A, O'hUigin C (1998) Evolution by gene duplication in the major histocompatibility complex. Cytogenet Cell Genet 80:123-127

Kirisits MJ, Vardimon D, Kunz HW, Gill TJ III (1994) Mapping of the TNFA locus in the rat. Immunogenetics 39:59-60

Krishnan BR, Jamry I, Chaplin DD (1995) Feature mapping of the HLA class I region: Localization of the POU5F1 and TCF19 genes. Genomics 30:53-58

Kunz HW, Gill TJ III, Dixon BD, Taylor FH, Greiner DL (1980) Growth and reproduction complex in the rat: genes linked to the major histocompatibility complex that affect development. J Exp Med 152:1506-1518

Lambracht D, Wurst W, Günther E, Wonigeit K (1991) Polymorphism of the rat TNF genes and genetic fine mapping in the rat MHC. Immunobiology 183:272

Lambracht D, Prokop C, Hedrich HJ, Fischer Lindahl K, Wonigeit K (1995) Mapping of H-2M homolog and MOG genes in the rat MHC. Immunogenetics 42:418-421

Lambracht-Washington D, Shisa H, Butcher GW, Fischer Lindahl K (1998) A polymorphic microsatellite marker in the rat major histocompatibility complex class I gene, RT1.M4, and a new recombinant RT1 haplotype, r39. Immunogenetics 48:420-421

Lambracht-Washington D, Wonigeit K, Fischer Lindahl K (1999) Gene order of the RT1.M region of the rat MHC exemplifies interspecies conservation with the mouse. Transplant Proc 31:1515-1516

Melhem MF, Kunz HW, Gill TJ III (1993) A major histocompatibility complex-linked locus in the rat critically influences resistence to diethylnitrosamine carcinogenesis. Proc Natl Acad Sci USA 90:1967-1971

Nair S, Dey R, Sanford JP, Doyle D (1992) Molecular cloning and analysis of an eIF-4A-related liver nuclear protein. J Biol Chem 267:12928-12935

Parker KE, Carter CA, Fabre JW (1990) A rat class I cDNA clone with Alu-like sequence and mapping to two genes in RT1.C/E. Immunogenetics 31:211-214

Parker KE, Carter CA, Murphy G, Fabre JW (1991) The rat RT.BM1 MHC class I cDNA shows a high level of sequence similarity to the mouse H-2T23^d gene. Immunogenetics 34:211-213

Pham-Dinh D, Jones EP, Pitiot G, Della Gaspera B, Daubas P, Mallet J, Le Paslier D, Fischer Lindahl K, Dautigny A (1995) Physical mapping of the human and mouse MOG gene at the distal end of the MHC class Ib region. Immunogenetics 42:386-391

Pietzsch A, Büchler C, Schmitz G (1998) Genomic organization, promoter cloning, and chromosomal localization of the Dif-2 gene. Biochem Biophys Res Commun 245:651-657

Rolstad B, Vaage JT, Naper C, Lambracht D, Wonigeit K, Joly E, Butcher GW (1997) Positive and negative recognition by rat NK cells. Immunol Rev 155:91-104

Rosner MH, Vigano MA, Ozato K, Timmons PM, Poirier F, Rigby PW, Staudt LM (1990) A POU-domain transcription factor in early stem cells and germ cells of the mammalian embryo. Nature 345:686-692

Salgar SK, Kunz HW, Gill TJ III (1998) Structural organization, sequence analysis, and physical mapping of the Grc-linked class Ib gene RT1.S3 in the rat. Immunogenetics 48:76-81

Shiina T, Tamiya G, Oka A, Takishima N, Inoko H (1999) Genome sequencing analysis of the 1.8 Mb entire human MHC class I region. Immunol Rev 167:193-199

Sonnhammer ELL, Durbin R (1995) A dot-matrix program with dynamic threshold control suited for genomic DNA and protein sequence analysis. Gene 167:1-10

Totaro A, Grifa A, Carella M, Rommens JM, Valentino MA, Roetto A, Zelante L, Gasparini P (1998) Cloning of a new gene (FB19) within HLA class I region. Biochem Biophys Res Commun 250:555-557

Uehara H (1991) Mouse Oct-3 maps between the Tcl12 embryonic lethal gene and the Qa gene in the H-2 complex. Immunogenetics 34:266-269

Vernet C, Ribouchon M-T, Chimini G, Pontarotti P (1994) Structure and evolution of a member of a new subfamily of GTP-binding proteins mapping to the human MHC class I region. Mamm Genome 5:100-105

Walter L, Günther E (1998) Identification of a novel highly conserved gene in the centromeric part of the major histocompatibility complex. Genomics 52:298-304

Walter L, Günther E (1999) Sequence analysis of the genomic interval between the Rps18 and RT1-A genes in the RT1[u] haplotype. Transplant Proc 31:1513-1514

Walter L, Tiemann C, Heine L, Günther E (1995) Genomic organization and sequence of the rat major histocompatibility complex class Ia gene RT1.A[u]. Immunogenetics 41:332

Walter L, Fischer K, Günther E (1996) Physical mapping of the Ring1, Ring2, Ke6, Ke4, Rxrb, Col11a2 and RT1.Hb genes in the rat major histocompatibility complex. Immunogenetics 44:218-221

Wang C-R, Lambracht D, Wonigeit K, Howard JC, Fischer Lindahl K (1995) Rat RT1 orthologs of mouse H-2M class Ib genes. Immunogenetics 42:63-67

Wang M, Stepkowski SM, Tian L, Langowski JL, Hebert JS, Kloc M, Yu J, Kahan BD (1996) Nucleotide sequences of three distinct cDNA clones coding for the rat class I heavy chain RT1[n] antigen. Immunogenetics 45:73-75

Woon PY, Osoegawa K, Kaisaki PJ, Zhao B, Catanese JJ, Gauguier D, Cox R, Levy ER, Lathrop GM, Monaco AP, de Jong PJ (1998) Construction and characterization of a 10-fold genome equivalent rat P1-derived artificial chromosome library. Genomics 50:306-316

Yuan X-J, Salgar S, Cortese Hassett AL, McHugh KP, Kunz HW, Gill TJ III (1996) Physical mapping of the E/C and grc regions of the rat major histocompatibility complex. Immunogenetics 44:9-18

Yuan X-J, Kunz HW, Gill TJ III (1999) Physical mapping and sequencing of class I genes in a 150-kb contig in the EC region. Transplant Proc 31:1507-1512

MHC gene organization of the bony fish, medaka

Kiyoshi Naruse[1], Akihiro Shima[1,2], and Masaru Nonaka[1]

[1]Department of Biological Sciences, Graduate School of Science, University of Tokyo, Bunkyo-ku, Tokyo 113-0033, Japan
[2]Department of Integrated Biosciences, Graduate School of Frontier Science, University of Tokyo, Bunkyo-ku, Tokyo 113-0033, Japan

Summary. We isolated the MHC *class I A*, *class II B*, *β2-microglobulin* (*B2m*), *LMP2, LMP7, TAP2*, and complement *B/C2, C3* and *C4* genes from the Japanese teleost fish, medaka, and analyzed the linkage relationships between these genes using two backcross typing panels. Multiple cDNA species were identified for the *MHC class I A* (3), *class II B* (3) and *C3* (2) genes, whereas only single cDNA species was isolated from the other genes. Two *class I A, LMP2, LMP7* and *TAP2* genes were closely linked to each other, and independent linkages were observed between two *class II B* genes and between two *C3* genes. The other genes were mapped on different linkage groups. These results indicate that the counterparts of mammalian MHC genes show dispersed distribution over the medaka genome, except for the two *class I A* genes and the genes involved in class I antigen presentation. A similar result has been reported for zebrafish, which is remotely related to medaka, suggesting that the common ancestor of teleosts may have had a dispersed MHC organization. Although it remains to be determined whether the dispersed MHC organization is ancestral to, or derived from, the centralized MHC that is characteristic of higher vertebrates, the medaka genes involved in class I antigen presentation seem to constitute the core of the MHC.

Key words. Medaka fish, MHC, Class I antigen presentation, Evolution

Introduction

The mammalian major histocompatibility complex (MHC) consists of three closely linked regions: class I, II and III (Klein 1986). Three kinds of genes encoding cell surface molecules that present antigen to T cells are present in the class I and II regions: the *class I A, class II A* and *class II B* genes. In addition, the class II region harbors genes (*LMP2, LMP7, TAP1* and *TAP2*) that encode proteins involved in class I antigen processing and transport. LMP2 and LMP7 are γ-interferon induced proteolytic subunits, which replace their constitutive counterparts and change proteasomal specificity, resulting in production of peptides suitable for presentation by class I molecules. TAP1 and TAP2 molecules form a heterodimer on the endoplasmic reticulum (ER) membrane, and transport class I antigenic peptides from the cytoplasm to the ER lumen. There is no structural similarity among class I A, LMP or TAP molecules. Thus, these mammalian MHC genes provide an example of linkage among functionally related, but structurally unrelated, genes. The human MHC class III region contains a great variety of genes including four complement genes, *C2*, *B*, *C4A* and *C4B* (Carroll et al. 1984). The *C2* and *C4* genes provide another example of a linkage between functionally related, but structurally unrelated, genes within the MHC: C2 and C4, which belong to completely different protein families, assemble to form the C3 convertase upon activation of the classical pathway.

To elucidate the origin and evolution of MHC gene organization, phylogenetic studies of some vertebrate classes are now in progress. In the chicken, two MHC regions (*B* and the *Rfp-Y*) are located on the same microchromosome (Miller et al. 1996) and the B-F/B-L region contains all genes belonging to all three classes (EMBL accession number AL023516). *Xenopus laevis* MHC also contains the genes belonging to three classes, although the organization of these three regions seems to differ from mammals (Nonaka et al. 1997). Thus, all tetrapods studied to date have a centrally organized MHC. In contrast, in all four teleost species currently analyzed (medaka, zebrafish, carp and salmon), the *class I A* genes are not linked to the *class II A* or *II B* genes (Flajnik et al. 1999). These results, and additional studies in zebrafish, suggest that teleosts have a dispersed MHC gene organization (Bingulac-Popovic et al. 1997; Postlethwait et al. 1998), which is very different from that of tetrapods.

To further clarify MHC gene organization in teleosts, we isolated MHC *class I, II* and *III* genes from the medaka, *Oryzias latipes,* and analyzed the linkage of these genes. The medaka belongs to Beloniformes, which is distantly related within the teleosts to Cypriniformes, to which zebrafish belong (Nelson 1994). In the present study, we report that medaka also have a dispersed MHC gene organization, suggesting that this is a common feature of teleosts.

Molecular cloning of the medaka counterparts of mammalian MHC-encoded genes

Class I A gene

The probes for library screening for MHC *class I A* genes were obtained using RT-PCR amplification using the degenerate primers: TGAGTGGSTRMAGAA-GTATGTG and RMWYYVYTGGGGTARAARCCTGT. Primers were designed by comparing *class I A* gene sequences from guppy, zebrafish, rainbow trout and *Xenopus*. The sequence of the amplified fragment was highly homologous with the MHC *class I A* sequences of guppy (Sato et al. 1995), zebrafish (Takeuchi et al. 1995) and rainbow trout (Hansen et al. 1996). The amplified sequences were extended by 5' and 3' rapid amplification of cDNA ends (RACE) and the extended fragments were used as probes for screening.

A cDNA library was constructed from 1μg of total RNA from a whole adult medaka body (HNI strain) using the SMART cDNA library construction kit (Clontech). After screening the cDNA library, three different species of *class I A*-like genes were isolated. We named these genes *Orla-UAA, -UBA* and *-UCA*, respectively. Several clones for the *Orla-UAA,* and only one clone each for *Orla-UBA* and *-UCA,* were identified.

The amino acid sequences of the α3 domain of *Orla-UAA, -UBA* and *-UCA* were aligned with those of MHC class I A amino acid sequences from various vertebrate species using the Clustal W software (Thompson et al. 1994). Figure 1 shows a phylogenetic tree based on this alignment that was constructed using the neighbor-joining method (Saitou and Nei 1987). *Orla-UAA* and *-UBA* have very similar amino acid sequences, whereas *Orla-UCA* shows less similarity. All three

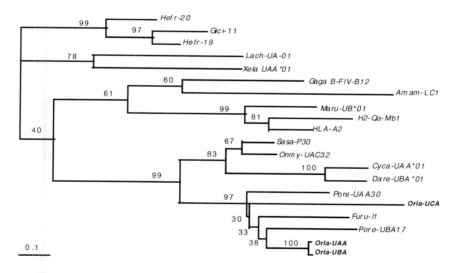

Fig. 1. A phylogenetic tree for the α3 domain of MHC class I A amino acid sequences from various vertebrate species. Numbers on each branch represent the bootstrap percentages for 1000 replications. Accession numbers used for the present study are as follows: rainbow trout (Onmy-UCA-C30, U55380), zebrafish (Dare-UBA*01, Z46777), carp (Cyca-UA1*1, X91015), pufferfish (Furu-I1, AF001215), nurse shark (Gici-11, AF028557), horn shark (Hefr-20, AF028559; Hefr-19, AF028558), coelacanth (Lach-UA-01, U08043), guppy (Pore-UAA30, Z54085; pore-UBA), Atlantic salmon (Sasa-P30, L07606), *A. ameiva* (Amam-LC1, M81094), chicken (Gaga-B12, M31012), human (HLA-A2, U18930), mouse (H2-Qa-Mb1, M20985), *Xenopus* (Xela-UAA*01, L20733), and *M. rufogriseus* (Maru-UB*01, L04952).

genes are members of a clade consisting of the medaka, pufferfish and guppy *class I A* genes. The medaka and guppy belong to Atherinomorpha and pufferfish to Percomorpha of Acanthopterigii. These results suggest that amplification of the *class I A* genes occurred in the common ancestor of Acanthopterigii. The *Orla-UAA* and *-UBA* genes seem to have arisen from a very recent gene duplication in the medaka lineage.

Class II B gene

Total RNA was isolated from the spleens of 1000 outbred medaka fish. RT-PCR was performed using primers corresponding to the two regions of the β2 domain that are most highly conserved among the teleost *class II β* sequences: TGCAGYGCCTAYGACTTCTACCC (CSAYDFYP) and GTGGATCTGGTA-GTACCAGTCCCCGTT (NGDWYYQIH). The nucleotide sequence of the amplified product (~120 bp) was homologous to teleost *class II β* sequences. 5' RACE was performed using two specific primers synthesized according to this sequence and resulted in the isolation of a clone covering the leader peptide, β1, and part of the β2 domain (data not shown).

A medaka cDNA library was constructed using a cDNA synthesis kit (Amersham) and λgt10. Screening of this library, using the 5' RACE product as a probe, resulted in the isolation of a cDNA clone that we named K14. Preliminary PCR analysis of the inbred medaka (HNI strain) genome indicated the presence of two highly homologous genes that we named *Orla-DAB* and *Orla-DBB*. To isolate these genes, a genomic DNA library was constructed from HNI DNA. DNA isolated from a single fish (100μg) was digested partially with *Sau*3AI, and the 16 to 23kb fragments were isolated using a 0.6% sea plaque agarose gel. These fragments were ligated with *Bam*HI digested λEMBL3 arms (Stratagene), and packaged *in vitro* using Gigapack II Gold (Stratagene).

The library, containing 5 x 10^5 independent clones, was screened using an insert of K14 as the probe. Twelve clones were isolated, and one contained the entire *Orla-DAB* gene. The other 11 clones contained another gene, which was neither *Orla-DAB* nor *-DBB*. We named this gene *Orla-DCB*. A part of the *Orla-DBB* gene encompassing exons encoding the β2 and TM/CT domains was amplified and isolated. The mRNAs from the *Orla-DAB* and *-DCB* genes were detected in HNI strain of medaka by RT-PCR analysis (data not shown), but *-DBB* mRNA was not detected. In an outbred orange-red variety of medaka, the mRNA from *Orla-DBB* gene were detected.

Figure 2 shows the phylogenetic tree for the 2 domains of MHC class II amino acid sequences constructed as described in Figure 1. Analysis of the phylogenetic tree indicated that these three genes are generated by gene duplication in the medaka fish lineage, and that *Orla-DAB* and *Orla-DBB* are sister genes.

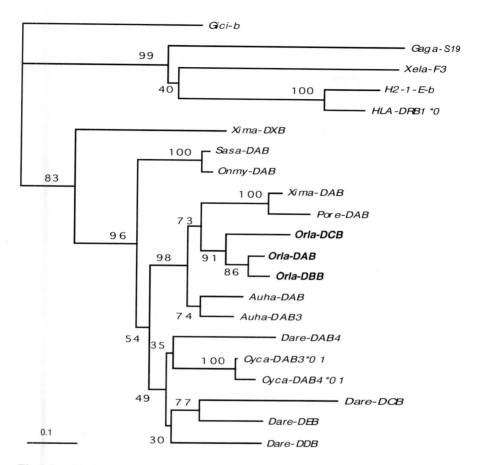

Fig. 2. Phylogenetic tree based on the amino acid sequences of β 2 and TM/CT domains of Mhc class II β proteins from the various vertebrates. Numbers on each branch are bootstrap percentage of 500 replications. The accession numbers used for the present study are as follows: nurse shark (Gici-b, L20275), A. hansbaenschi (Auha-DAB3, 2134016) (Auha-DAB, L13228) zebrafish (Dare-DAB4, U08870), (Dare-DDB, U08872), (Dare-DCB, U08873), (Dare-DEB, U08874), Carp (Cyca-DAB3*01, X95431), (Cyca-DAB4*01, X95435), guppy (Pore-DAB, Z54077-1), Xipophorus (Xima-DAB, AF040760), (Xima-DXB, AF40761), Atlantic salmon (Sasa-DAB, X70166), rainbow trout, (Onmy-DAB, U20943), chicken (Gaga-S19, S66480), toad (Xela-F3, D13684), human, (HLA-DRB1*0, A94681) mouse (H2-1-E-b, I48422).

B2m gene

The probe, β2mHNI, was obtained by PCR amplification of medaka (strain HNI) genomic DNA using the degenerate primers: 2mf, CBMSBCCSAARGTBCAK-GTGTAC; and 2mr, GCCWKYCSTBKTYGAAGGMCAKGTC. These primers were designed by comparing the previously reported β2m sequences from carp (L05536; Dixon et al. 1993), zebrafish (L05383; Ono et al. 1993), barbel (Z54151), rainbow trout (L63534; Shum et al. 1996) and chicken (M84767; Riegert et al. 1996). A single band of the expected size (130bp) was observed on a 12% polyacrylamide gel. Sequencing of the fragment showed a high degree of similarity with the *β2m* sequences from zebrafish, barbel, rainbow trout, and carp (data not shown). We named this fragment B2mHNI.

A liver AA2 cDNA library, previously described by Kuroda et al. (1996), was screened with the B2mHNI fragment labeled with ^{32}P by random priming. After screening, two positive clones were identified (clones 311 and 321, respectively). The sequences of these two clones were highly homologous.

Figure 3 shows the neighbor-joining tree that was constructed using the amino acid sequences from various vertebrates. Medaka β2m was included in the fish β2m clade. Bootstrap percentages supporting the monophyly of the fish β2m clade was sufficiently high (99%). These results strongly suggest that the isolated clones represent the β2m cDNA of medaka.

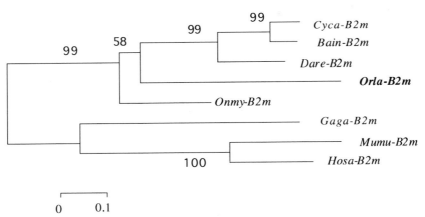

Fig. 3. A phylogenetic tree based on the amino acid sequences of β2-microglobulin from various vertebrates. The numbers on each branch represent the bootstrap percentages for 500 replications. The accession numbers used for the present study are as follows: zebrafish (Dare-B2m, L05383), barbel (Bain-B2m Z54151), carp (Cyca-B2m L05536), chicken (Gaga-B2m M84767), rainbow trout (Onmy-B2m L63534), mouse (Mumu-B2m X01838), and human (Hosa-B2m M17987).

Figure 4 shows the Southern blot of medaka (strain HNI) genomic DNA digested with 12 different restriction enzymes, and probed with the whole insert of clone 321, which contained the entire coding region as well as the 5' and 3' non-translated regions of *β2m* mRNA. Nine out of the 12 restriction enzyme digests produced a single hybridization band. Two bands were obtained with only three restriction enzymes, *Bg*III, *Dra*I and *Pst*I. These results strongly suggest that the medaka β2m molecule is encoded, not by multiple genes, but by a single gene in the haploid genome.

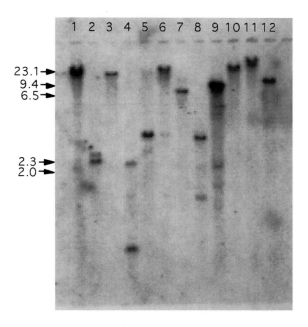

Fig. 4. Southern blot of medaka genomic DNA digested with 12 restriction enzymes. 1. *Bam*HI, 2. *Bg*III, 3. *Bst*XI, 4. *Dra*I, 5. *Eco*RI, 6. *Eco*RV, 7. *Hin*dIII, 8. *Pst*I, 9. *Pvu*II, 10. *Sca*I, 11. *Sma*I, and 12. *Xba*I. The size of the markers is indicated in kilobase pairs.

LMP and *TAP* genes

Molecular cloning of the medaka genes, *LMP2* and *LMP7*, has previously been described (Namikawa-Yamada et al. 1997). Briefly, RT-PCR amplification of *LMP2* and *LMP7* mRNA was performed from medaka liver RNA using degenerate primers based on amino acid sequences conserved between mammalian *LMP2* and δ, and *LMP7* and X. The entire primary structures of medaka *LMP2* and *LMP7* genes were deduced from a cDNA and genomic clone, respectively. The amino acid sequence of the mature medaka LMP2 peptide was aligned with the LMP2 and δ sequences of various vertebrates, ascidian δ, and the yeast PRE3, using Clustal W software (Thompson et al. 1994). The phylogenetic tree of LMP2 and δ, constructed using the neighbor-joining method and based on this alignment (Saitou and Nei 1987), is shown in Figure 5A.

When yeast *PRE3* is used as the outgroup, the possible gene duplication between δ and *LMP2* (shown by the dot) is located after the divergence of ascidians, but before the divergence of lampreys. If this cladogram is correct, the δ/*LMP2* gene duplication occurred at the very beginning of vertebrate evolution. Medaka LMP2 formed a cluster with other vertebrate LMP2, and this clustering was supported by a high bootstrap value.

A similar phylogenetic analysis of medaka LMP7 with X and LMP7 of various vertebrates, ascidian X and yeast PRE2, resulted in the cladogram shown in Figure 5B. Again, the possible *X/LMP7* gene duplication (shown by the dot) was located after the divergence of ascidians, but before the divergence of cyclostomes. This phylogenetic tree, however, has some interesting features: 1) The two cyclostome sequences formed a cluster with the vertebrate LMP7, rather than vertebrate X, even though these cyclostome sequences were identified as X, based on the identification of several diagnostic amino acid residues specific to this protein; 2) Two types of shark LMP7 were identified by diagnostic residues (S8 is characteristic of LMP7, whereas S1 has both X-specific and LMP7-specific residues); and 3) Chicken X is located outside of the mammalian, toad and shark X clustering. These results suggest that the evolution of X and LMP7 is not so simple, and implicate special genetic mechanisms such as gene conversion between *X* and *LMP7*.

Fig. 5. Phylogenetic trees for LMP2 and δ (**A**), and LMP7 and X (**B**). The trees were constructed by the neighbor-joining method using the amino acid sequences of the mature peptides. The medaka sequences are shown in bold. The numbers represent bootstrap percentages supporting the given partitioning. Dots indicate a possible branching point between LMP2 and δ (**A**), or LMP7 and X (**B**). The abbreviations and accession numbers used are as follows: SacePRE3, yeast, P38642; Haro δ, ascidian, Namikawa et al. unpublished data; Hosa δ, human, P28072; Rano δ, rat, P28073; Mumu δ, mouse, P28076; Xela δ, *Xenopus*, D87689; Laja δ, lamprey, Namikawa et al. unpublished data; HosaLMP2, human, P28065; RanoLMP2, rat, P28077; MumuLMP2, mouse, P28076; OrlaLMP2, medaka, D89724; SacePRE2, yeast, P30656; HaroX, ascidian, Namikawa et al. unpublished data; HosaX, human, P28074; RanoX, rat, P28075; XelaX, *Xenopus*, Namikawa et al. unpublished data; GiciX, shark, D64058; GagaX, chicken, X57210; HosaLMP7, human, P28062; RanoLMP7, rat, P28064; MumuLMP7, mouse, P28063; XelaLMP7A, *Xenopus*, D44540; XelaLMP7B, *Xenopus*, D44549; OrlaLMP7, medaka, D89724; GiciLMP7-S8, shark, D64057; GiciLMP7-S1, shark, D64056; EpbuX, hagfish, D64054; LajaX, lamprey, D64055.

A. δ/LMP2

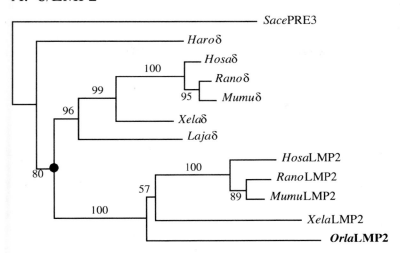

B. X/LMP7

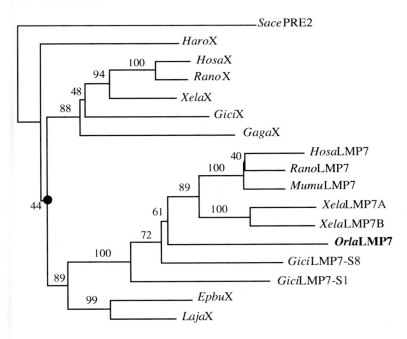

The medaka *TAP2* cDNA, corresponding to about 180 amino acid residues in the ATP-binding region, was obtained by RT-PCR using RNA extracted from the whole body of an outbred medaka fish. A BLAST search using this sequence picked up salmon and zebrafish *TAP2* with the highest scores, followed by *TAP2* sequences of other vertebrates (Terado et al., unpublished data). Although there is no doubt about the identity of medaka *TAP2*, its entire primary structure remains to be determined.

Complement B/C2, C3, C4

Isolation and characterization of medaka *B/C2* has been described previously (Kuroda et al. 1996). Degenerate PCR primers were designed based on the two amino acid sequences of the serine protease domain of B/C2, which are completely conserved in all vertebrate B and C2 sequences reported so far. RT-PCR amplification of medaka liver cDNA resulted in a single DNA band of the expected size (about 250 bp). After cloning into a plasmid, 19 clones were randomly chosen and the nucleotide sequences of their inserts were determined.

All inserts had the same nucleotide sequence that predicted an amino acid sequence showing 31% and 33% identity with the corresponding regions of the human *B* (Horiuchi et al. 1993) and human *C2* (Bentley 1986) genes, respectively. Using this sequence as a probe, a medaka liver cDNA library (containing 7 x 10^5 independent clones) was screened. Thirteen clones were isolated, and nucleotide sequencing of both termini of each insert indicated that all of them are derived from a single gene, and that three of the clones contained an entire protein coding sequence.

The insert of the longest clone was 2384 bp, and predicted a single long open reading frame of 754 amino acids. This amino acid sequence was aligned with the amino acid sequences of various vertebrate *B* and *C2* genes using Clustal W software (Thompson et al. 1994). The medaka sequence showed a significant similarity to other *B* and *C2* sequences throughout its entire length. This suggests that the medaka molecule has the same basic domain structure as the B and C2 proteins of other species, that is: three SCR domains, a von Willebrand domain and a serine protease domain from the N-terminus. The calculated amino acid identities based on this alignment of this medaka sequence were, 34.1% with human C2, 32.6% with mouse C2, 33.6% with human B, 32.5% with mouse B, 34.3% with *Xenopus* B, 37.1% with zebrafish B and 27.4% with lamprey B. The mammalian and *Xenopus* B and C2 sequences showed almost the same degree of identity to the medaka sequence, making it difficult to assign *B* or *C2* to the medaka sequence. Therefore, we tentatively termed this molecule, medaka B/C2 (Kuroda et al 1996).

The phylogenetic tree of the *B* and *C2* sequences was drawn using the neighbor joining method (Saitou and Nei 1987) (Fig. 6). Medaka B/C2 formed a cluster with the other teleost *B* sequences, and this clustering was supported by a high bootstrap value. The dot indicates a possible *B/C2* gene duplication event. If it is correct, shark *B* and the second carp *B* form a cluster with mammalian *C2*, suggesting that the *B/C2* gene duplication predated the emergence of cartilaginous fish. However, the bootstrap values to support this branching order (56, 28 and 41) are low, and the timing of *B/C2* gene duplication during evolution is still not clear. It also remains to be clarified if the teleost *B* cluster is closer to mammalian *B* or mammalian *C2*. The clustering of the two *Xenopus* sequences with mammalian B is supported by a high bootstrap value of 99%. Similarly, the clustering of all jawed vertebrate *B* and *C2* is supported by a high bootstrap value of 96%. Thus, we have concluded that the *B/C2* gene duplication occurred after the divergence of cyclostomes, but before the divergence of amphibians.

To isolate medaka *C3* and *C4* cDNA clones, a RT-PCR strategy was undertaken using the primers corresponding to the two highly conserved amino acid sequences of vertebrate C3 and C4 in the thioester region. Two *C3* clones and one *C4* cDNA clone were isolated from a liver library of inbred medaka (AA2 strain),

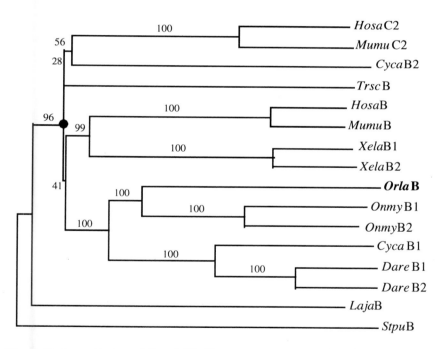

Fig. 6. Phylogenetic tree of B and C2. The tree was constructed as described in Figure 1. The dot indicates the possible B/C2 divergence. The abbreviations used in this figure are defined in Figure 5. The accession numbers are as follows: HosaC2, X04481; MumuC2, J05660; CycaB2, carp, AB007005; TrscB, shark, Terado et al., unpublished data; HosaB, X72875; MumuB, J05661; XelaB1, D29796; XelaB2, D49373; OrlaB, D84063; OnmyB1, trout, AF089861; OnmyB2, AF089860; CycaB1, AB007004; DareB1, zebrafish, U34662; LajaB, D13568; StpuB, sea urchin, AF059284.

and were named Orla-C3-1, -C3-2, and -C4. The Orla-C3-1, -C3-2, and -C4 amino acid sequences showed a significant similarity to other C3 and C4 sequences throughout their entire length. The functionally important residues in the thioester region, the β–α and α–γ processing sites, and the C3a and C4a regions, were well conserved in Orla-C3-1, -C3-2 and -C4. However, a catalytic histidine residue, which greatly increases the rate of reaction of the thioester, was replaced by alanine in Orla-C3-2.

The phylogenetic tree of the *C3, C4, C5*, and *α2M* sequences is shown in Figure 7. Orla-C3-1 and -C3-2 showed clustering with the C3 of teleosts and other vertebrates, including lampreys and hagfish. This was supported by a high bootstrap value of 93%. Orla-C4 showed clustering with human, mouse and *Xenopus* C4. One reliable conclusion that can be drawn from Figure 7 (supported by a bootstrap value of 92%), is that the *C3/C4/C5* gene duplication occurred in

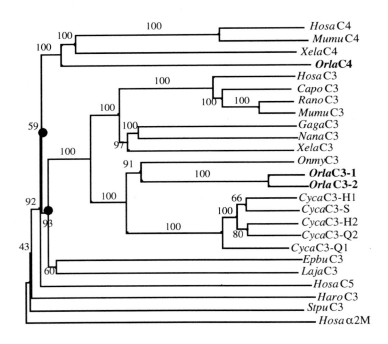

Fig. 7. Phylogenetic trees for C3, C4 and C5. Human α-2-macroglobulin was used as an outgroup. The dots indicate the possible divergence of C5 and C4 from C3. New abbreviations and accession numbers are: HosaC4, K02403; MumuC4, P01029; XelaC4, D78003; OrlaC4, Kuroda et al., unpublished data; HosaC3, P01024; CapoC3, guinea pig, P12387; RanoC3, X52477; MumuC3, P01027; GagaC3, I50711; NanaC3, cobra, Q01833; XelaC3, U19253; OnmyC3, L24433; OrlaC3-1 and OrlaC3-2, Kurada et al., unpublished data; CycaC3-H1, CycaC3-S, CycaC3-H2, CycaC3-Q2 and CycaC3-Q1, Nakao et al., unpublished data; EpbuC3, Z11595; LajaC3, Q00685; HosaC5, P01031; HaroC3, AB006964; StpuC3, AF025526; Hosa2M, P01023.

the vertebrate lineage after divergence of the urochordate. There are, however, still some ambiguities regarding the evolutionary history of the *C3, C4* and *C5* genes. First, disputes about the branching order of C3, C4 and C5 (Nonaka and Takahashi 1992; Hughes 1994) still remain to be resolved, since the bootstrap percentage which support the scenario that C5 diverged first is not high enough. Second, all C3, including hagfish and lamprey C3, formed a clade supported by a high bootstrap value of 93%, suggesting that the *C3/C4/C5* gene duplication predated the emergence of cyclostomes. However, functional analyses suggest that C4 and C5 are absent in cyclostomes, and it is possible that although the *C3/C4/C5* gene duplication predated the emergence of cyclostomes, the functional differentiation among C3, C4 and C5 did not occur, or the *C4* and *C5* genes were secondarily deleted in cyclostomes.

The presence of multiple C3 proteins or genes has been reported in several teleost species (Nakao et al. 1998; Gongora et al. 1998; Sunyer et al. 1996, 1997a), and differential binding specificities have been reported among multiple C3 proteins of trout and sea bream (Sunyer et al. 1996, 1997b). Subsequently, *C3* gene multiplication has been interpreted as a unique strategy undertaken by teleosts to expand their innate immune recognition capabilities and compensate for their inefficient adaptive immune responses (Sunyer et al. 1998). Interestingly, in all species analyzed so far, at least one of the multiple *C3* genes lacks the catalytic histidine residue that greatly increases the rate of reaction of the thioester.

Linkage analysis

To analyze the linkage relationships amongst medaka MHC genes, we used restriction fragment length polymorphisms (RFLP) analysis by Southern hybridization for the *class II B* genes and PCR-RFLP analysis for all the other genes. Table 1 shows the list of genes mapped by PCR-RFLP.

Mapping *class II B* genes by Southern hybridization

The linkage analysis was performed using 20 backcross offspring of Hd-Rr x (Hd-Rr x HNI) F1 crosses, which had been typed previously for many genetic markers including *LMP2/LMP7* and complement *B/C2*. DNA isolated from whole fish bodies (5g) was digested with *Hind*III, separated on a 1% agarose gel, and transferred to a nylon membrane. After ultraviolet crosslinking, hybridization with radio-labeled probes, prepared using the Rediprime kit (Amersham), was performed at 65°C for 16-20 h in 10x Denhardt's solution, 1M sodium chloride, 50mM Tris, 10mM EDTA, 0.1% SDS, and 0.1mg/ml denatured salmon sperm DNA.

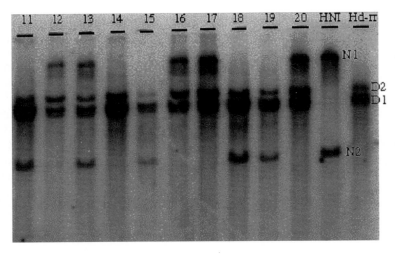

Fig. 8. Representative example of RFLP-typing of the *class II B* genes by Southern hybridization.

Medaka DNA from HNI strains, gave two bands (marked as N1 and N2 in Fig. 8). Referring to the genomic map of the *DAB, DBB* and *DCB* genes, it is obvious that N1 corresponds to doublets of *DAB* and *DBB*, whereas N2 corresponds to *DCB*. Medaka DNA from Hd-Rr strains also showed two apparent bands, a thin upper band and a thick lower band (named D2 and D1, respectively). Better resolution of the bands, obtained using a smaller amount of DNA, indicated that the D1 band of Hd-Rr is also a doublet. The hybridization pattern of the backcross offspring indicated the following points: 1) When both N1 and N2 bands are present or absent, the relative intensity of D1 to D2 is similar to that of Hd-Rr; 2) If the N1 band is present and the N2 band is absent, the relative intensity of D2 increases; and 3) In contrast, if the N2 band is present and the N1 band is absent, the relative intensity of D1 increases. These results indicate that N1 and D1 are alleles, and N2 and D2 are alleles, although N1 and D1 most probably represent two closely linked loci. These hybridization patterns also indicate that the three genes, designated *DAB, DBB* and *DCB,* are located at two different chromosomal loci.

From linkage analysis using PCR-RFLP analysis, the loci represented by the N1 band was linked with the *Tp53* gene, and the locus represented by the N2 band was linked with the *muscle actin* gene (data not shown). The PCR fragment amplified using the primers designed for the *DAB* gene sequence showed close linkage with the N1 band (data not shown). These results also indicate the N1 band represents the *DAB* locus.

Mapping of MHC genes by PCR-RFLP

Using 39 backcross progeny obtained by crossing AA2 fish with (AA2-HNIF1) fish, we mapped 13 MHC and related genes by PCR-RFLP analysis. Over 70 expressed gene loci, including the *Tp53* and *muscle actin* loci, were already mapped using this typing panel DNA (Naruse et al. 1999, submitted). Table 1 shows the MHC and related genes mapped in the present study.

Table 1. List of the genes mapped by PCR-RFLP analysis

No.	Gene	LG	Primers	Polymorphism
1	*B2m*	24	ATGAAAGAGCTTTTCTTCATTGC CTGGCCAGGGTCATGACTGTACAC	in/del
2	*B/C2*	20	GCTCAACATCTACATTCGCCT TGACAACTTCATACACATCA CCTCTTTGGAAACAGAC TTCACAGCTCAGCAGGT	*Rsa*I
3	*C3-1*	1	GCAGAGATGCTTTGAAT ATTTTATTCTGCACACATCTTTTACCA	*Hinf*I
4	*C3-2*	1	GCAGAGATGCTTTGAAT TTTCCACTGGGCCTTGGTAAAAT	*Hph*I
5	*C4*	14	TGTGCAGAAAGACCCTGCTATAAGG TGGTGTTATGGCGGAGGCAAG	in/del
6	*Hsc70*	20	ATGCCAGAGGGAATGCCAGGAG TTCAGCATTTATTGTAGAGTGACATC	*Dra*I
7	*LMP2*	7	GGCTCTGATTCCAGAGTGTCTGCA CAGAGTGGCAGCTGAGCGAACCT	*Bgl*II
8	*LMP7*	7	CTGTAYGRCTGAGAAACAA TCCTTGTCCCATCCACAGAT	*Pvu*II
9	*DAB*	9	GTCTGGAGAGAAAATCTCCTG TCTGACTCTGGCATGGACGGGT	in/del
10	*UAA*	7	TCCATCAGTGTCTCTCCTCCAGAAG GCAGACACAGAGAGAATGACAGCG	in/del
11	*UBA*	7	TCCATCAGTGTCTCTCCTCCAGAAG GCAGACACAGAGAGAATGACAGCG	in/del
12	*UCA*	23	TATTTTTGTTGCTGTCATTG AGCCTAAATCACAAAAGGGGTTT	allele specific
13	*Tap2*	7	AAATTCTTCTGGACGGGAATCCA TCACCACAAGAGTCTGGTTTGGG	in/del

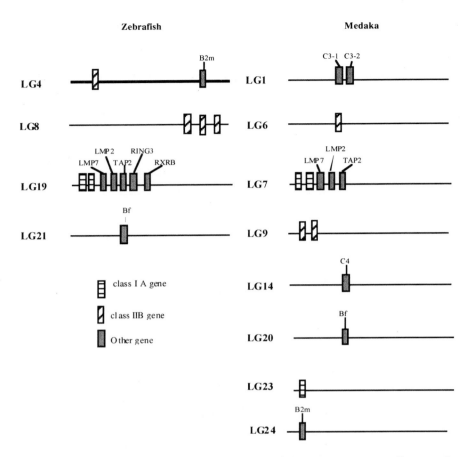

Fig. 9. Linkage maps of the medaka and zebrafish genes, corresponding to the mammalian MHC-encoded genes and their close relatives.

Figure 9 depicts the summary of the linkage relationships of the medaka MHC and related genes. Two MHC *class I A* genes (*UAA* and *UBA*) are linked to each other, and assigned to the same linkage group (LG7) as the *LMP2*, *LMP7* (Namikawa et al. 1997) and *TAP2* genes. Another *class I A* gene, *UCA*, was mapped to LG23. The *class II B* loci, *DAB* and *DBB*, were mapped to LG9. Genotyping of the *DCB* locus was not performed using this typing panel, but the *muscle actin* gene that is linked with the *DCB* locus, was assigned to LG6. Thus, by inference, the *DCB* locus was mapped to LG6. The β*2m* gene was mapped to LG24. The complement *B/C2* and *C4* genes were assigned to LG20 and LG14, respectively. Two *C3* genes, *C3-1* and *C3-2*, were mapped to LG1. The male sex determination gene, *Y*, was assigned to LG1. Thus, the *C3-1* and *C3-2* loci are sex-linked.

These results indicate that the medaka counterparts of the mammalian MHC genes are dispersed throughout the medaka genome, except for the cluster of the MHC *class I A* gene and the genes involved in class I antigen presentation. As shown in Figure 9, a similar result was obtained in zebrafish (Binglac-Popovic et al. 1997), suggesting that a common teleost ancestor may have had a dispersed MHC organization.

Conclusions

We have isolated medaka counterparts of the mammalian MHC-encoded genes; three *class I A*, three *class II B*, a *LMP2*, a *LMP7*, a *TAP2*, a complement *B* and a *C4* genes. In addition, a few functionally and structurally related genes, a *β2m* and two *C3* genes, were isolated. Phylogenetic tree analyses indicated that certain gene duplications which are believed to have played a critical role in development of vertebrate immune system such as between *δ* and *LMP2*, *X* and *LMP7* and *C3* and *C4* predated the appearance of teleost fish. In contrast, it was not clear if the *B/C2* gene duplication predated the appearance of teleost or not. Linkage analysis among these genes revealed the presence of three linkage groups. One contains two *class I A*, a *LMP2*, a *LMP7* and a *TAP2* gene, and the other two contains two *C3* genes and two *class II B* genes, respectively. There is no linkage among these three linkage groups and loci for the other isolated medaka MHC and related genes, indicating that the mammalian MHC-encoded genes are dispersed over the medaka genome. A similar dispersed MHC is also reported with zebrafish which is remotely related to medaka among teleost, indicating that the dispersed configuration is a common feature of teleost MHC. Since class I and II genes were reported to be linked in shark, it seems more probable that the dispersed MHC configuration in teleost is a derived character rather than an ancestral character. In either case, the tight linkage among the *class I A* gene and the genes involved in class I antigen processing and transportation in teleost strongly suggests that the core part of the MHC is composed of the genes essential for class I antigen presentation. Thus, the MHC is considered to have provided the place where tightly linked genes have co-evolved to establish the new biological system for class I antigen presentation.

Acknowledgements

We thank the following collaborators whose contributions are summarized here; K. Hasegawa, S. Fukamachi, H. Mitani, M. Kondo, H. Oku, A. Kojima, Y. Bessho, H. Hori, T. Terado, H. Kimura, N. Kuroda, C. Yamada-Namikawa and M. Sasaki.

References

Bentley DR (1986) Primary structure of human complement component C2. Homology to two unrelated protein families. Biochem J 239:339-345

Bingulac-Popovic J, Figueroa F, Sato A, Talbo WS, Johnson SL, Gates M, Postlethwait JH, Klein J (1997) Mapping of Mhc class I and class II regions to different linkage groups in the zebrafish, *Danio rerio*. Immunogenetics 46:129-134

Carroll MC, Belt T, Palsdottir A, Porter RR (1984) Structure and organization of the C4 genes. Philos Trans R Soc London B Biol Sci 306:379-388

Dixon B, Stet RJ, van Erp SH, Pohajdak B (1993) Characterization of beta 2-microglobulin transcripts from two teleost species. Immunogenetics 38:27-34

Flajnik MF, Ohta Y, Namikawa-Yamada C, Nonaka M (1999) Insight into the primordial MHC from studies in ectothermic vertebrates. Immunol Rev 167:59-67

Gongora R, Figueroa F, Klein J (1998) Independent duplications of Bf and C3 complement genes in the zebrafish. Scand J Immunol 48:651-658

Hansen JD, Strassburger P, Du Pasquier L (1996) Conservation of an alpha 2 domain within the teleostean world, MHC class I from the rainbow trout *Oncorhynchus mykiss*. Dev Comp Immunol. 20:417-425

Horiuchi T, Kim S, Matsumoto M, Watanabe I, Fujita S, Volanakis JE (1993) Human complement factor B: cDNA cloning, nucleotide sequencing, phenotypic conversion by site-directed mutagenesis and expression. Mol Immunol 30:1587-1592

Hughes AL (1994) Phylogeny of the C3/C4/C5 complement-component gene family indicates that C5 diverged first. Mol Biol Evol 11:417-425

Klein J (1986) Natural History of the Major Histocpmpatibility Complex. Wiley, New York

Kuroda N, Wada H, Naruse K, Simada A, Shima A, Sasaki M, Nonaka M (1996) Molecular cloning and linkage analysis of the Japanese medaka fish complement Bf/C2 gene. Immunogenetics 44:459-467

Miller MM, Goto RM, Taylor RL Jr, Zoorob R, Auffray C, Briles RW, Briles WE, Bloom SE (1996) Assignment of Rfp-Y to the chicken major histocompatibility complex/NOR microchromosome and evidence for high-frequency recombination associated with the nucleolar organizer region. Proc Natl Acad Sci USA 93:3958-3962

Nakao M, Yano T (1998) Structural and functional identification of complement components of the bony fish, carp (*Cyprinus carpio*). Immunol Rev 166:27-38

Namikawa-Yamada C, Naruse K, Wada H, Shima A, Kuroda N, Nonaka M, Sasaki M (1997) Genetic linkage between the LMP2 and LMP7 genes in the medaka fish, a teleost. Immunogenetics 46:431-433

Nelson JS (1994) Fishes of the World, 3rd ed. John Wiley and Sons Inc., New York

Nonaka M, Namikawa-Yamada C, Sasaki M, Salter-Cid L, Flajnik MF (1997) Evolution of proteasome subunits delta and LMP2: complementary DNA cloning and linkage analysis with MHC in lower vertebrates. J Immunol 159:734-740

Nonaka M, Takahashi M (1992) Complete complementary DNA sequence of the third component of complement of lamprey. Implication for the evolution of thioester containing proteins. J Immunol 148:3290-3295

Ono H, Figueroa F, O'hUigin C, Klein J (1993) Cloning of the beta 2-microglobulin gene in the zebrafish. Immunogenetics 38:1-10

Postlethwait JH, Yan YL, Gates MA, Horne S, Amores A, Brownlie A, Donovan A, Egan ES, Force A, Gong Z, Goutel C, Fritz A, Kelsh R, Knapik E, Liao E, Paw B, Ransom D, Singer A, Thomson M, Abduljabbar TS, Yelick P, Beier D, Joly JS, Larhammar D, Rosa F, et al. (1998) Vertebrate genome evolution and the zebrafish gene map. Nat Genet 18:345-349

Saitou N, Nei M (1987) The neighbor-joining method: a new method for reconstructing phylogenetic trees. Mol Biol Evol 4:406-425

Sato A, Figueroa F, O'hUigin C, Reznick DN, Klein J (1996) Identification of major histocompatibility complex genes in the guppy, *Poecilia reticulata*. Immunogenetics 43:38-49

Shum BP, Azumi K, Zhang S, Kehrer SR, Raison RL, Detrich HW, Parham P (1996) Unexpected beta2-microglobulin sequence diversity in individual rainbow trout. Proc Natl Acad Sci USA 93:2779-2784

Sunyer JO, Tort L, Lambris JD (1997a) Structural C3 diversity in fish: characterization of five forms of C3 in the diploid fish *Sparus aurata*. J Immunol 158:2813-2821

Sunyer JO, Tort L, Lambris JD (1997b) Diversity of the third form of complement, C3, in fish: functional characterization of five forms of C3 in the diploid fish *Sparus aurata*. Biochem J 326:877-881

Sunyer JO, Zarkadis IK, Lambris JD (1998) Complement diversity: a mechanism for generating immune diversity? Immunol Today 19:519-523

Sunyer JO, Zarkadis IK, Sahu A, Lambris JD (1996) Multiple forms of complement C3 in trout that differ in binding to complement activators. Proc Natl Acad Sci USA 93:8546-8551

Takeuchi H, Figueroa F, O'hUigin C, Klein J (1995) Cloning and characterization of class I Mhc genes of the zebrafish, *Brachydanio rerio*. Immunogenetics 42:77-84

Thompson JD, Higgins DG, Gibson TJ (1994) CLUSTAL W: improving the sensitivity of progressive multiple sequence alignment through sequence weighting, position-specific gap penalties and weight matrix choice. Nucleic Acids Res 22:4673-4680

Polymorphic olfactory receptor genes and HLA loci constitute extended haplotypes

Andreas Ziegler[1], Anke Ehlers[1], Simon Forbes[2], John Trowsdale[2], Barbara Uchanska-Ziegler[1], Armin Volz[1], Ruth Younger[3], and Stephan Beck[3]

[1]Institut für Immungenetik, Universitätsklinikum Charité, Humboldt-Universität zu Berlin, Spandauer Damm 130, 14050 Berlin, Germany
[2]Department of Pathology, Division of Immunology, University of Cambridge, Tennis Court Road, Cambridge CB2 1QP, United Kingdom
[3]The Sanger Centre, Wellcome Trust Genome Campus, Hinxton, Cambridge CB10 1SA, United Kingdom

Summary. Olfactory receptor (OR) genes are often clustered and are known to be located on most human chromosomes. In the largest OR gene cluster so far analyzed and sequenced in any organism, we have identified 27 OR genes between HLA-F and HFE, of which so far 23 are located within about 650 kilobasepairs between HLA-F and RFP. Their products could be involved in the recognition of individual-specific, MHC-dependent odours and be tailored to perform this function. In this first systematic OR gene polymorphism study, twelve HLA-linked OR genes were analyzed on ten HLA-homozygous or – hemizygous cell lines with different HLA haplotypes. All potentially functional HLA-linked OR genes were found to exhibit polymorphism, although the degree differed considerably. The nucleotide changes *within a given gene* occur at defined positions, and may be part of putative transmembrane domains or occur at other residues. It has been possible so far to define 12 haplotypes for the HLA-linked OR genes on the 18 chromosomes 6 analyzed. The strong linkage disequilibrium, which is known to extend from HLA-A to HFE, is expected to conserve these extended HLA/OR haplotypes. If HLA and HLA-linked polymorphic OR genes also turn out to be functionally connected, a view supported by our finding of polymorphism of these OR genes, the entire region might be considered to constitute an extended gene complex which could, as we have already suggested, be designated "Immuno-Olfactory Supercomplex" (IOS).

Key words. HLA complex, MHC, Olfactory receptors, Linkage disequilibrium, Immuno-Olfactory Supercomplex

Introduction

Biomedical research is heading for an exciting new phase in which behavioural features can be linked with genes and finally understood at the molecular level. It has recently been demonstrated that olfaction plays a role in determining the degree to which individuals are attracted to each other, and polymorphisms of genes belonging to, or closely linked to, the major histocompatibility complex (MHC) of mice and rats have been implicated in this phenomenon (Yamazaki et al. 1976; Yamazaki et al. 1979; Singh et al. 1987; Brown et al. 1989; reviewed by Penn and Potts 1998). Remarkably, it has been found that mice are able to "learn" their own MHC identity and that of potential mates by a process which has been termed familial imprinting (Yamazaki et al. 1988; Eklund 1997).

These studies have recently been extended also to humans, showing that a human odor was rated the more "attractive", the fewer HLA class I antigens were shared by the provider of the odor and its recipient (Wedekind et al. 1995; Wedekind and Füri 1997). In support of these observations, Herz and Cahill (1997) have found that, for choosing their partners, women relied on odors more than on other sensory cues. Furthermore, the studies of Ober (1995) on a Hutterite population may suggest the involvement of odor cues in explaining longer interbirth intervals in couples sharing one HLA haplotype: this might be due to lower copulation rates (Penn and Potts 1998). In addition, Hutterite couples share HLA haplotypes less frequently than expected on the basis of random mating expectations (Ober et al. 1997). Obviously, olfaction-based attraction may also provide the explanation for this phenomenon. However, a study on South American Indians came to the opposite conclusion, namely lack of MHC-dependent mating preferences in humans (Hedrick and Black 1997). In this context, it is remarkable that unexplained fertility as well as recurrent spontaneous abortions seems to occur more frequently in couples sharing several HLA specificities (Ober et al. 1988; Weckstein et al. 1991; Laitinen 1993; Ho et al. 1994). In spite of these partly conflicting results, there is mounting evidence that humans may be subject to MHC-dependent olfactory preferences and dislikes as well, and that sharing of MHC specificities by couples could negatively influence reproductive performance.

The evidence available to date points to MHC class I molecules or substances associated with them as sources of olfactory individuality (reviewed by Penn and Potts 1998). A particularly attractive theory to explain olfaction-influenced mate choice under natural, i.e. not germfree, conditions (the "peptide-microflora hypothesis", Penn and Potts 1998) involves the unique peptide binding specificity of MHC molecules, which contribute to the individuality of an organism, in conjunction with the microflora harboured by certain glands, such as sweat glands or preputial glands. Peptides originating from MHC protein complexes could be broken down and metabolized by microorganisms to small, volatile molecules, which could then be excreted in urine or sweat, thereby endowing an individual with a characteristic, MHC-dependent odor. Unfortunately, attempts to provide evidence for this or any other hypothesis seeking to explain the existence of olfactory individuality have met with limited success so far (see e.g. Singer et al. 1997).

It has been pointed out before that next to nothing is currently known on how MHC-dependent odors are detected by mice, rats or humans (Ziegler 1997; Penn and Potts 1998). Principally, there seem to be at least two different routes by which odorants could be perceived by mammals: via the main olfactory epithelium (MOE) or the vomeronasal organ (VNO). While it is believed that the MOE serves in the detection of environmental odors (Zhao et al. 1998), the VNO seems to recognise molecules defining the reproductive and social status of individuals within a species, leading to neuroendocrine and corresponding behavioural responses (Wysocki 1989), although this organ has also been shown to be crucial for the localization of prey, e.g. in snakes. Since MHC-dependent odorants appear to be volatile (see above), their target is most likely the MOE and not the VNO which is thought to interact with nonvolatile pheromones (Dulac and Axel 1995). In both types of epithelia, olfactory receptors are carried by specialized cilia (MOE) or microvilli (VNO). However, there are further important distinctions: the MOE and the VNO are anatomically separated from each other (in humans, the MOE in the upper part of the nasal cavity, the VNO in the frontal part of the nasal septum), and the axonal projections from the olfactory neurons are sent either to glomeruli within the olfactory bulb (MOE) (reviewed by Farbman 1992) or form glomerular neuropils within the accessory olfactory bulb (VNO) (Halpern 1987; see also Belluscio et al. 1999 and Rodriguez et al. 1999). In contrast to information on odors received by the MOE, the VNO pathways bypass higher cognitive centers.

Furthermore, the two organs harbour different types of receptors (Fig. 1). Although all of these are G-protein-coupled receptors with seven transmembrane domains, the MOE-type olfactory receptors (M-OR) are encoded by a huge family of up to 1,000 (rodents), 400 (dog) and 500 (human) genes (Buck and Axel 1991; reviewed by Mombaerts 1999a,b), while little is currently known on the number of the two types of VNO genes (V1-OR and V2-OR); in rodents, both V-OR families appear to include about 50-100 genes (Dulac and Axel 1995; Herrada and Dulac 1997; Matsunami and Buck 1997; Ryba and Tirindelli 1997; Belluscio et al. 1999 and Rodriguez et al. 1999). The three families share no significant sequence homology. The large number of M-OR and V-OR genes together with the finding that at least some of the genes seem to be monoallelically expressed (Chess et al. 1994; Chess 1998; Rodriguez et al. 1999) has so far precluded an understanding of the molecular mechanism by which an individual olfactory neuron is able to express only a single type of olfactory receptor molecule on its cilia or microvilli (Ressler et al. 1994; Vassar et al. 1994; Malnic et al. 1999; Rodriguez et al. 1999). The reason for this selective expression is to ensure that a given neuron will only be able to transmit the signal obtained from one receptor species, so that the quality of an olfactory stimulus is recognized by a specific combination of

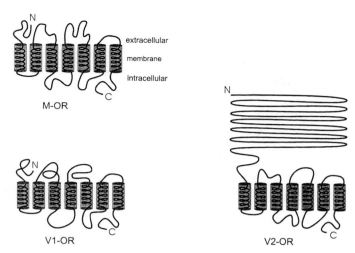

Fig. 1. Schematic structural representation of the three types of olfactory receptor molecules known to exist in mammals. M-OR, MOE-type olfactory receptor; V1-OR and V2-OR, the two types of VNO-expressed olfactory receptors.

activated glomerular neuropils (VNO) or glomeruli within the olfactory bulb (MOE) (see e.g. Malnic et al. 1999; Belluscio et al. 1999; Rodriguez et al. 1999).

While no data are available for V-OR genes regarding expression in other tissues, it is noteworthy that the expression of some M-OR genes is clearly not restricted to the MOE. This has been studied in considerable detail in the testes of humans, dog, mouse and rat (Parmentier et al. 1992; Vanderhaeghen et al. 1993; Vanderhaeghen et al. 1997a). In humans, 36 different cDNA-derived PCR products have been found so far which are very similar to those of the other species. The significance of this expression remains, however, unclear. Since one M-OR type protein has been found to be expressed on the midpiece of dog spermatozoa, these molecules might play a role in sperm maturation, sperm motility or sperm-egg interaction (Vanderhaeghen et al. 1993; Walensky et al. 1995, 1998). For example, spermatozoa need chemical cues in order to reach their target within the female genital tract. Chemoattractants for spermatozoa are present in follicular fluid (Ward and Kopf 1993; Cohen-Dayag et al. 1995), and these substances might presumably even be peptides (Nandedkar et al. 1996; see Eisenbach 1999, for a review), as in several invertebrate species (Garbers et al. 1986). Apart from a role in male fertility, Dreyer (1998) has pointed out that M-OR and V-OR might fulfill a crucial function in developing embryos, where they could play a key role in cell recognition and targeting, not only in the brain, but also in other organs (the "area code hypothesis"). In this context, it is also worth mentioning that apoptosis of various cellular components of the olfactory system seems to be an essential feature of the development of this sense, not only in mammals (Voyron et al. 1999), but also in invertebrates (Harzsch et al. 1999).

This feature, the monoallelic expression of at least some receptors, and their high degree of ligand specificity (Wojtasek et al. 1998; Touhara et al. 1999; Hatt et al. 1999; Duchamp-Viret et al. 1999) provide remarkable parallels to the immune system of mammals, where the proper selection of T and B cells are prerequisites for a correct interplay of cellular and humoral immunity.

Within the human genome, many M-OR genes have already been assigned a chromosomal location (Ben-Arie et al. 1994; Fan et al. 1995; Vanderhaeghen et al. 1997b; Rouquier et al. 1998a, 1998b; Trask et al. 1998a, 1998b; Brand-Arpon 1999). As a rule, M-OR genes occur in clusters (Ben-Arie et al. 1994; Buettner et al. 1998; Trask et al. 1998a, 1998b), some of which are close to the telomeres of several human chromosomes and exhibit polymorphic duplications (Trask et al. 1998a). There are already some examples where the basic structure of OR gene clusters is similar in mouse and man (see e.g. Bulger et al. 1999). Since it has already been pointed out that MHC-dependent mating preferences might be controlled by linked genes, one of which is responsible for the signal and another for the receptor (Yamazaki et al. 1976), a cluster of OR genes could be located in close vicinity of the HLA region. There are also theoretical considerations arguing that genetic kin recognition systems require "detecting", "matching", and "using" genes, although these loci are not necessarily linked (Grafen 1990; Penn and Potts 1998). A few expressed M-OR genes have previously been found near the HLA complex (Fan et al. 1995; Gruen et al. 1996) as well as near the MHC of the mouse (reviewed by Amadou et al. 1999), lending support to the suggestion of Yamazaki and coworkers (1976). The finding of an exceptionally strong linkage disequilibrium across the ~5,000 kilobasepair (kb) region directly telomeric of HLA-F (Malfroy et al. 1997) is in line with this idea as well.

To provide a basis for further functional work, our study of HLA-linked olfactory receptor genes is currently aimed at answering the following questions (Ziegler 1997):
1. How many HLA-linked OR genes exist?
2. To which of the three gene families do these OR loci belong?
3. What is the extent of their polymorphism, if any?
4. Is it possible to define extended HLA/M-OR/V-OR gene haplotypes?

We have mapped and cloned the region from HLA-F to HFE (about 5,000 kb), and determined most of its sequence. Of 27 olfactory loci identified so far, 25 are of the MOE type, while two belong to the V1-OR gene family. Twelve M-OR genes have been analyzed for polymorphisms using a panel of HLA homozygous or hemizygous cell lines of widely varying ethnic origin, representing ten different HLA haplotypes. We find that all potentially expressable genes are polymorphic, resulting in at least eleven different M-OR gene haplotypes in the ten cell lines analyzed. Our results also show that HLA-linked M-OR genes and HLA loci constitute extended haplotypes.

Materials and methods

All experimental procedures are described in detail elsewhere (Ehlers et al. manuscript submitted; Younger et al. manuscript in preparation). In brief, the human cell lines employed in this study were derived from donors belonging to different ethnic groups; they were either HLA class I homozygous or hemizygous and represented ten different HLA haplotypes on 18 chromosomes 6.

For each of the M-OR genes analyzed in these cell lines, two to three primers were designed, based on the genomic sequence determined previously by us. The specificity of the primers was evaluated by alignment with all known human OR sequences. As a rule, PCR using these highly specific primers resulted in overlapping PCR products, which were then sequenced. In those cases where an unambiguous allele assignment proved impossible due to heterozygosity of one of the OR loci in the HLA homozygous cell lines, PCR was used to amplify a DNA segment encoding the full-length receptor. These fragments were cloned and independent clones sequenced to define the respective alleles.

Results and discussion

A novel nomenclature for OR genes

Large numbers of similar genes can only be described and discussed when a meaningful, logical designation for them is at hand. Although this notion may seem superfluous for MHC enthusiasts, it is not commonplace in the field of OR genetics. However, the urgent need for an entirely novel nomenclature has already been recognized (Mombaerts 1999a). It should ideally take care of the different types of genes (currently three), the enormous number of the loci, their chromosomal assignment, allow for an allele designation, indicate the functional status, and finally also the species to which the OR belongs. Presently, no proposal has been published which attempts to integrate all of this information.

We would like to suggest a unifying nomenclature which we find flexible, easy to comprehend and handle. If we take the allele hs6M1-6*01 as an example, the designation consists of a two letter abbreviation of the species (hs, *Homo sapiens*), followed by the number of the chromosome on which the gene is located (in this case chromosome 6, although "Z" is used where the location is unknown). This is followed by a letter and number indicating whether the gene belongs to the M-OR, V1-OR or V2-OR families (M1, V1 or V2, respectively; here M1 [M1, because an M2 family might be found eventually]). This description is followed by a dash and an arbitrary gene identity number (here, -6). Finally, if alleles of the gene are known, an asterisk and an additional number (here, *01) describes the corresponding allele. As a rule, the original sequence, e.g. from a PAC, should receive the allele designation "*01". A pseudogene could additionally be designated with a "P", but this may turn out to be a two-edged sword, because a given gene could be non-functional in one individual and thus qualify as a pseudogene, while it may be perfectly active in another member of a species. The alleles which have been found for the hs6M1-4 locus (see below) prove that this caveat is not without justification. This nomenclature has all the flexibility which is needed because of the huge numbers of OR genes in all mammalian species investigated so far, and the indication of a chromosomal assignment is clearly also very useful. In this context, we find it practical to refer to olfactory receptors as "OR" (even if they are unlikely to "smell" anything, as in the case of a testis-expressed protein), and add a prefix indicating the gene family (i.e. M-, V1-, or V2-), when we address a family of OR as a whole at the protein or gene level.

Genomic organization and phylogenetic analysis

The genomic organization of the 6p22.1-p21.3 region has been elucidated by sequencing of overlapping PACs and BACs, most of which we had previously isolated and mapped (see e.g. Volz et al. 1997; Peters et al. 1998). The location of the various OR genes is schematically presented in Figure 2. In total, 27 such genes were identified to date. Of the four OR genes telomeric of *RFP*, but still in the region of strong linkage disequilibrium with the HLA complex, only one (hs6M1-10) has all features of a functional gene, while the second M-OR gene and the two V1-OR loci are probably non-functional. Our work has in fact led to the first chromosomal assignment for any human V1-OR gene. Since the region around these V1-OR loci has not yet been completely sequenced, it remains unknown whether only M-OR genes occur in large clusters, or whether this feature extends to V-OR loci as well. Like M-OR genes (Ben-Arie et al. 1994), the two V1-OR loci are each characterized by the presence of a single exon.

Between the *GABABR1* and *RFP* loci, i.e. within about 650 kb, 23 OR loci, all of the M-OR type, were found. These as well as the two M-OR genes identified telomeric of RFP reveal the typical genomic organization of M-OR genes (see above). Eleven of the 23 M-OR genes contained an open reading frame, rendering them potentially functional. Very few additional genes were unequivocally identified in this region, among them two (*GABABR1* and *MAS*-related G-protein-coupled receptor) which encode other proteins exhibiting seven transmembrane domains and structural similarity with V2-OR genes in rodents (Herrada and Dulac 1997; Matsunami and Buck 1997; Ryba and Tirindelli 1997). The *GABABR1* gene has been mapped previously to this region (Peters et al. 1998). Its genomic organization is completely different from that of the M-OR genes, but it exhibits polymorphism as well (Peters et al. 1998; Sander et al. 1999). This gene must be regarded as a candidate locus for various neurological diseases, among them juvenile myoclonic epilepsy (Sander et al. 1999) and the "phonological awareness phenotype" of dyslexia (Grigorenko et al. 1997; Gayan et al. 1999). The *MAS*-related locus bears similarity to *MAS* (on 6q27), which has been described as an oncogene (Young et al. 1986). Nothing is known regarding the

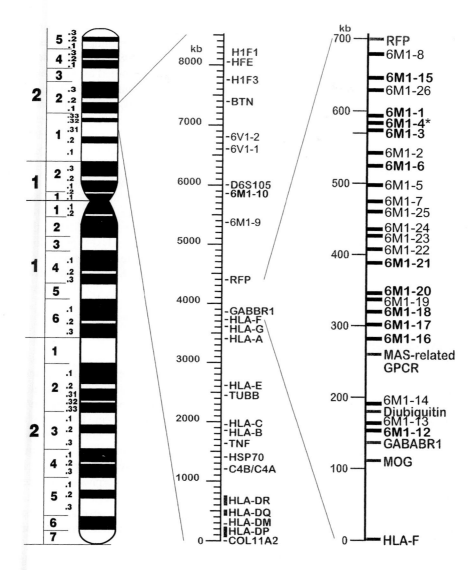

Fig. 2. Chromosome 6 ideogram depicting the cytogenetic and approximate physical locations (distances in kilobasepairs, kb) of several loci belonging to the HLA complex and of various OR genes (hs6M1-1 to –25, hs6V1-1, -2). OR genes in bold script are potentially active at least in some individuals; the others are likely to be not expressed at the protein level.

function of the gene on 6p, but *MAS* has recently been found to be an important modulating factor in the electrophysiology of the hippocampus, possibly connected with the repression of anxiety (Walther et al. 1998). It remains to be seen whether the locus telomeric of HLA-F must also be regarded as a candidate gene for any of the neurological or psychiatric disorders associated with this segment of chromosome 6 (for a discussion, see Ziegler 1997). Another gene in the sequenced region is diubiquitin. The product of this gene is apparently involved in modulating the growth of B cells and dendritic cells (Bates et al. 1997; Liu et al. 1999); it is thus unlikely to have a function within the brain or in olfactory perception.

Using the olfactory database in conjunction with multiple alignment and phylogenetic tree software (Thompson et al. 1997; Galtier et al. 1996), we have also begun to study the relationship of the M-OR genes found on human chromosome 6 and other complete M-OR genes (results not shown). From this, with varying degrees of confidence based on bootstrap analysis, provisional subgroups could be delineated. Within these subgroups, there often appear to be collections of M-OR genes from the same chromosome or chromosomal cluster. The genomic and phylogenetic analyses will be presented in more detail elsewhere (Younger et al. manuscript in preparation).

Polymorphism of HLA-linked M-OR genes

A characteristic feature of HLA class I and II loci is their extreme genetic polymorphism, which to some degree appears to extend also to other genes in close vicinity (*GABABR1*, see above). As pointed out before, one of the objectives of our study was a systematic analysis to determine for the first time the degree of polymorphism exhibited by any cluster of M-OR genes. We chose to analyse HLA-homozygous or hemizygous cell lines from widely varying ethnic backgrounds in order to maximize our chances to detect nucleotide changes. In retrospect, this was probably a good idea, because the polymorphisms detected were totally different from those found in HLA class I and II genes. A polymorphism analysis of HLA-linked M-OR genes from several hundred individuals from Berlin, mostly of Caucasian origin, is currently in progress (our unpublished results).

With the exception of hs6M1-1, where only a single silent nucleotide change was observed which distinguished the two alleles, at least two alleles encoding different amino acid sequences were found for all other M-OR genes analyzed. The amino acid substitutions were detected within all receptor domains, the fourth extracellular region being the only exception. Figure 3 gives a schematic representation of the polymorphic amino acid residues found in three of the OR genes. We regard it as extremely unlikely that the corresponding nucleotide substitutions occurred by chance, since *within a given gene*, only certain positions

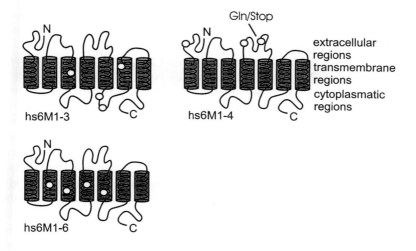

Fig. 3. Schematic structural representation of three HLA-linked M-OR proteins with an indication of the location of polymorphic residues. (open circles)

Table 1. Domain assignment of amino acid substitutions within HLA-linked MOE-type olfactory receptors

Gene	Domains and observed substitutions
hs6M1-3	TM3 (Thr-Ala), CP3 (Arg-Gln, Val-Ile)
hs6M1-4	EC1 (Ile-Leu), EC3(Ala-Val, STOP-Gln)
hs6M1-6	TM2 (His-Tyr), TM3 Ala-Thr), TM4 (Ala-Val), TM5 (Ala-Thr)
hs6M1-10	CP3 (Gln-Arg)
hs6M1-12	TM1 (Phe-Leu), CP1 (Ala-Val)
hs6M1-15	EC2 (Met-Val), CP4 (Asp-Asn)
hs6M1-16	TM2 (Asp-Asn)

Designation of receptor domains:
EC1,2,3: 1st, 2nd, and 3rd extracellular domain
TM1,2,3,4,5: 1st, 2nd, 3rd, 4th, and 5th transmembrane domain
CP1,3,4: 1st, 3rd, and 4th cytoplasmic domain

turned out to be variable in all cell lines investigated.

To our surprise, the analysis was complicated by the finding that some of the HLA-homozygous cell lines, many of them homozygous typing cells used in HLA Workshops, exhibited heterozygosity for some of the M-OR loci. In these cases, the unequivocal definition of alleles was thus possible only after cloning and sequencing of the respective alleles from both haplotypes. The sequences 5' and 3' of each of the genes exhibited no polymorphism Table 1 gives a summary of the observed amino acid substitutions. It is obvious that the changes within these genes do not occur only within particular domains. On the contrary, they seem to be scattered, occurring in extracellular and cytoplasmic domains as well as in putative transmembrane stretches of the protein. For example, it is difficult to imagine how a change in the EC2 domain could be connected in any way with a substitution involving CP4 (hs6M1-15). At least two of the receptors exhibit polymorphisms that immediately suggest a functional importance, namely hs6M1-6 and hs6M1-4. In the first case, only putative transmembrane regions which might be involved in ligand binding (see below) are affected by the substitutions. However, the hs6M1-4 gene contains an open reading frame in four alleles, while only the hs6M1-4*01 allele exhibits a stop codon, which leads to premature termination of translation and an inactive product. Among the few cells analyzed so far, the latter allele appears to be quite frequent. Furthermore, it is evident that it is not restricted to any ethnic group. The HLA-linked M-OR genes are thus clearly polymorphic, although the extent of this polymorphism in different populations remains to be determined.

Functional considerations

In the absence of a detailed 3-dimensional structure for any receptor with seven transmembrane domains, it is currently only possible to speculate with regard to a functional significance of any of the observed changes. Amino acid substitutions within TM2, 3, 5, and 6 of the β_2-adrenergic receptor are connected with agonist binding and receptor activation (reviewed by Strader et al. 1994), and studies of further G-protein-coupled receptors support this notion, as is the finding of high variability within TM3, 4, and 5 of M-OR (Buck and Axel 1991). Pilpel and Lancet (1999) have aligned 197 M-OR sequences from several species and deduced 36 hypervariable residues, which are all part of transmembrane domains. 17 of these residues are located in the extracellular two-thirds of the respective segments, which are thought to bind ligands in other G-protein-coupled receptors. Twelve of these residues, mostly in the third and fifth transmembrane domains, correspond to residues that are known to be contacted by ligands in other receptors. At first glance, the hs6M1-6 gene (Table 1, Fig. 3) might be thought to be a candidate for a change in ligand recognition. However, according to the predictions of Lancet and Pilpel (1999), the amino acid substitions found for any of the HLA-linked M-OR are unlikely to lead to a change in ligand binding specificity. In our opinion, these predictions should be treated with extreme caution. In line with this, Krautwurst and coworkers (1998) observed that a Val-Ile substitution at position 206 of the protein (transmembrane domain 5) led to a change in ligand preference from octanal to heptanal. The position of this replacement does not correspond to any of the hypervariable residues defined by Lancet and Pilpel (1999). The study by Krautwurst et al. (1999) demonstrates also that seemingly minor changes in the protein sequence can have profound consequences in terms of odor recognition. The involvement of residue 206 in heptanal binding has been observed also by Zhao et al. (1998). In the case of the HLA-linked receptor hs6M1-4, this residue is Ile as well, and the corresponding nucleotide triplet is subject to a nucleotide change in the third position, which, however, remains without consequences at the amino acid level.

Similar considerations apply with regard to amino acid substitutions within the extra- or intracellular domains, which we observed in some of the M-OR (Table 1, Fig. 3). The intracytoplasmic domains 2 and 3 are thought to be involved in G-protein binding. Drastic amino acid changes as observed in hs6M1-3 and hs6M1-10 could presumably influence signal transduction after ligand binding, but currently this suggestion lacks any experimental proof. Changes of extracellular residues may not have any functional consequence, but again, this is a matter of pure speculation.

However, the hs6M1-4 alleles clearly deserve further study, because the lack of the corresponding protein in individuals being homozygous for the *01 allele could lead to anosmia with regard to the odorant recognized by this receptor. The best-known case of specific anosmia is that of androstenone, which cannot be smelled by some people, while it appears sandelwood-like or urine-like to others. The ability to recognize this substance has a genetic component (Wysocki and Beauchamp 1984). It is of course possible that homozygous carriers of the hs6M1-4*01 allele do not exhibit any specific anosmia, because the hs6M1-4-ligand might be detected also by other OR. Examples for redundancy within the olfactory system have been described (Malnic et al. 1999; Touhara et al. 1999; Duchamp-Viret et al. 1999). Rouquier and coworkers (1998a) have described an OR gene (912-83 in their nomenclature, in chromosomal region 11q11-12) which is active in nonhuman primates, but inactivated in all human individuals tested. This case is of course different from that of the hs6M1-4 gene, because the latter appears to be subject to dynamic fluctuations *within* a species, not just *among* species.

Nevertheless, we are convinced that a real understanding of the specificities and functions of OR relies on a determination of the 3-dimensional structure at atomic resolution of one of these proteins, although it remains doubtful whether an extrapolation from one structure to that of another OR is really justified, given the extreme variability of the OR gene family on one hand and the difficulties encountered even in the case of protein families in which the structures of many more members have already been solved. This latter point is illustrated by comparing the structures of HLA-B*3501 complexed with an octamer and a nonamer peptide (Smith et al. 1997; Menssen et al. 1999), where the binding of the two peptides to the same "backbone" leads to considerable changes not only in the positioning of the peptides, but also of the HLA class I heavy chains. This peptide-induced repositioning of identical heavy chains was quite unexpected and casts doubt on the validity of any specificity predictions for related proteins with different amino acid sequences.

Regarding the expression of HLA-linked M-OR genes, we have obtained preliminary data, mainly from dot blot studies, that selected loci can be expressed in various tissues (Ehlers, Volz and Ziegler, unpublished observations). These experiments are hampered by the principal difficulty of obtaining fresh biopsies from the most likely site of expression, namely the human olfactory epithelium. In the small samples which might be available in exceptional cases, it could well be that those olfactory neurons expressing the receptor under investigation are missing, while they might be present in a second sample of olfactory mucosa from the same individual. This problem could probably be solved when monoclonal antibodies are available which detect specific olfactory receptors, and which work also on tissue samples excised some time after death. However, it is evident that such studies are more easily done in other primates or rodents. So far, our analyses have indicated expression of the hs6M1-16 gene e.g. in testicular tissue, but in the absence of *in situ*-hybridization experiments, we do not know whether the expression is confined to germ cells. It is an interesting aspect of testis expression studies of M-OR genes reported previously (Parmentier et al. 1992; Vanderhaeghen et al. 1993, 1997a; Walensky et al. 1995, 1998), that the PCR-primers used to amplify M-OR transcripts are not expected to amplify transcripts of any of the HLA-linked OR genes identified so far. Therefore, potential expression of these genes in the testis will have gone undetected.

It is clearly also of the utmost importance to define ligands for HLA-linked M-OR. This might be done using the expression systems employed by Krautwurst and coworkers (1998) or Hatt and colleagues (1999). Such experiments would open the possibility to study animals with deletions of their own MHC-linked M-OR, but transgenic for the corresponding human M-OR.

Extended HLA/M-OR haplotypes

The detection of different alleles for the potentially active HLA-linked M-OR loci has also allowed us to define M-OR haplotypes. In the case of the two HLA-hemizygous cell lines BM19.7 and BM28.7, this was straightforward, but in the case of the other eight cell lines investigated, three presented difficulties, because for more than one M-OR gene, two alleles were observed. Therefore, an unambiguous assignment of each allele to one of the two haplotypes was not possible. This can only be achieved by generating cell lines hemizygous for the M-OR loci or by carrying out family studies. A summary of the data obtained is presented in Table 2, together with preliminary haplotype designations.

Table 2. HLA-A and -B antigens as well as M-OR haplotypes in ten HLA-homozygous or –hemizygous cell lines

Cell line	HLA A	B	1	3	4	6	10	12	15	16	M-OR haplotype
BM28.7	1	35	*01	*02	*05	*02	*01	*01	*01	*02	2
BM19.7	2	13	*02	*02	*05	*02	*01	*02	*01	*03	3
LG2	2	27	*01	*01	*01	*01	*01	*03	*01	*01	1
KR3598	2	44	*01	*02	*01+*02	*01+*02	*02	*02+*03	*01	*01	5+6
H2LCL	3	7	*01	*01	*01	*01	*01	*03	*01	*01	1
WT51	23	65	*01	*03	*03	*03	*01	*01+*04	*02	*01	7+8
SA	24	7	*01	*01+*02	*01	*01+*02	*01	*02+*03	*01	*01	9+10
YAR	26	38	*01	*01	*01	*01	*01	*03	*01	*01	1
OLGA	31	62	*01	*01+*02	*01+*02	*01+*02	*01	*01+*04	*01	*01	11+12
AMAI	68	53	*01	*04	*04	*02	*01	*01	*03	*03	4

A total of 18 chromosomes 6 were analyzed in the ten cell lines, and 12 OR haplotypes could be delineated. In three of the cell lines, the same OR haplotype (designated as "1") appears to be present, at least on the basis of the loci investigated so far, although the respective cell lines exhibit very different HLA antigens. However, the opposite situation was also observed, since three cell lines expressed HLA-A2, but exhibited four different OR haplotypes, in spite of the fact that one of the cell lines was hemizygous for the respective chromosomal region (BM19.7). It seems possible that the number of OR haplotypes will increase, even within the small panel of cell lines investigated, when more OR loci will have been analyzed.

Fourteen different extended haplotypes can be distinguished if HLA class I haplotypes and OR haplotypes are considered together. This procedure appears quite justified since the linkage disequilibrium, which is known to characterize many common HLA haplotypes (e.g. A1-B35-Cw4 as in BM28.7, or A3-B7 as in H2LCL), extends several thousand kb telomeric of HLA-F and includes even the HFE locus (Malfroy et al. 1997; Tay et al. 1997; see also Fig. 2). All OR genes detected during this study are part of this region and therefore subject to suppression of recombination, leading to "extended HLA haplotypes" (Awdeh et al. 1983; Alper et al. 1986; Naruse et al. 1998) which may also be termed "ancestral HLA haplotypes" (Tay et al. 1997; Dawkins et al. 1999). For example, Tay and coworkers (1997) have found that nine of fourteen HLA-A3/B7 haplotypes shared the same microsatellite alleles between HLA-A and D6S105 (see Fig. 2), and even nine of ten HLA-A1/B8 haplotypes shared the microsatellite alleles, although a different set of alleles to those with HLA-A3/B7 was detected. These results resemble the findings reported here (see Table 2). Therefore, evidence from independent sources, employing entirely different loci, supports the notion that extended HLA/M-OR haplotypes exist. These haplotypes tend to be preserved because of linkage disequilibrium and are likely to be transmitted intact to the offspring.

An "Immuno-Olfactory Supercomplex"?

This brings us back to the suggestion of Yamazaki and colleagues (1976), that MHC-dependent mating preferences might be controlled by linked genes, one of which is responsible for the signal and another for the receptor. Our results clearly demonstrate that the suggestion of Yamazaki et al. (1976) might basically be correct, since a large number of OR genes is located in the immediate vicinity of the telomeric end of the HLA complex. In addition, that part of the region harbouring most of the OR genes, i.e. between the GABABR1 and RFP genes, is to a large extent syntenic between mouse and human (see e.g. Amadou et al. 1999; Beck et al. unpublished observations).

As an example, sequencing and subsequent analyses of the clones dJ271M21 (from human) and bK573K1 (from mouse) revealed the two clones to be syntenic, since both contained the 22 exons of the alternatively spliced GABABR1 neurotransmitter gene (see also Peters et al. 1998) and the two exons of the diubiquitin locus. The human M-OR genes in dJ271M21 also appear to be the orthologues of the mouse M-OR loci found in bK573K1: on the DNA level, hs6M1-12 is closely related to mm17M1-1, mm17M1-2, mm17M1-3, and mm17M1-4. The similarity ranges from 84.1% to 81.1%. On the other hand, hs6M1-14 is very similar to mm17M1-6, with a conserved region from amino acid residue 77 to the end of the mouse M-OR. Within this region, 79.4% of the nucleotides are shared; however, the start of the protein is not retained in humans, and where there is a CAG (Gln) at position 77 of the mouse M-OR, the corresponding human codon is a STOP codon (TAG). This situation is of course resembling the finding of Rouquier et al. (1998a) which we have already discussed above. Based on this very limited comparison between MHC-linked M-OR genes in two distantly related species, it would appear that some OR genes underwent additional duplications in the rodent, although we cannot exclude the possibility of deletion of extra copies in the human genome. When the complete sequences of the two syntenic regions in the human and mouse genomes will finally be available, a detailed comparison of MHC-linked OR genes can be carried out, which might also lead to the detection of OR regulatory sequences shared by the two species.

Therefore, the genetic prerequisites of the suggestion (Yamazaki et al. 1976) appear to be met. However, in order to justify the designation of this genomic region, both in human and mouse, as "Immuno-Olfactory Supercomplex" (IOS) which Ziegler (1997) has already proposed, a functional relationship of the two sets of loci and their products will have to be demonstrated as well. Clearly, it would have to be shown that at least some MHC-dependent odors can be recognized by the products of MHC-linked M-OR genes. Furthermore, the potential expression of these MHC-linked receptor proteins on spermatozoa deserves attention, since Rülicke and coworkers (1998) have presented preliminary evidence in the mouse that parents are able to promote specific combinations of MHC haplotypes during fertilization, since virus-infected mice produced more MHC-heterozygous embryos than sham-infected animals. The reasons for these findings are unknown, and the lack of expression of MHC antigens on germ cells and preimplantation embryos (reviewed by Desoye et al. 1991) does not facilitate an explanation. However, given the fact that already a considerable number of OR have been detected on spermatozoa, together with various signal-transducing proteins (see above), there might be a role for MHC-linked olfactory receptors in reproduction as well.

Acknowledgements

The authors thank the European Union, the Volkswagen Foundation, and the Wellcome Trust for supporting the research described in this article.

References

Alper CA, Awdeh ZL, Yunis EJ (1986) Complotypes, extended haplotypes, male segregation distortion, and disease markers. Hum Immunol 15:366-373

Amadou C, Kumanovics A, Jones EP, Lambracht-Washington D, Yoshino M, Lindahl KF (1999) The mouse major histocompatibility complex: some assembly required. Immunol Rev 67:211-221

Awdeh ZL, Raum D, Yunis EJ, Alper CA (1983) Extended HLA/complement allele haplotypes: evidence for T/t-like complex in man. Proc Natl Acad Sci USA 80:259-263

Bates EE, Ravel O, Dieu MC, Ho S, Guret C, Bridon JM, Ait-Yahia S, Briere F, Caux C, Banchereau J, Lebecque S (1997) Identification and analysis of a novel member of the ubiquitin family expressed in dendritic cells and mature B cells. Eur J Immunol 27:2471-2477

Belluscio L, Koentges G, Axel R, Dulac C (1999) A map of pheromone receptor activation in the mammalian brain. Cell 97:209-220

Ben Arie N, Lancet D, Taylor C, Khen M, Walker N, Ledbetter DH, Carrozzo R, Patel K, Sheer D, Lehrach H, North MA (1994) Olfactory receptor gene cluster on human chromosome 17: possible duplication of an ancestral receptor repertoire. Hum Mol Genet 3:229-235

Brand-Arpon V, Rouquier S, Massa H, de Jong PJ, Ferraz C, Ioannou PA, Demaille JG, Trask BJ, Giorgi D A (1999) A genomic region encompassing a cluster of olfactory receptor genes and a myosin light chain kinase (MYLK) gene is duplicated on human chromosome regions 3q13-q21 and 3p13. Genomics 56:98-110

Brown RE, Roser B, Singh PB (1989) Class I and class II regions of the major histocompatibility complex both contribute to individual odors in congenic inbred strains of rats. Behav Genet 19:659-674

Buck L, Axel R (1991) A novel multigene family may encode odorant receptors: A molecular basis for odor recognition. Cell 65:175-187

Buettner JA, Glusman G, Ben-Arie N, Ramos P, Lancet D, Evans GA (1998) Organization and evolution of olfactory receptor genes on human chromosome 11. Genomics 53:56-68

Bulger M, von Doorninck JH, Saitoh N, Telling A, Farrell C, Bender MA, Felsenfeld G, Axel R, Groudine M (1999) Conservation of sequence and structure flanking the mouse and human beta-globin loci: the beta-globin genes are embedded within an array of odorant receptor genes. Proc Natl Acad Sci USA 96:5129-134

Chess A, Simon I, Cedar H, Axel R (1994) Allelic inactivation regulates olfactory receptor gene expression. Cell 78:823-834

Chess A (1998) Olfactory receptor gene regulation. Adv Immunol 69:437-447

Cohen-Dayag A, Tur-Kaspa I, Dor J, Mashiach S, Eisenbach M (1995) Sperm capacitation in humans is transient and correlates with chemotactic responsiveness to follicular factors. Proc Natl Acad Sci USA 92:11039-11043

Dawkins R, Leelayuwat C, Gaudieri S, Tay G, Hui J, Cattley S, Martinez P, Kulski J (1999) Genomics of the major histocompatibility complex: haplotypes, duplication, retroviruses and disease. Immunol Rev 167:275-304

Desoye G, Dohr GA, Ziegler A (1991) Expression of human major histocompatibility antigens on germ cells and early preimplantation embryos. Lab Invest 64:306-312

Dreyer WJ (1998) The area code hypothesis revisited: olfactory receptors and other related transmembrane receptors may function as the last digits in a cell surface code for assembling embryos. Proc Natl Acad Sci USA 95:9072-9077

Duchamp-Viret P, Chaput MA, Duchamp A (1999) Odor response properties of rat olfactory receptor neurons. Science 284:2171-2174

Dulac C, Axel R (1995) A novel family of genes encoding putative pheromone receptors in mammals. Cell 83:195-206

Eisenbach M (1999) Sperm chemotaxis. Rev Reprod 1:56-66

Eklund A (1997) The effect of early experience on MHC-based mate preferences in two B10.W strains of mice (Mus domesticus). Behav Genet 27:223-229

Fan W, Liu Y-C, Parimoo S, Weissman SM (1995) Olfactory receptor-like genes are located in the human major histocompatibility complex. Genomics 27:119-123

Farbman AI (1992) Cell Biology of Olfaction. Cambridge University Press, New York

Galtier N, Gouy M, Gautier C (1996) SEAVIEW and PHYLO_WIN: two graphic tools for sequence alignment and molecular phylogeny. Comput Appl Biosci 12:543-548

Garbers DL, Bentley JK, Dangott LJ, Ramarao CS, Shimomura H, Suzuki N, Thorpe D (1986) Peptides associated with eggs: mechanisms of interaction with spermatozoa. Adv Exp Med Biol 207:315-357

Gayan J, Smith SD, Cherny SS, Cardon LR, Fulker DW, Brower AM, Olson RK, Pennington BF, DeFries JC (1999) Quantitative-trait locus for specific language and reading deficits on chromosome 6p. Am J Hum Genet 64:157-164

Grafen A (1990) Do animals really recognize kin? Anim Behav 39:42-54

Grigorenko EL, Wood FB, Meyer MS, Hart LA, Speed WC, Shuster A, Pauls DL (1997) Susceptibility loci for distinct components of developmental dyslexia on chromosomes 6 and 15. Am J Hum Genet 60:27-39

Gruen JR, Nalabolu SR, Chu TW, Bowlus C, Fan WF, Goei VL, Wei H, Sivakamasundari R, Liu Y, Xu HX, Parimoo S, Nallur G, Ajioka R, Shukla H, Bray-Ward P, Pan J,

Weissman SM (1996) A transcription map of the major histocompatibility complex (MHC) class I region. Genomics 36:70-85

Halpern M (1987) The organization and function of the vomeronasal system. Annu Rev Neurosci 10:325-362

Harzsch S, Miller J, Benton J, Beltz B (1999) From embryo to adult: persistent neurogenesis and apoptotic cell death shape the lobster deutocerebrum. J Neurosci 19:3472-3485

Hatt H, Gisselmann G, Wetzel CH (1999) Cloning, functional expression and characterization of a human olfactory receptor. Cell Mol Biol 45:285-291

Hedrick PW, Black FL (1997) HLA and mate selection: no evidence in South Amerindians. Am J Hum Genet 61:505-511

Herrada G, Dulac C (1997) A novel family of putative pheromone receptors in mammals with a topographically organized and sexually dimorphic distribution. Cell 90:763-773

Herz RS, Cahill ED (1997) Differential use of sensory information in sexual behavior as a function of gender. Hum Nat 8:275-289

Ho HN, Yang YS, Hsieh RP, Lin HR, Chen SU, Chen HF, Huang SC, Lee TY, Gill TJ (1994) Sharing of human leukocyte antigens in couples with unexplained infertility affects the success of in vitro fertilization and tubal embryo transfer. Am J Obstet Gynecol 170:63-71

Krautwurst D, Yau KW, Reed RR (1998) Identification of ligands for olfactory receptors by functional expression of a receptor library. Cell 95:917-926

Laitinen T (1993) A set of MHC haplotypes found among Finnish couples suffering from recurrent spontaneous abortions. Am J Reprod Immunol 29:148-154

Liu YC, Pan J, Zhang C, Fan W, Collinge M, Bender JR, Weissman SM (1999) A MHC-encoded ubiquitin-like protein (FAT10) binds noncovalently to the spindle assembly checkpoint protein MAD2. Proc Natl Acad Sci USA 96:4313-4318

Malfroy L, Roth MP, Carrington M, Borot N, Volz A, Ziegler A, Coppin H (1997) Heterogeneity in rates of recombination in the 6-Mb region telomeric to the human major histocompatibility complex. Genomics 43:226-231

Malnic B, Hirono J, Sato T, Buck LB (1999) Combinatorial receptor codes for odors. Cell 96:713-723

Matsunami H, Buck LB (1997) A multigene family encoding a diverse array of putative pheromone receptors in mammals. Cell 90:775-784

Menssen R, Orth P, Ziegler A, Saenger W (1999) Decamer-like conformation of a nona-peptide bound to HLA-B*3501 due to non-standard positioning of the C terminus. J Mol Biol 285:645-653

Mombaerts P (1999a) Molecular biology of odorant receptors in vertebrates. Annu Rev Neurosci 22:487-509

Mombaerts P (1999b) Odorant receptor genes in humans. Curr Opin Genet Dev 9:315-320

Nandedkar TD, Parkar SG, Iyer KS, Mahale SD, Moodbidri SB, Mukhopadhyaya RR, Joshi DS (1996) Regulation of follicular maturation by human ovarian follicular fluid peptide. J Reprod Fertil Suppl 50:95-104

Naruse TK, Nose Y, Ando R, Araki N, Shigenari A, Ando A, Ishihara M, Kagiya M, Nabeya N, Isshiki G, Inoko H (1998) Extended HLA haplotypes in Japanese homozygous typing cells. Tissue Antigens 51:305-308

Ober C, Elias S, O'Brien E, Kostyu DD, Hauck WW, Bombard A (1988) HLA sharing and fertility in Hutterite couples: evidence for prenatal selection against compatible fetuses. Am J Reprod Immunol Microbiol 18:111-115

Ober C, Weitkamp LR, Cox N, Dytch H, Kostyu D, Elias S (1997) HLA and mate choice in humans. Am J Hum Genet 61:497-504

Parmentier M, Libert F, Schurmans S, Schiffmann S, Lefort A, Eggerickx, D, Ledent C, Mollereau C, Gérard C, Perret J, Grootegoed A, Vassart G (1992) Expression of members of the putative olfactory receptor gene family in mammalian germ cells. Nature 355:453-455

Penn D, Potts W (1998) How do major histocompatibility complex genes influence odor and mating preferences? Adv Immunol 69:411-436

Peters HC, Kammer G, Volz A, Kaupmann K, Ziegler A, Bettler B, Epplen JT, Sander T, Riess O (1998) Mapping, genomic structure, and polymorphisms of the human GABABR1 receptor gene: evaluation of its involvement in idiopathic generalized epilepsy. Neurogenetics 2:47-54

Pilpel Y, Lancet D (1999) The variable and conserved interfaces of modeled olfactory receptor proteins. Protein Sci 8:969-977

Ressler KJ, Sullivan SL, Buck LB (1993) A zonal organization of odorant receptor gene expression in the olfactory epithelium. Cell 73:597-609

Rodriguez I, Feinstein P, Mombaerts P (1999) Variable patterns of axonal projections of sensory neurons in the mouse vomeronasal system. Cell 97:199-208

Rouquier S, Friedman C, Delettre C, van den Engh G, Blancher A, Crouau-Roy B, Trask BJ, Giorgi D (1998a) A gene recently inactivated in human defines a new olfactory receptor family in mammals. Hum Mol Genet 7:1337-1345

Rouquier S, Taviaux S, Trask BJ, Brand-Arpon V, van den Engh G, Demaille J, Giorgi D (1998b) Distribution of olfactory receptor genes in the human genome. Nat Genet 18:243-250

Rülicke T, Chapuisat M, Homberger FR, Macas E, Wedekind C (1998) MHC-genotype of progeny influenced by parental infection. Proc R Soc Lond B Biol Sci 265:711-716

Ryba NJ, Tirindelli R (1997) A new multigene family of putative pheromone receptors. Neuron 19:371-379

Sander T, Peters C, Kammer G, Samochowiec J, Zirra M, Mischke D, Ziegler A, Kaupmann K, Bettler B, Epplen JT, Riess O (1999) Association analysis of exonic variants of the gene encoding the GABAB receptor and idiopathic generalized epilepsy. Am J Med Genet 88:305-310

Singer AG, Beauchamp GK, Yamazaki K (1997) Volatile signals of the major histocompatibility complex in male mouse urine. Proc Natl Acad Sci USA 94:2210-2214

Singh PB, Brown RE, Roser B (1987) MHC antigens in urine as olfactory recognition cues. Nature 327:161-164

Strader CD, Fong TM, Tota MR, Underwood D, Dixon RA (1994) Structure and function of G protein-coupled receptors. Annu Rev Biochem 63:101-132

Tay GK, Cattley SK, Chorney MJ, Hollingsworth PN, Roth MP, Dawkins RL, Witt CS (1997) Conservation of ancestral haplotypes telomeric of HLA-A. Eur J Immunogenet 24:275-285

Thomas MB, Haines SL, Akeson RA (1996) Chemoreceptors expressed in taste, olfactory and male reproductive tissues. Gene 178:1-5

Thompson JD, Gibson TJ, Plewniak F, Jeanmougin F, Higgins DG (1997) The CLUSTAL_X windows interface: flexible strategies for multiple sequence alignment aided by quality analysis tools. Nucleic Acids Res 25:4876-4882

Touhara K, Sengoku S, Inaki K, Tsuboi A, Hirono J, Sato T, Sakano H, Haga T (1999) Functional identification and reconstitution of an odorant receptor in single olfactory neurons. Proc Natl Acad Sci USA 96:4040-4045

Trask BJ, Friedman C, Martin-Gallardo A, Rowen L, Akinbami C, Blankenship J, Collins C, Giorgi D, Iadonato S, Johnson F, Kuo WL, Massa H, Morrish T, Naylor S, Nguyen OT, Rouquier S, Smith T, Wong DJ, Youngblom J, van den Engh G (1998a) Members of the olfactory receptor gene family are contained in large blocks of DNA duplicated polymorphically near the ends of human chromosomes. Hum Mol Genet 7:13-26

Trask BJ, Massa H, Brand-Arpon V, Chan K, Friedman C, Nguyen OT, Eichler E, van den Engh G, Rouquier S, Shizuya H, Giorgi D (1998b) Large multi-chromosomal duplications encompass many members of the olfactory receptor gene family in the human genome. Hum Mol Genet 7:2007-2020

Vanderhaeghen P, Schurmans S, Vassart G, Parmentier M (1993) Olfactory receptors are displayed on dog mature sperm cells. J Cell Biol 123:1441-1452

Vanderhaeghen P, Schurmans S, Vassart G, Parmentier M (1997a) Specific repertoire of olfactory receptors in the male germ cells of several mammalian species. Genomics 39:239-246

Vanderhaeghen P, Schurmans S, Vassart G, Parmentier M (1997b) Molecular cloning and chromosomal mapping of olfactory receptor genes expressed in the male germ line: Evidence for their wide distribution in the human genome. Biochem Biophys Res Commun 237:283-287

Vassar R, Chao SK, Sitcheran R, Nunez JM, Vosshall LB, Axel R (1994) Topographic organization of sensory projections to the olfactory bulb. Cell 79:981-991

Volz A, Davies A, Ragoussis I, Laun K, Ziegler A (1997) Dissection of the 5.5 Mbp region directly telomeric of HLA-B including a long range restriction map, YAC and PAC contigs. DNA Seq 8:181-187

Voyron S, Giacobini P, Tarozzo G, Cappello P, Perroteau I, Fasolo A (1999) Apoptosis in the development of the mouse olfactory epithelium. Brain Res Dev Brain Res 115:49-55

Walensky LD, Roskams AJ, Lefkowitz RJ, Snyder SH, Ronnett GV (1995) Odorant receptors and desensitization proteins colocalize in mammalian sperm. Mol Med 1:130-141

Walensky LD, Ruat M, Bakin RE, Blackshaw S, Ronnett GV, Snyder SH (1998) Two novel odorant receptor families expressed in spermatids undergo 5'-splicing. J Biol Chem 273:9378-9387

Walther T, Balschun D, Voigt JP, Fink H, Zuschratter W, Birchmeier C, Ganten D, Bader M (1998) Sustained long term potentiation and anxiety in mice lacking the Mas protooncogene. J Biol Chem 273:11867-11873

Ward CR, Kopf GS (1993) Molecular events mediating sperm activation. Dev Biol 158:9-34

Weckstein LN, Patrizio P, Balmaceda JP, Asch RH, Branch DW (1991) Human leukocyte antigen compatibility and failure to achieve a viable pregnancy with assisted reproductive technology. Acta Eur Fertil 22:103-107

Wedekind C, Seebeck T, Bettens F, Paepke AJ (1995) MHC-dependent mate preferences in humans. Proc R Soc Lond B Biol Sci 260:245-249

Wedekind C, Füri S (1997) Body odour preferences in men and women: do they aim for specific MHC combinations or simply heterozygosity? Proc R Soc Lond B Biol Sci 264:1471-1479

Wojtasek H, Hansson BS, Leal WS (1998) Attracted or repelled?--a matter of two neurons, one pheromone binding protein, and a chiral center. Biochem Biophys Res Commun 250:217-222

Wysocki CJ (1989) Vomeronasal chemorecognition: its role in reproductive fitness and physiology. In: Neural Control of Reproductive Function, Alan R. Liss, New York, pp. 545-566

Wysocki CJ, Beauchamp GK (1984) Ability to smell androstenone is genetically determined. Proc Natl Acad Sci USA 81:4899-4902

Yamazaki K, Boyse EA, Miké V, Thaler HT, Mathieson BJ, Abbott J, Boyse J, Zayas ZA, Thomas L (1976) Control of mating preference in mice by genes in the major histocompatibility complex. J Exp Med 144: 1324-1335

Yamazaki K, Yamaguchi M, Baranoski L, Bard J, Boyse EA, Thomas L (1979) Recognition among mice. Evidence from the use of a Y-maze differentially scented by congenic mice of different major histocompatibility types. J Exp Med 150:755-760

Yamazaki K, Beauchamp GK, Kupniewski D, Bard J, Thomas L, Boyse EA (1988) Familial imprinting determines H-2 selective mating preferences. Science 240:1331-1332

Young D, Waitches G, Birchmeier C, Fasano O, Wigler M (1986) Isolation and characterization of a new cellular oncogene encoding a protein with multiple potential transmembrane domains. Cell 45:711-719

Zhao H, Ivic L, Otaki JM, Hashimoto M, Mikoshiba K, Firestein S (1998) Functional expression of a mammalian odorant receptor. Science 279:237-242

Ziegler A (1997) Biology of chromosome 6. DNA Seq 8:189-202

Distribution of polypurine/polypyrimidine tract sequences in the human MHC region and their possible functions

Shigehiko Kanaya[1,2], Tatsuo Fukagawa[3], Asako Ando[4], Hidetoshi Inoko[4], Yoshihiro Kudo[1], and Toshimichi Ikemura[3]

[1]Department of Electrical and Information Engineering, Faculty of Engineering, Yamagata University, Yonezawa, Yamagata-ken 992-8510, Japan
[2]Division of Physiological Genetics, Department of Ontogenetics, National Institute of Genetics, Mishima, Shizuoka-ken 411-8540, Japan
[3]Division of Evolutionary Genetics, Department of Population Genetics, National Institute of Genetics, Mishima, Shizuoka-ken 411-8540, Japan
[4]Department of Genetic Information, Division of Molecular Life Science, Tokai University School of Medicine, Bohseidai, Isehara, Kanagawa-ken 259-1193, Japan

Summary. To investigate the biological significance of polypurine/poly-pyrimidine (pur/pyr) tract sequences in the human MHC region, we searched both DNA strands for sequences longer than 100 nt with an A+G% (AG%) higher than 85% (A+G tract). Among all human genome sequences registered in DDBJ (ca. 32 Mb in total), we obtained 6247 A+G tracts. There exists one tract per 51 kb and therefore, roughly one tract per one replicon size. One hundred seventeen tracts (one tract per 33 kb) were found in the MHC region, which is a significantly higher level than that usually found in human genome segments. One of the A+G tracts is found in the DNA-replication switch region at the junction of MHC classes II and III. Other tracts were examined in connection with polymorphism levels of several MHC genes. We also investigated apparently long A+G tracts in the entire human genome. The longest A+G tract found so far is the 2798-nt tract found in the 3' downstream region of rRNA genes. The biological significance of the long A+G tracts was investigated in connection with triplex formation, pausing of DNA replication, and enhancement of recombination.

Key words. Triplex, Recombination, Polypurine/polypyrimidine tract, G+C%, A+G%, Replication timing, Chromosome band boundary, Human MHC

Introduction

The genes in the MHC region are extremely polymorphic. Some loci have heterozygosities as high as 80% (Yeager and Hughes 1999). Furthermore, a large number of recombination events are known to have accumulated in the MHC region, which suggests that the region has DNA structures with an intrinsically high potential for mutation and recombination. In an attempt to elucidate molecular mechanisms that produce these characteristics, we investigated polypurine/polypyrimidine tract sequences (pur/pyr tracts) in the human MHC region. There are a large number of pur/pyr tracts scattered over the human genome, each several hundred bp long (Siedlaczck et al. 1993). Interest in the pur/pyr tracts has increased due to the *in vitro* observation that these sequences can form a triple-helix (triplex) even under physiological conditions (Hoffman-Liebermann et al. 1986; Hampel et al. 1991). Triplex formation is predicted to be important as a molecular mechanism for modulating DNA-fork movement, transcription, and recombination, as well as in organizing the subnuclear arrangement of chromosomal DNAs within an interphase nucleus. During DNA replication, a DNA polymerase is known to stall in the center of pur/pyr-tracts with triplex-forming potential (Lapidot et al. 1989; Baran et al. 1991; Samadashwily et al. 1993). The pausing of a replication fork has been associated with a high frequency of mutagenesis (Bebenek et al. 1989). The pur/pyr-tracts are also associated with recombination activation due to triplex formation (Collier et al. 1988; Weinreb et al. 1990; Kohwi and Panchenko 1993) and/or single-strand displacement mediated by the triplex formation, which are enhanced by torsional stress produced during transcription- and replication-progression (Yancopoulos and Alt 1985; Kim and Wang 1989; Thomas and Rothstein 1989). One of the pur/pyr tracts of interest in the present study is that found in the DNA-replication switch region described below.

Pur/Pyr sequences in the DNA-replication switch region at the junction between MHC classes II and III

Chromosome band zones have been connected to various genome characteristics such as DNA replication timing during S phase and long-range G+C% (GC%) mosaic structures (Bernardi et al. 1985; Ikemura 1985; Ikemura and Aota 1988; Ikemura and Wada 1991; Holmquist 1992; Craig and Bickmore 1994). Therefore, band boundaries are thought to correspond both to switching points of DNA replication timing (i.e., from early to late replication) and to GC% mosaic boundaries. The boundary may have control signals for replication timing. The 4-megabase (Mb) human MHC is composed of multiple bands at a high-resolution level. We previously found a boundary for Mb-sized GC% mosaic domains at the junction between MHC classes II and III, and proposed this boundary as a chromosome band boundary (Ikemura et al. 1990; Fukagawa et al. 1995). Considering genome behaviors connected with chromosome bands, band boundaries are presumably detectable at a molecular level by identifying the early-to-late switch point for replication timing. We reported the replication timing for a 500-kb region harboring the junction of MHC classes II and III, for which we completed bidirectional chromosome walking, and a large portion of this junction has been sequenced (Matsumoto 1992; Sugaya et al. 1994, 1997). Many sets of closely spaced PCR primers (e.g. even within 1 kb) were designed for clarifying the switching zone of replication timing (Tenzen et al. 1997). DNA replication timing changed precisely at the GC% boundary with a 2-hour time difference between the GC- and AT-rich sides (Figs. 1a and b). A 209-bp pur/pyr tract with triplex-forming potential was found at exactly the midpoint of the replication switch zone. Its purine strand can be expressed mostly as palindromic sequences composed of two tetranucleotides, (GGAA) and (AGAA). Under proper superhelical stress, this type of palindromic pur/pyr sequence (termed H-palindrome) undergoes a distinct structural transition involving an intrastrand disproportionation in which half the mirror repeat sequence dissociates into single strands and one of the two single strands folds back on the other half of the mirror repeat duplex to form an intramolecular triplex (known as H-DNA) (Mirkin and Frank-Kamenetskii 1994). Figure 1c shows examples of such H-DNA triplexes formed for the 209-bp pur/pyr tract when perfectly consecutive Hoogsteen-pairing was assumed. Replication fork movement is known to pause *in vitro* at triplexes formed in pur/pyr tracts (Baran et al. 1991; Brinton et al. 1991). While the function of triplexes in the *in vivo* replication process has not yet been clarified, the 209-nt pur/pyr tract was shown to be able to form the triplex structure in the human interphase nucleus as explained later. It is worthwhile to note that in the pig MHC, a centromere is located exactly at the junction region of MHC classes II and III (Chardon et al. 1999).

Distribution of pur/pyr tracts in the human MHC

As noted above, triplex-forming pur/pyr tracts are believed to play various biological roles in basic functions such as replication, recombination, transcription, and chromosomal folding. We examined the distribution of pur/pyr tracts in the human MHC by searching both DNA strands for 100-nt sequences with an A+G% (AG%) higher than 85%. When the detected 100-nt sequences overlapped, they merged into one sequence (called A+G tract). Among all human genome sequences registered in DDBJ (release 37; ca. 32 Mb in total), we obtained 6247 A+G tracts. There was one tract per 51 kb and therefore, roughly one tract per one replicon size. One hundred seventeen tracts (i.e., one tract per 33 kb) were found in MHC, which is a significantly higher level than that usually found in human genome segments. The distribution of A+G tracts in the MHC (Fig. 2) shows four high-frequency regions (peaks in %Freq in Fig. 2; 0.1-0.2 Mb,

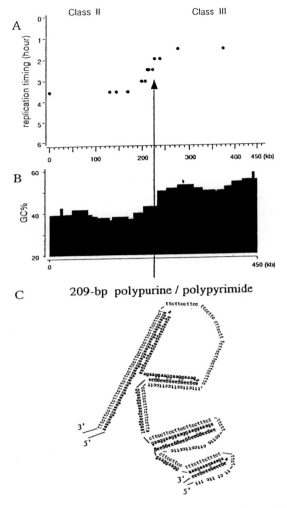

Fig. 1. The 209-bp pur/pyr tract found at the exact mid point of the switching region of replication timing during the S phase. Replication timing **(a)** and GC% distribution **(b)** in the junction between MHC classes II and III is from Tenzen et al. (1997). **(a)** Y-axis means the replication timing; the start time of the S phase was taken as 0, and replication timings for individual loci were plotted as reported by Tenzen et al. (1997). **(b)** For the GC% distribution in this and other MHC regions, see Fukagawa et al. (1995). **(c)** An example of H-DNA triplexes for the 209-bp pur/pyr tract leaving the longest single-strand, when only the perfectly consecutive Hoogsteen-pairing longer than 10 bps was assumed. The 209-bp sequence corresponds to that from 5301 to 5510 nucleotide in GenBank Locus HSMHC3A5 (Accession No. U89335), and its location in **a**, **b** is indicated by a vertical arrowhead line.

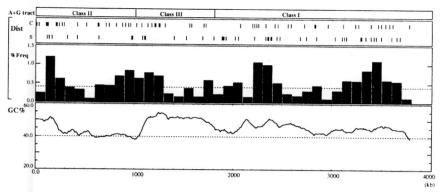

Fig. 2. Distribution of A+G tracts and GC% in the MHC region. Bars in Dist represent A+G tracts in standard (S) and complement (C) strands. %Freq represents the % level of A+G tracts per each 10-kb sequence, and GC% represents GC content in a 100-kb window length.

0.9-1.0 Mb, 2.2-2.3 Mb, and 3.4-3.5 Mb). A+G tracts tend to be clustered in GC%-transition regions. The evident GC% transition around 1.0 Mb corresponds to the DNA-replication switch region at the junction between MHC classes II and III as described above.

Figure 3 shows relative positions of genes and A+G tracts. Twenty tracts (among the 117 in the MHC) are located within genes (Fig. 4). It is reported that, in the class II region of the MHC, the HLA-DRA gene is monomorphic in all mammalian species examined to date, whereas HLA-DQA1 is polymorphic (Hughes and Nei 1990; Yeager and Hughes 1999). A 261-nt A+G tract is located between exons 1 and 2 in DQA1 (Fig. 4), whereas no tracts are found in HLA-DRA (Fig. 3). Yeager and Hughes (1999) reported that HLA-DQB1 has low levels of recombination, and HLA-DPB1 has a high level. A 119-base A+G tract is located between exons 2 and 3 in HLA-DPB1 (Fig. 4), whereas no tracts are detected in HLA-DQB1. These results are consistent with the view that the existence of A+G tracts may result in enhancement of recombination.

Long A+G tracts in the human genome

One way to elucidate the possible biological significance of A+G tracts is to investigate the currently known long tracts. The longest A+G tract in the human genome sequences available in the current DNA database is a 2798-nt tract located in the 3' downstream region of rRNA genes (Table 1). In this 17-kb downstream region, there are 12 A+G tracts including the longest 2798-nt tract; 4

136

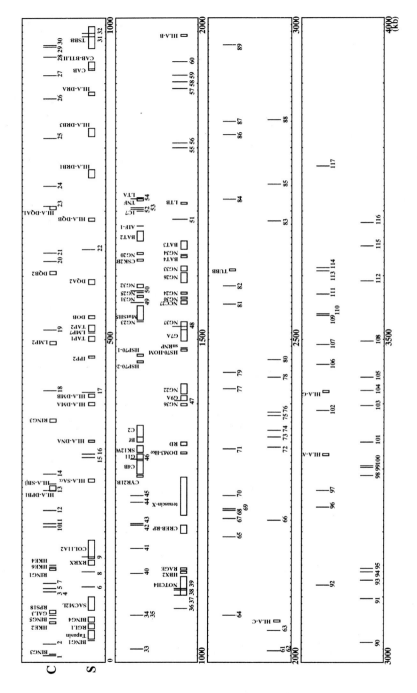

Fig. 3. Organization of A+G tracts (bars numbered from 1 to 117) in the MHC region.

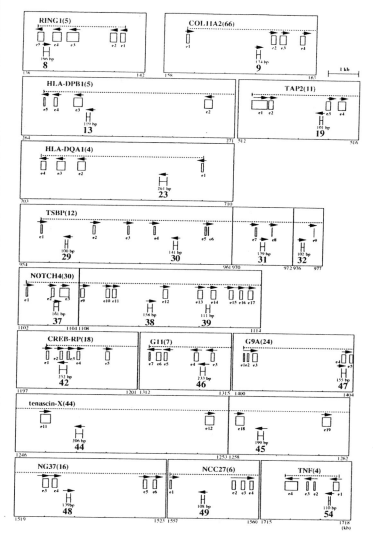

Fig. 4. A+G tracts within genes. Exons are numbered from 5' to 3' (e1, e2, ...). The number of exons for respective genes is described in parentheses. Numbers of A+G tracts correspond to those in Figure 3. Two A+G tracts (ID = 42 and 47) completely cover protein-coding regions. Interestingly, in a 155-nt A+G tract (ID = 42), H-DNA can be formed by 24-nt consecutive Hoogsteen-pairing. Ruskin and Green (1985) reported that changes from a pyrimidine to purine tract at the 3' end of transcripts can interfere with mRNA splicing. Thus, the 20 A+G-tracts may be involved in the regulation of splicing.

Table 1. Very long A+G tracts in human genome sequences

LOCUS	A+G%	Length (nt)	Note
HSU13369*	87.5	2798	rDNA
HUMPKD1GEN	96.3	2544	autosomal dominant polycystic kidney disease
HSU91325	91.9	1228	15p13.11
AC004216	87.8	1167	12q24
AF011889	91.4	1084	Xq28; Hunter syndrome
HSU66083	90.8	1050	X; MAGE-9 antigen
AC005347	92.2	1037	16p13.3
HSL241B9C	86.9	1030	4p16.3; Huntington's disease region
AC005570	87.5	1026	
HSU75931	96.5	996	Xp22
HS389A20	95.6	958	X
HSCAMF3X1	94.8	932	Yp; pseudoautosomal boundary (PAB1)
HSU13369*	86.8	920	rDNA
HSU13369*	86.4	913	rDNA
AC002102	89.2	910	9q34
AC005254	91.9	908	19p12
HS229041	95.0	856	21q22.2
U82668*	92.6	853	Xp22; pseudoautosomal region
D87013	85.2	835	22q11.2; immunoglobulin light chain
HUAC003119	93.8	830	16p13.11
U82668*	95.4	809	Xp22; pseudoautosomal region
AC002402	94.8	808	11
AC002478	95.1	787	
AC003668	94.7	783	Xp22
AC005035	87.4	783	2
AC004772	87.0	779	Y
HSSG1SG2	94.4	772	semenogelin I and II
HSQ4H9	91.1	765	21q22.3
HUAC003661	93.8	764	16p12
HS164C20	94.8	759	6q16.1-22.1
AC004258	91.4	749	19p13.3; CIRBP
HSB13C9	95.7	748	22
HUAC003049	93.8	746	16p11.2-p12
AC006134	95.2	742	19q12
AC004756	93.8	729	17
AC004084	90.7	712	7q22-q31.1
HSU13369*	87.2	708	rDNA
AC003967	87.8	702	19p12
AC005803	94.1	701	17

* represents LOCUS consisting of more than one A+G tract with longer than 700 nt.

others are longer than 700 nt (Fig. 5). DNA fork movement from the 5' side of rDNAs (from right to left in Fig. 5) is known to stall in this pur/pyr-tract cluster and thus evade collision between DNA fork movement and RNA polymerase movement for transcribing rRNA genes. Triplex structures formed in these A+G tracts are thought to contribute to pausing of DNA fork movement. Interestingly, all 12 A+G tracts are on one strand. This may relate to the predicted function of the A+G tracts.

To further study the global organization of A+G tracts in the entire human genome, we examined the histogram distribution of A+G tract length (Fig. 6). For example, there were 125 A+G-tracts longer than 500 nt. Table 1 lists tracts longer than 700 nt, along with biological notes for their respective genome segments. Biological notes in some sequences are not described in DNA databases because these sequences were determined by large scale-sequencing projects. Concerning biological notes so far known, the respective genome segments appears to be connected with enhanced recombination. Characteristics connected with the enhanced recombination are as follows. For example, the size of rDNA-clusters among the human population is known to be variable, so enhanced recombination of this region should be common. The 17-kb downstream cluster of A+G tracts may be responsible, at least in part, for the high rate of rDNA-cluster recombination. Pseudoautosomal regions are also known to have a high recombination level during meiosis and were shown to have long pur/pyr tracts (932 nt in pseudoautosomal boundary HSCAMF3X1; two A+G tracts, 853 and 809 nt, in pseudoautosomal region U82668). Genome segments responsible for certain genetic diseases, such as autosomal dominant polycystic kidney disease (PKD1), Hunter syndrome, and Huntington's disease, also tend to have long pur/pyr-tracts (Table 1). The genomic fragment containing a 2544-nt A+G tract in PKD1 is duplicated in 16p13.1, and this region is known to participate in translocations and deletions (Burn et al. 1995). A common cause of Hunter syndrome was predicted to be due to inversion of the IDS gene resulting from recombination with an IDS-related sequence (Bondeson et al. 1995) and a 1030-nt A+G tract is located near the IDS gene (Fig. 5). Huntington's disease, known to be a (CAG)n-expansion disease, also has a 1030-nt A+G tract. It is thus thought that pur/pyr-tracts are associated with enhanced recombination and that their organization is an important factor for understanding dynamics of human chromosomes.

In situ visualization of triplexes in human interphase nuclei

Using monoclonal antibodies specific for triplex DNAs, Lee et al. (1987), Burkholder et al. (1988), and Agazie et al. (1994, 1996) reported that the mammalian nucleus shows a characteristic distribution of foci-type signals and thus a certain portion of chromosomal DNA is naturally organized into a triplex configuration. In triplex formation, single-stranded DNA is always left unpaired, and it can hybridize with other single-stranded DNA with significant complemen-

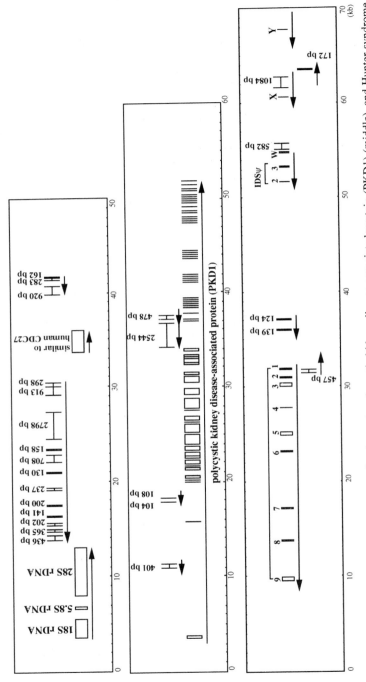

Fig. 5. A+G tract organization. rDNA region (upper), polycystic kidney disease-associated protein (PKD1) (middle), and Hunter syndrome region (bottom).

tarity. If individual signals stained by monoclonal antibodies contained a sufficient amount of triplex DNAs and the resultant single-stranded DNAs, sequence-dependent foci-type signals can be visualized in the nondenatured nucleus *in situ* by choosing proper DNA probes. FISH has been successfully developed for detecting RNA in the interphase nucleus (Lawrence et al. 1989; Johnson et al. 1991; Dirks et al. 1993). For detecting RNA in the presence of double-stranded genome DNA, nondenatured nuclei were used because the RNAs of interest are mostly present in a single-stranded form. This nondenatured FISH method may detect the single-stranded DNA displaced by triplex-formation. The 209-bp pur/pyr tract in the MHC gives clear foci-type signals (Fig. 7a). This pattern can also be obtained by analyzing nuclei pre-treated by a variety of RNase, indicating that the respective binding sites are not RNAs. Recently, we observed that a wide range of DNA probes with triplex-forming potential, e.g., (AG/CT)n, (GAA/TTC), and (GGAA/TTCC)n, give sequence-specific foci-type signals. Figure 7b shows an example of the (AG/CT)n probe. These signals, including the 209-bp tract, showed an exact overlap with triplexes immunolocalized by the known antitriplex monoclonal antibodies (Ikemura et al. 1998; Ohno et al. 1999).

Judging from signal intensities and probe sizes (the 209-bp pur/pyr tract), nondenatured FISH signals of interest should not be due to a single copy genome DNA. When GenBank was searched for the 209-bp sequence using the FASTA program, more than 50 sequences had greater than 80% nucleotide identity over

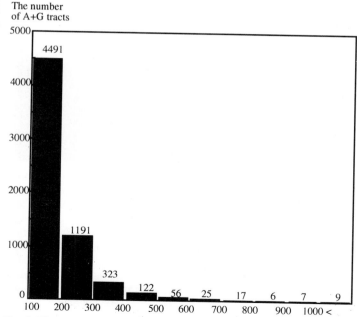

Fig. 6. Histograms of A+G tracts in human genome sequences.

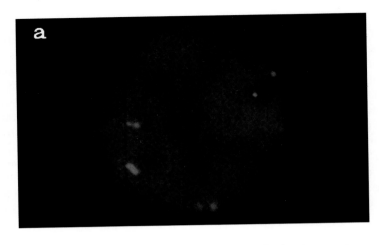

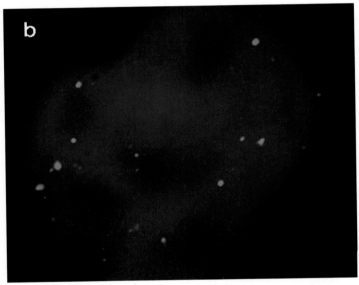

Fig. 7. Nondenatured FISH assay with triplex-forming DNA probes. Human monocytic leukemia cells were analyzed. (**a**) Cy3-labeled 209-bp pur/pyr tract probe. (**b**) FITC-labeled (AG/CT)n probe. RNase-treated nuclei gave similar results. The DNA probe mix for the nondenatured FISH was composed of 50% formamide in 2 x SSC, 5% dextran sulfate, 100 µg/ml shared salmon sperm DNA, and 0.1 – 1 µg/ml fluorescent DNA probe (Ohno et al. 1999). After 5-min denaturation at 80°C followed by cooling on ice, 30 µl of the mixture was applied to each slide covered with a glass coverslip and incubated for 4 h at 37°C in a humid chamber. The slides were then washed according to a standard procedure for FISH for RNA detection (Lawrence et al. 1989). Cy3- and FITC-labeled probes gave red and green signals, respectively, and nuclei were counterstained by DAPI (blue).

200-nt stretches. This means that there are several hundred copies of highly homologous sequences in the human genome. It should be noted that the number of nondenatured FISH signals (Fig. 7a) is far less than the number of the predicted sequences. Each signal is thought to visualize a supramolecular structure composed of multiple sequences that are spatially clustered within an interphase nucleus through interstrand triplexes and/or base-pairing between single strands being displaced by intramolecular triplex formation. Recently, we found that all these nondenatured FISH signals closely associate spatially with centromeres in interphase nuclei (Ikemura et al. 1998; Ohno et al. 1999).

References

Agazie YM, Lee JS, Burkholder GD (1994) Characterization of a new monoclonal antibody to triplex DNA and immunofluorescent staining of mammalian chromosomes. J Biol Chem 269:7019-7023

Agazie YM, Burkholder GD, Lee JS (1996) Triplex DNA in the nucleus: direct binding of triplex-specific antibodies and their effect on transcription, replication and cell growth. Biochem J 316:461-466

Baran N, Lapidot A, Manor H (1991) Formation of DNA triplexes accounts for arrests of DNA synthesis at d(TC)n and d(GA)n tracts. Proc Natl Acad Sci USA 88: 507-511

Barch MJ (1991) The ACT Cytogenetics Laboratory Manual. 2nd ed. Raven Press, Ltd., New York, pp 31

Bebenek K, Abbots J, Roberts JD, Wilson SH, Kunkel TA (1989) Specificity and mechanism of error-prone replication by human immunodeficiency virus-1 reverse transcriptase J Biol Chem 264:16948-16956

Bernardi G, Olofsson B, Filipski J, Zerial M, Salinas J, Cuny G, Meunier-Rotival M, Rodier F (1985) The mosaic genome of warm-blooded vertebrates. Science 228:953-958

Bondeson ML, Dahl N, Malmgren H, Kleijer WJ, Tonnesen T, Carlberg BM, Pettersson U (1995) Inversion of the IDS gene resulting from recombination with IDS-related sequences is a common cause of the Hunter syndrome. Hum Mol Genet 4:615-621

Brinton BT, Caddle MS, Heintz NH (1991) Position and orientation-dependent effects of a eukaryotic Z-triplex DNA motif on episomal DNA replication in COS-7 cells. J Biol Chem 266:5153-5161

Burkholder GD, Latimer LJP, Lee JS (1988) Immunofluorescent staining of mammalian nuclei and chromosomes with a monoclonal antibody to triplex DNA. Chromosoma 97:185-192

Burn TC, Connors TD, Dackowski WR, Petry LR, Van Raay TJ, Millholland J, Venet M, Miller G, Hakim RM, Doggett NA, Landes GM, Klinger KW, Qian F, Onuchic LF, Wanick T, Germino GG (1995) Analysis of the genomic sequence for the autosomal dominant polycystic kidney disease (PKD1) gene predicts the presence of a leucine-rich repeat. Hum Mol Genet 4:575-582

Chardon P, Renard C, Vaiman M (1999) The major histocompatibility complex in swine. Immunol Rev 167:179-192

Collier DA, Griffin JA, Wells RD (1988) Non-B right-handed DNA conformations of homopurine-homopyrimidine sequences in the murine immunoglobin Ca switch region. J Biol Chem 263:7397-7405

Craig JM, Bickmore WA (1994) The distribution of CpG islands in mammalian chromosomes. Nat Genet 7:376-382

Dirks RW, Van de Rijke FM, Fujishita S, Van der Ploeg M, Raap AK (1993) Methodologies for specific intron and exon RNA localization in cultured cells by haptenized and fluorochromized probes. J Cell Sci 104:1187-1197

Fukagawa T, Sugaya K, Matsumoto K, Okumura K, Ando A, Inoko H, Ikemura T (1995) A boundary of long-range G + C% mosaic domains in the human MHC locus: pseudoautosomal boundary-like sequence exists near the boundary. Genomics 25:184-191

Hampel KJ, Crosson P, Lee JS (1991) Polyamines favor DNA triplex formation at neutral pH. Biochemistry 30:4455-4459

Hoffman-Liebermann B, Liebermann D, Troutt A, Kedes LH, Cohen SN (1986) Human homologs of TU transposon sequences: polypurine/polypyrimidine sequence elements that can alter DNA conformation in vitro and in vivo. Mol Cell Biol 6:3632-3642

Holmquist GP (1992) Chromosome bands, their chromatin flavors, and their functional features. Am J Hum Genet 51:17-37

Hughes AL, Nei M (1990) Evolutionary relationships of class II major histocompatibility complex genes in mammals. Mol Biol Evol 6:559-579

Ikemura T (1985) Codon usage and tRNA content in unicellular and multicellular organisms. Mol Biol Evol 2:13-34

Ikemura T, Aota S (1988) Global variation in G+C content along vertebrate genome DNA: Possible correlation with chromosome band structures. J Mol Biol 203:1-13

Ikemura T, Ohno M, Uemura T, Sasaki H, Tenzen T (1998) Genome and subnuclear organization of non-B forming DNAs in mammalian chromosomes. Cytogenet Cell Genet 81:101-102

Ikemura T, Wada K (1991) Evident diversity of codon usage patterns of human genes with respect to chromosome banding patterns and chromosome numbers; relation between nucleotide sequence data and cytogenetic data. Nucleic Acids Res 19:4333-4339

Ikemura T, Wada K, Aota S (1990) Giant G+C% mosaic structures of the human genome found by arrangement of GenBank human DNA sequences according to genetic positions. Genomics 2:207-216

Johnson CV, Singer RH, Lawrence JB (1991) Fluorescent detection of nuclear RNA and DNA: implications for genome organization. Methods Cell Biol 35:73-99

Kohwi Y, Panchenko Y (1993) Transcription-dependent recombination induced by triple-helix formation. Genes Dev 7:1766-1778

Kim RA, Wang JC (1989) A subthreshold level of DNA topoisomerases leads to the excision of yeast rDNA as extrachromosomal rings. Cell 57:975-985

Lapidot A, Baran N, Manor H (1989) (dT-dC)n and (dG-dA)n tracts arrest single-stranded DNA replication in vitro. Nucleic Acids Res 17:883-900

Lawrence JB, Singer RH, Marselle LM (1989) Highly localized tracts of specific transcripts within interphase nuclei visualized by in situ hybridization. Cell 57:493-502

Lee JS, Burkholder GD, Latimer LJP, Haug BL, Braun RP (1987) A monoclonal antibody to triplex DNA binds to eucaryotic chromosomes. Nucleic Acid Res 15:1047-1061

Matsumoto K, Arai M, Ishihara N, Ando A, Inoko H, Ikemura T (1992) Cluster of fibronectin type III repeats found in the human major histocompatibility complex class III region shows the highest homology with the repeats in an extracellular matrix protein, Tenascin. Genomics 12:485-491

Mirkin SM, Frank-Kamenetskii MD (1994) H-DNA and related structures. Annu Rev Biophys Biomol Struct 23:541-576

Ohno M, Tenzen T, Watanabe Y, Yamagata S, Kanaya S, Ikemura T (1999) Non-B DNA structures spatially and sequence-specifically associated with individual centromeres in the human interphase nucleus. Chromosome Today 13:in press

Ruskin B, Green MR (1985) The role of the 3' splice site consensus sequence in mammalian pre-mRNA splicing. Nature 317:732-734

Samadashwily GM, Dayn A, Mirkin SM (1993) Suicidal nucleotide sequences for DNA polymerization. EMBO J 12:4975-4983

Siedlaczck I, Epplen C, Rieb O, Epplen JT (1993) Simple repetitive (GAA)n loci in the human genome. Electrophoresis 14:973-977

Stein CA (1995) Does antisense exist? Nat Med 1:1119-1121

Sugaya K, Fukagawa T, Matsumoto K, Mita K, Takahashi E, Ando A, Inoko H, Ikemura T (1994) Three genes in the human MHC class III region near the junction with the class II: gene for receptor of advanced glycosylation end products, PBX2 homeobox gene and a Notch homolog, human counterpart of mouse mammary tumor gene int-3. Genomics 23:408-419

Sugaya K, Sasanuma S, Nohata J, Kimura T, Fukagawa T, Nakamura Y, Ando A, Inoko H, Ikemura T, Mita K (1997) Gene organization of human NOTCH4 and (CTG)n polymorphism in this human counterpart gene of mouse proto-oncogene Int3. Gene 189: 235-244

Tenzen T, Yamagata T, Fukagawa T, Sugaya K, Ando A, Inoko H, Gojobori T, Fujiyama A, Okumura K, Ikemura T (1997) Precise switching of DNA replication timing in the GC content transition area in the human major histocompatibility complex. Mol Cell Biol 17:4043-4050

Thomas BJ, Rothstein R (1989) Elevated recombination rates in transcriptionally active DNA. Cell 56:619-630

Weinreb A, Collier DA, Birshtein BK, Wells RD (1990) Left-handed Z-DNA and intramolecular triplex formation at the site of an unequal sister chromatid exchange. J Biol Chem 265:1352-1359

Yancopoulos GD, Alt FW (1985) Developmentally controlled and tissue-specific expression of unrearranged VH gene segments. Cell 40:271-281

Yeager M, Hughes AL (1999) Evolution of the mammalian MHC: natural selection, recombination and convergent evolution. Immunol Rev 167:45-58

Potential for paralogous mapping to simplify the genetics of diseases and functions associated with MHC haplotypes

Roger Dawkins, Jemma Berry, Patricia Martinez, Silvana Gaudieri, Jennie Hui, Sonia Cattley, Natalie Longman, Jerzy Kulski, and Patrick Carnegie

The Centre for Molecular Immunology and Instrumentation, The University of Western Australia, Nedlands, Western Australia 6907, Australia

Summary. The major histocompatibility complex (MHC) region contains several hundred genes in addition to the well-known HLA Class I and II loci. Some of these will explain MHC-disease associations. We have used paralogous and syntenic mapping to short list candidates and conclude that non-HLA genes present on human 6,1 and 19 and murine 17, 4 and 7 regulate immune responses and autoimmune disease. These genes are likely to be components of the early pre-HLA MHC and include tenascin and possibly notch *inter alia*.

Key words. Major histocompatibility complex, Immunoregulation, Psoriasis, Haemochromatosis and autoimmunity

Complexity

There are now several hundred human diseases which are associated with HLA antigens. These diseases are extremely diverse, ranging from some which appear to be immunologically mediated, through others such as haemochromatosis and narcolepsy which are clearly quite different in pathogenesis. The sheer number and extreme diversity of these diseases suggest that there must be different mechanisms involved and that at least several different gene families will be implicated. In fact, there are now clear-cut examples where mechanistic explanations involve non-HLA genes scattered throughout the major histocompatibility complex (MHC) region.

This manifest complexity is exacerbated by the fact that many MHC-associated diseases are influenced by genes in other genomic regions. In some diseases, ten or more different genomic locations have been implicated in different studies. This polygenic nature of MHC-associated diseases is also found in the mouse and probably other species.

Classifications

In an attempt to reduce some of the complexity, attempts have been made to classify these diseases into several categories. It is quite clear that there is one group of diseases in which there are isolated mutations within a functional gene. For example, 21OH deficiency can be explained in terms of specific mutations. The same may be partly true for C2 deficiency, possibly haemochromatosis and conceivably other disorders such as narcolepsy.

A second group of diseases is characterised by autoimmunity. Diseases such as systemic lupus erythematosus (SLE) and insulin dependent diabetes mellitus (IDDM) are associated with specific ancestral haplotypes (AHs) which, however, differ in different ethnic groups. It would appear likely that there are immunoregulatory genes carried by these haplotypes and that quantitative differences in expression or regulation may prove to be responsible for the autoimmunity or at least the associated pathology. Unfortunately, however, many of these autoimmune diseases are associated with different haplotypes, again raising the prospect of substantial complexity. This impression is supported by the fact that a disease such as multiple sclerosis (MS) is positively associated with the 7.1 AH in Caucasoids, whereas IDDM is negatively associated with the same haplotype. Thus, in addition to regulatory genes there must be others which are more or less disease specific.

Many other MHC-associated diseases are more difficult to classify. Some appear to involve abnormalities of differentiation (eg psoriasis), dysregulated inflammation (psoriasis, ankylosing spondylitis, rheumatoid arthritis) and some appear to involve an aberrant response to infection (rate of progression after infection with human immunodeficiency virus).

Simplifying polygenicity

In a further attempt to provide insights and strategies for dealing with this complexity, we asked whether the extra-MHC genes might represent unlinked members of large gene families, but there appear to be no clear examples to date.

Elsewhere we have suggested that one possible simplification is provided by a mechanism based on the presence or absence of human endogenous retroviruses (HERVs) or their derivatives (Dawkins et al. 1999). Here, we ask whether some of the complexity might be reduced by applying the principle of paralogous mapping. Katsanis et al. (1996), Kasahara (1997) and others suggested that different regions of the genome may be similar by virtue of early chromosomal duplications. Irrespective of the explanation for paralogy, it is clear that there are genomic regions which are similar in that each possesses distinctive and therefore recognisable clusters of heterogeneous genes. For example, clusters of MHC genes are found on Chromosomes 1, 9 and 19 as well as 6 (Dawkins et al. 1999).

One region which resembles the MHC in some respects is located on 1q21-25, including the epidermal differentiation complex (EDC). Our interest in this region was provoked by the discovery of the PERB11 (MIC) family within the MHC (Leelayuwat et al. 1994; Bahram et al. 1994). Although PERB11 is similar to HLA Class I genes and HFE – also located within the MHC – there are also similarities to MR1 and CD1, which are located within 1q21-25. The likelihood of substantial and functionally important sharing between these two genomic regions increased recently when it was shown that Pg8 and S, which are tightly clustered just telomeric of HLA-C, have paralogues within the EDC. Pg8 is similar but not identical to trichohyalin (Guillaudeux et al. 1998). S is similar and possibly identical to corneodesmosin (Guerrin et al. 1998). Independently, several groups had suggested that the EDC might be relevant to the pathogenesis of psoriasis (Zhao and Elder 1997; Hardas et al. 1996). Polymorphism of EDC genes might result in abnormalities of epidermal differentiation. Interestingly, Capon et al. (1998) have claimed that 1q21 may be the second most important region controlling psoriasis. Yet further but unexpected evidence in favour of the potential power of paralogous mapping has been provided by Roetto et al. (1999). Juvenile haemochromatosis (JH) is similar to the adult form in many respects but is not MHC associated and mutations of the HFE gene have not been found. In view of the possibilities of a similar gene elsewhere in the genome, Roetto et al. mapped JH to chromosome 1q between D1S442 and D1S2347. CD1 is within this area but could not be implicated. Since the paralogous region has been subjected to multiple duplication events – as in the MHC – it is very likely that there will be copies of other similar genes. In fact, MR1 – which is closer in sequence to HFE than CD1 – has already been mapped nearby (Hashimoto et al. 1999).

The relationships between HFE, MR1 and PERB11 (members of the PHFZ or "1.5" family) are shown in Figure 1 together with 2 examples of CD1. It can be seen that the sequences are almost identical in some sections but each pairwise comparison yields a different conclusion, as would be expected given that the three genes appear to have diverged functionally. We predict that the 1q21-25 paralogous region will contain polymorphic copies of the "1.5" family and that these may explain the associations with psoriasis and juvenile haemochromatosis (Dawkins et al. 1999).

The model

These possibilities suggest to us that it may be possible to simplify the complexity faced by those attempting to explain MHC – disease associations. For example, if the extra-MHC regions actually contain genes also represented to various degrees within the MHC, our working hypothesis would focus on the genes which are shared and would explain how the similar genomic regions could be involved,

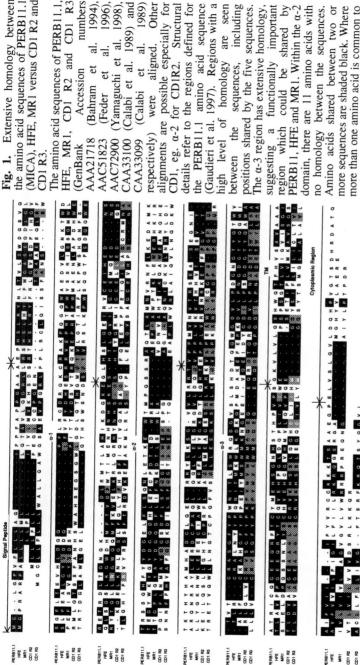

Fig. 1. Extensive homology between the amino acid sequences of PERB11.1, HFE, MR1 versus CD1 R2 and CD1 R3.

The amino acid sequences of PERB11.1, HFE, MR1, CD1 R2 and CD1 R3 (GenBank Accession numbers AAA21718 (Bahram et al. 1994), AAC51823 (Feder et al. 1996), AAC72900 (Yamaguchi et al. 1998), CAA33100 (Calabi et al. 1989) and CAA33099 (Calabi et al. 1989) respectively) were aligned. Other alignments are possible especially for CD1, eg. α-2 for CD1R2. Structural details refer to the regions defined for the PERB11.1 amino acid sequence (Gaudieri et al. 1997). Regions with a high level of homology are seen between the sequences, including positions shared by the five sequences. The α-3 region has extensive homology, suggesting a functionally important region which could be shared by PERB11, HFE and MR1. Within the α-2 domain, there are 11 amino acids with no homology between the sequences. Amino acids shared between two or more sequences are shaded black. Where more than one amino acid is common to at least two sequences, those that are shared with PERB11.1 are shaded black, while those shared between other sequences are shaded gray. Dashes indicate insertions and/or deletions (indels). Blank sections indicate that amino acid sequence is not available.

although not necessarily to the same degree. Indeed, we would predict that in some of the families one paralogous region may be more important than another, possibly explaining why some genome-wide surveys have yielded somewhat conflicting results. Pursuing the model in relation to psoriasis, there is some evidence that the MHC is most important in those patients with an early onset and relatively severe disease (Burden et al. 1998). In some families, genes on chromosome 1q21 appear to be almost as important (Capon et al. 1998). In other families the secondary genes appear to be on chromosome 17 and perhaps elsewhere (Nair et al. 1995). Such findings are consistent with the hypothesis that there is polymorphism within each of the paralogous regions and that the importance of one region or another is a feature of a particular family and possibly the associated environment factors relevant to that particular family.

Paralogy and synteny

If paralogous mapping is valid, it should be possible to extend the strategy by mapping syntenic regions in other species. For example the paralogous clusters on human 6, 9 and 19 are syntenic to mouse 17, rat 20, pig 7 and cow 23. If the shared genes are important in disease, it could be expected that these chromosomes would have been implicated in the corresponding diseases in animals.

We have addressed this approach by comparing multiple sclerosis (MS) with experimental autoallergic encephalomyelitis (EAE), systemic lupus erythematosus (SLE) with the spontaneous NZB disease and insulin dependent diabetes mellitus (IDDM) with the spontaneous diabetes which occurs in the non-obese diabetic (NOD) mouse.

Initially we simply recorded the apparent sharing of genes (putative synteny) between human and mouse chromosomes (Fig. 2). We then indicated which chromosomes have been implicated in disease in either species. It can be seen that numerous chromosomes could be important but that, in man, only six (1, 2, 6, 11, 14, 19) appear to be important in all three diseases. In the mouse, three (4, 7, 17) are relevant to the equivalent models and all three relate to the previously selected human chromosomes (4 to 1, 7 to 19 and 17 to 6). Interestingly, these pairs had already been recorded as being syntenic. The fact that human 6 and murine 17 share MHC genes suggests that the approach is valid. Furthermore, human 1 and 19 are known to be paralogous to the MHC. Kasahara (1999) has also shown a paralogous relationship between chromosomes 1, 9 and 19 which is independent of the MHC (Fig. 3).

This simple approach can be developed but already shows promise as a method for identifying genes and genomic regions which regulate autoimmune diseases.

Given the conservation of the paralogous and syntenic regions (Kasahara et al. 1999) and their prominence, as shown in Figure 2, we propose that there are clusters of genes which regulate or otherwise influence protective immune

| | Mouse | | | | | | | | | | | | | | | | | | | | | | | |
	1	2	3	4	5	6	7	8	9	10	11	12	13	14	15	16	17	18	19	X	Y	MS	SLE	IDDM
H 1	79	1	70	87	2			5		1	1	1	6					1				+	+	+
2	67	41		6	23						7	14			1		9	1				+	+	+
3	1	1	17	1		24		1	47				10			31	1					+		
4		23		63	3		17			1		1						1					+	
5	1										34		49		12		2	25				+		
6	5			6	1		1		6	26			17				97					+	+	+
7		1			49	51	1			10	10	5					1					+		
u 8	5		7	9			21						14	26	3				1					
9		39		41								7						10						
m 10		10			3	10			11				16				3	38						+
a 11		31		1		59		44		1								39				+	+	+
n 12		1	1		15	62	1			36					40								+	
13	2				7			10					18										+	
14					1		1	1	1	1		44	22									+	+	+
15		23				30		30				1												+
16						17	45			7				1	11	14						+	+	
17			1			1		1		158	2		1									+		
18	3				2									1	4	26							+	+
19						1	91	21	8	20	1	1					15					+	+	+
20		70																					+	
21					1					15			1		25	6								
22	1				6	2				8	9				26	26								
X			1							2			1					1	150			+		+
Y																					6			
EAE		+	+	+			+	+			+	+	+	+		+	+	+						
SLE (NZB)	+			+	+	+	+				+		+				+	+						
IDDM (NOD)	+		+	+	+	+		+				+					+							

Fig. 2. Paralogous and syntenic mapping may identify genomic regions which regulate autoimmune responses. This is a modified version of the Oxford grid provided by the Jackson Laboratories at:

http://www.informatics.jax.org/searches/oxfordgrid_form.shtml.

Here we initially highlighted the chromosomes that had greater than 80 similarities between them. A literature search was then performed to determine the chromosomal regions in both the mouse and human that are associated with autoimmune diseases. These regions were added to the grid and the regions associated with all three diseases highlighted. Some of the regions of high chromosomal similarity correlate with the chromosomes that show associations with all three autoimmune diseases.

References used to generate this figure: Alliel et al. 1998; Awata et al. 1997; Barcellos et al. 1997; Butterfield et al. 1999; Butterfield et al. 1998; Chataway et al. 1998; Concannon et al. 1998; Cordell et al. 1998; Croxford et al. 1997; Cucca et al. 1998; Ebers et al. 1996; Encinas et al. 1996; Field et al. 1996; Gaffney et al. 1998; Kono et al. 1994; Kuokkanen et al. 1996; McAleer et al. 1995; Mein et al. 1998; Melanitou et al. 1998; Morel et al. 1994; Moser et al. 1998; Nakagawa et al. 1998; Podolin et al. 1998; Rodrigues et al. 1994; Rothe et al. 1997; Rozzo et al. 1996; Sawcer et al. 1996; Shai et al. 1999; She et al. 1998; Sundvall et al. 1995; Takeoka et al. 1999; Tsao et al. 1997; Vyse et al. 1997; Yui et al. 1996.

6p21-p25	19p13.1-p13.3	1q21-25	9q33-q34
OLFR2[2]	OLFR[2]		OLFR3[2]
ZNF184[2]	ZNF85[2]		
	ZNF91[2]		
Histone family[2]		Histone family[2]	
MOG[1]		MPZ[1]	
S[1]		LOR[1]	
POU5F1[3,4]	POU2F2[3,4]	POU2F1[3,4]	
PERB11[1]		MR1[1]	
MHC class I[1,2]		CD1[1]	
GNL1[3,4]	GNA11/15[3,4]		
TUBB[2]	TUBB5[2]		TUBB2[2]
NTRK4[2]		NTRK1[2]	NTRK2[2]
BAT1[2]	BAT1-like[2]		
LTA[2]	CD70[2]	APT1LG1[2]	CD30L[2]
TNF[2]	CD137L[2]		
LTB[2]	4-1BBL[2]		
AIF1[2]			AIF1-like[2]
CDKN1A[3,4]	CDKN2D[3,4]		
VARS2[1]			VARS1[1]
BAT2[1,2]			BAT2-like[1,2,3]
HSPA1[1,2]		HSPA6/7[1,2]	HSPA5[1,2]
C2[3,4]	C3[1,2]		C5[1,2]
C4A/C4B[1,2]			C8G[3,4]
TNX[1,2]		TNR[1,2]	HXB[1,2]
CREBL1[2]		ATF6[2]	
LPAATA[2]			LPAATB[2]
PBX2[1,2]		PBX1[1,2]	PBX3[1,2]
NOTCH4[1,2]	NOTCH3[1,2]	[NOTCH2][1,2]	NOTCH1[1,2]
TAP1[1,2]			ABC1[3,4]
TAP2[1,2]			ABC2[1,2]
PSMB8/9[1,2]	PSMC1[3,4]		PSMB4[1,2]
RING3[1,2]	HUNKI[2]	[BRDT][2]	RING3L[1,2]
COL11A2[1,2]		[COL11A1][1,2]	COL5A1[1]
RXRB[1,2]		RXRG[1,2]	RXRA[1,2]
RALGDS2[2]			RALGDS[2]
BING1[2]			ZNF-X[2]
RPL12L[2]			RPL[2]
	LMNB2[1]	LAMA[1]	
	CACNL1A4[1]	CACNL1A5[1]	
	TPM4[1,3]	TPM3[1,3]	TPM2[1,3]
	CAPN4[3,4]	CAPN2[3,4]	
	EDG6[3,4]	EDG1[3,4]	
	FCER2[3,4]	FCER1[3,4]	
	LY94[3,4]	LY9[3,4]	
	VAV1[1]		VAV2[1]
	PCSK4[3,4]		PCSK5[3,4]
	STXPB2[3,4]		STXPB1[3,4]
		AK2[1]	AK1/AK3[1,3]
		LMX1A[1,3]	LMX1B[1,3]
		PTGS2[1,3]	PTGS1[1,3]
		TAL1[1,3]	TAL2[1,2]
		SPTA1[1]	SPTAN1[1]
		ABL2[1]	ABL1[1]

Fig. 3. Chromosome location of paralogous MHC genes

Footnotes:

Genes on chromosome 6 are ordered in a telomere to centromere fashion. The order has been derived from the references listed below and is not necessarily in the exact order. Genes on chromosomes 1, 9 and 19 are ordered relative to their paralogous genes on chromosome 6. Where there is no known paralog on 6p21-25 (eg LMNB2) the order is inconsequential. Genes in parentheses are located elsewhere on chromosome 1.

References used to generate this figure :

1 - Dawkins et al. (1999)

2 - Kasahara M (1999)

3 - The Online Mendelian Inheritance in Man (http://www.ncbi/nlm/nih/gov/Omim)

4 - Mouse Genome Informatics (http://www.informatics.jax.org)

responses. These clusters must be polymorphic, at least in vertebrates. Certain haplotypes associated with vigorous responses have been selected and are now found to be important in the development of autoimmune disease. It remains to be determined which components are directly important but the data summarised here suggest that they will be components of, or close to, the gamma (or complement) block in the human MHC and that they will be represented in the primordial pre HLA MHC.

A corollary of this hypothesis is that other genes, including other MHC genes, are important in determining <u>which</u> autoimmune disease actually develops. We include indels, such as HERVs, within this category.

Genomic sequencing of haplotypes

In attempting to determine the importance of paralogous regions, it has become abundantly clear that insufficient attention has been given to complete genomic sequencing of alternative haplotypes. This work is progressing within the MHC, and has revealed very substantial differences. For example, two Caucasoid haplotypes, 7.1 and 8.1, are similar in some respects but vastly different in others. This, when taken together with the different disease associations, raises possible candidates. Unfortunately, similar detailed studies have not been done in other paralogous regions and until haplospecific sequences are available, substantial reservations are necessary.

Conclusion

We have considered the possibility that paralogous regions may explain some of the apparent complexity surrounding explanations for MHC disease associations. There appear to be some plausible examples of the phenomenon, but further information on the genetic control of relevant functions is required. The extent of polymorphism within paralogous regions must be defined. Distinct haplotypes must be identified and extensive genomic sequencing of these haplotypes will be necessary. In the meanwhile, we propose paralogous and syntenic mapping as a potentially useful approach to the development of novel working hypotheses.

Acknowledgements

We thank M. Kasahara, H. Inoko, T. Gojobori and P. Pontarotti for helpful discussions. Supported by the National Health and Medical Research Council, the Immunogenetics Research Foundation, Royal Perth Hospital, Sir Charles Gairdner Hospital and PathCentre. Publication 9915 of the Centre for Molecular Immunology and Instrumentation of the University of Western Australia.

References

Alliel PM, Perin JP, Belliveau J, Pierig R, Nussbaum JL, Rieger F (1998) [Endogenous retroviral sequences analogous to that of the new retrovirus MSRV associated with multiple sclerosis (part 1)]. Comptes Rendus de l Academie des Sciences - Serie Iii, Sciences de la Vie 321:495-499

Awata T, Kurihara S, Kikuchi C, Takei S, Inoue I, Ishii C, Takahashi K, Negishi K, Yoshida Y, Hagura R, Kanazawa Y, Katayama S (1997) Evidence for association between the class I subset of the insulin gene minisatellite (IDDM2 locus) and IDDM in the Japanese population. Diabetes 46:1637-1642

Bahram S, Bresnahan M, Geraghty DE, Spies T (1994) A second lineage of mammalian major histocompatibility complex class I genes [see comments]. Proc Nat Acad Sci USA 91:6259-6263

Barcellos L, Klitz W, Field L, Tobias R, Bowcock A, Wilson R, Nelson M, Nagatomi J, Thomson G (1997) Association mapping of disease loci, by use of a pooled DNA genomic screen. Am J Hum Genet 61:734-747

Burden AD, Javed S, Bailey M, Hodgins M, Connor M, Tillman D (1998) Genetics of psoriasis: Paternal inheritance and a locus on chromosome 6p. Invest Dermatol 110:958-960

Butterfield RJ, Blankenhorn EP, Roper RJ, Zachary JF, Doerge RW, Sudweeks J, Rose J, Teuscher C (1999) Genetic analysis of disease subtypes and sexual dimorphisms in mouse experimental allergic encephalomyelitis (EAE): relapsing/remitting and monophasic remitting/nonrelapsing EAE are immunogenetically distinct. J Immunol 162:3096-3102

Butterfield RJ, Sudweeks JD, Blankenhorn EP, Korngold R, Marini JC, Todd JA, Roper RJ, Teuscher C (1998) New genetic loci that control susceptibility and symptoms of experimental allergic encephalomyelitis in inbred mice. J Immunol 161:1860-1867

Calabi, F, Jarvis J, Martin L, Milstein C (1989) Two classes of CD1 genes. Eur J Immunol 19:285-292

Capon F, Novelli G, Semprini S, Clementi M, Nudo M, Vultaggio P, Mazzanti C, Gobello T, Botta A, Fabrizi G, Dallapiccola B (1998) Searching for psoriasis susceptibility genes in Italy: genome scan and evidence for a new locus on chromosome 1. J Invest Dermat 112:101-104

Chataway J, Feakes R, Coraddu F, Gray J, Deans J, Fraser M, Robertson N, Broadley S, Jones H, Clayton D, Goodfellow P, Sawcer S, Compston A (1998) The genetics of multiple sclerosis: principles, background and updated results of the United Kingdom systematic genome screen. Brain 121:1869-1887

Concannon P, KJ. G-E, Hinds D, Wapelhorst B, Morrison V, Stirling B, Mitra M, Farmer J, Williams S, Cox N, Bell G, Risch N, Spielman R (1998) A second-generation screen of the human genome for susceptibility to insulin-dependent diabetes mellitus. Nat Genet 19:292-296

Cordell HJ, Todd JA, Lathrop GM (1998) Mapping multiple linked quantitative trait loci in non-obese diabetic mice using a stepwise regression strategy. Genet Res 71:51-64

Croxford JL, O'Neill JK, Baker D (1997) Polygenic control of experimental allergic encephalomyelitis in Biozzi ABH and BALB/c mice. J Neuroimmunol 74:205-211

Cucca F, Guy J, Kawaguchi Y, Esposito L, Merriman M, Wilson A, Cordell H, Bain S, Todd J (1998) A male-female bias in type 1 diabetes and linkage to chromosome Xp in MHC HLA-DR3-positive patients. Nat Genet 19:301-302

Dawkins R, Leelayuwat C, Gaudieri S, Tay G, Hui J, Cattley S, Martinez P, Kulski J (1999) Genomics of the major histocompatibility complex: haplotypes, duplication, retroviruses and disease. Immunol Rev 167:275-304

Ebers GC (1996) Genetic epidemiology of multiple sclerosis. Curr Opin Neurol 9:155-158

Feder, JN, Gnirke A, Thomas W, Tsuchihashi Z, Ruddy DA, Basava A, Dormishian F, Domingo R Jr, Ellis MC, Fullan A, Hinton LM, Jones NL, Kimmel BE, Kronmal GS, Lauer P, Lee VK, Loeb DB, Mapa FA, McClelland E, Meyer NC, Mintier GA, Moeller N, Moore T, Morikang E, Wolff RK, et al. (1996) A novel MHC class I-like gene is mutated in patients with hereditary haemochromatosis. Nat Genet 13:399-408

Field L, Tobias R, Thomson G, Plon S (1996) Susceptibility to insulin-dependent diabetes mellitus maps to a locus (IDDM11) on human chromosome 14q24.3-q31. Genomics 33:1-8

Gaffney P, Kearns G, Shark K, Ortmann W, Selby S, Malmgren M, Rohlf K, Ockenden T, Messner R, King R, Rich S, Behrens T (1998) A genome-wide search for susceptibility genes in human systemic lupus erythematosus sib-pair families. Proc Natl Acad Sci USA 95:14875-14879

Gaudieri S, Leelayuwat C, Townend DC, Mullberg J, Cosman D, Dawkins RL (1997) Allelic and interlocus comparison of the PERB11 multigene family in the MHC. Immunogenetics 45:209-216

Guerrin M, Simon M, Montezin M, Haftek M, Vincent C, Serre G (1998) Expression cloning of human corneodesmosin proves its identity with the product of the S gene and allows improved characterization of its processing during keratinocyte differentiation. J Biol Chem 273:22640-22647

Guillaudeux T, Janer M, Wong GK, Spies T, Geraghty DE (1998) The complete genomic sequence of 424,015 bp at the centromeric end of the HLA class I region: gene content and polymorphism. Proc Natl Acad Sci USA 95:9494-9499

Hardas BD, Zhao X, Zhang J, Longqing X, Stoll S, Elder JT (1996) Assignment of psoriasin to human chromosomal band 1q21: coordinate overexpression of clustered genes in psoriasis. J Invest Dermatol 106:753-758

Hashimoto K, Okamura K, Yamaguchi H, Ototake M, Nakanishi T, Kurosawa Y (1999) Conservation and diversification of MHC class I and its related molecules in vertebrates. Immunol Rev 167:81-100

Kasahara M (1997) New insights into the genomic organization and origin of the major histocompatibility complex: role of chromosomal (genome) duplication in the emergence of the adaptive immune system. Hereditas 127:59-65

Kasahara M (1999) The chromosomal duplication model of the major histocompatibility complex. Immunol Rev 167:17-32

Katsanis N, Fitzgibbon J, Fisher EMC (1996) Paralogy mapping: identification of a region in the human MHC triplicated onto human chromosomes 1 and 9 allows the prediction and isolation of novel PBX and NOTCH loci. Genomics 35:101-108

Kono DH, Burlingame RW, Owens DG, Kuramochi A, Balderas RS, Balomenos D, Theofilopoulos AN (1994) Lupus susceptibility loci in New Zealand mice. Proc Natl Acad Sci USA 91:10168-10172

Kuokkanen S, Sundvall M, Terwilliger J, Tienari P, Wikstrom J, Holmdahl R, Pettersson U, Peltonen L (1996) A putative vulnerability locus to multiple sclerosis maps to 5p14-p12 in a region syntenic to the murine locus Eae2. Nat Genet 13:477-480

Leelayuwat C, Townend DC, Degli-Esposti MA, Abraham LJ, Dawkins RL (1994) A new polymorphic and multicopy MHC gene family related to nonmammalian class I [published erratum appears in Immunogenetics 41:174]. Immunogenetics 40:339-351

McAleer M, Reifsnyder P, Palmer S, Prochazka M, Love J, Copeman J, Powell E, Rodrigues N, Prins J, Serreze D, et al. (1995) Crosses of NOD mice with the related NON strain. A polygenic model for IDDM. Diabetes 44:1186-1195

Mein C, Esposito L, Dunn M, Johnson G, Timms A, Goy J, Smith A, Sebag-Montefiore L, Merriman M, Wilson A, Pritchard L, Cucca F, Barnett A, Bain S, Todd J (1998) A search for type 1 diabetes susceptibility genes in families from the United Kingdom. Nat Genet 19:297-300

Melanitou E, Joly F, Lathrop M, Boitard C, Avner P (1998) Evidence for the presence of insulin-dependent diabetes-associated alleles on the distal part of mouse chromosome 6. Genome Res 8:608-620

Morel L, Rudofsky UH, Longmate JA, Schiffenbauer J, Wakeland EK (1994) Polygenic control of susceptibility to murine systemic lupus erythematosus. Immunity 1:219-229

Moser K, Neas B, Salmon J, Yu H, Gray-McGuire C, Asundi N, Bruner G, Fox J, Kelly J, Henshall S, Bacino D, Dietz M, Hogue R, Koelsch G, Nigtingale L, Shaver T, Abdou N, Albert D, Carson C, Petri M, Treadwell E, James J, Harley J (1998) Genome scan of human systemic lupus erythematosus: evidence for linkage on chromosome 1q in African-American pedigrees. Proc Natl Acad Sci USA 95:14869-14874

Nair RP, Guo SW, Jenisch S, Henseler T, Lange EM, Terhune M, Westphal E, Christophers E, Voorhees JJ, Elder JT (1995) Scanning chromosome 17 for psoriasis susceptibility: lack of evidence for a distal 17q locus. Hum Hered 45:219-230

Nakagawa Y, Kawaguchi Y, Twells R, Muxworthy C, Hunter K, Wilson A, Merriman M, Cox R, Merriman T, Cucca F, McKinney P, Shield J, Tuomilehto J, Tuomilehto-Wolf E, Ionesco-Tirgoviste C, Nistico L, Buzzetti R, Pozzilli P, Joner G, Thorsby E, Undlien D, Pociot F, Nerup J, Ronningen K, Todd J, et al (1998) Fine mapping of the diabetes-susceptibility locus, IDDM4, on chromosome 11q13. Am J Hum Genet 63:547-556

Podolin PL, Denny P, Armitage N, Lord CJ, Hill NJ, Levy ER, Peterson LB, Todd JA, Wicker LS, Lyons PA (1998) Localization of two insulin-dependent diabetes (Idd) genes to the Idd10 region on mouse chromosome 3. Mamm Genome 9:283-286

Rodrigues N, Cornall R, Chandler P, Simpson E, Wicker L, Peterson L, Todd J (1994) Mapping of an insulin-dependent diabetes locus, Idd9, in NOD mice to chromosome 4. Mamm Genome 5:167-170

Roetto A, Totaro A, Cazzola M, Cicilano M, Bosio S, D'Ascola G, Carella M, Zelante L, Kelly AL, Cox TM, Gasparini P, Camaschella C (1999) Juvenile Hemochromatosis Locus Maps to Chromosome 1q. Am J Hum Genet 64:1388-1393

Rothe H, Jenkins NA, Copeland NG, Kolb H (1997) Active stage of autoimmune diabetes is associated with the expression of a novel cytokine, IGIF, which is located near Idd2. J Clin Invest 99:469-474

Rozzo SJ, Vyse TJ, Drake CG, Kotzin BL (1996) Effect of genetic background on the contribution of New Zealand black loci to autoimmune lupus nephritis. Proc Natl Acad Sci USA 93:15164-15168

Sawcer S, Jones HB, Feakes R, Gray J, Smaldon N, Chataway J, Robertson N, Clayton D, Goodfellow PN, Compston A (1996) A genome screen in multiple sclerosis reveals susceptibility loci on chromosome 6p21 and 17q22 [see comments]. Nat Genet 13:464-468

Shai R, Quismorio FJ, Li L, Kwon O, Morrison J, Wallace D, Neuwelt C, Brautbar C, Gauderman W, Jacob C (1999) Genome-wide screen for systemic lupus erythematosus susceptibility genes in multiplex families. Hum Mol Genet 8:639-644

She J, Marron M (1998) Genetic susceptibility factors in type 1 diabetes: linkage, disequilibrium and functional analyses. Curr Opin Immunol 10:682-689

Sundvall M, Jirholt J, Yang HT, Jansson L, Engstrom A, Pettersson U, Holmdahl R (1995) Identification of murine loci associated with susceptibility to chronic experimental autoimmune encephalomyelitis. Nat Genet 10:313-317

Takeoka Y, Taguchi N, Kotzin BL, Bennett S, Vyse TJ, Boyd RL, Naike M, Konishi J, Ansari AA, Shultz LD, Gershwin ME (1999) Thymic microenvironment and NZB mice: the abnormal thymic microenvironment of New Zealand mice correlates with immunopathology. Clin Immunol 90:388-398

Tsao B, Cantor R, Kalunian K, Chen C, Badsha H, Singh R, Wallace D, Kitridou R, Chen S, Shen N, Song Y, DA. I, Yu C, Hahn B, Rotter J (1997) Evidence for linkage of a candidate chromosome 1 region to human systemic lupus erythematosus. J Clin Invest 99:725-731

Vyse TJ, Rozzo SJ, Drake CG, Izui S, Kotzin BL (1997) Control of multiple autoantibodies linked with a lupus nephritis susceptibility locus in New Zealand black mice. J Immunol 158:5566-5574

Yamaguchi H, Kurosawa Y, Hashimoto K (1998) Expanded genomic organization of conserved mammalian class I-related genes, human MR1 and its murine ortholog. Biochem Biophys Res Commun 250:558-564

Yui M, Muralidharan K, Moreno-Altamirano B, Perrin G, Chestnut K, Wakeland E (1996) Production of congenic mouse strains carrying NOD-derived diabetogenic genetic intervals: an approach for the genetic dissection of complex traits. Mamm Genome 7:331-334

Zhao XP, Elder JT (1997) Positional cloning of novel skin-specific genes from the human epidermal differentiation complex. Genomics 45:250-258

Transposable elements and the metamerismatic evolution of the HLA class I region

Jerzy K. Kulski[1], Silvana Gaudieri[2], and Roger L. Dawkins[1]

[1]Centre for Molecular Immunology and Instrumentation, The University of Western Australia, Nedlands 6009, Western Australia
[2]Centre for Information Biology, National Institute of Genetics, Mishima 411-8540, Japan

Summary. We have analysed continuous genomic sequence of 1.8 Mb from the HLA class I region of the MHC with a view to understanding the evolution, organisation and sequential interrelationships between members of the multicopy HLA class I and PERB11 (MIC) gene families, human endogenous retroviruses (HERVs) and retroelements that are distributed within this region. Analysis and mapping of genomic sequence from PERB11.2 (MICB) to HLA-F has revealed that the multicopy HLA and PERB11 (MIC) sequences, HERV-16 (P5 gene family) and associated retroelements such as Alu, LTR, MER and L1 are contained within repeated segments that can be classified into at least 6 groups based on the distinctive features of paralogous transposable elements. Most of these segments appear to have evolved from a basic duplication unit or duplicon composed of a HLA class I, HERV-16 (P5) and PERB11 (MIC) sequence, and the associated retroelements. Exponential amplification of duplicons by diversifying single and multisegmental duplications has resulted in many copies of pseudogenes and gene fragments, and three subgenomic blocks (alpha, beta, and kappa) that differ in the number, orientation and complexity of duplicons. Retroelements, particularly HERV-16, are closely associated with the breakpoints within and between duplicons (Kulski et al 1999b), suggesting that they have had a major role in the spread and diversity of the multicopy gene families.

From our analyses we conclude that the HLA class I genomic region is a metameric design of three distinct subgenomic blocks that are characterised by the presence of HLA class I, HERV-16 and PERB11 (MIC) sequences and distinctive retroelements. The blocks have evolved metamerismatically by the expansion and contraction of duplicons involving retroelements and basic recombination processes such as duplications, insertions, deletions and translocations. In this context, we also consider the distribution of many transposable elements within the MHC as "bandaids' or "scars" brought about in response to genomic stress or damage.

Key words. Transposable elements, HLA, Evolution, Duplication, Duplicons

Introduction to metameres, metamerism and duplicons in the MHC

> The basic principle of arthropod design is <u>metamerism</u>, the construction of the body from an extended series of repeated segments. The key to arthropod diversification lies in recognizing that an initial form composed of numerous nearly identical segments can evolve by reduction and fusion of segments, and by specialization of initially similar parts on different segments, into the vast array of divergent anatomies seen in advanced arthropods. Stephen Jay Gould (1989, page 103).

Different analogies and models have been used to describe the complex genomic organisation and evolution of the multicopy gene families within the MHC. These include the expansion and contraction of MHC genes like the bellows of an accordion (Klein et al. 1998), the explosive outcome of a big bang (Abi Rached et al. 1999), the "birth and death" process (Nei et al. 1997), or the wear and tear of "native beads" (Dawkins et al. 1999). Although each of the models has its own particular merit, in this paper we examine and explain the organisation and evolution of the HLA genes within the class I region according to the basic principle of metamerism. The design is metameric where the genomic region is largely composed of an extended series of duplicated segments analogous to the body design of arthropods (Gould 1989). The metameres in zoology are structures of a longitudinal series of similar segments making up the body of an arthropod, whereas in genomics they are a varying number of common repeating units or segments that make up one specific region or different chromosomal regions. In this regard, the evolution of the HLA class I region can be seen as having originated by expansions (duplications, insertions, and translocations) and contractions (deletions and translocations) of nearly identical genomic segments termed here as <u>duplicons</u>. Therefore, the design of the HLA class I region is <u>metameric</u> while its evolution is <u>metamerismatic</u> when each duplicon is further expanded or contracted by different DNA processes. Micromeric and macromeric events involving small (<10 kb) and large (>10 kb) genomic sequences such as introns, exons, retroelements, pseudogenes and genes have added substantially to the diversity of the genomic regions already containing multimeric duplications (Gaudieri et al. 1997a, b, 1999; Kulski et al. 1997, 1999a, b).

DNA sequence and analysis

The entire genomic sequence of the HLA class I region of the MHC encompassing 1.8 Mb from PERB11.2 (MICB) to HLA-F was obtained from WWW, the University of Washington Genome Center (Geraghty DE and Olson MV 1998) at http://www.genome.washington.edu/UWGC/hla/hlaseqsum.asp for the analysis of the alpha, beta and kappa blocks. Sequences for analysis of the beta and alpha blocks were also obtained from DDBJ/EMBL/GenBank as accession numbers AB000882 and D84394 (Mizuki et al. 1997; Shiina et al. 1998) and AF55066 (submitted in 1998 by A. Hampe from CNRS, Rennes, France), respectively.

The genomic lineages and maps presented in this paper are based on dot plot matrix comparisons and phylogenetic analyses of sequences using computer programs previously described (Gaudieri et al. 1997a, b, 1999; Kulski et al. 1997, 1999a, b). Repeat elements, such as HERVs, Alu, L1, LTRs, and MERs, within the contiguous sequences were identified using the programs CENSOR (Jurka et al. 1996) or RepeatMasker (A.F.A. Smit and P. Green) that are available on the World Wide Web (http://charon.girinst.org/~server/censor.html, and http://ftp.genome.washington.edu/cgi-bin/RepeatMasker, respectively).

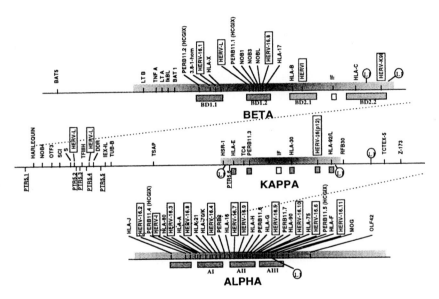

Fig. 1. Genomic map of the location of members of different multicopy gene families and HERVs within subgenomic blocks, beta, kappa and alpha. The map is modified from those presented by Dawkins et al. (1999) and Shiina et al. (1999) and is based in part on the ancestral haplotype 62.1 (HLA-A2, -B62, -Cw10, -DR4). The location of HERVs, full-length L1 and pTRC5 sequences are boxed, circled and underlined, respectively. The duplicons (extended regions of duplication) produced by different duplication events, such as BD1 and BD2 in the beta block, are shown as rectangles and squares below the blocks. The relationship between duplicons within the kappa block are unknown and therefore are unlabelled. The duplicons within the alpha block have probably originated by multisegmental duplication and are shown as tripartite segments (AI to AIII) where each segment contains 3 duplicons arranged according to their predicted histories (see Kulski et al. 1999b).

Organisation of HLA class I region into three genomic blocks

The HLA class I region of the MHC is composed of duplicated members of different multicopy gene families including HLA (human leukocyte antigen) class I (Geraghty et al. 1992), PERB11 (Perth beta block transcript 11) or MIC (MHC class I –related chain) (Bahram et al. 1994; Leelayuwat et al. 1994), human endogenous retrovirus type 16 (Kulski and Dawkins 1999) or P5 (Vernet et al. 1993), and hemochromatosis candidate genes (HCGs) II, IV, VIII and IX (Pichon et al. 1996) distributed over 1.8 Mb of continuous sequence (Shiina et al. 1999). It was previously recognised that these multicopy gene families are organised into two subgenomic regions or polymorphic frozen blocks, designated as the alpha and beta blocks, with each block (250 to 350 kb in length) reflecting substantial polymorphism and a common ancestral and duplicative history (Marshall et al. 1993; Leelayuwat et al. 1995; Pichon et al. 1996; Dawkins et al. 1999). Another region, provisionally referred to as the kappa block, is located between the alpha and beta blocks, and it contains the PERB11.3 (MICC) and HLA-E and –92 gene sequences. The kappa block has received less attention in the past, possibly due in part to a lack of sufficient continuous genomic sequence for DNA analysis, and it is not clear that there is sufficient polymorphism to justify the designation as a polymorphic frozen block.

Figure 1 is a genomic map of the members of the different multicopy gene families and HERVs located within the blocks, alpha, beta and kappa. The subgenomic regions harbouring duplicons produced by two different duplication events within the beta block, and multisegmental duplication events within the alpha block are highlighted. Together, the alpha and beta blocks contain at least 6 copies of PERB11 (MIC), 12 copies of HERV-16 (P5), and 15 copies of full-length and fragmented HLA class I sequences within about 574 kb of highly polymorphic sequence of the haplotypes that have been examined so far (Gaudieri et al. 1999). In comparison, the kappa block contains HLA-E, PERB11.3 (MICC), HLA-30 and -92 within a region of about 251 to 235 kb separated from the alpha and beta blocks by 247 kb and 737 kb of sequence, respectively (Shiina et al. 1999). Numerous coding and non-coding sequences of transcriptional factors, RNA helicases and ribosomal proteins are found between the blocks including OTF3, SC1, S, DDR, and TNF-alpha stimulated ABC protein genes (Shiina et al. 1999). The organisation of the duplicated sequences within the beta block is in the opposite orientation to the duplicated sequences within the alpha and the kappa blocks. It is evident from the organisation and location of the members of different multicopy gene families that each block has evolved separately from intra-block duplications of one or more duplicons.

The basic unit of duplication

Analysis of genomic sequence from the alpha and beta blocks of the MHC has revealed that the PERB11 (MIC), HCG and HLA class I gene sequences, pseudogenes and gene fragments are contained within distinct segments or duplicons. When each duplicon was compared for gene and retroelement patterns, at least 4 distinct groups (A to D) were recognised to have evolved from a single ancestral unit or primordial HLA duplicon containing an HLA and PERB11 (MIC) gene, an endogenous retrovirus (HERV-16) sequence, HCG genes II, IV, VIII, and IX, and distinctive Alu sequences and other retroelements (Kulski et al. 1999b). The structure of the primordial duplicon reconstructed from the paralogous regions of the rearranged duplicons is shown in Figure 2.

Classification of HLA class I duplicons by structural categories

Large indels, mainly transposable elements such as Alu, MER, LTR, L1 and HERVs, are a major distinguishing feature between duplicons (Gaudieri et al. 1997a, b; Kulski et al. 1997, 1999b). Many transposable elements, since their integration into primordial or ancient duplicons, have been retained whereas others have been deleted either during or following a series of duplication events. Different transposable elements that have been inserted after each new round of duplication can be used as "specific" markers to differentiate between duplicons. For example, the phylogenetically related category A duplicons (HLA-80, -16, -90) contain insertions and deletions of Alu J1 and MER21B, category B duplicons (HLA-J, -70, -G, -75, -F) have Alu J3 and J4, category C duplicons (HLA-A and –H) have gained Alu Y and MER9 elements, and retained Alu J3 but lost Alu J4, while category D duplicons (HLA-X and –17) have features characteristic of their position within the beta block (Kulski et al. 1999b) as well as having retained PERB11 (MIC) and some of the paralogous Alu F elements found within category B and C duplicons (Fig. 2). Although HLA-16 is part of the group A genomic category (Fig. 2), it has lost Alu J1 at some stage during or following its origin.

We have now classified the HLA duplicons, including those shown in Figure 2, into 6 structural categories A to F on the basis of retroelement patterns and shared Alu elements (Kulski et al. 1999b). The HLA-B and -C duplicons were placed into category E because they have sequence and retroelement patterns that are markedly different to the other HLA duplicons illustrated in Figure 2. However, the HLA-B and –C duplicons can still be considered to have evolved from a primordial duplication unit by simply proposing an early deletion of the PERB11 (MIC) and HERV-16 sequences from the HLA-B/HLA-C preduplication lineage. The characteristic features of the HLA-B and -C duplicons shown in Figure 3 have been previously described by Kulski et al. (1997) and Gaudieri et al. (1997b).

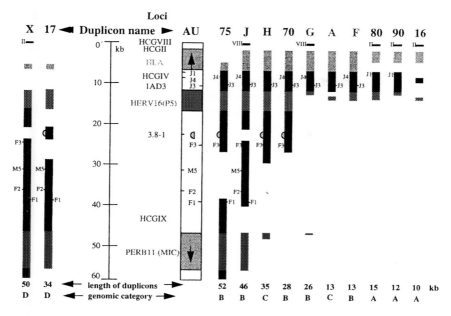

Fig. 2. HLA class I duplicons of categories A to D are duplication and deletion products originating from the same ancestral unit (AU). The hypothetical AU sequence is shown as a column labelled AU with the location of genes, pseudogenes and HERV-16 (P5) listed on the left side of the column under "Loci". The positions of the different paralogous Alu J and Alu F elements, and MER5B (M5) sequences are shown within the column. The direction of transcription for the HLA class I and PERB11 (MIC) genes is shown by the vertical arrows within the shaded blocks of the AU column. The duplicons are shown as vertical lines in relation to the AU. The duplicons from the alpha and the beta blocks are positioned on the right and left side of the AU, respectively. Sequences deleted from duplicons during evolution are represented by the gaps within the vertical lines. The gaps are positioned relative to the sequences that have been retained by some duplicons and therefore are included as part of the AU. The HLA duplicons are also presented according to genomic categories (A to D) and sequence length (10 to 52 kb) and not according to their sequential position in the genome (see Fig. 1). The estimated length of 60 kb for the AU is hypothetical and calculated from a composite of the duplicons HLA-X, -17, -75, -J and -H. The paralogous Alu J elements J1, J3, and J4 have been labelled using the same nomenclature as Kulski et al. (1999b). The paralogous Alu F elements, which probably preceded the origin of Alu J elements (Kapitonov and Jurka 1996) are shown as F1 to F3 within duplicons and the AU. The half circles represent the duplicated 3.8-1 sequence, and M5 represents the DNA transposon MER5B. Many other transposable elements contained within the duplicons have been omitted from the figure for simplicity of presentation.

The HLA-X duplicon fits into both category D and E. The HLA-X duplicon was placed into category D because it contains the PERB11 (MIC) gene as well as the paralogous Alu F elements found within duplicons of the alpha block (Fig. 2). However, the HLA-X duplicon has also retained Alu Jr2AN, Alu Jr3AN, LTR9 and MER21B sequences that are features of the category E duplicons (Fig. 3).

The HLA duplicons within the kappa block are less easily recognised as having arisen from the primordial duplicon because of the disruptions within the linear sequence of the HLA class I genes, HERV-16, PERB11 (MIC) and associated retroelements. If PERB11 (MIC) and HERV16 sequences were originally associated with HLA-E, -92 and -30 then most of them have been deleted from the kappa block with only the fragments of PERB11.3 (MICC) and HERV-16 (P5.12)

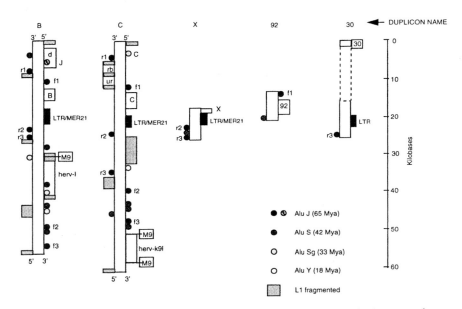

Fig. 3. Paralogous Alu J elements show the evolutionary relationship between the category E duplicons. The HLA-B, -C, -X, -92, and -30 duplicons are in category E on the basis of shared Alu J elements shown as solid circles and labelled f1, f2, f3, r1, r2 and r3 that correspond to the paralogous Alu J elements reported previously by Kulski et al. (1997) as f1AN, f2AN, f3AN, r1AN, r2AN and r3AN, respectively. The HLA-X duplicon fits into a mixed category because it also shows the characteristics of a category D duplicon in Figure 2. The shaded and open circles represent post duplication Alu elements of different subfamilies as indicated in the figure key. The shaded rectangles are fragmented or full-length L1 sequences. The open rectangles are the locations of pseudogenes, dihydrofolate reductase (d), human endogenous retrovirus type I (HERV-I) on duplicon HLA-B, ribosomal protein L3 (rb), and ubiquitin protease (ur) and HERV-K9I on duplicon HLA-C. The solid rectangles include LTR9 and in some cases MER21B.

remaining. Alternatively, the 3 HLA class I sequences, and the single PERB11 (MIC) and HERV-16 fragments within the kappa block have evolved by different mechanisms to those described for the alpha and beta blocks. We have found that HLA-92 contains at least one Alu J element (AluJ.f1AN) that is paralogous to those found within the HLA-B and -C duplicons, thereby linking the kappa block more closely to the beta than to the alpha block (Fig. 3) and the category E duplicons. It is difficult to categorize HLA-30 into any one of the five duplicon groups because all that remains of this sequence is a genomic fragment of exon 3. However, 10-14 kb of genomic sequence (IF) that is located telomeric of HLA-30 has homology to part of a 30-kb intervening fragment that we had previously found between the HLA-B and -C duplicons (Kulski et al. 1997). In addition, the duplicated AluJ.r3AN and LTR26 retroelements within the HLA-B, -C and –X duplicons of the beta block were found within a position 15 kb from HLA-30 and 51 kb from PERB11.3 (MICC) within the kappa block (see Figs. 1 and 3).

The HLA-E gene sequence has a number of features that are different to the other HLA duplicons, and therefore, it has been placed alone in category F. The most striking features of the HLA-E gene in comparison to other classical and non classical class I HLA genes, and shown in Figure 4, are the Alu Sg element in intron 5 and the Alu Y element and MSTc retroelements within the 3'UTR coding region. Other features associated with the HLA-E gene region include variable numbers and subtypes of Alu elements within the 5' enhancer/promoter region that may have caused the deletion of enhancer A and interferon response sequences, and the insertion of a full-length L1 sequence within close proximity. The coding regions of some other HLA class I sequences are also closely associated with retroelements (Fig. 4). Alu J1 has been associated with the deletion of exons 1 to 3 from HLA-80, -16 and –90, an Alu Y element is found close to the 5'UTR of HLA-F and the 3'UTR of HLA-G, and retroelements are present within some of PERB11 (MIC) coding regions. The insertion of retroelements near to and within HLA-E coding regions and the fragmentary nature of the other HLA class I, PERB11 (MIC) and HERV-16 sequences suggests that the kappa block has been subjected to intense sequence rearrangements and recombinations different from those of the beta and alpha blocks. In addition, although the effects of the intra-gene retroelement insertions on the expression of non classical class I genes HLA-E, -F and -G have not been investigated (see Fig. 4), retroelements could be expected a priori to contribute to transcription and translation patterns and regulatory responses different to those observed for the classical class I genes, HLA-A, -B, and -C.

Metamerismatic expansion of duplicons within the genomic blocks

If the primordial HLA duplicon had been duplicated and maintained perfectly in tandem, an array of repeat units within a single block would be expected to have

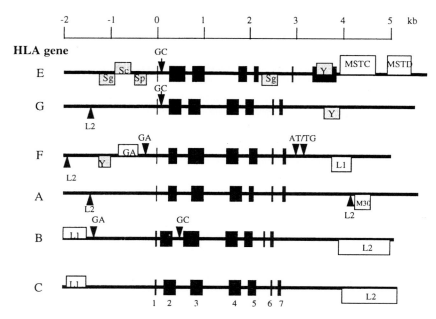

Fig. 4. Alu repeats and other retroelements associated with classical and nonclassical HLA class I genes. HLA-A, -F, -G genes were taken from accession number AF55066, HLA-B and -C from accession number D84394, and HLA-E from accession number AC006139. All classical (HLA-A, -B, -C) and non classical (HLA-E, -F and -G) HLA class I gene sequences were compared against HLA-A26 mRNA (accession number D32131), HLA-B55 mRNA (accession number M77778) and HLA-E (accession number M20022) to determine the position of the exons (shown as solid vertical lines and numbered 1 to 7 for HLA-C). The position of Alu elements are shown as shaded boxes labelled Sc, Sp, Sg or Y according to their subfamily designations. The location of other retroelements are shown as open boxes or vertical arrows labelled L1, L2, M30 (MER30) or MST. Vertical arrows also show the position of simple repeat regions labelled as GA, AT, TG and GC rich regions.

Table 1. Different properties of the alpha, beta and kappa blocks

Block	Pseudogene/ functional[a]	Segments (No.)	Retroelements %[b]	Paralogous retroelements (%)	Pattern of duplication	Human indel polymorphisms[c]
Alpha	4	At least 10	39.8	57.7	Single, multisegmental and transpositions	Several
Beta	0.5	4	49.7	28.5	Single	At least one
Kappa	3	At least 3	50.6	10.3	Single	?

[a] For HLA and MIC (PERB11) members.
[b] From RepeatMasker output.
[c] Venditti and Chorney (1992); Watanabe et al (1997); Komatsu-Wakui et al (1999).

occurred as shown in Fig. 5A. By contrast, the organisation of the HLA class I duplicons within the class I region reflects a more complex and imperfect process that has resulted from duplication, translocation and inversion of duplicon sequences, and numerous deletions/insertions within duplicons, as illustrated in Fig. 5B. Three distinct blocks have been generated with each block reflecting a separate evolutionary history. A summary of the characteristics of each block is presented in Table 1 that has been expanded from the one published by Gaudieri et al. (1999) in order to include the kappa block.

The metamerismatic evolution of the HLA class I region can be explained by four basic steps shown in the simplified scheme of Fig. 5B:

1. A HERV-16 sequence was inserted into a progenitor duplicon composed of an HLA-class I and PERB11 (MIC) gene probably early in primate history about the time when Alu J elements were emerging. Although we cannot exclude the possibility that some duplicons lacking an HERV-16 insertion also existed at this stage of evolution, we have omitted them from our model for simplicity. The origin of HERV-16 is presently not known, and so far has been identified only in the human genome. However, HERV-16 is closely related in sequence to HERV-L and MERV-L that are widely distributed within the human and mouse genomes, respectively (Kulski and Dawkins 1999; Kulski et al. 1999a). The PERB11 (MIC) and HLA-class I sequences are structurally and phylogenetically related but their progenitor, and their time of origin and divergence is unknown (Dawkins et al. 1999).

2. The primordial duplicon containing the HLA class I, HERV-16 and PERB11 (MIC) -like sequences was then duplicated and the products transposed in the same orientation to form the "seeds" for the generation of the kappa and alpha block regions.

3. The HLA class I, HERV-16 and PERB 11 (MIC) sequences within the beta block are in the opposite orientation to those in the alpha and kappa blocks suggesting that one of the ancient duplicons had undergone an inversion and translocation. This step is likely to have involved a translocation from the kappa to the beta block because there are paralogous Alu J elements shared between the duplicons of HLA-X, -17, -B, -C, -30 and –92 but not with any of the duplicons of the alpha block (compare Figs. 2 and 3). The evidence for the different stages of duplication within the beta block has been presented in more detail elsewhere (Gaudieri et al. 1997a, 1999; Kulski et al. 1997, 1999b).

4. A model for the exponential amplification of duplicons within the alpha block by a series of diversifying multisegmental duplications starting during the period of Alu S activity has been presented elsewhere (Kulski et al. 1999b). This model was proposed to account for the present physical separation of closely related sequences and genes elucidated in part from phylogenetic comparisons and by using Alu sequences and other transposable elements as evolutionary markers. The HLA-A and –H duplicons are similar in that they share paralogous Alu Y and MER 9 sequences and were probably generated

during or after the period of Alu Y amplification. However, the HLA-A duplicon has completely lost the genomic region between HERV-16 and PERB11 (MIC) which has still been partly retained within the HLA-H duplicon. The fragment of HLA-21 that is located between the HLA-A and –70 duplicons may have been duplicated from the progenitor HLA-A duplicon.

The scheme presented in Fig. 5B is not limited to any particular haplotype and remains fluid as indicated by reports of a 40-50 kb deletion between HLA-A and –G in HLA-A9 (Vendetti et al. 1992) and in HLA-A24 (Watanabe et al. 1997), an insertion of 70 kb between HLA-J and PERB11.4 (MICD) in haplotypes HLA-A30, -A31 and -A33 (El Kahloun et al. 1992; Watanabe et al. 1997) and a 100-kb deletion of the PERB11.1 (MICA) associated with HLA-B48 (Komatsu-Wakui et al. 1999).

The metamerismatic process outlined in Fig. 5B has generated not only numerous pseudogenes and gene fragments, but at least six HLA class I genes and two PERB11 (MIC) genes that have remained transcriptionally active. Of the many copies of HERV-16 sequence, at least one is transcriptionally active producing a mRNA that is antisense to the *pol* gene region (Kulski and Dawkins 1999). However, Klein et al. (1999) have presented a strong case for a diminution or alteration in the functions of the HLA class I genes whereby the A locus is being replaced by the functionally dominant B locus, HLA-C is becoming dormant, and HLA-E, -F and -G "are being pushed into the pseudogene category". In addition, the recent report of the PERB11(MIC) null haplotype (Komatsu-Wakui et al. 1999) may raise questions about the relevance of this gene in humans. The functions of PERB11 (MIC) that has an HLA class I like structure and sequence (Nei et al. 1998; Dawkins et al. 1999) have not been fully established, although in the epithelial cells of intestine and skin the PERB11 (MIC) molecule may interact with gamma/delta T cells to protect epithelial surfaces (Groh et al. 1998). The deletion of the entire PERB11.1 (MICA) coding gene in an HLA-B*4801 haplotype (Komatsu-Wakui et al. 1999) further illustrates the plasticity of the HERV-16 and PERB11(MIC) subgenomic region and the susceptibility of this region to deletions and rearrangements that are shown in Figure 2.

Since the generation of duplication products can be explained by DNA recombination processes, the metamerismatic model can also accommodate gene conversion-like events. Despite a few examples (Hughes 1995; Hoegstrand and Boehme 1999), gene conversion is a relatively minor process for the evolution of genes and polymorphisms in the MHC. The existence of considerable locus specificity in polymorphic sequences suggests that the exchange of sequences between HLA-A, -B, and -C loci has not been a major contributor to diversity of these genes (Parham et al. 1988; Nei et al. 1997).

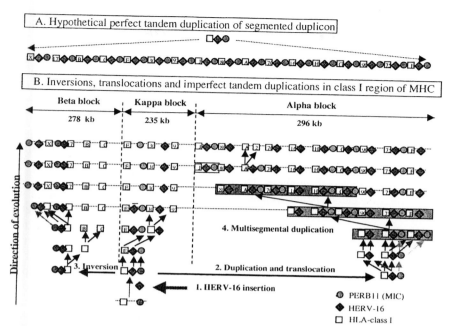

Fig. 5. Metamerismatic evolution of HLA class I region. The metamerismatic model is a scheme for the origin of the alpha, beta and kappa blocks by coevolution of the HLA class I, HERV-16 (P5) and PERB11 (MIC) sequences starting early in primate history from the primordial duplicon or ancestral unit shown in Fig. 2. (A) shows the hypothetical organisation of the class I region if the primordial duplicon had undergone perfect tandem duplications to produce 17 known HLA-class I sequences (squares) linked to HERV-16 (diamonds) and PERB 11 (MIC) (circles) sequences. (B) proposes that the actual evolutionary process has been imperfect involving inversions, translocations and a series of expansions (duplications) and contractions (deletions) resulting in the present subgenomic organisation of the alpha, beta and kappa blocks. The direction of evolution is indicated by the arrow on the left-handed side of the figure. The 4 basic steps of HERV-16 insertion, duplication and translocation, inversion and translocation, and multisegmental duplication are described in the text. A tripartite arrangement of duplicons are boxed and shaded to show how the alpha block may have expanded exponentially rather than linearly, resulting in the physical separation of closely related duplicons such as HLA -80, -16 and -90 from each other by less related duplicons such as HLA-A, HLA-70, etc. Deletions have also played a major part in the evolution of the class I region with the loss of components such as PERB11 (MIC), HCGVIII, HERV-16 and other transposable elements from duplicons during their expansion and contraction within the alpha, beta and kappa blocks.

Genomic stress, "bandaids", "scars" and diversity

In addition to the forces of natural selection such as balancing and purifying selection (Grimsley et al. 1998; Satta et al. 1998), genomic damage and repair are important driving forces in the generation of sequence diversity and polymorphism. During evolutionary time, stress or damage to MHC genomic sequences could have come from many and varied sources including infections, chemicals or even from cellular processes such as replication, transcription, translation and transposition in response to cellular stresses. The enormous number of transposable elements such as HERVs, LTRs, MERs, Alu and L1 sequences distributed throughout the genome are most often described as "junk DNA" but they can be better seen as genomic "scars" or "bandaids" involved with recombination processes in response to genomic stress and damage (Teng et al. 1996). This view is not new and corresponds closely to the insights originally proposed so eloquently and convincingly by Barbara MacClintock (1984) in her Nobel prize acceptance speech entitled "The significance of responses of the genome to challenge". Undoubtedly, transposable elements have participated actively in the contraction and expansion of duplicated regions as a response to the environmental challenges imposed on germ cells and their genomes. Therefore, the influence of genomic stress and transposable elements on the generation of diversity and polymorphism within duplicons of the MHC and other regions of the human genome is an integral part of the metamerismatic model. Because of the large number of potentially mobile transposable elements, retrotransposition and integration of some Alu, L1 and HERV sequences also have contributed to gene inactivation and disease (Nouvel 1994; Miki 1998; Larsson and Andersson 1998).

Transposable elements

Transposable elements such as HERV-16, L1 sequences, and Alu elements have been found closely associated with sequential breakpoints within and between different HLA duplicons suggesting a role for transposable elements in deletion events and the contraction of duplicons (Kulski et al. 1999b). In addition, numerous transposable elements including Alu elements, HERVs, L1 sequences, MERs, LTRs have been inserted after duplication thereby contributing to the expansion of duplicons (Gaudieri et al. 1997a, b; Kulski et al. 1997, 1999b). The role and distribution of retroelements within the class II region of the MHC have been previously reviewed by Andersson et al. (1998). The following is an overview of the transposable elements associated with the metamerismatic evolution of the HLA class I region of the MHC.

Alu elements

These elements (about 290 bp in length) are found only in primates and they contribute up to 10% of the human genome (Schmid 1996; Smit 1996). Different subtypes of Alu sequence amplified through successive waves of fixation (Kapitonov and Jurka 1996) are associated with primate lineage history (Fig. 6) that suggest some role in speciation or fitness, although the more specific functions of Alu elements appear to be regulatory responses to cellular stress such as viral infections, heat shock and translational inhibition by cyclohexamide (Schmid 1996; Chu et al. 1998). Like L1 sequences (Teng et al. 1996), Alu elements may act as genomic "bandaids' (Boeke 1997) possibly reflecting DNA repair mechanisms when normal DNA "patch repair" mechanisms are not able to cope with the damage or loss of continuous sequence from either single or double strands (Kornberg 1980). Paralogous or duplicated Alu elements have coevolved with HLA duplicons and they can be readily distinguished from other Alu elements within the human genome by distinct insertion sites and flanking sequence (Kulski et al. 1997, 1999b). Most paralogous Alu elements found within HLA duplicons by BLAST searches have been restricted to the intra block regions. Of 16 paralogous Alu J elements found so far within the HLA duplicons, 4 are within the beta and alpha blocks, 3 within the kappa and beta blocks, 3 and 6 are restricted to the alpha and the beta blocks, respectively.

We have used the estimated evolutionary age of different Alu subfamily members (Fig. 6) to model the lineages of the duplicons (Fig. 5B). For example, the HLA-B and -C duplicons share only Alu J elements and consequently have emerged before the HLA-X/PERB11.2 (MICB) and HLA-17/PERB11.1 (MICA) duplicons that share both Alu J and the older Alu S elements (Gaudieri et al. 1997a; Kulski et al. 1997, 1999b; Yamazaki et al. 1999). HLA-A and HLA-H duplicons appear to be relatively recent duplication products because they share Alu Y sequences (Kulski et al. 1999b). Whereas Alu Ya5 elements are found more frequently in humans than in gorillas or chimpanzees, Alu Yb8 elements appear to be human specific (Schmid 1996). Some Alu Yb8 elements are still transpositionally mobile and they have been associated with different human genetic diseases including those causing mutations in low density lipoprotein receptor and neurofibromatosis type I genes (Miki 1998). Alu Y elements also have been used in population studies to differentiate between ethnic groups and to trace the migration patterns of modern humans (Schmid 1996), and therefore could be useful markers for studying and characterising the lineages and history of HLA ancestral haplotypes.

L1

The L1 sequences are known to have evolved before the emergence of primates. They have been classified into different mammalian and primate subfamilies on the basis of sequence divergence (Smit et al. 1995), and used as evolutionary markers to estimate the ages of HLA duplications within the class I region of the MHC (Gaudieri et al. 1999; Yamazaki et al. 1999). Approximately 15 % of the human genome content is composed of L1 sequences but mostly as highly truncated structures rather than as full-length retroelements of 9 kb with intact helicase and reverse transcriptase coding sequences (Kazazian and Moran 1998). We and others have observed that many of the older L1 elements act as receptors or integration sites for other L1 family members, HERVs, LTRs and Alu elements (Beck et al. 1996; Kulski et al. 1997, 1999a). Complete and mobile L1 sequences may amount to only 30-60 active elements per human diploid cell (Sassaman et al. 1997). The transposition and insertion of new L1 sequences may occur as a consequence of DNA damage (Teng et al. 1996) and be part of a DNA sequence repair mechanism possibly in association with Alu elements (Boeke 1997). We have found 6 almost full-length primate-specific L1 sequences (PA2) in various stages of initial decay located near HLA-75, -E, -92 and -C, HERV-K9I and HTEX-4 (Fig. 1). The relatively intact structures of these L1 elements suggest that they are recent insertions in primates. Further analysis of these L1 sequences in the orthologous location of human and non-human primates may help to establish the evolutionary time of insertion and elucidate how dimorphic these L1 sequences are in human ancestral haplotypes.

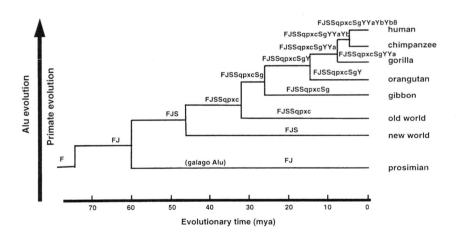

Fig. 6. Coevolution of Alu subfamily members and primates. The Alu nomenclature and molecular ages are taken from Batzer et al. (1996) and Kapitonov and Jurka (1996), respectively. The approximate dating of primate origins is according to Martin (1993).

HERVs

Recent reassessments of HERV sequences within the human genome have highlighted their role in disease generation and protection (Larsson and Andersson 1998). In this regard, we have found that the *pol* gene region of the HERV-16 (P5.1) sequence in the beta block of the MHC appears to be transcribed as an antisense sequence in activated lymphocytes and therefore may have a protective role against retroviral infections (Kulski and Dawkins 1999). The alpha and beta block of the MHC is extremely rich in HERVs occurring at a ratio that is at least 10 times higher than usually observed in the human genome (Kulski et al. 1999a). This high content is mostly due to the duplication of at least 11 copies of HERV-16 sequence in association with the HLA class I and PERB11 (MIC) genes (Fig. 1). Four other HERV families have been identified within the alpha and beta blocks (Fig. 1) including HERV-L (1 copy), HERV-I (2 copies), HERV-K9I (1 copy) and HARLEQUIN (1 copy), and two partial HERV-L sequences were found between the beta and kappa blocks. Moreover, there are at least 5 copies of a pTRC sequence (also called pHE.1, see accession number M85205), that is a transcript of HERV-9 (Lania et al. 1992), in the region between the beta and kappa blocks. In view of the current evidence on HERVs, one or other of these HERV sequences may have a major affect on disease generation, susceptibility and resistance.

LTRs, DNA transposons and microsatellites

Many different LTRs, DNA transposons and microsatellites are present within the class I region of the MHC (Shiina et al. 1999). A number of simple nucleotide repeats in the alpha and beta block have been generated by Alu elements and L1 fragments (Gaudieri et al. 1997a; Kulski et al. 1997) whereas others occur independently of annotated transposable elements. Because of different dynamics and evolution of micro- and mini-satellites, we have not covered them extensively in our analysis. We have noted that the HLA class I genomic region is rich in purine/pyrimidine repeats some of which may be involved in DNA secondary structural alterations such as triple helix formation (Soyfer and Potaman 1996; Maueler et al. 1998), and others in the deletion of 100 kb of genomic sequence including the complete PERB11.1 (MICA) gene and HERV-L sequence (Komatsu-Wakui et al. 1999). In addition, purine/pyrimidine repeats at different locations of the class I region, one called the CL element (Abraham et al. 1992) and another located within the 5'UTR of the HLA-F gene (Santos et al. 1998), have been used as markers of polymorphism.

Conclusion

DNA sequences from different laboratories during the past decade have revealed that the MHC genomic region is composed of many duplications containing a diverse array of multicopy gene families, transposable elements, and simple repeat sequences. No single model such as the big bang, accordion, beads, life and death or metamerism seems adequate to describe or explain the intricate sequence of events that have been involved in the evolution of the highly polymorphic and duplicated regions of the MHC. Nevertheless, the metamerismatic model provides some insight into the evolution of the multimeric design of the HLA class I region. Transposable elements represent an additional and important driving force in MHC evolution with respect to generating diversity, plasticity, adaptability and regulatory configurations and structures. Moreover, the metamerismatic model can be developed further by comparative genomics, population studies, biochemical experimentation and genetic manipulations for a better understanding of the role of transposable elements and the different DNA processes involved in the evolution of the human MHC.

Acknowledgments

We thank Natalie Longman and Sonia Cattley for their assistance with the figures, and Takashi Gojobori and Joerg Epplen for critical discussions and suggestions. Silvana Gaudieri is the recipient of a fellowship from the Japan Society for the Promotion of Science (97111). This work was supported by grants from the National Health and Medical Research Council (No 981719) and the Immunogenetics Research Foundation. This paper is publication No. 9921 of the Centre for Molecular Immunology and Instrumentation, The University of Western Australia, Perth, Western Australia.

References

Abi Rached L, McDermott MF, Pontarotti P (1999) The MHC big bang. Immunol Rev 167:33-45

Abraham LJ, Leelayuwat C, Grimsley G, Degli-Esposti MA, Mann A, Zhang WJ, Christiansen FT, Dawkins RL (1992) Sequence differences between HLA-B and TNF distinguish different MHC ancestral haplotypes. Tissue Antigens 39:117-121

Andersson G, Svensson AC, Setterblad N, Rask L (1998) Retroelements in the human MHC class II region. Trends Genet 14:109-114

Bahram S, Bresnahan M, Geraghty DE, Spies T (1994) A second lineage of mammalian major histocompatibility complex class I genes. Proc Natl Acad Sci USA 91:6259-6263

Batzer MA, Deininger PL, Hellmann-Blumberg U, Jurka J, Labuda D, Rubin CM, Schmid CW, Zietkiewicz E, Zuckerkandl E (1996) Standardized nomenclature for Alu repeats. J Mol Evol 42:3-6

Beck S, Abdulla S, Alderton RP, Glynne RJ, Gut IG, Hosking LK, Jackson A, Kelly A, Newell WR, Sanseau P, Radley E, Thorpe KL, Trowsdale J (1996) Evolutionary dynamics of non-coding sequences within the class II region of the human MHC. J Mol Biol 255:1-13

Boeke JD (1997) LINEs and Alus-the polyA connection. Nat Genet 16:6-7

Chu W-M, Ballard R, Carpick BW, Williams BRG, Schmid CW (1998) Potential Alu function: Regulation of the activity of double-stranded RNA-activated kinase PKR. Mol Cell Biol 18:58-68

Dawkins R, Leelayuwat C, Gaudieri S, Tay G, Hui J, Cattley S, Martinez P, Kulski J (1999) Genomics of the Major Histocompatibility Complex: Haplotypes, retroviruses and disease. Immunol Rev 167:275-304

El Kahloun A, Vernet C, Jouanolle A-M, Boretto J, Mauvieux V, LeGall J-Y, David V, Pontarotti P (1992) A continuous restriction map from HLA-E to HLA-F. Structural comparison between different HLA-A haplotypes. Immunogenetics 35:183-189

El Kahloun A, Chauvel B, Mauvieux V, Dorval I, Jouanolle A-M, Gicquel I, Le Gall J-Y, David V (1993) Localization of seven new genes around the HLA-A locus. Hum Mol Genet 2:55-60

Gaudieri S, Giles K, Kulski J, Dawkins R (1997a) Duplication and polymorphism in the MHC: Alu generated diversity and polymorphism within the PERB11 gene family. Hereditas 127:37-46

Gaudieri S, Kulski JK, Balmer L, Giles KM, Inoko H, Dawkins RL (1997b) Retroelements and segmental duplications in the generation of diversity within the MHC. DNA Seq 8:137-141

Gaudieri S, Kulski JK, Dawkins RL, Gojobori (1999) Different evolutionary histories in two subgenomic regions of the Major Histocompatibility Complex. Genome Res 9:541-549

Geraghty DE, Koller BH, Hansen JA, Orr HT (1992) The HLA class I gene family includes at least six genes and twelve pseudogenes and gene fragments. J Immunol 149:1934-1946

Gould SJ (1989) Wonderful Life. The Burgess Shale and the Nature of History. Penguin Books LTD, London, England

Grimsley C, Mather KA, Ober C (1998) HLA-H: a pseudogene with increased variation due to balancing selection at neighboring loci. Mol Biol Evol 15:1581-1588

Groh V, Steinle A, Bauer S, Spies T (1998) Recognition of stress-induced MHC molecules by intestinal epithelial gamma delta T cells. Science 279:1737-1740

Hoegstrand K, Boehme J (1999) Gene conversion can create new MHC alleles. Immunol Rev 167:305-317

Hughes AL (1995) Origin and evolution of HLA class I pseudogenes. Mol Biol Evol 12:247-258

Jurka J, Klonowski P, Dagman V, Pelton P (1996b) CENSOR- a program for identification and elimination of repetitive elements from DNA sequences. Comp Chem 20:119-121

Kapitonov V, Jurka J (1996) The age of Alu subfamilies. J Mol Evol 42:59-65

Kazazian HH, Moran JV (1998) The impact of L1 retrotransposons on the human genome. Nat Genet 19:19-24

Klein J, Sato A, O'hUigin C (1998) Evolution by gene duplication in the major histocompatibility complex. Cytogenet Cell Genet 80:123-127

Klein J, Zhu Z, Gutknecht J, Figueroa F, Kasahara M (1991) Lessons in evolution. In: Srivastava R, Ram BP, Tyle P (eds) Immunogenetics of the Major Histocompatibility Complex. VCH, New York, pp 18-38

Komatsu-Wakui M, Tokunaga K, Ishikawa Y, Kashiwase K, Moriyama S, Tsuchiya N, Ando H, Shiina T, Geraghty DE, Inoko H, Juji T (1999) MIC-A polymorphism in Japanese and a MIC-A-MIC-B null haplotype. Immunogenetics 49:620-628

Kornberg A (1980) DNA Replication. WH Freeman and Co., New York

Kulski JK, Dawkins RL (1999) The P5 multicopy gene family in the MHC is related in sequence to human endogenous retroviruses HERV-L and HERV-16. Immunogenetics 49:404-412

Kulski JK, Gaudieri S, Bellgard M, Balmer L, Giles K, Inoko H, Dawkins RL (1997) The evolution of MHC diversity by segmental duplication and transposition of retroelements. J Mol Evol 45:599-609

Kulski JK, Gaudieri S, Inoko T, Dawkins RL (1999a) Comparison between two HERV-rich regions within the Major Histocompatibility Complex. J Mol Evol 48:675-683

Kulski JK, Gaudieri S, Martin A, Dawkins RL (1999b) Coevolution of PERB11 (MIC) and HLA class I genes with HERV-16 and retroelements by extended genomic duplication. J Mol Evol 49:84-97

Lania L, Di Cristofano A, Strazzullo M, Majello B, La Mantia G (1992) Structural and functional organization of the human endogenous retroviral ERV9 sequences. Virology 191:464-468

Leelayuwat C, Townsend DC, Degli-Esposti MA, Abraham LJ, Dawkins RL (1994) A new polymorphic and multicopy MHC gene family related to non-mammalian class I. Immunogenetics 40:339-351

Leelayuwat C, Pinelli M, Dawkins RL (1995) Clustering of diverse replicated sequences in the MHC. Evidence for en bloc duplication. J Immunol 155:692-698

Marshall B, Leelayuwat C, Degli-Esposti MA, Pinelli M, Abraham LJ, Dawkins RL (1993) New major histocompatibility complex genes. Hum Immunol 38:24-29

Martin RD (1993) Primate origins: plugging the gaps. Nature 363:223-234

Maueler W, Kyas A, Keyl H-G, Epplen J (1998) A genome-derived (gaa.ttc)24 trinucleotide block binds nuclear protein(s) specifically and forms triple helices. Gene 215:389-403

McClintock B (1984) The significance of responses of the genome to challenge. Science 226:792-801

Miki Y (1998) Retrotransposal integration of mobile genetic elements in human diseases. J Hum Genet 43:77-84

Mizuki N, Ando H, Kimura M, Ohno S, Miyata S, Yamazaki M, Tashiro H, Watanabe K, Ono A, Taguchi S, Sugawara C, Fukuzumi Y, Okumura K, Goto K, Ishihara M, Nakamura S, Yonemoto J, Kikuti YY, Shiina T, Chen L, et al. (1997) Nucleotide sequence analysis of the *HLA* class I region spanning the 237-kb segment around the *HLA-B* and -*C* genes. Genomics 42:55-66

Nei M, Gu X, Sitnikova T (1997) Evolution by the birth-and-death process in multigene families of the vertebrate immune system. Proc Natl Acad Sci USA 94:7799-7806

Nouvel P (1994) The mammalian genome shaping activity of reverse transcriptase. Genetica 93:191-201

Parham P, Lomen CE, Lawlor DA, Ways JP, Holmes N, Coppin HL, Salter RD, Wan AM, Ennis PD (1988) Nature of polymorphism in HLA-A, -B, and -C molecules. Proc Natl Acad Sci USA 85:4005-4009

Pichon L, Carn G, Bouric P, Giffon T, Chauvel B, Lepourcelet M, Mosser J, Legall J-V, David V (1996) Structural analysis of the HLA-A/HLA-F subregion: precise localization of two new multigene families closely associated with the HLA class I sequences. Genomics 32:236-244

Santos EJM, Epplen JT, Epplen C, Guerreiro JF (1998) Microsatellite evolution in the 5'UTR of the HLA-F gene. Human Evol 13:57-64

Sassaman DM, Dombroski BA, Moran JV, Kimberland ML, Naas, TP, DeBerardinis RJ, Gabriel A, Swegold GD, Kazazian Jr HH (1997) Many human L1 elements are capable of retrotransposition. Nat Genet 16:37-43

Satta Y, Li YJ, Takahata N (1998) The neutral theory and natural selection in the HLA region. Front Biosci 27:d459-467

Schmid CW (1996) Alu structure, origin, evolution, significance, and function of one-tenth of human DNA. Prog Nucleic Acids Res Mol Biol 53:283-319

Shiina T, Tamiya G, Oka A, Takishima N, Inoko H (1999) Genome sequence analysis of the 1.8 Mb entire human MHC class I region. Immunol Rev 176:193-199

Shiina T, Tamiya A, Oka A, Yamagata T, Yamagata N, Kikkawa E, Goto A, Mizuki N, Watanabe K, Fukuzumi Y, Taguchi S, Sugawara C, Ono A, Chen L, Yamazaki M, Tashiro H, Ando A, Ikemura T, Kimura M, Inoko H (1998) Nucleotide sequencing analysis of the 146-kilobase segment around the *IkBL* and *MICA* genes at the centromeric end of the *HLA* Class I region. Genomics 47:372-382

Smit AFA (1996) The origin of interspersed repeats in the human genome. Curr Opin Genet Dev 6:743-748

Smit AFA, Toth G, Riggs AD, Jurka J (1995) Ancestral, mammalian-wide subfamilies of LINE-1 repetitive sequences. J Mol Biol 246:401-417

Soyfer VN, Potaman VN (1996) Triple-helical Nucleic Acids. Springer-Verlag, New York

Teng S-C, Kim B, Gabriel A (1996) Retrotransposon reverse-transcriptase-mediated repair of chromosomal breaks. Nature 383:641-644

Venditti CP, Chorney MJ (1992) Class I gene contraction within the HLA-A subregion of the human MHC. Genomics 14:1003-1009

Venditti CP, Harris JM, Geraghty DE, Chorney MJ (1994) Mapping and characterization of non-*HLA* multigene assemblages in the human MHC class I region. Genomics 22:257-266

Vernet C, Ribouchon MT, Chimini G, Jouanolle AM, Sidibe I, Pontarotti P (1993) A novel coding sequence belonging to a new multicopy gene family mapping within the human MHC class I region. Immunogenetics 38:47-53

Watanabe Y, Tokunaga K, Geraghty DE, Tadokoro K, Juji T (1997) Large-scale comparative mapping of the MHC class I region of predominant haplotypes in Japanese. Immunogenetics 46:135-141

Yamazaki M, Tateno Y, Inoko H (1999) Genomic organisation around the centromeric end of the HLA class I region: large-scale sequence analysis. J Mol Evol 48:317-327

Polymorphism in the *HLA* class I region

Yoko Satta and Naoyuki Takahata

Department of Biosystems Science, School of Advanced Sciences, Graduate University for Advanced Studies, Hayama, Kanagawa 240-0193, Japan

Summary. We compared a pair of approximately 650 kb DNA sequences in the human leukocyte antigen (*HLA*) class I region located on chromosome 6p21.3. It is immediately apparent that the extent of polymorphism varies greatly within the class I region of 1.8 Mb. As repeatedly noted, DNA polymorphism is extremely high at the three class Ia loci (*HLA-A*, *-B*, and *-C*) owing to balancing selection. However, the intergenic subregion between the *HLA-B* and *HLA-C* loci is even more polymorphic than the class Ia loci themselves. Although there are a number of repetitive sequences and retro-transposable insertions in the *HLA-B/C* intergenic subregion, none of them appear to be particularly responsible for the elevated polymorphism in the subregion. Nevertheless, the hitch-hiking effect of balanced class Ia loci on linked subregions is evident: as the physical map distance from a class Ia locus increases, the extent of polymorphism decreases rather sharply to the level observed at non-*HLA* loci. The decreasing rate is subregion-specific, suggesting that the recombination rate is also subregion-specific. Based on a population genetics model of the human demographic history, we have inferred the recombination rate in a subregion either telomeric or centromeric to each of the three class Ia loci. Some subregions are consistent with the genome-wide average rate of recombination (1 cM per Mb), but two other subregions, telomeric to the *HLA-A* locus and centromeric to the *HLA-B* locus, are considerably suppressed in recombination.

Key words. DNA polymorphism, *HLA* class I region, Recombination, Human demography, Hitch-hiking effect

Introduction

Both linkage analyses and direct observations of the number of chiasmata per cell have suggested that recombination may occur at the average rate of 1 % (or 1 cM) per Mb per generation in the human genome (Vogel and Motulsky 1997). However, detailed studies for specified genomic regions such as the β-globin gene cluster of about 70 kb on chromosome 11 have raised a possibility that the rate may vary substantially even within a relatively small genomic region (Antonarakis et al. 1982). At present, it may not be infeasible to conduct family analyses for a Mb-long genomic region, but such analyses are still scanty.

An alternative approach for estimating the recombination rate is to use the pattern and extent of DNA polymorphism (Satta et al. 1999). Six *HLA* class Ia (*HLA-A, -B, -C*) and class IIb (*DRB, DQB, DPB*) loci are shown to be highly polymorphic. The per-site nucleotide differences in pairwise comparisons of alleles at these loci are more than 50 times greater than those at non-*HLA* loci (Satta 1992; Klein et al. 1993a). As noted, large *HLA* nucleotide differences are not confined to nonsynonymous sites in exons which encode the peptide binding region (PBR) and occur at linked synonymous sites and introns as well (Cereb et al. 1997; Bergström et al. 1998). These elevated silent polymorphisms are most likely caused by hitch-hiking effects of balancing selection responsible for the PBR diversity so that the tighter the linkage of silent sites to the PBR encoding exons, the more polymorphic (Satta 1997; Takahata and Satta 1998). This relationship was previously examined with the synonymous nucleotide differences at 37 loci in the *HLA* region (Satta et al. 1999).

There are two limitations in the above approach. One is statistical and concerned with the number and length of individual linked loci near a classical *HLA* locus we can compare. The number of linked loci must be large enough to encompass a genomic region of some 100 kb and the length of individual DNA sequences must be long enough to estimate accurately the synonymous nucleotide differences at each of linked loci. The other is historical and concerned with the use of DNA polymorphism. Polymorphism is compounded by various evolutionary forces and affected by the demographic history of organisms. Hence, it is necessary to develop a realistic demographic model that is compatible with all available data on silent DNA polymorphism in a particular organism.

Humans seem to be the best organism in these regards. Extensive studies of mitochondrial and non-*HLA* nuclear polymorphism have shown that the effective population size has been 10,000 for the last 1 or 2 million years (Li and Sadler 1991; Takahata 1993, 1995), whereas the trans-specific mode of *HLA* polymorphism as well as the ancestral polymorphism prior to the divergence between the human and the African apes (Klein et al. 1990, 1993b; Takahata and Satta 1997; Takahata 1999) has suggested that the effective population size had been 100,000 before *Homo* emerged. In relation to the ongoing human genome project, a new type of sequence data pertinent to our purpose has been accumulating. The sequencing of the whole *HLA* class I region of 1.8 Mb (Shiina et al. 1999) and other chromosomal regions of the same scale has been completed. In the case of the class I region, two other independent sequencing projects were carried out: one encompassing *HLA-B/C* loci (Guillaudeux et al. 1998) and the other *HLA-A* locus (GenBank accession no. AF055066). Although these are partly redundant when the same YAC clones were sequenced, the majority can be used for evaluating the extent of DNA polymorphism. Here we use this information on

DNA polymorphism along a fairly long stretch of the human genome and estimate the recombination rate based on a population genetics model.

Materials and methods

Sequence alignment and analysis

We aligned two allelic DNA sequences of 337,690 bp *HLA-B/C* and 317,954 bp *HLA-A* subregions by using MegAlign and further modifying by eye. We counted the number of silent and non-silent nucleotide differences in a 1 kb non-overlapping sliding window. All indels were excluded so that some windows were smaller than 1 kb in size. The per-site nucleotide differences were then computed in 650 windows and plotted against their locations on the physical map.

Population genetics model

For the theoretical expectation of the per-site nucleotide differences in a subregion linked to a class Ia locus, we used the formula given in Appendix of Takahata and Satta (1998). It is assumed that the effective population size reduced from 100,000 to 10,000 at 50,000 generations or 1 million years ago (Fig. 1), the neutral mutation rate is 2×10^{-8} per site per generation, the recombination rate is in proportion to the physical map distance from a class Ia locus, and the selection intensity at the class Ia locus is 0.02.

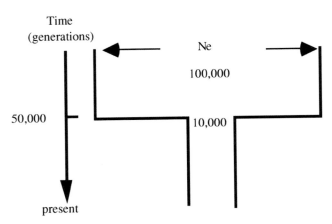

Fig. 1. A crane neck model of human demography so as to be consistent with the nucleotide diversity at both *HLA* and non-*HLA* loci.

Results and discussion

The extent of nucleotide differences per site at each of the three class Ia loci is 4 to 5 %. This is nearly the same as the average nucleotide diversity at the synonymous sites of the three class Ia loci (Satta 1992) and more than 50 times higher than the standard level (0.08%) averaged over 49 non-*HLA* loci (Li and Sadler 1991). Although a number of functional loci are located in the *HLA* class I region (Shiina et al. 1999), it seems that there is no clear-cut evidence of balancing selection at non-classical class I or any other loci in the class I region (but see Castro et al. 1996). It is evident from Figure 2 that balancing selection operating at a class Ia locus can influence polymorphism strongly in nearby subregions within some 100 kb. The extent of polymorphism in a linked neutral subregion is sensitive to the recombination rate with the nearest selected locus (Satta 1997). This sensitivity may enable us to estimate the recombination rate fairly accurately from linked neutral polymorphism. It is noted in Figure 2 that the observed level of polymorphism is ragged in one window to another. One reason is a rather small number of nucleotide differences observed in each 1 kb window. However, an alteration of window size does not erase the ragged pattern (data not shown). It is more likely that the raggedness reflects a number of recombination events because multiple recombinations with single but different breakpoints can lead to such a pattern of the nucleotide differences between a pair of alleles. If this is the case and if we can examine a large number of alleles to compute all pairwise nucleotide differences, the raggedness may become small.

Despite such a ragged distribution, the level of nucleotide differences tends to decrease in windows as they are more distant from the nearest class Ia locus and the overall decreasing pattern may fit to a theoretical curve with an increasing recombination rate (Satta et al. 1999). Either the centromeric or telomeric subregion of *HLA-C* as well as the centromeric subregion of *HLA-A* is consistent with the genome-wide inference of the recombination rate, 1 cM per Mb (Vogel and Motulsky 1997). By contrast, the centromeric subregion of *HLA-B* and the telomeric subregion of *HLA-A* show significantly greater nucleotide differences than expected, suggesting that these subregions have undergone infrequent recombination. The observed level of nucleotide differences in the centromeric subregion of *HLA-B* becomes consistent with that in the case where 0.1 cM corresponds to 1 Mb and that in the telomeric subregion of *HLA-A* becomes consistent with that in the case where 0.01-0.1 cM corresponds to 1 Mb. In fact, the linkage analysis of CEPH family supports the latter estimate of the recombination rate. The direct estimate in the 4 Mb subregion from *HLA-A* to *HFE* (hemochromatosis locus) is 0.05 cM per Mb and the 2 Mb subregion telomeric to *HFE* is 0.5 cM per Mb (Malfroy et al. 1997). Based on single-sperm typing of markers that flank the entire *HLA* region, Carrington (1999) observed that the recombination rate depends on haplotypes used and ranges from 0.2 cM to 0.9 cM per Mb. Unfortunately, however, the molecular mechanism for recombination suppression is unknown.

Per-site nucleotide differences
(1 kb non-overlapping windows)

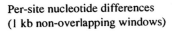

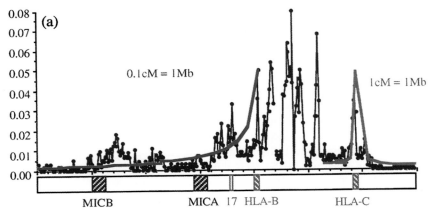

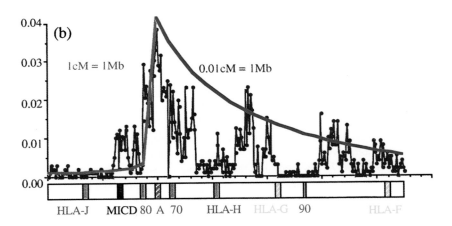

Fig.2. Polymorphism measured by the nucleotide differences per site in a 1 kb long non-overlapping sliding window in the *HLA-B/C* subregion (**a**) and in the *HLA-A* subregion (**b**). The red and green curves indicate the expected extents of polymorphism at the linked neutral subregion with 1 cM = 1 Mb and 0.1 cM = 1 Mb (**a**) or 0.01 cM = 1 Mb (**b**), respectively. The bottom bar indicates the approximate positions of bona fide class I loci or their homologs in these subregions. The locus positions are after Shiina et al. (1999) and Guillaudeux et al. (1998).

The intergenic subregion between *HLA-B* and *HLA-C* shows an unusual extent of nucleotide differences (Gaudieri et al., this volume). This subregion is largely devoid of functional loci. One possibility for the high variation may be due to the fact that the subregion is sandwiched by the highly polymorphic *HLA-B* and *HLA-C* loci. However, even if balancing selection acts at both loci, it is unlikely that the extent of polymorphism in a sandwiched subregion becomes higher than that at the selected loci unless some synergetic interaction operates. Alternatively, the paralogy between two telomeric subregions of *HLA-B* and *HLA-C* (Kulski et al. 1997; Mizuki et al. 1997; Guilaudeux et al. 1998) may confuse the allelic comparison. To examine this possibility, we aligned about 30 kb subregions telomeric to the two loci and examined the per-site nucleotide differences in both allelic and paralogous comparisons. It is clear that the per-site nucleotide differences in the allelic comparison are significantly smaller than those in the paralogous comparison (data not shown). The sequence divergence between *HLA-B* and *HLA-C* telomeric subregions is estimated as 0.105 ± 0.003 based upon Kimura's (1980) two parameter model. This extent of sequence divergence suggests that the *HLA-B* and *HLA-C* loci were duplicated about 50 million years ago (see also Kulski et al. 1997) and no subsequent sequence exchanges have occurred since that time. This estimate of the *HLA-B* and -*C* duplication may be used to resolve the controversial divergence time between Platyrrhini and Catarrhini: there should not exist any ortholog of *HLA-C* in Platyrrhini if it diverged more than 50 million years ago (Martin 1993; Takahata 1999).

One feature in the class I region is the high density of insertion sequences (Kulski et al. 1997; Mizuki et al. 1997; Shiina et al. 1999). In the 40 kb intergenic subregion between the *HLA-B* and *HLA-C* loci, there are 13 *Alu* and one truncated *L1* sequences. Nucleotide substitutions tend to be concentrated near some *Alu* sequences. However, this does not fully account for the unusual polymorphism and there are a number of variable sites outside *Alu* sequences. In any event, the causality between *Alu* insertion and high mutability remains to be studied (Roger et al. 1998; Aho et al. 1999).

The extent of polymorphism at selectively neutral regions is proportional to the divergence time of genes which is in turn proportional to the population size (Kingmann 1982; Kimura 1983). As mentioned, in the human population, the extent of silent nucleotide differences is 0.08% per site averaged over 49 non-*HLA* loci (Li and Sadler 1991) and corresponds to the average sequence divergence time of 400,000 years if the silent substitution rate is 10^{-9} per site per year. In other words, silent nucleotide differences reflect the population or species history during the last 400,000 years on average, and they are therefore used for inquiring the origin of modern humans (Takahata 1995; Harding et al. 1997; Harpending et al. 1998). The high extent of silent polymorphism in the *HLA* region reflects more ancient history. To address the early hominid evolution 2 to 5 million years ago, genomic regions with hitch-hiked silent polymorphism of 0.4% to 1% are appropriate (Klein et al. 1993b; Takahata 1995). There are a number of such candidate regions including non-classical class I loci and class I related

184

pseudogenes. When the hitch-hiking effect is quantified on the basis of accurate recombination rates, relatively ancient polymorphism observed in some *HLA* subregion can provide unique opportunities for studying the demographic history at any critical phase in human evolution.

Acknowledgments

We thank Sang-Hee Lee for her comments. This work was supported in part by a Promotion-of-Science grant from NOVARTIS Foundation (Japan). This paper is contribution no. 15 from the Department of Biosystems Science.

References

Aho S, Rothenberger K, Tan EML, Ryoo YW, Cho BH, McLean WHI, Uitto J (1999) Human periplakin: genomic organization in a clonally unstable region of chromosome 16p with an abundance of repetitive sequence elements. Genomics 56:160-168

Antonarakis SE, Boehm CD, Giardina PJV, Kazazian HH Jr (1982) Nonrandom association of polymorphic restriction sites in the _-globin gene cluster. Proc Natl Acad Sci USA 79:137-141

Bergström TF, Josefsson A, Erlich HA, Gyllensten U (1998) Recent origin of *HLA-DRB1* alleles and implications for human evolution. Nat Genet 18:237-242

Carrington M (1999) Recombination within the human MHC. Immunol Rev 167:245-256

Castro ML, Morales P, Fernandez-Soria V, Suarez B, Recio MJ, Alvarez M, Martin-Villa M, Arnaiz-Villena A (1996) Allelic diversity at the primate *Mhc-G* locus: exon 3 bears stop codons in all Cercopithecinae sequences. Immunogenetics 44:327-336

Cereb N, Hughes AL, Yang SO (1997) Locus-specific conservation of the *HLA* class I introns by intra-locus homogenization. Immunogenetics 47:30-36

Guillaudeux T, Janer M, Ka-Shu G, Spies T, Geragthy DE (1998) The complete genomic sequence of 424,015 bp at the centromeric end of the *HLA* class I region: gene content and polymorphism. Proc Natl Acad Sci USA 95:9494-9499

Harpending HC, Batzer MA, Gurven M, Jorde LB, Rogers AR, Sherry ST (1998) Genetic traces of ancient demography. Proc Natl Acad Sci USA 95:1961-1967

Harding RM, Fullerton SM, Griffiths RC, Bond J, Cox MJ, Schneider JA, Moulin DS, Clegg JB (1997) Archaic African and Asian lineages in the genetic ancestry of modern humans. Am J Hum Genet 60:772-789

Kimura M (1980) A simple method for estimating evolutionary rates of base substitutions through comparative studies of nucleotide sequences. J Mol Evol 16:111-120

Kimura M (1983) The Neutral Theory of Molecular Evolution. Cambridge University Press, Cambridge, pp 214-229

Kingmann JFC (1982) On the genealogy of large populations. J Appl Prob 19A:27-43

Klein J, Gutknecht J, Fischer N (1990) The major histocompatibility complex and human evolution. Trends Genet 6:7-11

Klein J, Satta Y, O'hUigin C, Takahata N (1993a) The molecular descent of the major histocompatibility complex. Annu Rev Immunol 11:269-295

Klein J, Takahata N, Ayala FJ (1993b) *MHC* polymorphism and human origins. Sci Am 269:78-83

Kulski JK, Gaudieri S, Bellgard M, Balmer L, Giles K, Inoko H, Dawkins RL (1997) The evolution of *MHC* diversity by segmental duplication and transposition of retroelements. J Mol Evol 45:599-609

Li W-H, Sadler LA (1991) Low nucleotide diversity in man. Genetics 129:513-523

Malfroy L, Roth MP, Carrington M, Borot N, Volz A, Ziegler A, Coppin H (1997) Heterogeneity in rates of recombination in the 6-Mb region telomeric to the human major histocompatibility complex. Genomics 43:226-231

Martin RD (1993) Primate origins, plugging the gaps. Nature 363:223-234

Mizuki N, Ando H, Kimura M, Ohno S, Miyata S, Yamazaki M, Tashiro H, Watanabe K, Ono A, Taguchi S, Sugawara C, Fukuzumi Y, Okumura K, Goto K, Ishihara M, Nakamura S, Yonemoto J, Kikuti YY, Shiina T, Chen L, Ando A, Ikemura T, Inoko H (1997) Nucleotide sequence analysis of the *HLA* class I region spanning the 237-kb segement around the *HLA-B* and *-C* genes. Genomics 42:55-66

Roger M, Sanchez FO, Schurr E (1998) Comparative study of the genomic organization of DNA repeats within the 5'-flanking region of the natural resistance-associated macrophage protein gene (NRAMP1) between humans and great apes. Mamm Genome 9:435-439

Satta Y (1992) Balancing selection at *HLA* loci. In: Takahata N, Clark AG (eds) Mechanisms of Molecular Evolution. Japan Scientific Societies Press, Sinauer Associates, Tokyo, pp 129-149

Satta Y (1997) Effects of intra-locus recombination on *HLA* polymorphism. Hereditas 127:105-112

Satta Y, Kupfermann H, Li Y-J, Takahata N (1999) Molecular clock and recombination in primate *Mhc* genes. Immunol Rev 167:367-379

Shiina T, Tamiya G, Oka A, Takishima N, Inoko H (1999) Genome sequencing analysis of the 1.8 Mb entire human *MHC* class I region. Immunol Rev 167:193-199

Takahata N (1993) Allelic geneaolgy and human evolution. Mol Biol Evol 10:2-22

Takahata N (1995) A genetic perspective on the origin and history of humans. Annu Rev Ecol Syst 26:343-372

Takahata N (1999) Molecular phylogeny and demographic history of humans. The Proceedings of the Dual Congress, in press

Takahata N, Satta Y (1997) Evolution of the human lineage: Phylogenetic and demographic inferences from primate DNA sequences. Proc Natl Acad Sci USA 94:4811-4815

Takahata N, Satta Y (1998) Footprints of intragenic recombination at *HLA* loci. Immunogenetics 47:430-441

Vogel F, Motulsky AG (1997) Human Genetics. Springer, Berlin, pp 37-44

Nucleotide diversity within the human major histocompatibility complex: function of hitch-hiking effect, duplications, indels and recombination

Silvana Gaudieri[1], Roger L. Dawkins[2], Kaori Habara[1], Jerzy K. Kulski[2], and Takashi Gojobori[1]

[1]Center for Information Biology, National Institute of Genetics, Mishima, Shizuoka-ken 411-8540, Japan
[2]Centre for Molecular Immunology and Instrumentation, University of Western Australia, Nedlands, 6008, Western Australia, Australia

Summary. The recent availability of genomic sequences within the human Major Histocompatibility Complex (MHC) has allowed a detailed study of its organisation, gene content and nucleotide diversity. The MHC is characterised by polymorphic multicopy gene families including the Human Leucocyte Antigens (HLA) Class I and II genes (Geraghty et al. 1992), large kilobase insertions and deletions (indels) (Venditti and Chorney 1992; Watanabe et al. 1997), and uneven rates of recombination (Carrington 1999). Balancing selection has been reported to operate on the antigen presenting HLA Class I genes (Hughes and Nei 1988), with elevated silent differences at linked neutral sites attributed to a hitch-hiking effect (Grimsley et al. 1998; Horton et al. 1998; Satta et al. 1998). To examine the extent of nucleotide diversity at non-HLA Class I loci, we compared different human haplotypes in 380 kb of sequence from IkBL to telomeric of HLA-C, including HLA-B. We show that the level of nucleotide diversity within the MHC is several fold greater than elsewhere in the genome with peaks not always corresponding to the HLA Class I loci. This unusually high level of nucleotide diversity is contained within a region of the MHC described as a polymorphic frozen block (PFB) (beta block). A region associated with limited meiotic recombination and high non-coding nucleotide diversity. A hitch-hiking effect from balancing selection operating on the HLA Class I genes is likely to be contributing to the nucleotide diversity pattern within the MHC, however, other evolutionary processes such as duplications and indels also appear to play an important role.

Key words. MHC, Nucleotide diversity, Selection, Duplication, Indel

MHC and nucleotide diversity

The human MHC spans approximately 4 Mb on the short arm of chromosome six and is characterised by polymorphism, sequence duplication, indels, and disease association (reviewed in Dawkins et al. 1999). There is substantial evidence to indicate that these characteristics are closely related (Dawkins et al. 1999).

Nucleotide diversity within the human genome has been estimated to be between 0.08-0.2% (Li and Saddler 1991; Rowen et al. 1996; Horton et al. 1998; Lai et al. 1998). A large increase in nucleotide diversity has been observed within the coding regions of the peptide presenting HLA Class I genes in the MHC (Parham et al. 1995), and has been attributed to balancing selection (Hughes and Nei 1988), with surrounding nucleotide diversity at neutral sites associated with a hitch-hiking effect (Grimsley et al. 1998; Horton et al. 1998; Satta et al. 1998).

Recently, the comparison of human haplotypes within the MHC has shown a high level (greater than 10%) of nucleotide diversity flanking the polymorphic HLA Class I and II genes. Horton et al. (1998) compared approximately 80 kb of sequence between two haplotypes within the Class II region, obtaining an average of 5% nucleotide diversity, and demonstrating that high levels of nucleotide diversity extend beyond the HLA-DQB1 locus and includes retroelement indels. Similarly, Guillaudeux et al. (1998) used the overlap between cosmids from the heterozygous cell line CGM1 to show extensive nucleotide diversity over approximately 40 kb of sequence between the HLA Class I genes, HLA-B and -C. The level of nucleotide diversity peaked at 7% in a region containing a retroviral sequence (Guillaudeux et al. 1998). These two examples of nucleotide diversity within the MHC, as expected (Abraham et al. 1993; Gaudieri et al. 1997c), differ greatly with the level of nucleotide diversity detected in other regions of the genome (0.08-0.2%) (Li and Saddler 1991; Rowen et al. 1996; Horton et al. 1998; Lai et al. 1998).

In *Drosophila*, it has been shown that the hitch-hiking effect on neutral sites from balancing selection acting upon a nearby gene is affected by the mutation and recombination rate (Kreitman and Hudson 1991; Aquadro 1992). Given the variation in the recombination rate in the 4 Mb of sequence within the MHC (Carrington 1999), we examined a region of the MHC termed a PFB; specifically the beta block containing HLA-B and -C (Marshall et al. 1993; Dawkins et al. 1999).

Polymorphic frozen blocks

The MHC has been described as a composite of PFBs (Marshall et al. 1993). These blocks are approximately 200-300 kb in length, and their combination is observed in a population as MHC haplotypes (Degli-Esposti et al. 1992b), however, the blocks themselves have changed little, or remained frozen (Gaudieri et al. 1997c). Recombination has been observed between the blocks but rarely, if ever, within the blocks (Dawkins et al. 1999). The blocks are thought to be characterised by continuous nucleotide diversity and recombination suppression, and evidence from sequencing and family studies supports these features (Degli-Esposti et al. 1992b; Abraham et al. 1993; Gaudieri et al. 1997c; Dawkins et al. 1999).

Given that recombination is rare within the blocks (Dawkins et al. 1999), we could examine the expected nucleotide diversity pattern based on the assumptions of balancing selection acting on the HLA Class I loci. The expectation would be an elevated pattern of nucleotide diversity within the region of low recombination rate with a single peak in nucleotide diversity at the locus under balancing selection.

We examined the pattern of nucleotide diversity in continuous sequence within the beta block from IkBL to telomeric of HLA-C. The nucleotide diversity was much greater than reported elsewhere in the genome and had several peaks and troughs (Fig. 1).

Nucleotide diversity of continuous sequence within the beta block of the MHC

Approximately 380 kb of continuous sequence from two sequencing groups spanning the beta block region from IkBL to telomeric of HLA-C (Mizuki et al. 1997; Guillaudeux et al. 1998; Shiina et al. 1998) allowed an extension of our earlier analyses of the level of nucleotide diversity between two haplotypes distant from the HLA Class I loci (Abraham et al. 1993; Gaudieri et al. 1997c). Sequences were aligned using the program ClustalW (http://www.ddbj.nig.ac.jp/E-mail/clustalw-e.html) and the resultant outputs were used in the program CLTOSS (http://193.50.234.246/ ~beaudoin/anrs/cgi-bin/Pre_align_process2.cgi). CLTOSS removed all gaps from the alignments to normalize the number of nucleotides examined in each window. The gap-stripped alignments were used in the in-house program Window6.pl using a window size of 100 nucleotides.

The sequence alignment was broken into three different haplotype comparisons. From IkBL to MICA (PERB11.1), cosmids from the Mann cell line (HLA-A29; -B44; -Cw4; -DR7) (AC004181; AC006046; AC004183; AC004184; AC004215; AC004214) (Guillaudeux et al. 1998) were compared to the Boleth cell line (HLA-A2; -B62; -Cw10; -DR4) (AB000882) (Shiina et al. 1998). From MICA (PERB11.1) to HERV-I the Mann cell line (AC004180; and AC004182) was compared to the heterozygous CGM1 cell line (in this comparison HLA-A3; -B8; -Cw-; -DR3) (D84394) (Mizuki et al. 1997). The region from HERV-I to telomeric of HLA-C was compared between the two haplotypes in CGM1 (HLA-A3,29; -B8,14; -Cw-,-; -DR3,7) (AC004205; AC004204; AC006048; AC004185 ; and AC006047 were compared against D84394) (Mizuki et al. 1997; Guillaudeux et al. 1998).

To determine the level of sequence error between the sequencing groups from

Japan (Tokai University) and the US (University of Washington), we compared sequence from the same haplotype. In this case, cosmid Y5C028 (AC004210) was compared against D84394 with a resultant substitution and indel error rate of less than 0.05% (Mizuki et al. 1997; Guillaudeux et al. 1998).

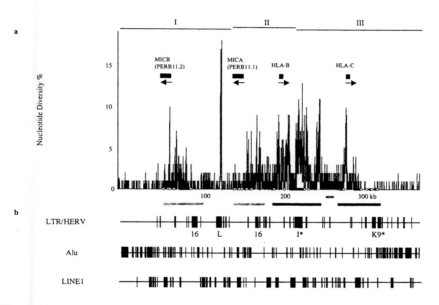

Fig. 1. Nucleotide diversity profile from IkBL to telomeric of HLA-C. **a.** The graph depicts the nucleotide diversity pattern from IkBL to telomeric of HLA-C determined from the pairwise comparison of three cell lines Mann, Boleth and CGM1. The comparison between Mann and Boleth is indicated by I, Mann and CGM1 by II and between the two CGM1 haplotypes by III. The level of the nucleotide diversity varies greatly along the length of the comparison from 0-18% (18% within a (TA)n rich region in the LTR sequence of HERV-L). The HLA-B, -C, and MICA (PERB11.1) and MICB (PERB11.2) genes are indicated by black boxes with their transcription direction indicated by an arrowhead. The duplication segments of HLA-B and -C are indicated by black horizontal lines, whereas the MIC (PERB11) duplication segments are indicated by grey horizontal lines. The horizontal double line corresponds to approximately 10 kb of sequence between the HLA duplication segments, which shares similarity with a telomeric region between HLA-30 and MICC (PERB11.3). Indels are indicated by grey circles. **b.** Retroelement sequences were identified by RepeatMasker2 (http://ftp.genome.washington.edu/cgi-bin/RepeatMasker) and its output was illustrated using an in-house program called DrawRep.pl. The HERV-I and HERV-K91 sequences have inserted within the duplicated segments and are indicated by an asterisks.

Nucleotide diversity greater than elsewhere in the genome with peaks at HLA Class I and non-HLA Class I loci

The nucleotide diversity pattern we obtain within the beta block of the MHC is extreme and interrupted with several peaks (Fig. 1). High nucleotide diversity (greater than 1%) is maintained at nucleotide sites distant from the highly polymorphic Class I loci. With the addition of indels, the number is even greater (Table 1, Fig. 2). Large indels greater than 100 bp contain retroelement sequences (Table 1, Fig. 2) and these have proven particularly useful as molecular clocks in human evolution, and in disease associations (Kulski et al. 1997).

The level of nucleotide diversity peaks at 18% within a (TA)n rich region in the 5' LTR of HERV-L (Fig. 1). This 2.5 kb (TA)n rich region is difficult to align and alternative alignments may alter the level of nucleotide diversity. However, G and C interruptions within the (TA)n pattern in both sequences contribute the bulk of the observed differences. The HERV-I and surrounding L1 sequences, and another retroelement cluster of L1 and Alu sequences form other peaks at greater than 10% nucleotide diversity. The level of nucleotide diversity is several fold greater than previously reported elsewhere in the genome: 0.2% in the TCR complex (Rowen et al. 1996); 0.09% over 138 kb in a non-MHC linked region on 6p23 containing the polymorphic SCA1 gene (Horton et al. 1998); 0.08% from a set of autosomal sequences (Li and Saddler 1991); and 0.09% in a 4 Mb region containing APOE (Lai et al. 1998).

Table 1. Nucleotide diversity within the centromeric Class I region of the MHC (beta block)

Region	Cell line	HLA alleles	Length kb[1]	Nucleotide diversity[2] %	Indels[3] (<100bp) %	No of indels (>100bp)	Indels (>100bp) composition
MHC (6p21.3)	Mann vs Boleth	A29; B44; Cw4; DR7 vs A2; B62; Cw10; DR4 (I)[4]	138.7	0.45	0.07	4	2-Alu;1-SVA; 1-simple repeat
	Mann vs CGM1	A29; B44; Cw4; DR7 vs A3; B8; Cw-; DR3 (II)[4]	74.2	1.3	0.12	0	
	CGM1 vs CGM1	A3; B8; Cw-; DR3 vs A29; B14; Cw-; DR7 (III)[4]	160.7	0.9	0.04	4	2-Alu; 1-SVA; 1-L1+Alu

[1] Length of comparison minus indels
[2] Average nucleotide diversity per 100 nucleotides corrected for multiple substitutions by Kimura's two parameter model (Kimura 1981) based on actual transition:transversion ratios
[3] Average number of indels per 100 nucleotides, consecutive indel sites counted as a single event
[4] I-III corresponds to Figure 1

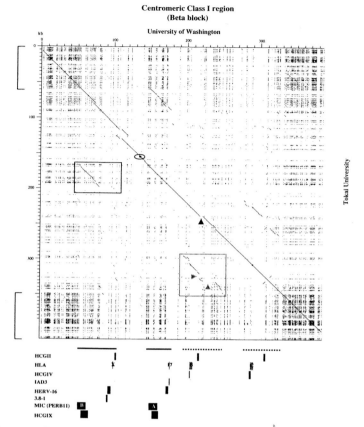

Fig. 2. Segmental duplication of linked multicopy gene families. Dotplot analysis of the region from IkBL to telomeric of HLA-C between the sequences from the Japanese (Tokai University) and US (University of Washington) groups. The large diagonal line corresponds to the alignment between the two sequences. An example of an indel between the sequences is indicated by a black arrowhead. Diagonal lines within the dotplots indicate regions of duplication, with interruptions corresponding to indels. Within the dotplot, the two duplications sets are indicated by black and dashed boxes, and examples of indels between the segments are indicated by grey arrowheads. The position of the multicopy gene families HLA Class I, HCG (II, IV and IX), 1AD3, HERV-16 (P5 + PERB3), 3.8-1 and MIC (PERB11) are shown below the x-axis. Based on the duplication endpoints in the dotplot and the ordering of the multicopy gene families, the duplicated segments are shown as black and dashed horizontal lines corresponding to the boxes. For the well characterised HLA Class I and MIC (PERB11) members, transcribed genes are indicated by grey boxes. The HCG, 1AD3 and HERV-16 families are shown by black boxes. The two Alu-rich regions are indicated by open brackets. The (TA)n rich region within the 5' LTR of HERV-L is shown by a black circle.

Although the HLA-B and -C loci exhibit nucleotide diversity of approximately 10%, they do not correspond to the highest peaks (Fig. 1). The haplotype comparisons are different between HLA-B and the peak containing HERV-I and its surrounding sequence, but a comparison of the HLA-B alleles from corresponding haplotypes shows a similar level of nucleotide diversity to that obtained in Figure 1 (Guillaudeux et al. 1998; Gaudieri et al., in press).

PFB boundaries correspond to low levels of nucleotide variability

The variation in the level of nucleotide diversity does not appear to be random, but rather forms distinct troughs and peaks (Fig. 1). Two regions of low nucleotide diversity (0-2%) centromeric of MICB (PERB11.2) and telomeric of HLA-C coincide with the proposed centromeric and telomeric boundaries of the beta block (Marshall et al. 1993; Dawkins et al. 1999). These two regions do not show greater than 2% nucleotide diversity within a single 100 nucleotide window, and on average the nucleotide diversity is well below 1% (Fig. 1). The endpoints of the regions of low nucleotide diversity coincide with the centromeric and telomeric ends of the available sequence, and therefore, is likely to extend on either side. From Figure 1, it can be seen that these regions are Alu-rich and contain the old and more recent Alu J and Alu Y subtypes (Kapitonov and Jurka 1996), respectively. Alus, particularly the 3' poly A tail, are associated with microsatellites and polymorphism (Epstein et al. 1990; Gaudieri et al. 1997a). Given that older Alu J and S subtypes are within the region, some form of sequence exchange, recombination may be maintaining the low diversity.

A decrease in nucleotide diversity is expected at the ends of the PFBs where recombination may occur, this is reflected in the pattern observed in Figure 1. Similarly, hitch-hiking from balancing selection acting on the HLA loci would also predict a similar a decrease in nucleotide diversity when the recombination rate increases. However, a hitch-hiking effect would expect a single peak at the loci reported to be under balancing selection, but this is clearly not the case. The nucleotide diversity pattern within the beta block is extreme with several peaks and troughs. Although a hitch-hiking effect from balancing selection operating on the HLA loci is likely to contribute to the level of nucleotide diversity observed in the beta block, the peaks and troughs suggest that other factors affect nucleotide diversity within the MHC. The beta block is subject to duplications and indels (Gaudieri et al. 1997a, b; Kulski et al. 1997) and we will examine the influence these processes have on the pattern of nucleotide diversity.

Duplication within the MHC

Multicopy gene families and duplication within the MHC

The organisation and evolution of the HLA Class I sequences within the MHC has been explained by chromosomal rearrangements involving a series of duplications and deletions (Klein et al. 1998; Geraghty et al. 1992; Vernet et al. 1993) and interlocus recombinations (Hughes and Nei 1989). More recently several groups favour the occurrence of repeated HLA Class I gene duplications whereby some genes have been deleted or become nonfunctional (Nei et al. 1997; Klein et al. 1998). The recent availability of genomic sequence for the central and telomeric region of the MHC, containing the alpha (spanning HLA-A, -G and -F) and beta blocks, has allowed the examination of the genomic organisation of the multicopy gene families HLA Class I (Geraghty et al. 1992), MIC (PERB11) (Bahram et al. 1994; Leelayuwat et al. 1994), P5 (Vernet et al. 1993), PERB3 (Marshall et al. 1993), 3.8-1 (Pichon et al. 1997), HCGII, IV, VIII and IX (Pichon et al. 1996b) and 1AD3 (Totaro et al. 1995). The organisation of the multicopy gene families in the two blocks reflect a common origin and duplicative history (Leelayuwat et al. 1995; Pichon et al. 1996a).

Dotplot analysis of continuous sequence within the MHC

The sequences from the Japanese and US groups spanning IkBL to telomeric of HLA-C were compared against each other using the dot matrix program Dotter (Sonnhammer and Durbin 1996). Duplicated sequences could be identified as diagonal lines corresponding to consecutive dots, which represent windows of similarity based on an assigned level of stringency (Fig. 2). The diagonal lines are interrupted by indels within the x- or y-axis of the sequence.

The locations of the multicopy gene family members of HLA Class I, HERV-16 (P5+PERB3), MIC (PERB11), 1AD3, 3.8-1, and HCGII, IV and IX on the sequences were determined using the program BLASTn (National Center for Biotechnology Information, Bethesda, Maryland) and the annotation information of the database entry.

Segment duplications

The arrangement of the multicopy gene families and the lines within the dotplots indicate they are linked and form two sets of duplicated segments. These duplications contain segments that share approximately 30 kb of sequence, including coding and intergenic sequences. The two large segmental duplications contain the transcribed genes HLA-B and -C, and MICA (PERB11.1) and MICB (PERB11.2) (Gaudieri et al. 1997b; Kulski et al. 1997). Based on L1 and Alu dating, the two duplications occurred as separate events resulting in four distinct segments (Gaudieri et al. 1997a; Kulski et al. 1997; Yamazaki et al. 1999). The duplications are imperfect and duplication/truncation sites of the segments are associated with retroelement sequences. An important consequence of the duplications is the increased possibility of recombination between paralogous retroelements and other repetitive sequences resulting in genomic changes (Mazzarella and Schlessinger 1997).

Within the beta block, there is little sequence that can be regarded as unique (Fig. 1). Furthermore, when the alpha and beta blocks are compared there are extensive similarities, suggesting that each block has been subject to cycles of duplication following the initial event that created each block. However, the blocks differ in the mode of duplication, the alpha block is characterised by multisegment duplications creating an exponential increase in the multicopy gene families, whereas in the beta block two separate duplication events have given rise to four segments (Gaudieri et al. 1999; Kulski et al. 1999c).

Retroelement indels

The major differences between the duplications are large indels (greater than 100 bp) containing retroelements (Alu, L1, HERV, MaLR and processed pseudogenes) and other repetitive sequences (MIR, MER) (Gaudieri et al. 1997a, b; Kulski et al. 1997). The insertion sites of sequences within the duplications of the beta block are closely associated with retroelement sequences (Gaudieri et al. 1997a, b; Kulski et al. 1997). There is increasing evidence that retroelements contribute to the recombination and duplications observed in the MHC (Kulski et al. 1997; Andersson et al. 1998) and in other gene clusters in the genome (Erickson et al. 1992). A HERV-K (C4) in the C4 region of the MHC has contributed to interlocus and interallelic length heterogeneity of the complement gene locus (*C4*) by integration into intron 9 of C4A genes and in some C4B genes (Chu et al. 1995). Furthermore, instability of operon structures in bacteria has been shown to be associated with transposon insertion sequences (Itoh et al. 1999).

Nucleotide diversity and duplications

Except for the peak in nucleotide diversity in the (TA)n-rich region within the 5'LTR of HERV-L, all other peaks are contained within the duplicated segments in the beta block (Fig. 1). Although the HLA duplication occurred prior to the MIC (PERB11) duplication (Gaudieri et al. 1997a; Yamazaki et al. 1999), all segments contain high levels of nucleotide diversity (Fig. 1). In fact, each segment appears to have at least one peak over 5% and in the MICB (PERB11.2) segment there is a bell shaped curve of nucleotide diversity (Fig. 1). However, there are prominent troughs within the duplicated segments, that in some cases may be explained in relation to indels within the duplicated segments. If there is no selective pressure acting on the indels, the level of nucleotide diversity within the indels that occurred subsequent to the duplication events should be related to the age of the insertion event. There have been several retroelement indels and transpositions within the beta block and we will examine these sequences.

To determine the age of the inserted HERV sequences we examined their open reading frames (ORFs) and internal insertions. The dot plot in Figure 2 shows that both HERV-16 elements are part of the MIC (PERB11) duplication segments. In contrast, HERV-I and HERV-K91 appear to have been inserted after the duplication of the HLA-B and -C genomic segments (Fig. 2). Analysis of ORFs revealed the absence of intact coding sequence within the putative gag, pol and env gene regions of HERV-I and HERV-L with the exception of HERV-K91 that had two large ORFs within the region of the putative protease and pol genes (Kulski et al. 1999b). HERV-I appears to have been a receptors for the insertion of other retrotransposons including Alu elements and fragments of L1 and THE1 (Kulski et al. 1999b).

Although the HERV-I sequence is likely to be an old insertion into the HLA-B segment, the indel event occurred subsequent to the segment duplication and would be expected to exhibit a lower level of nucleotide diversity than HLA-B. However, the level of nucleotide diversity within HERV-I and flanking L1 sequences is higher or equivalent to that observed at HLA-B and -C (Fig. 1). It remains to be shown what role, if any, HERV-I has within this region. The recently inserted HERV-K91 sequence, expectedly, exhibits a low level of nucleotide diversity (less than 1%). The duplicated HERV-16 sequences differ in the level of nucleotide diversity, however, this may be related to the function of the transcribed copy (see below).

A 10 kb region exhibiting a low level of nucleotide diversity between the segments containing HLA-B and -C is duplicated in a telomeric region between HLA-30 and MICC (PERB11.3) (Kulski et al. submitted), suggesting this may be a recent translocation.

If the indels and recent translocations are removed from Figure 1, the pattern of nucleotide diversity still contains peaks and troughs that suggest that factors besides a hitch-hiking effect from balancing selection operating on the antigen presenting HLA Class I loci is shaping the pattern of nucleotide diversity within the beta block. Other factors may include a mix of selective forces acting on other non-HLA Class I loci. The duplication time of segments also appear to affect the pattern of nucleotide diversity. Furthermore, when the indels are removed almost all troughs in nucleotide diversity occur outside of the duplication segments.

The pattern of nucleotide diversity is extreme within the boundaries of the proposed beta block. It is possible there is a suppression of recombination within the beta block due to the high level of nucleotide diversity. The level of nucleotide diversity may be maintaining the organisation of the beta block, especially as the region contains large duplicated segments on which each segment contains at least one polymorphic and transcribed gene. Moreover, recombination has been shown to be nucleotide homology dependent in yeast and mouse (Radman 1991; Yoshino et al. 1995).

Nucleotide diversity, duplications, indels and disease association

The beta block of the MHC has been associated with several diseases including susceptibility to autoimmune diseases such as Myasthenia Gravis (Degli-Esposti et al. 1992a), and rapid progression to AIDS following HIV-1 infection (Cameron et al. 1992). In contrast to humans, chimpanzees can be actively infected with HIV-1 and can mount an antibody response but in almost all cases, they do not progress to AIDS (Heeney et al. 1994). Within the beta block, chimpanzees exhibit a large 130 kb deletion containing at least one copy of MIC (PERB11) and retroelements such as HERV-16 and most likely HERV-L (Leelayuwat et al. 1993). The duplicated HERV-16 sequences (P5-1 and P5-8) differ in the level of nucleotide diversity (Fig. 1). P5-1 is transcribed in certain lymphoid cells and tissues in an antisense direction to the HERV-16 sequence (Vernet et al. 1993) and it has been suggested that this transcript may have an antiviral role (Kulski et al. 1999a). Retroelement sequences (including HERVs) within the beta block can exhibit high levels of nucleotide diversity, including indels, which may be involved in the variation of the host's response to itself and to infectious agents.

The role of HERVs and retroelements in the organisation and nucleotide diversity within the MHC has been shown, however, their role in disease is not as well established in the MHC, but is certainly an area of increasing interest. While HERVs within the MHC predominantly contain many stop codons suggesting that they are incapable of producing infectious virions, they may still be transcriptionally and translationally active. The processes of interaction may involve competition for cell receptors, modulation of nearby genes and induction of antibody responses to viral antigens (Lower et al. 1996; Kulski et al. 1999b).

Conclusion

(1) We show that nucleotide diversity within the beta block of the MHC containing HLA-B and -C is several fold greater than elsewhere within the genome.

(2) The pattern of nucleotide diversity shows several peaks and troughs. Several of the peaks in nucleotide diversity do not correspond to the HLA Class I loci under balancing selection, suggesting that other factors besides hitch-hiking is operating on the pattern of nucleotide diversity.

(3) Nucleotide diversity is influenced by other evolutionary processes within the MHC including duplications and indels.

(4) The pattern of nucleotide diversity reflects the existence of PFBs within the MHC, whereby extreme nucleotide diversity exists within the block, flanked by low levels of nucleotide diversity.

(5) Nucleotide diversity may act to repress recombination within the block with potential recombination areas most likely occurring outside the block in the

regions of low nucleotide diversity; centromeric of MICB (PERB11.2) and telomeric of HLA-C.

(6) Nucleotide diversity, duplications and indels, and disease association are closely related.

Acknowledgments

We thank Dr Kazuho Ikeo and Professor Joergen Epplen for helpful suggestions with the manuscript. SG is supported by a Japanese Society for the Promotion of Science (JSPS) fellowship (97119). JKK and RLD are grateful for support from a grant (981719) awarded by the National Health and Medical Research Council, Australia.

References

Abraham LJ, Grimsley G, Leelayuwat C, Townend DC, Pinelli M, Christiansen FT, Dawkins RL (1993) A region centromeric of the major histocompatibility complex class I region is as highly polymorphic as HLA-B. Implications for recombination. Hum Immunol 38:75-82

Andersson G, Svensson A, Setterblad N, Rask L (1998) Retroelements in the human MHC class II region. Trends Genet 14:109-114

Aquadro CF (1992) Why is the genome variable? Insights from *Drosophila*. Trends Genet 8:355-362

Bahram S, Bresnahan M, Geraghty DE, Spies T (1994) A second lineage of mammalian major histocompatibility complex class I genes. Proc Natl Acad Sci USA 91:6259-6263

Cameron PU, Mallal SA, French MAH, Dawkins RL (1992) Central MHC genes between HLA-B and complement C4 confer risk for HIV-1 disease progression. In: Tsuji K, Aizawa M, Sasazuki T (eds) HLA 1991. Vol 2. Oxford University Press, New York, pp 544-547

Carrington M (1999) Recombination within the human MHC. Immunol Rev 167:245-256

Chu X, Rittner C, Schneider PM (1995) Length polymorphism of the human complement C4 gene is due to an ancient retroviral integration. Immunogenetics 12:74-81

Dawkins RL, Leelayuwat C, Gaudieri S, Tay GK, Hui J, Cattley S, Martinez P, Kulski JK (1999) Genomics of the major histocompatibility complex: Haplotypes, duplications, retroviruses and disease. Immunol Rev 167:275-304

Degli-Esposti MA, Andreas A, Christiansen FT, Schalke B, Albert E, Dawkins RL (1992a) An approach to the localization of the susceptibility genes for generalised myasthenia gravis by mapping recombinant ancestral haplotypes. Immunogenetics 35:355-364

Degli-Esposti MA, Leaver AL, Christiansen FT, Witt CS, Abraham LJ, Dawkins RL (1992b) Ancestral haplotypes: conserved population MHC haplotypes. Hum Immunol 34:242-252

Epstein N, Nahor O, Silver J (1990) The 3' ends of Alu repeats are highly polymorphic. Nucleic Acids Res 18:4634

Erickson LM, Kim HS, Maeda N (1992) Junctions between genes in the haptoglobin gene cluster of primates. Genomics 14:948-958

Gaudieri S, Giles K, Kulski JK, Dawkins RL (1997a) Duplication and polymorphism in the MHC: Alu generated diversity and polymorphism within the PERB11 gene family. Hereditas 127:37-46

Gaudieri S, Kulski JK, Balmer L, Inoko H, Dawkins RL (1997b) Retroelements and segmental duplications in the generation of diversity within the MHC. DNA Seq 8:137-141

Gaudieri S, Leelayuwat C, Tay GK, Townend DC, Dawkins RL (1997c) The Major Histocompatibility Complex (MHC) contains conserved polymorphic genomic sequences that are shuffled by recombination to form ethnic specific haplotypes. J Mol Evol 45:17-23

Gaudieri S, Kulski JK, Dawkins RL, Gojobori T (1999) Different evolutionary histories in two subgenomic regions of the Major Histocompatibility Complex. Genome Res 9:541-549

Gaudieri S, Kulski JK, Dawkins RL, Gojobori T (1999) Extensive nucleotide diversity within a 370 kb sequence from the central region of the Major Histocompatibility Complex. Gene:in press

Geraghty DE, Koller BH, Hansen JA, Orr HT (1992) The HLA class I gene family includes at least six genes and twelve pseudogenes and gene fragments. J Immunol 149:1934-1946

Grimsley C, Mather KA, Ober C (1998) HLA-H: a pseudogene with increased variation due to balancing selection at neighboring loci. Mol Biol Evol 15:1581-1588

Guillaudeux T, Janer M, Wong GK, Spies T, Geraghty DE (1998) The complete genomic sequence of 424,015 bp at the centromeric end of the HLA class I region: Gene content and polymorphism. Proc Natl Acad Sci USA 95:9494-9499

Heeney JL, Van Els C, De Vries P, Ten Haaft P, Otting N, Koornstra W, Boes J, Dubbes R, Niphuis H, Dings M, Cranage M, Norley S, Jonker M, Bontrop RE, Osterhaus A (1994) Major histocompatibility complex class I-associated vaccine protection from simian immunodeficiency virus-infected peripheral blood cells. J Exp Med 180:769-774

Horton R, Niblett D, Milne S, Palmer S, Tubby B, Trowsdale J, Beck S (1998) Large-scale sequence comparisons reveal unusually high levels of variation in the HLA-DQB1 locus in the Class II region of the human MHC. J Mol Biol 282:71-97

Hughes AL, Nei M (1988) Pattern of nucleotide substitution at major histocompatibility complex class I loci reveals overdominant selection. Nature 335:167-170

Hughes AL, Nei M (1989) Ancient interlocus exchange in the history of the HLA-A locus. Genetics 122:681-686

Itoh T, Takemoto K, Mori H, Gojobori T (1999) Evolutionary instability of operon structures disclosed by sequence comparisons of complete microbial genomes. Mol Biol Evol 16:332-346

Kapitonov V, Jurka J (1996) The age of Alu subfamilies. J Mol Evol 42:59-65

Kimura M (1981) Estimation of evolutionary distances between homologous nucleotide sequences. Proc Natl Acad Sci USA 78:454-458

Klein J, Sato A, O'hUigin C (1998) Evolution by gene duplication in the major histocompatibility complex. Cytogenet Cell Genet 80:123-127

Kreitman M, Hudson RR (1991) Inferring the evolutionary histories of the Adh and Adh-dup loci in *Drosophila melanogaster* from patterns of polymorphism and divergence. Genetics 127:565-582

Kulski JK, Gaudieri S, Bellgard M, Balmer L, Giles K, Inoko H, Dawkins RL (1997) The evolution of MHC diversity by segmental duplication and transposition of retroelements. J Mol Evol 45:599-609

Kulski JK, Dawkins RL (1999a) The P5 multicopy gene family in the MHC is related in sequence to human endogenous retroviruses HERV-L and HERV-16. Immunogenetics 49:404-412

Kulski JK, Gaudieri S, Inoko H, Dawkins RL (1999b) Comparison between two HERV-rich regions within the Major Histocompatibility Complex. J Mol Evol 49:84-97

Kulski JK, Gaudieri S, Martin A, Dawkins RL (1999c) Coevolution of MIC (PERB11) and HLA Class I Genes with HERV-16 and retroelements by extended genomic duplication. J Mol Evol 48:675-683

Lai E, Riley J, Purvis I, Roses A (1998) A 4-Mb high-density single nucleotide polymorphism-based map around human APOE. Genomics 54:31-38

Leelayuwat C, Zhang WJ, Townend DC, Gaudieri S, Dawkins RL (1993) Differences in the central MHC between humans and chimpanzees: Implications for development of autoimmunity and acquired immune deficiency syndrome. Hum Immunol 38:30-41

Leelayuwat C, Townend DC, Degli-Esposti MA, Abraham LJ, Dawkins RL (1994) A new polymorphic and multicopy MHC gene family related to nonmammalian class I. Immunogenetics 40:339-351

Leelayuwat C, Pinelli M, Dawkins RL (1995) Clustering of diverse replicated sequences in the MHC: Evidence for en bloc duplication. J Immunol 155:692-698

Li W-H, Saddler LA (1991) Low nucleotide diversity in man. Genetics 129:513-523

Lower R, Lower J, Kurth R (1996) The viruses in all of us: Characteristics and biological significance of human endogenous retrovirus sequences. Proc Natl Acad Sci USA 93:5177-5184

Marshall B, Leelayuwat C, Degli-Esposti MA, Pinelli M, Abraham LJ, Dawkins RL (1993) New major histocompatibility complex genes. Hum Immunol 38:24-29

Mazzarella R, Schlessinger D (1997) Duplication and distribution of repetitive elements and non-unique regions in the human genome. Gene 205:29-38

Mizuki N, Ando H, Kimura M, Ohno S, Miyata S, Yamazaki M, Tashiro H, Watanabe K, Ono A, Taguchi S, Sugawara C, Fukuzumi Y, Okumura K, Goto K, Ishihara M, Nakamura S, Yonemoto J, Kikuti YY, Shiina T, Chen L, Ando A, Ikemura T, Inoko H (1997) Nucleotide sequence analysis of the HLA class I region spanning the 237 kb segment around the HLA-B and -C genes. Genomics 42:55-66

Nei M, Gu X, Sitnikova T (1997) Evolution by the birth-and-death process in multigene families of the vertebrate immune system. Proc Natl Acad Sci USA 94:7799-7806

Parham P, Adams EJ, Arnett KL (1995) The origins of HLA-A,B,C polymorphism. Immunol Rev 143:141-180

Pichon L, Carn G, Bouric P, Giffon T, Chauvel B, Lepourcelet M, Mosser J, Legall J, David V (1996a) Structural analysis of the HLA/HLA-F subregion: Precise localization of two new multigene families closely associated with the HLA Class I sequences. Genomics 32:236-244

Pichon L, Hampe A, Giffon T, Carn G, Legall JY, David V (1996b) A new non-HLA multigene family associated with the PERB11 family within the MHC class I region. Immunogenetics 44:259-267

Pichon L, Venditti C, Harris J, Elshof A, Pinelli M, Chorney M (1997) Studies of the 6.7 family of dispersed genomic fragments within the MHC class I region. Exp Clin Immunogenet 14:131-140

Radman M (1991) Avoidance of inter-repeat recombination by sequence divergence and a mechanism of neutral evolution. Biochimie 73:357-361

Rowen L, Koop BF, Hood L (1996) The complete 685-kilobase DNA sequence of the human b T cell receptor locus. Science 272:1755-1762

Satta Y, Li YJ, Takahata N (1998) The neutral theory and natural selection in the HLA region. Front Biosci 27:d459-467

Shiina T, Tamiya G, Oka A, Yamagata T, Yamagata N, Kikkawa E, Goto A, Mizuki N, Watanabe K, Fukuzumi Y, Taguchi S, Sugawara C, Ono A, Chen L, Yamazaki M, Tashiro H, Ando A, Ikemura T, Kimura M, Inoko H (1998) Nucleotide sequencing analysis of the 146-kilobase segment around the IkBL and MICA genes at the centromeric end of the HLA Class I region. Genomics 47:372-382

Sonnhammer ELL, Durbin R (1996) A dot-matrix program with dynamic threshold control suited for genomic DNA and protein sequence analysis. Gene 167:GC1-10

Totaro A, Grifa A, Roetto A, Lunardi C, D'Agruma L, Sbaiz L, Zelante L, De Sandre G, Camaschella C, Gasparini P (1995) New polymorphisms and markers in the HLA class I region: relevance to hereditary hemochromatosis (HFE). Hum Genet 95:429-434

Venditti CP, Chorney MJ (1992) Class I gene contraction within the HLA-A subregion of the human MHC. Genomics 14:1003-1009

Venditti CP, Harris JM, Geraghty DE, Chorney MJ (1994) Mapping and characterization of non-HLA multigene assemblages in the human MHC class I region. Genomics 22:257-266

Vernet C, Ribouchon MT, Chimini G, Jouanolle AM, Sidibe I, Pontarotti P (1993) A novel coding sequence belonging to a new multicopy gene family mapping within the human MHC class I region. Immunogenetics 38:47-53

Watanabe Y, Tokunaga K, Geraghty DE, Tadokoro K, Juji T (1997) Large-scale comparative mapping of the MHC class I region of predominant haplotypes in Japanese. Immunogenetics 46:135-141

Yamazaki M, Tateno Y, Inoko H (1999) Genomic organisation around the centromeric end of the HLA Class I region: Large-scale sequence analysis. J Mol Evol 48:317-327

Yoshino M, Sagai T, Lindahl KF, Toyoda Y, Moriwaki K, Shiroishi T (1995) Allele-dependent recombination frequency: homology requirement in meiotic recombination at the hot spot in the mouse major histocompatibility complex. Genomics 27:298-305

3. Function and evolutionary dynamics of MHC genes

Proteasomes and MHC class I-peptide generation

Keiji Tanaka[1], Nobuyuki Tanahashi[1], and Naoki Shimbara[2]

[1]The Tokyo Metropolitan Institute of Medical Science, and CREST, Japan Science and Technology Corporation (JST), 18-22 Honkomagome, Bunkyo-ku 3-chome, Tokyo 113-8613, Japan
[2]Department of Biomedical R&D, Sumitomo Electric Industries, 1 Taya-cho, Sakae-ku, Yokohama 244, Japan

Summary. Proteasomes are processing enzymes capable of generating major histocompatibility complex (MHC) class I-ligands from endogenous antigens. Thus far, immunoproteasomes and the proteasome activator PA28 induced by γ-interferon have been shown to play a central role in the production of immuno-dominant MHC class I-ligands in cells. However, the antigen processing mechanism, particularly the mechanism of how proteasomes correctly excise class I-ligands without destroying them, is poorly understood. In this article, we discuss the role of the regions adjacent to a class I epitope in its dual-cleavage excision. Our results suggest that such regions may function as anchors to trap target peptides for proteasomal degradation. Furthermore, we propose that proline residue(s) within epitopic sequences presumably contributes to efficient production of MHC class I-ligands through prevention of their random cleavage by proteasomes.

Key words. Antigen processing, Immunoproteasome, MHC class I, PA28, Proteasome

Introduction

Cytotoxic T lymphocytes (CTLs) recognize small peptides derived from intracellular (also called endogenous) antigens in association with polymorphic MHC class I molecules. Production of immunodominant epitopes destined for presentation by MHC class I molecules occurs mainly in the cytosol by limited digestion of antigenic proteins. The resulting peptides are then transported into the endoplasmic reticulum (ER) by the transporter associated with antigen processing (TAP), where they become associated with MHC class I molecules and are then delivered to the cell surface to activate CTLs (Momburg and Hämmering 1998;

Rock and Goldberg 1999).

Most cellular proteins are covalently modified by attachment with multiple ubiquitins (Ubs) to form a poly-Ub chain functioning as a degradation signal (Hershko and Ciechanover 1998). Subsequent destruction of poly-ubiquitinylated proteins is catalyzed by the 26S proteasome, a eukaryotic ATP-dependent, 2-2.5 MDa protease complex (Fig. 3, panel A), composed of a core proteinase, known as the 20S proteasome, and a pair of symmetrically disposed PA700 regulatory particles (also called a 19S complex) (Coux et al. 1996; Rechsteiner 1998; Baumeister et al. 1998). The 20S proteasome and the PA700/19S complex consist of 14 and at least 20 different subunits, respectively (Tanaka 1998). The regulator PA700 is attached to both ends of the central 20S proteasome in opposite orientations to form the enzymatically active 26S proteasome. Accumulated evidence indicates that the 26S proteasome contributes not only to almost complete degradation of most cellular proteins but also to generation of such antigenic peptides as MHC class I-ligands, because the selective inhibitors of proteasomes almost completely block MHC class I-mediated antigen presentation (Tanaka et al. 1997; Rock and Goldberg 1999). A simplified model of this pathway is depicted in Figure 1.

MHC class I molecules expressed on the plasma membranes of cells are tightly associated with antigenic peptides with sizes of mostly 8-10 amino acid residues, indicating that processing of antigenic proteins by the proteasome must be occurring strictly in the cytoplasm. However, how the proteasome correctly generates immunodominant MHC class I-peptides is poorly understood. In this review, we briefly summarize our current understanding of the antigen processing pathway mediated by the proteasome (for further information, see recent reviews by Tanaka and Kasahara 1998, and Rock and Goldberg 1999).

Proteasomes and γ-Interferon regulation

The 20S proteasome acts as protein-destroying machinery with a variety of catalytic centers that presumably contribute to the hydrolysis of multiple peptide bonds in single polypeptide substrates by a coordinated mechanism (Coux et al. 1996; Baumeister 1998). The 20S proteasome with a molecular mass of 700-750 kDa is a barrel-like particle appearing as a stack of four rings made up of two outer α-rings and two inner β-rings, associated in the order of αββα. The α- and β-rings are each made up of seven structurally similar α- and β-subunits, respectively.

Immunoproteasomes

To date, cDNAs or genes encoding all subunits of human and budding yeast 20S proteasomes have been isolated by molecular-biological techniques (Tanaka

1998). As expected, 7 α- and 7 β-subunits have been found in yeast; however, 7 α- and 10 β-subunits were found in humans. The existence of three extra β-subunits was initially puzzling, because the β-ring is composed of seven subunits. Recently, γ-interferon (γ-IFN), a major immunomodulatory cytokine, was shown to induce replacement of 3 constitutively expressed β-type subunits (X, Y and Z) by γ-IFN-inducible subunits (termed LMP7, LMP2, and MECL1, respectively), with high amino acid sequence similarity to the subunits replaced (Tanaka et al. 1997; Früh and Yang 1999). Because the proteasome containing the γ-IFN-inducible subunits appears to be specialized for the production of MHC class I-

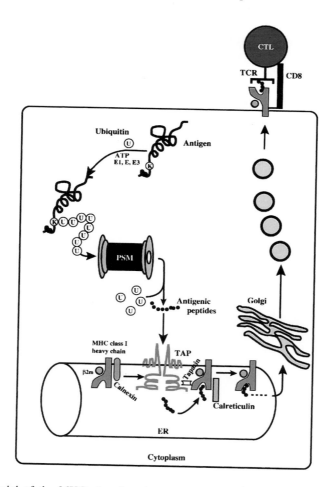

Fig. 1. Model of the MHC class I antigen processing pathway. U, ubiquitin; E1, Ub-activating enzyme; E2, Ub-conjugating enzyme; E3, Ub-protein ligase; PSM, proteasome; TAP, transporter associated with antigen processing; ER, endoplasmic reticulum; β2m, β2-microglobulin; CTL, cytotoxic T lymphocytes; TCR, T cell receptor.

ligands, we have proposed that such proteasomes should be called "immuno-proteasomes" (see the hypothetical model in Fig. 2) to distinguish them from those containing constitutively expressed subunits (Tanaka and Kasahara 1998).

Most β-type proteasomal subunits are synthesized as proproteins and processed to the mature forms by removal of their N-terminal pro-sequences (Schemidt and Kloetzel 1997). Recent evidence indicates that this precursor processing occurs by an autocatalytic mechanism. In addition, maturation of catalytically inactive β-type subunits appears to be exerted by catalytically active β-type subunits and to occur in the fully assembled 20S particle.

Accumulated evidence indicates that the three γ-IFN-inducible subunits replace the constitutive catalytic 20S subunits during proteasome biogenesis. Indeed, γ-IFN-inducible subunits are preferentially assembled into the 20S proteasome. Furthermore, MECL1 requires LMP2 for its efficient incorporation into preproteasomes, and the preproteasomes containing LMP2 and MECL1 require LMP7 for efficient maturation (Griffin et al. 1998). Thus, a mechanism exists that favors the assembly of homogeneous immunoproteasomes containing all three γ-IFN-inducible subunits (Tanaka and Kasahara 1998).

Of the 10 β-type subunits, only three pairs of γ-IFN-regulated subunits have proteolytically active threonine residues at their N-termini. Thus, the subunit exchanges induced by γ-IFN are likely to confer functional alterations upon proteasomes. In fact, γ-IFN alters the proteolytic specificity of the proteasome, increasing the tryptic and chymotryptic activities for cleavage of peptide bonds on

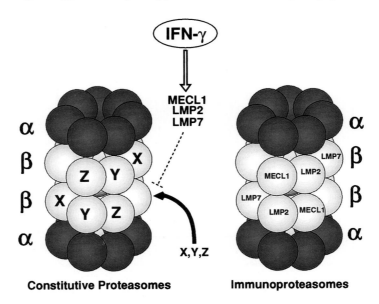

Fig. 2. Hypothetical model of the constitutive 20S proteasome and the γ-IFN-induced immunoproteasome containing LMP2, LMP7, and MECL1. For details, see text.

the carboxyl side of basic and hydrophobic amino acid residues, respectively, of fluorogenic substrates, but decreasing the caspase-like activities for peptides containing acidic amino acid residues (Rock and Goldberg 1999). These changes in peptidase activities suggest that, during protein breakdown, the proteasome from γ-IFN-treated cells should generate more peptides that have hydrophobic or basic carboxyl termini and fewer peptides with acidic carboxyl termini. Thus, the formation of immunoproteasomes should facilitate the production of MHC class I-binding peptides, because hydrophobic or basic carboxyl terminal residues normally serve as anchors for binding to MHC class I molecules (Rock and Goldberg 1999).

γ-IFN-inducible activator PA28

PA28 (also known as the 11S regulator) was identified as an activator protein of the 20S proteasome. Analysis of cDNAs showed that PA28 is composed of two related proteins, termed PA28α and PA28β, with an overall identity of approximately 50%. Although the PA28 complex greatly stimulates multiple peptidase activities of the 20S proteasome, it fails to enhance hydrolysis of large protein substrates with native or denatured structures, even though the proteins have already been ubiquitinylated. Thus, PA28 does not play a central role in the initial cleavage of protein substrates. It presumably has a stimulating effect on the degradation of polypeptides of intermediate size that are generated by the 26S proteasome, implying that the 26S proteasome and the PA28-proteasome complex may function sequentially or cooperatively. Intriguingly, γ-IFN has been found to increase the transcripts for PA28α and PA28β greatly and coordinately (Tanahashi et al. 1997; Tanaka and Kasahara 1998; Früh and Yang 1999). Moreover, PA28 has been shown to be involved in the generation of immunodominant MHC class I-ligands *in vitro* and *in vivo*. These findings strongly indicate potential involvement of PA28 in the immune response.

PA28 is a ring-shaped protein of approximately 200 kDa composed of hetero-hexameric or heteroheptameric PA28α and PA28β subunits (Song et al. 1996; Zhang et al. 1999) and binds directly to both ends of the cylindrical 20S molecule to form a conical or football-like structure (see Fig. 3, panel B) (Gray et al. 1994). Association of PA28 with the 20S proteasome does not require ATP-hydrolysis, in contrast to ATP-dependent assembly of the 26S proteasome. Recently, we found that an antibody against p45 ATPase, a subunit of the PA700 complex, immunoprecipitates not only a set of 26S proteasome components, but also PA28α and PA28β, indicating simultaneous binding of PA28 and PA700 to the 20S proteasome to form the PA700-20S-PA28 complex (here we use the simple abbreviation 20S for the 20S proteasome) (Hendil et al. 1998). Thus, this complex has been named provisionally the "hybrid-type" proteasome (Fig. 3, panel C) (Tanaka and Kasahara 1998). This type of proteasomes may contribute to more efficient proteolysis; perhaps intact substrate proteins are recognized first by

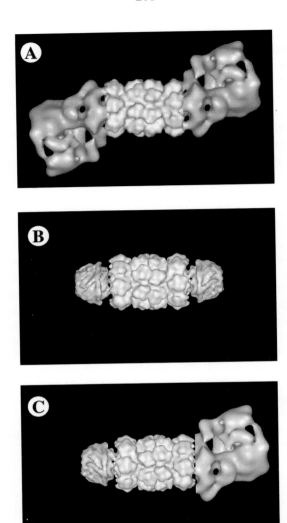

Fig. 3. Three-dimensional models of the proteasome. Panel A, the 26S proteasome made up of PA700-20S core-PA700; Panel B, the football-like proteasome made up of PA28-20S core-PA28; and Panel C, a "hybrid-type" proteasome made up of PA700-20S core-PA28. The structures in panels A and B were modeled based on the electron microscopic observation. The structure in panel C is not based on electron micrographs; it was created by synthesizing the structures shown in panels A and B. (Courtesy of Dr. W. Baumeister, Max-Planck-Institut für Biochemie, Martinsried, Germany)

PA700 and fed into the cavity of the 20S proteasome whose cleavage ability is greatly stimulated by PA28. Because PA28 occupies the same site on the 20S core particle as does PA700 of the 26S proteasome, when judged by electron microscopic analysis, it does not seem so surprising that both activators associate with the same 20S proteasome simultaneously in opposite orientations. Quite recently, we found that the "hybrid-type" proteasome is induced in response to γ-IFN stimulation (Tanahashi et al., submitted), implying that it presumably acts as an important processing enzyme responsible for the generation of MHC class I-ligands.

Mechanisms for production of immunodominant MHC class I-ligands

Proteasomes have multicatalytic protease functions that may be of advantage not only for rapid destruction of target proteins but also for selective cutting (Coux et al. 1996). To define the role of the proteasome in the antigen processing pathway, it is necessary to determine whether it is capable of yielding antigenic peptides correctly from antigenic proteins. There is now abundant evidence that the proteasome has the ability to yield MHC class I-ligands from various model substrates (Boes et al. 1994; Niedermann et al. 1995; Dick et al. 1996). However, there is as yet no satisfactory explanation of how the proteasome selects an epitope region within a protein sequence. In addition, little is known about how epitope sequences resist proteasomal digestion.

Double-cleavage production of the CTL epitope by proteasomes

By *in vitro* analyses, we revealed that the proteasome produces the tumor rejection antigen precursor peptide pRL1b (SIIPGLPLSL), but not pRL1a (IPGLPLSL) bound to the H-2L^d molecule, from the 29-mer synthetic peptides covering the CTL epitope that is derived from the c-*akt* proto-oncogene (Shimbara et al. 1997). The pRL1a peptide contains Pro as the second residue from the N-terminus, and certain other peptides containing Pro at this position appear to be poorly transported by the TAP system (Momburg and Hämmering 1998). Thus, the proteasome appears to generate peptide fragments that are efficiently transported by TAP, but that do not bind to class I molecules efficiently. It is interesting that certain peptides, such as pRL1a, that bind optimally to MHC class I molecules are poorly transported by TAP, but that when they are elongated N-terminally, their transport is improved. Consistent with this observation, it has been suggested that an unidentified protease(s) in the ER contributes to trimming of the N-terminus of peptides to optimize their binding to MHC class I molecules (Rock and Goldberg 1999).

The proteasome degrades polypeptides in a processive manner so that the pattern

of the products including pRL1b does not alter as a function of time during digestion (Shimbara et al. 1997). Interestingly, the double cleavage production of pRL-
1b by the proteasome seems to depend on the lengths of the flanking regions
adjacent to either end of the CTL epitope, because their successive deletions
caused almost complete prevention of pRL1b excision (Shimbara et al. 1997).
Thus, the flanking regions of suitable length appear to be essential for immunodominant peptide production, suggesting that the flanking regions presumably get
into the β-ring cavity of the proteasome complex and function as anchors for the
proteasome to trap the peptide. Based on these findings, we have proposed the
model called the "flanking-anchor hypothesis", which emphasizes the roles of the
flanking regions in correct CTL epitope production. According to this model, the
proteasome should catalyze efficient, one-step generation of an immunodominant
peptide by a double-cleavage mechanism, but should not generate peptides by a
sequential process through single-cleavage reactions. There is currently no direct
evidence that the proteasome has any specific structural motif(s) functioning as
peptide-binding anchors. Thus, it is also possible that specific residues or local
conformations of the flanking regions, rather than their lengths, are important for
anchoring peptides in the cavity of the proteasome complex.

Dick et al. (1996) reported that coordinated dual cleavages induced by the
proteasome regulator PA28 lead to predominant production of MHC class I-
ligands. Consistent with this report, we found that PA28 promotes dual cleavage
production of appropriate class I-ligands that could not be produced by the
proteasome alone (Shimbara et al. 1997). However, our result seems to differ
somewhat from that of Dick et al. (1996), because, in our experiments, PA28 also
promoted single cleavage of certain peptides without generating pRL1b. Thus,
PA28 collaborates with the proteasome for efficient production of
immunodominant peptides, by promoting not only single cleavage of all
susceptible peptides, but also dual cleavage in some peptides harboring certain
characteristic lengths.

Contribution of proline residues to efficient production of MHC class I-ligands by proteasomes

Accumulating evidence indicates that proteasomes can generate various CTL
epitopes or their precursor peptides *in vitro* from synthetic peptides, suggesting
the existence of the mechanism that prevents random cleavage of epitopic
sequences. However, it is not known why epitopic sequences, which contain
numerous amino acid residues susceptible to proteasomal digestion, are resistant
to proteasomal attack. Recently, we proposed that the Pro residue(s) within an
epitopic sequence plays an essential role in producing some epitopes (Shimbara et
al. 1998). The reason why we thought Pro residue(s) important is based on our
previous finding that most products of ornithine decarboxylase degraded *in vitro*
by the 26S ATP-dependent proteasome contained one or two Pro residues

(Tokunaga et al. 1994). This finding led us to the idea that the Pro residue has a role in the escape from random cleavage by the proteasome. Consistent with this assumption, replacement of the first or second Pro residue by alanine within two CTL epitopic precursor peptides, SIIPGLPLSL (pRL1b) and DMYPHFMPTNL (an epitope of the pp89 protein of mouse cytomegalovirus), resulted in markedly reduced production of epitopic peptides (Shimbara et al. 1998). Thus, Pro residue(s) presumably contributes to efficient production of MHC class I-ligands through prevention of their random cleavage by the proteasome. To emphasize the role of Pro residues in generating MHC class I-ligands, we shall tentatively designate our proposal as the "Pro-rule".

It is notable that approximately 40-50% of peptides bound to MHC class I molecules contain one or more Pro residues. However, over 50% of the peptides associated with MHC class I molecules do not contain Pro. Therefore, the proteasome has multiple, as yet unknown mechanisms for preventing complete degradation of target proteins. For example, an alternative mechanism for preventing random cleavage may be the recognition of the tertiary conformation of an epitope sequence. There may be a particular configuration or a polarity distribution that is resistant to proteasomal attack.

Perspectives

In this article, we summarized our recent studies. There appear to be two major unsolved questions concerning the roles of the proteasome as an antigen processing enzyme. First, it is totally unknown why the proteasome is capable of generating peptides that are efficiently transported by the TAP system and that finally fit in the pocket of the peptide-binding groove of MHC class I molecules. Second, how PA28 collaborates with the proteasome *in vivo* is an enigma. We proposed here two models that account for the generation of MHC class I-ligands by the proteasome: the "flanking-anchor hypothesis" and the "Pro-rule". However, other mechanisms are likely to exist that contribute to the generation of MHC class I-ligands. Further studies are required to uncover such mechanisms.

References

Baumeister W, Walz J, Zühl F, Seemüller E (1998) The proteasome: Paradigm of a self-compartmentalizing protease. Cell 92:367-380

Boes B, Hengel H, Ruppert T, Multhaup G, Koszinowski UH, Kloetzel PM (1994) Interferon γ stimulation modulates the proteolytic activity and cleavage site preference of 20S mouse proteasome. J Exp Med 179:901-909

Coux O, Tanaka K, Goldberg AL (1996) Structure and functions of the 20S and 26S proteasomes. Annu Rev Biochem 65:801-847

Dick TP, Ruppert T, Groettrup M, Kloetzel PM, Kuehn L, Koszinowski UH, Stevanovi´c

S, Schild H, Rammensee H-G (1996) Coordinated dual cleavages induced by the proteasome regulator PA28 lead to dominant MHC ligands. Cell 86:253-262

Früh K, Yang Y (1999) Antigen presentation by MHC class I and its regulation by interferon γ. Curr Opinion Immunol 11:76-81

Gray CW, Slaughter CA, DeMartino GN (1996) PA28 activator protein forms regulatory caps on proteasome stacked rings. J Mol Biol 236:7-15

Griffin TA, Nandi D, Cruz M, Fehling HJ, Kaer LV, Monaco JJ (1997) Immunoproteasome assembly: cooperative incorporation of interferon γ-inducible subunits. J Exp Med 187:97-104

Hendil KB, Khan S, Tanaka K (1998) Simultaneous binding of PA28 and PA700 activators to proteasomes. Biochem J 332:749-754

Hershko A, Ciechanover A (1998) The ubiquitin system. Annu Rev Biochem 67:425-479

Momburg F, Hämmerling G (1998) Generation and TAP-mediated transport of peptides for major histocompatibility complex class I molecules. Adv Immunol 68:191-256

Niedermann G, Butz, S, Ihlenfeldt HG, Grimm R, Maier B, Eichmann K (1995) Contribution of proteasome mediated proteolysis to the hierarchy of epitopes presented by major histocompatibility complex class I molecules. Immunity 2:289-299

Rechsteiner M (1998) The 26S proteasome. In: Peters JM, Harris JR, Finley D (eds) Ubiquitin and the Biology of the Cell. Plenum Press, New York, pp 147-189

Rock KL, Goldberg AL (1999) Degradation of cell proteins and the generation of MHC class I-presented peptides. Annu Rev Immnunol 17:739-779

Schmidt M, Kloetzel PM (1997) Biogenesis of eukaryotic 20S proteasomes: The complex maturation pathway of a complex enzyme. FASEB J 11:1235-1243

Shimbara N, Nakajima H, Tanahashi N, Ogawa K, Niwa S, Uenaka A, Nakayama E, Tanaka K (1997) Double-cleavage production of the CTL epitope by proteasomes and PA28: Role of the flanking region. Genes to Cells 2:785-800

Shimbara N, Ogawa K, Nakajima H, Yamasaki N, HidakaY, Niwa S, Tanahashi N, Tanaka K (1998) Contribution of proline residue for efficient production of MHC class I-ligands by proteasomes. J Biol Chem 273:23062-23071

Song X, Mott JD, von Kampen J, Pramanik B, Tanaka K, Slaughter CA, DeMartino DN (1996) A model for the quaternary structure of the proteasome activator, PA28. J Biol Chem 42:26410-26417

Tanaka K (1998) Molecular biology of proteasomes. Biochem Biophys Res Commun 247:537-541

Tanaka K, Tanahashi N, Tsurumi C, Yokota K, Shimbara N (1997) Proteasomes and antigen processing. Adv Immunol 64:1-38

Tanaka K, Kasahara M (1998) The MHC class I ligand generating system: Roles of immunoproteasomes and INF-γ inducible PA28. Immunol Rev 163:161-176

Tanahashi N, Yokota K, Ahn JW, Chung CH, Fujiwara T, Takahashi E, DeMartino GN, Slaughter CA, Toyonaga T, Yamamura K, Shimbara N, Tanaka K (1997) Molecular properties of the proteasome activator PA28 family proteins and γ-interferon regulation. Genes to Cells 2:195-211

Tokunaga F, Goto T, Koide T, Murakami Y, Hayashi S, Tamura T, Tanaka K, Ichihara A (1994) ATP- and antizyme-dependent endoproteolytic breakdown of ornithine decarboxylase to oligopeptides by the 26S proteasome. J Biol Chem 269:17382-17385

Zhang Z, Krutchinsky A, Endicott S, Realini C, Rechsteiner M, Standing KG (1999) Proteasome activator 11S REG or PA28: Recombinant REGα/REGβ hetero-oligomers are heptamers. Biochemistry 38:5651-5658

Comparative aspects of the MHC class I-related MR1, CD1D, and MIC genes in primates

Lutz Walter, Jung Won Seo, and Eberhard Günther

Division of Immunogenetics, University of Göttingen, Heinrich-Düker-Weg 12, D-37073 Göttingen, Germany

Summary. Homologous sequences of the MHC class I-related genes *MR1*, *CD1D*, and MIC have been isolated from different nonhuman primates. *MR1* pseudogenes were found in the genomes of gibbon and chimpanzee. In *MR1* exons 2 and 3 the frequency of nonsynonymous substitutions does not exceed that of synonymous substitutions in codons of the putative peptide-binding residues indicating that potential ligands of MR1 are probably conserved and not diverse. Gene trees constructed from MR1 and CD1D sequences show branching of Catarrhini and Platyrrhini species and relatively high degree of sequence conservation. Homologs of *MICA/MICB* as well as orthologs of *MICC*, *MICD*, and *MICE* genes could be identified in nonhuman primates. Gene tree analysis indicates that orthologs of human *MICA* and *MICB* genes are found in chimpanzee, gorilla, orangutan, and gibbon, whereas orthology is lost in Old World monkey species like rhesus macaques, baboon, African green monkey and guereza. It is concluded that the evolution of *MICA/MICB*-related genes is different from MHC class Ia genes because orthology is maintained for a shorter time.

Key words. Primates, MR1, CD1D, MIC, Orthology

In the genome of various mammalian species several genes have been identified that are more distantly related to classical and nonclassical MHC class I genes. Among these are the MHC-linked MIC genes as well as the *MR1* and CD1 genes, which both map outside the MHC.

The CD1 family encompasses 5 members in the human genome which are closely linked on chromosome 1q22-23 (Calabi and Milstein 1986; Yu and Milstein 1988). Based on evolutionary relationship the CD1 genes can be ordered into 3 groups: 1) *CD1B*, *CD1A* and *CD1C*, 2) *CD1D*, 3) *CD1E* (Hughes 1991). Only homologs of human *CD1D* have been identified in mouse (Balk et al. 1991) and rat (Ichimiya et al. 1994). CD1 molecules could be shown to present glycolipids or hydrophobic peptides to cytotoxic T cells (Castaño et al. 1995; Sieling et al. 1995; Burdin et al. 1998).

The recently identified *MR1* gene is linked to the CD1 cluster in human and maps to 1q25 (Hashimoto et al. 1995), whereas in mouse and rat *Mr1* is found on chromosomes 1 and 13, respectively, and thus, maps to a different chromosome than *Cd1* (Yamaguchi et al. 1997; Walter and Günther 1998; Walter et al. 1998). The human, mouse, and rat *MR1* genes are ubiquitously transcribed (Hashimoto et al. 1995; Yamaguchi et al. 1997; Walter and Günther 1998) and show alternative splicing (Riegert et al. 1998; Walter and Günther 1998). The function of the MR1 molecule is not known so far.

The family of MIC (Bahram et al. 1994) or PERB11 (Leelayuwat et al. 1994) genes encompasses 6 members which are dispersed over the human MHC class I region (Shiina et al. 1999). The stress-inducible *MICA* and *MICB* genes are expressed in epithelial cells of the small intestine and their products are recognized by γδT cells (Groh et al. 1996, 1998). In contrast to other class Ib genes, *MICA* and *MICB* show substantial polymorphism (Fodil et al. 1996, 1999; Ando et al. 1997; Komatsu-Wakui et al. 1999; Petersdorf et al. 1999; Visser et al. 1999).

In order to identify *MR1*, *CD1D*, and MIC genes in primates and to analyze their evolutionary conservation, oligonucleotides derived from the 5' and 3' ends of exons 2 and 3, respectively, of human *MR1* and *CD1D*, as well as oligonucleotides derived from the 5' ends of exon 4 and intron 5 of human *MICA*, and the 5' ends of exon 4 and intron 4 of human *MICC*, *MICD*, and *MICE* were designed and used in polymerase chain reaction (PCR). The sequences of these PCR products were determined and gene trees were constructed.

Figure 1 shows a gene tree constructed from *MR1* exons 2 and 3. The *Mr1* sequence of the rat was included as outgroup. The tree is split into two major branches corresponding to the Catarrhini and Platyrrhini species. Interestingly, the gibbon and chimpanzee sequences turned out to be derived from *MR1* pseudogenes. A deletion and consequent frame shift is found in exon 3 of the gibbon gene and a mutation in exon 3 resulting in a premature stop codon (TGG →TAG) in the chimpanzee gene. In the database also a human *MR1* pseudogene is found (EMBL/GenBank accession number AF073486), which significantly clusters with the chimpanzee pseudogene and not the functional human *MR1* gene (Fig. 1). Due to the presence of introns the *MR1* pseudogenes in human, chimpanzee and gibbon do not represent processed pseudogenes. Since *MR1* appears to be highly conserved in evolution (Yamaguchi et al. 1997; Riegert et al. 1998; Walter and Günther 1998), it is unlikely that chimpanzees and gibbons do not have a functional *MR1* gene. Our repeated failure to isolate a functional *MR1* gene in chimpanzee and gibbon is most probably due to mismatch of primers used in PCR.

In MHC class Ia genes nonsynonymous (d_N) versus synonymous (d_S) substitutions usually show a remarkably increased d_N/d_S ratio in the peptide-binding region of the α1 and α2 domains (Hughes and Nei 1988). Since among MHC class I-related genes *MR1* shows the highest similarity to MHC class Ia genes in exons 2 and 3 which code for the α1 and α2 domain, respectively, d_N and

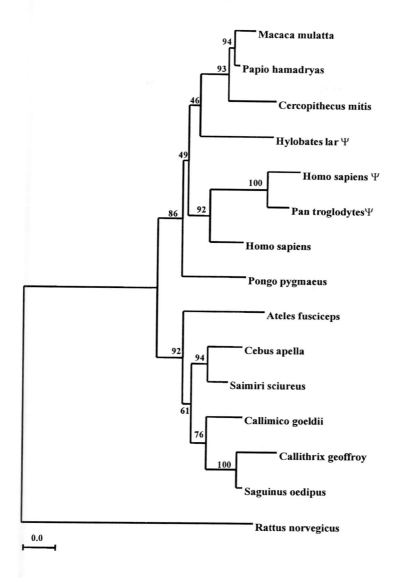

Fig. 1. Gene tree analysis of primate MR1 exon 2 and 3 sequences. Human *MR1* (Hashimoto et al. 1995), human MR1 pseudogene (database accession number AF073486) and rat *Mr1* (Walter and Günther 1998) are included. Ψ indicates a pseudogene. The Clustal W software (version 1.7) was used to align the sequences and to construct the neighbor-joining tree. Reliability of branching patterns is given as percentage and is based on 1000 bootstrap replications.

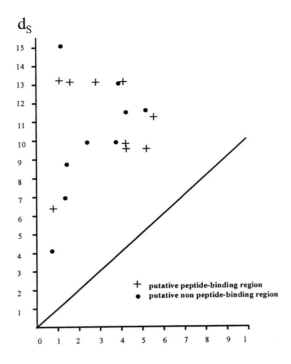

Fig. 2. Analysis of synonymous (d_S) and nonsynonymous (d_N) substitutions in primate MR1 exons 2 and 3. Values for d_S and d_N were calculated with the MEGA software, version 1.01 (Kumar et al. 1993).

d_S were determined for the primate MR1 sequences mentioned in Figure 1. In contrast to class Ia genes, d_N is found not to exceed d_S in the putative peptide-binding region of MR1 (Fig. 2). This indicates that a putative ligand of MR1 is probably monomorphic and conserved.

A further example of non-MHC linked class I-related genes is the cluster of CD1 genes. Oligonucleotides specific for human *CD1D* were used to isolate *CD1D* orthologs in non-human primates. A gene tree constructed from the sequences determined shows that as expected all nonhuman sequences cluster with *CD1D* and not the other human CD1 genes (Fig. 3). Similar to MR1, a significant branching into Catarrhini and Platyrrhini species and a relatively high degree of sequence conservation can be observed.

For the analysis of homologs of the human MHC-linked MIC genes, corresponding sequences have been determined in nonhuman primates and a gene tree containing exon 4 sequences of primate MIC genes was constructed (Fig. 4).

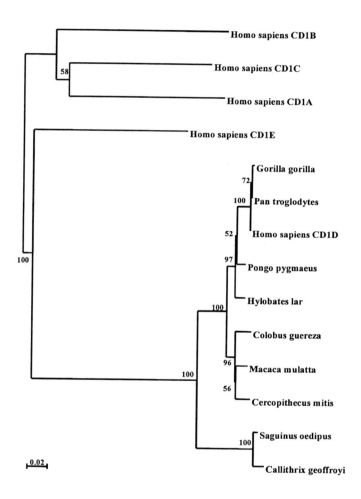

Fig. 3. Gene tree analysis of primate CD1 exon 2 and 3 sequences. The human *CD1A*, *CD1B*, *CD1C*, *CD1D*, *CD1E* sequences (Martin et al. 1987; Balk et al. 1989; Calabi et al. 1989) are included. For details of gene tree analysis see Figure 1.

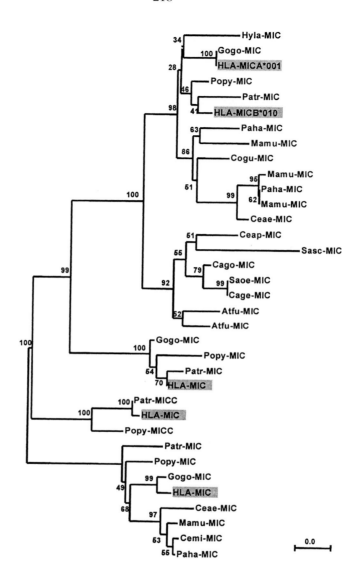

Fig. 4. Gene tree analysis of primate MIC exon 4 sequences. The human *MICA, MICB, MICC, MICD, MICE* sequences (Shiina et al. 1999) are included and marked. The designation of MIC genes includes the MHC designation of the respective species. For details of gene tree analysis see Figure 1.

Four major branches can be distinguished. The lower three represent the *MICC*, *MICD*, and *MICE* pseudogenes and respective orthologs of nonhuman primates. This indicates that in nonhuman primates orthologs of *MICC*, *MICD*, and *MICE* are present. Since in rhesus macaques, baboon, blue and African green monkey *MICD* orthologs are identified, these genes must have existed at least in a common ancestor of Hominoidea and Cercopithecoidea. The first major branch contains *MICA*, *MICB* and *MICA/MICB*-like genes. Interestingly, the *MICA/MICB*-related sequences of Old World monkeys do not cluster with either *MICA* or *MICB*. In particular we isolated three different *MICA/MICB*-related sequences from rhesus macaques, designated *Mamu-MIC1*, *Mamu-MIC2*, *Mamu-MIC3* (Seo et al. 1999). These three rhesus MIC genes cluster significantly with baboon, guereza and African green monkey MIC sequences, but not with either *MICA* or *MICB*. This indicates that orthology between the human *MICA/MICB* genes and the rhesus *MIC1/MIC2/MIC3* genes is lost. However, these rhesus MIC genes still cluster in the *MICA/MICB* group and, thus, appear to be derived from the same MIC ancestor gene which duplicated and diversified in the various primate species. In contrast to the *MICA/MICB*-related genes of macaques, baboon, guereza and African green monkey, orthologs of *HLA-A* and *HLA-B* are still present in these species indicating that the evolution of MIC genes is different from class Ia genes, because their orthology is obviously maintained for a shorter time.

Acknowledgments

J. W. Seo is a fellow of the Graduiertenkolleg "Perspektiven der Primatologie" that is financed by the DFG. The authors thank Dr. U. Sauermann, German Primate Center Göttingen, and C. Roos, Gene Bank of Primates supported by the German Primate Center, for DNA samples.

References

Ando H, Mizuki N, Ota M, Yamazaki M, Ohno S, Goto K, Miyata Y, Wakisaka K, Bahram S, Inoko H (1997) Allelic variants of the human MHC class I chain-related B gene (MICB). Immunogenetics 46:499-508

Bahram S, Bresnahan M, Geraghty DE, Spies T (1994) A second lineage of mammalian major histocompatibility complex class I genes. Proc Natl Acad Sci USA 91:6259-6263

Balk SP, Bleicher PA, Terhorst C (1989) Isolation and characterization of a cDNA coding for a fourth CD1 molecule. Proc Natl Acad Sci USA 86:252-256

Balk SP, Bleicher PA, Terhorst C (1991) Isolation and expression of cDNA encoding the murine homologues of CD1. J Immunol 146:768-774

Burdin N, Brossay L, Koezuka Y, Smiley ST, Grusby MJ, Gui M, Taniguchi M, Hayakawa K, Kronenberg MJ (1998) Selective ability of mouse CD1 to present glycolipids: alpha-galactosylceramide specifically stimulates V alpha 14+ NK T lymphocytes. J Immunol 161:3271-3281

Calabi F, Milstein C (1986) A novel family of human major histocompatibility complex-related genes not mapping to chromosome 6. Nature 323:540-543

Calabi F, Jarvis JM, Martin L, Milstein C (1989) Two classes of CD1 genes. Eur J Immunol 19:285-292

Castaño AR, Tangri S, Miller JE, Holcombe HR, Jackson MR, Huse WD, Kronenberg M, Peterson PA (1995) Peptide binding and presentation by mouse CD1. Science 269:223-226

Fodil N, Laloux L, Wanner V, Pellet P, Hauptmann G, Mizuki N, Inoko H, Spies T, Theodorou I, Bahram S (1996) Allelic repertoire of the human MHC class I MICA gene. Immunogenetics 44:351-357

Fodil N, Pellet P, Laloux L, Hauptmann G, Theodorou I, Bahram S (1999) MICA haplotypic diversity. Immunogenetics 49:557-560

Groh V, Bahram S, Bauer S, Herman A, Beauchamp M, Spies T (1996) Cell stress-regulated human major histocompatibility complex class I gene expressed in gastrointestinal epithelium. Proc Natl Acad Sci USA 93:12445-12450

Groh V, Steinle A, Bauer S, Spies T (1998) Recognition of stress-induced MHC molecules by intestinal epithelial $\gamma\delta$ T cells. Science 279:1737-1740

Hashimoto K, Hirai M, Kurosawa Y (1995) A gene outside the human MHC related to classical HLA class I genes. Science 269:693-695

Hughes AL (1991) Evolutionary origin and diversification of the mammalian CD1 antigen genes. Mol Biol Evol 8:185-201

Hughes AL, Nei M (1988) Pattern of nucleotide substitution at major histocompatibility complex loci reveals overdominant selection. Nature 355:167-170

Ichimiya S, Kikuchi K, Matsuura A (1994) Structural analysis of the rat homologue of CD1. Evidence for evolutionary conservation of the *CD1D* class and widespread transcription by rat cells. J Immunol 153:1112-1123

Komatsu-Wakui M, Tokunaga K, Ishikawa Y, Kashiwase K, Moriyama S, Tsuchiya N, Ando H, Shiina T, Geraghty DE, Inoko H, Juji T (1999) MIC-A polymorphism in Japanese and a MIC-A-MIC-B null haplotype. Immunogenetics 49:620-628

Kumar S, Tamura K, Nei, M (1993) MEGA: Molecular evolutionary genetics analysis, version 1.01. The Pennsylvania State University, University Park, PA 16802

Leelayuwat C, Townend DC, Degli-Esposti MA, Abraham LJ, Dawkins RL (1994) A new polymorphic and multicopy MHC gene family related to nonmammalian class I. Immunogenetics 40:339-351

Martin LH, Calabi F, Lefevre F-A, Bilsland CAG, Milstein C (1987) Structure and expression of the human thymocyte antigens CD1a, CD1b, and CD1c. Proc Natl Acad Sci USA 84:9189-9193

Petersdorf EW, Shuler KB, Longton GM, Spies T, Hansen JA (1999) Population study of allelic diversity in the human MHC class I-related MIC-A gene. Immunogenetics 49:605-612

Riegert P, Wanner V, Bahram S (1998) Genomics, isoforms, expression, and phylogeny of the MHC class I-related MR1 gene. J Immunol 161:4066-4077

Seo JW, Bontrop R, Walter L, Günther E (1999) Major histocompatibility complex-linked MIC genes in rhesus macaques and other primates. Immunogenetics, in press

Shiina T, Tamiya G, Oka A, Takishima N, Inoko H (1999) Genome sequencing analysis of the 1.8 Mb entire human MHC class I region. Immunol Rev 167:193-199

Sieling PA, Chatterjee D, Porcelli SA, Prigozy TI, Mazzaccaro RJ, Soriano T, Bloom BR, Brenner MB, Kronenberg M, Brennan PJ, Modlin R (1995) CD1-restricted T cell recognition of microbial lipoglycan antigens. Science 269:227-230

Visser CJT, Tilanus MGJ, Tatari Z, van der Zwan AW, Bakker R, Rozemuller EH, Schaeffer V, Tamouza R, Charron D (1999) Sequencing-based typing of MICA reveals 33 alleles: a study on linkage with classical HLA genes. Immunogenetics 49:561-566

Walter L, Günther E (1998) Isolation and molecular characterization of the rat MR1 homologue, a non-MHC linked class I-related gene. Immunogenetics 47:477-482

Walter L, Levan G, Günther E (1998) Mapping of the rat *Cd1* gene to chromosome 2. Rat Genome 4:8-12

Yamaguchi H, Hirai M, Kurosawa Y, Hashimoto K (1997) A highly conserved major histocompatibility complex class I-related gene in mammals. Biochem Biophys Res Commun 238:697-702

Yu CY, Milstein C (1988) A physical map linking the five CD1 human thymocyte differentiation antigen genes. EMBO J 8:3727-3732

Rat TL and CD1

Akihiro Matsuura and Miyuki Kinebuchi

Department of Pathology, School of Medicine, Sapporo Medical University
South-1, West-17, Chuo-ku, Sapporo 060-8556, Japan

Summary. Mouse *TL* and human *CD1* were once thought to be genetic homologues because of their thymus-specific expression. To elucidate their functional significance we studied their equivalents in the rat. Two rat class *Ib* genes, containing 3' sequences very similar to mouse *TL*, were identified and designated *RT1.P*. Neither of them, however, can encode ordinary class I molecules due to accumulation of harmful mutations in the 5' regions that are unique to RT1.P, while the 3' TL-like regions still retain protein-coding capacity. On the other hand, rat CD1 is functional. Rat has a single CD1D class gene conserved through mammalian evolution. Rat CD1 products are expressed in a wide variety of organs and tissues. The lymphoid cells isolated from thymus, spleen, lymph nodes, and liver were tested for their reactivity to CD1 transfected cells. CD1-reactive cells were found in the liver. Analysis of surface markers and T cell receptors revealed the similarity and difference among CD1-reactive cells.

Key words. Rat, TL, CD1, Evolution, Function, Unconventional T cells

Introduction

Mouse *TL* and human *CD1* were originally assumed to be genetic equivalents in different species because of their thymus-specific expression and biochemical characteristics such as molecular weight and β2m association (Old et al. 1963; McMichael et al. 1979). However, after their primary structures were determined by molecular cloning, it was revealed that they are only distantly related and coexist in the mouse. It became clear that such a simple interpretation could not be correct (Fisher et al. 1985; Calabi and Milstein 1986; Bradbury et al. 1988). Thymus is the major site of T cell development and differentiation, and therefore plays an important role in establishing the acquired immune system (AIS, see the chapter by J. Klein, this volume). To clarify the biological significance of thymus-specific class Ib molecules, we asked whether their structure and pattern of expression are conserved in rodents. We used rats as they have been widely used as models for the study of many complex diseases, including hypertension, autoimmune disorders, and cancer.

Rat TL

Characterization of rat class Ib genes related to mouse TL

Two rat class *Ib* genes containing 3' sequences very similar to mouse *TL* were isolated and characterized. The genes were designated as *RT1.P1* and *RT1.P2*, respectively (Matsuura et al. 1997). They were similar to each other. The *RT1.P1* appeared to be a pseudogene according to several sequence criteria. The 5' region sequences corresponding to exons 2 and 3 could not encode ordinary $\alpha 1$ and $\alpha 2$ domains due to many nucleotide insertion and deletion mutations. In any reading frame, there were stop codons in both exons. On the other hand, the 3' region sequences corresponding to exon 4 retained coding capacity of the $\alpha 3$ domain of ordinary class I molecules, but the sequences corresponding to exon 5 had a 19 bp deletion causing early termination in the transmembrane domain. The regions corresponding to exon-like sequences of *RT1.P2* were also sequenced, but they did not have functional 5' exons like *RT1.P1*. Therefore, neither *RT1.P1* nor *RT1.P2* could encode ordinary class I molecules. Figure 1 shows the genomic organization of *RT1.P1*.

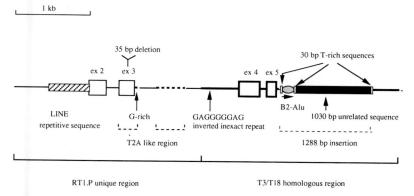

Fig. 1. Schematic diagram of the *RT1.P1* gene. Based on Haar plot analysis and close observation of the sequences, the structure of *RT1.P1* is depicted. Large boxes indicate exon-like regions. Line and dotted lines indicate introns. For details, see Matsuura et al. (1997). Database accession numbers: AB002169-AB002172.

Comparison of structural organization of *TL* family genes

Percent similarity of the exon 5 sequences of *RT1.Ps* with other known class I genes was investigated. Extremely high similarity (over 85 %) was observed with some of them including *T1d, T3b, T3d,* and *T18d*. Comparison of the genomic organization of these genes revealed three types of TL family genes (Fig. 2). *RT1.P1* and *T1/T16* were pseudogenes, and only *T3/T18* was functional and encoded TL antigens. While the 3' regions showed high similarity among TL family genes, their 5' region was unique to each type. In intron 3, there was a clear demarcation between 3' TL common and 5' type-specific regions.

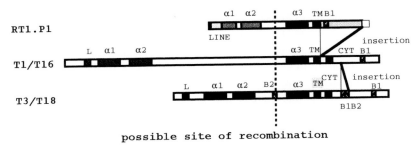

Fig. 2. Three types of *TL* family genes in rats and mice. Rat *RT1.P1* and mouse *T1/T16* can not encode ordinary class I molecules. Mouse *T3/T18* encodes TL antigens expressed preferentially on thymocytes (*T18*) and on intestinal epithelial cells (*T3*). T3 is expressed on some of T cell leukemias which are thought to be derived from immature thymocytes. We use the term "TL (thymus-leukemia)" for this gene family to respect the original description of TL antigen, historically the first class Ib molecules, by Old and Boyse (Old et al. 1963).

A recombination model of *TL* family gene evolution in *Muridae*

Comparison of the organization enabled us to deduce a model of *TL* gene family evolution from the ancestral rodent to the present-day rat and mouse (Fig. 3). This model is consistent with the molecular evolutionary genetic analysis of class I coding regions by Hughes (1991), in that frequent duplication and duplication played important roles in evolution of class Ib genes. Only the T3-type gene acquired protein-coding capacity. Once TL protein products were produced, they were under the influence of natural selection. Obata et al. (1994) found a significant positive correlation between α1 and α2 domains. Thus both exons appear to have evolved under tight linkage. The correlation between α1 + α2 and α3 is positive but rather weak. Furthermore, the correlation between α1 + α2 and TM + CYT is negative. These results agree with our model in which recombination in intron 3 is assumed to be important for creating *TL* family class Ib genes.

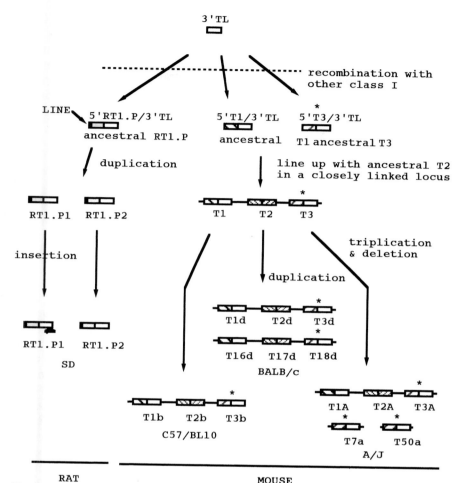

Fig. 3. A recombination model of *TL* family gene evolution in the rat and mouse. In an ancestral rodent common to the rat and mouse, perhaps even common to the *Cricetidae* family and *Muridae* family, the 3' TL sequence that includes part of intron 3, exon 4, intron 4, exon 5 and downstream sequences emerged. Whether it arose independently or as part of a class I gene is not known. By recombination with other class I genes at intron 3, 3' TL acquired 5' sequences of ancestral *T1, T3* and *RT1.P*. In ancestral *RT1.P*, a LINE repetitive sequence should exist in the 5' region. Putative evolutionary intermediates are also shown. After duplication and subsequent insertion of a TL-unrelated sequence in intron 5, present-day *RT1.P1* and *RT1.P2* were generated. In the mouse, ancestors of *T1, T2,* and *T3* were lined up in a closely linked chromosomal locus. This *T1-T2-T3* set was conserved in C57BL/10 (b, H2b-haplotype), duplicated in BALB/c (d), and triplicated and deleted in A/J (a) mice. Therefore, the numbers of structural genes for TL antigens are one (*T3b*) in B10, two (*T3d* and *T18d*) in BALB, and three (*T3A, T7a, T50a*) in A/J. Functional genes are shown by asterisks. *T7 a* and *T50a* are *Tla*[a] genes isolated from EARAD1 cell lines by Fung-Win Shen and Aki Matsuura (Horie et al. 1989, and unpublished observation).

It is not known why 3' *TL* has been conserved even in the pseudogenes. Based on the observation that *Pele-A* class I genes encode three distinct sizes of transmembrane domains (TMs) due to tandem duplications of sequences similar to human hypervariable minisatellite repeats and the λ chi site, it was hypothesized that MHC class I genes may have recruited "selfish" DNA as a TM sequence in their evolution to encode cell surface proteins (Crew et al. 1991). 3' *TL* might be a selfish DNA sequence.

Rat CD1

Rat CD1 typifies evolutionarily conserved CD1D

The non-MHC-encoded CD1 family has recently emerged as a novel antigen-presenting system that is distinct from MHC class I and class II molecules. The *CD1* family genes have been found in many mammalian species such as mice, rats, rabbits and sheep (Bradubury et al. 1988; Calabi et al. 1989a; Calabi et al. 1989b; Ferguson et al. 1996; Ichimiya et al. 1994; Rhind et al. 1999). We determined the genomic organization of the rat CD1 locus (Katabami et al. 1998) and compared it in detail with those of previously published CD1 genes. Structural organizations of CD1D, classic CD1 and H2-K are shown in Figure 4.

All *CD1* genes contain six exons; exon 1 for the 5' untranslated region (5'UN) and the leader peptide; exons 2-4 for the α1-3 domains, respectively; exon 5 for the transmembrane region and part of the cytoplasmic tail; and exon 6 for the remainder of the short cytoplasmic tail and the 3' untranslated region (3'UN). These results indicate that the overall exon-intron organization of the *CD1* gene is similar to that of the MHC class I gene, in that individual functional domains are encoded by separate exons. Comparison of the introns provides an interesting view as to the evolution of *CD1* genes found in different species. Overall organization, including intronic length, can be divided into two types, which are correlated with the previous categorization of two classes of *CD1* genes, classic *CD1* (including *hCD1A, hCD1B, and hCD1C*) and *CD1D* (including *mCD1D1, mCD1D2* and *rCD1*). For example, the length of intron 1, about 280 bp in *CD1D*, is shorter than that (about 350 bp) in classic *CD1*. The length of intron 2 in *CD1D*, about 280 bp, is also shorter than that (about 600 bp) in classic *CD1*. The lengths of intron 3 and intron 4 of *CD1D* class genes are longer than those of classic *CD1* with only a few exceptions; *mCD1D2* has a deletion in intron 3 and *hCD1A* has an *Alu* repeat in intron 3. Similarly, other intron lengths are also typical of the two classes. Compared to *H2-K*, all *CD1s* have short intron 3. It may explain why we did not find chimeric CD1 genes, such as *CD1D*/classic *CD1* or classic *CD1/CD1D*, which recombined at intron 3 as has been shown for *TL* family genes (Fig. 3).

The cytoplasmic portion of the CD1 molecule is encoded by the 3' end of the fifth exon and short sixth exon; a stretch of charged amino acid residues, RRR in

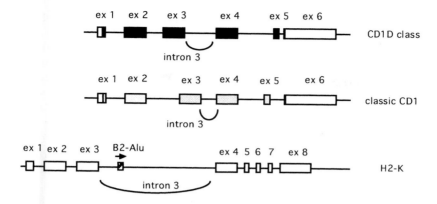

Fig. 4. Genomic organization of *CD1D*, classic *CD1* and *H2-K*. Exons and introns are indicated by boxes and straight lines, respectively. Comparison of *CD1* genes categorized them into two classes based on composition of exonic sequences as well as lengths of introns. The *CD1D* class genes include rat *CD1*, mouse *CD1D1* and *CD1D2*, and human *CD1D*. The classic *CD1* class genes include human *CD1A*, *CD1B*, and *CD1C*. *CD1E* is in an intermediate position. Organization of *H2-K* is shown as a representative of the classical MHC class I gene. Database accession numbers are D26439 and AB029486.

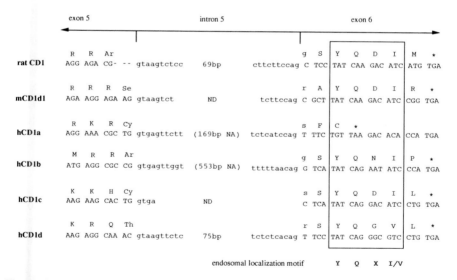

Fig. 5. Comparison of nucleotide and amino acid sequences of the cytoplasmic portion of CD1. *mCD1D2* is the same as *mCD1D1*. A conserved consensus sequence, YQXI/V, shown by a box is encoded by the sixth exon. Absence of this motif in CD1a is due to a nonsense mutation in the first nucleotide for codon Q. ND; not determined. NA; not available from the GenBank/EMBL/DDBJ database.

rat CD1, is followed by consensus sequence YQXI/V, YQDI in rat CD1 (Fig. 5). These features are well conserved in all CD1 molecules except for CD1a. This tyrosine-based motif, YQNI, was recently reported to be a signal for internalization and targeting of human CD1b antigens to endosomal compartments (MIICs, MHC class II compartments) where antigen-loading of class II molecules by endocytosed peptide occurs (Sugita et al. 1996). CD1a molecules lost this motif by a point mutation at the codon for Q (CAA) to a stop codon (TAA) in the sixth exon, indicating that prototypical CD1 might carry this motif. It may be the case for a pig CD1 which also lost this motif by probable acquisition of MR1-like cytoplasmic portion, WKHCDPSSALHRLE (Chun et al. 1999). The underlined sequences are very similar to a serine-containing motif conserved in the cytoplasmic tails of human and mouse MHC class Ia molecules (SD/EXSL). The latter serine residue (position 355) in the SD/EXSL motif is a site of phosphorylation in HLA-B7 molecules (Guild and Strominger 1984) and is required for constitutive endocytosis of HLA class I molecules (Vega and Strominger 1989).

Rat CD1 defines syntenic chromosomal segments

Despite the structural and functional resemblance to MHC, all *CD1* genes have been mapped outside the MHC; human CD1 genes on chromosome 1 (*1q22-q23*) in the order *CD1D-CD1A-CD1C-CD1B-CD1E*, and two mouse *CD1* genes (*mCd1d1* and *mCd1d2*) on chromosome 3 (Albertson et al. 1988; Yu et al. 1989; Kingsmore et al. 1989; Moseley and Seldin 1989). Besides being outside of the MHC, *CD1* defines a conserved linkage group border between human chromosome 1 and mouse chromosomes 1 and 3. A large linkage group of human chromosome 1 appears to be split into mouse chromosomes 1 and 3 (Fig. 6). By using fluorescence in situ hybridization, we have localized the rat *Cd1* to chromosome 2q34 which defines syntenic segments in the rat genome (Matsuura et al. 1997; Walter et al. 1998; Matsuura et al. 1999).

A model of CD1 evolution in mammals

Based on previously published and our observations, a model of CD1 evolution is depicted in Figure 7. Ancestral *CD1* gene might be *CD1D*-like and contain endosomal localization signal (shown by asterisk). Gene duplication resulted in generation of two prototypical CD1 classes. The *CD1D* has been conserved through mammalian evolution and is implicated in a role in innate immunity as CD1d molecules act as the ligand for NKT cells with an invariant T cell receptor. The other classic *CD1* has been deleted in rodents but expanded in other species and is implicated in infectious immunity, especially for protection against

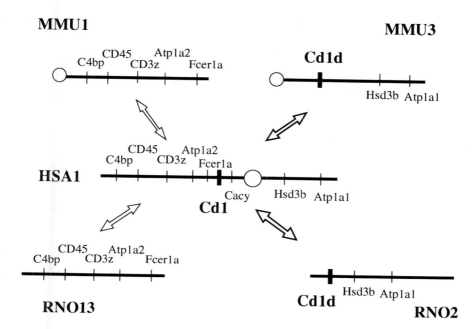

Fig. 6. Chromosomal location of CD1 in mice, humans, and rats. Location of rat Cd3z was recently defined (Matsuura et al. 1999)

Mycobacterium, etc. In humans, three classic *CD1* showed tissue-specific expression in thymus and antigen-presenting cells. Such a characteristic feature has also been observed for classic *CD1* in different species. Thymus-specific expression might be due to acquisition of the 5' regulatory elements relatively early in CD1 evolution. As indicated in a previous section, no chimeric CD1 gene between extracellular exons has been noticed to date. In contrast, exclusion of CD1a from endocytic pathways by truncation of cytoplasmic portion due to a single nucleotide alteration occurred recently in the primate radiation. Such "endosomal localization signal"-deficient CD1 has also been found in pig (Chun et al. 1999). The pig CD1 extracellular domains are similar to those of human CD1a. It is important to determine whether the acquisition of the novel cytoplasmic portion occurred by recombination at the relatively long fourth intron between *CD1* and *MR1* genes. Furthermore, it is of particular interest to ask whether functional differences exist amoug prototype CD1d group 2 proteins, CD1b group 1 proteins, and apparently newly arose CD1a-like protein with respect to both the spectrum of ligand binding and the paths to cell surfaces.

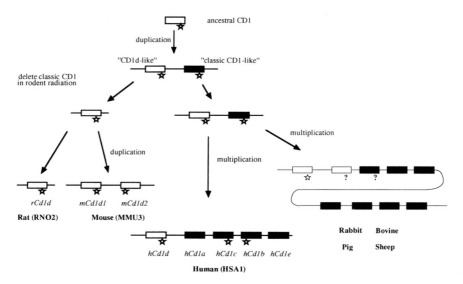

Fig. 7. A model of CD1 family gene evolution in mammals. Open and shaded boxes indicate *CD1D*-like and classic *CD1*-like genes, respectively. Asterisk indicates an endosomal localization signal near the carboxyl terminal of the cytoplasmic domain which was encoded by a single short exon 6. *CD1A* does not contain this signal. All species harbor single or only few copies of *CD1D*. Rodents lack classic *CD1*, but most species have multiple copies of classic *CD1*.

Function of rat CD1

A distinct function for mouse CD1 appears to serve as a ligand directly recognized by a population of T cells that express the T cell receptor (TCR) and NK1.1 (NKR-P1C) antigens (Bendelac et al. 1995). Similar NKR-P1 positive T cells (NKT cells) have also been identified in humans and rats, and recognize evolutionarily conserved CD1d molecules in the absence of exogenously added antigens (Davodeau et al. 1998; Exley et al. 1997; Knudsen et al. 1997). Mouse and human CD1d molecules can also present glycolipids such as α-galactosylceramide to NKT cells (Brossay et al. 1998; Spada et al. 1998). A striking feature of these cells is that they use an invariant TCRα chain (Vα14-Jα281 in mice, Vα24-JαQ in humans) paired preferentially with particular Vβ chains (Vβ8, 7 or 2 in mice and Vβ11 in humans) (Dellabona et al. 1994; Imai et al. 1986; Lantz and Bendelac 1994). We recently identified rat homologues of mouse TRAV14s and invariant TRAV14-J281s. Of particular interest, significant diversity and tissue-specific expression of these invariant TCRα chains were observed (Fig. 8; Matsuura et al. 1999, and unpublished data). From a phylogenetic standpoint, therefore, we and other investigators hypothesized that

```
           1                                              40
rTRAV14S1   MGKHLSACWVLLWLHHQWVAGRTQVEQSPQSLVVHEGESC
mTRAV14S1   MKKRLSACWVVLWLHYQWVAGKTQVEQSPQSLVVRQGENC
hTRAV24     MKKHLTTFLVILWLYFYRGNGKNQVEQSPQSLIILEGKNC

           41                                             80
rTRAV14S1   VLQCNYTVTPFNNLRWYKQDRGRAPVSLTVLTNKEEKTSR
mTRAV14S1   VLQCNYSVTPDNHLRWFKQDTGKGLVSLTVLVDQKDKTSN
hTRAV24     TLQCNYTVSPFSNLRWYKQDTGRGPVSLTIMTFSENTKSN

           81                              113
rTRAV14S1   GRYSATLDADAKHSTLHITASLLDDAATYICVV
mTRAV14S1   GRYSATLDKDAKHSTLHITATLLDDTATYICVV
hTRAV24     GRYTATLDADTKOSSLHITASOLSDSASYICV-
```

Fig. 8. Comparison of amino acid sequences of rat TRAV14, mouse TRAV14, and human TRAV24. This V gene encodes a part of invariant TCRα chain in each species. Three additional rat TRAV14s were identified to date (unpublished data).

CD1d molecules have similar functional significance in the mammalian immune system (Bendelac et al. 1997; Ichimiya et al. 1994; Porcelli 1995).

We then investigated CD1-reactive lymphocytes in the rat organs. Strong cytotoxic and proliferative responses against CD1 transfectants were observed with hepatic lymphocytes. The phenotypic analysis of these cells indicated that rat NKR-P1A positive, TCRαβ positive, NKT cells were in cell cycle progression with high DNA content. The rat invariant TRAV14-J281 chains were expressed abundantly by hepatic sinusoidal lymphocytes.

Concluding remarks

All species are believed to contain a number of class I genes and usually only a few of them are restriction elements for nominal antigens. As clearly shown for the *TL* gene family, variability of numbers and species of MHC-linked class Ib genes suggests that frequent duplications and deletions occurred in this chromosomal region. Amadou (1999) called this situation the "crazy butterfly". In contrast, the CD1 chromosomal region does not appear to have been subjected to such expansive forces in rodent radiation. In other mammals, CD1s expanded their number and species. What can we generally conclude from our studies and previous studies? Image of restriction elements and reactive cells (T cells and NKT cells) is illustrated in Figure 9. Most striking feature of class Ib and restricted T cells is evolutionary conservation of both CD1d and an invariant

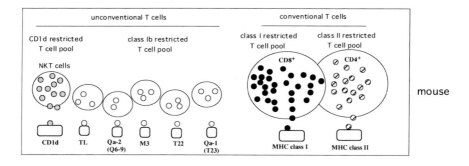

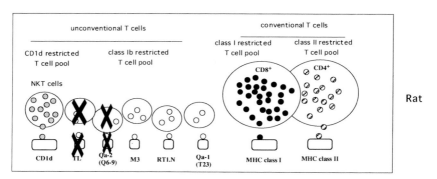

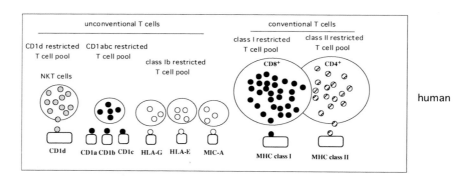

Selecting molecules and T cells

Fig. 9. Selecting molecules and selected T cells in the mouse, rats, and humans. Conventional T cells important for acquired immunity are selected in thymus by classical class I molecules (for CD8 T cells) and class II molecules (for CD4 T cells), both of which are extremely polymorphic in these species. Conventional T cells have extremely diverse T cell receptors. In contrast, unconventional T cells including NKT cells, γδ T cells, and perhaps previously undescribed populations are selected by class Ib molecules, such as CD1 and TL.

Vα14-Jα281 chain. Interaction of CD1d-bearing cells and CD1d-reactive lymphocytes in thymic and extrathymic tissues is important for the rodent and human immune system. Whether class Ib molecules uniquely found in each species restrict T cell pools with specific functions in each species or T cell pools with similar functions needs to be determined. If the genetic homologues do not exist, other class Ib molecules may take over their function by convergent evolution, as in the case of the wings of birds and insects. Is it true for class Ib?

References

Albertson DG, Fishpool R, Sherrington P, Nacheva E, Milstein C (1988) Sensitive and high resolution in situ hybridization to human chromosomes using biotin labelled probes: assignment of the human thymocyte CD1 antigen genes to chromosome 1. EMBO J 7:2801-2805

Amadou C (1999) Evolution of the Mhc class I region: the framework hypothesis. Immunogenetics 49:362-367

Bendelac A, Lantz O, Quimby ME, Tewdell JW, Bennink J, Brutkiewicz RR (1995) CD1 recognition by mouse NK1[+] T lymphocytes. Science 268:863-865

Bradbury A, Belt KT, Neri TM, Milstein C, Calabi F (1988) Mouse CD1 is distinct from and co-exists with TL in the same thymus. EMBO J 7:3081-3086

Brossay L, Chioda M, Burdin N, Koezuka Y, Giulia C, Dellabona P, Kronenberg M (1998) CD1d-mediated recognition of an α-galactosylceramide by natural killer T cells is highly conserved through mammalian evolution. J Exp Med 188:1521-1528

Calabi F, Milstein C (1986) A novel family of human major histocompatibility complex-related genes not mapping to chromosome 6. Nature 323:540-543

Calabi F, Belt KT, Yu CY, Bradbury A, Mandy WJ, Milstein C (1989a) The rabbit CD1 and the evolutionary conservation of the CD1 gene family. Immunogenetics 30:370-377

Calabi F, Jarvis JM, Martin L, Milstein C (1989b) Two classes of CD1 genes. Eur J Immunol 19:285-292

Chun T, Wang K, Zuckermann FA, Gaskins HR (1999) Molecular cloning and characterization of a novel CD1 gene from the pig. J Immunol 162:6562-6571

Crew MD, Filipowsky ME, Neshat MS, Smith GS, Walford RL (1991) Transmembrane domain length variation in the evolution of major histocompatibility complex class I genes. Proc Natl Acad Sci USA 88:4666-4670

Davodeau F, Peyrat MA, Necker A, Dominici R, Blanchard F, Leget C, Gaschet J, Costa P, Jacques Y, Godard A, Vie H, Poggi A, Romagne F, Bonneville M (1997) Close phenotypic and functional similarities between human and murine $\alpha\beta$ T cells expressing invariant TCR α-chains. J Immunol 158:5603-5611

Dellabona P, Padovan E, Casorati G, Brockhaus M, Lanzavecchia A (1994) An invariant Vα24-JαQ/Vβ11 T cell receptor is expressed in all individuals by clonally expanded CD4[-]CD8[-] T cells. J Exp Med 180:1171-1176

Exley M, Garcia J, Balk SP, Porcelli SA (1997) Requirements for CD1d recognition by human invariant Vα14[+] CD4[-]CD8[-] T cells. J Exp Med 186:109-120

Ferguson ED, Dutia BM, Hein WR, Hopkins J (1996) The sheep CD1 gene family contains at least four CD1B homologues. Immunogenetics 44:86-96

Fisher DA, Hunt III SW, Hood L (1985) Structure of a gene encoding a murine thymus leukemia antigen, and organization of *Tla* genes in the BALB/c mouse. J Exp Med 162:528-545

Horie M, Matsuura A, Chang KJ, Niikawa J, Shen FW (1991) Properties of the promotor region of the $T18^d$ $(T13^c)$ *Tla* gene. Immunogenetics 33:171-177

Guild BC, Strominger JL (1984) Human and murine class I MHC antigens share conserved serine 335, the site of HLA phosphorylation in vivo. J Biol Chem 259:9235-9242

Hughes AL (1991) Independent gene duplications, not concerted evolution, explain relationships among class I MHC genes of murine rodents. Immunogenetics 33:367-373

Ichimiya S, Kikuchi M, Matsuura A (1994) Structural analysis of the rat homologue of CD1: Evidence for evolutionary conservation of the CD1D class and widespread transcription by rat cells. J Immunol 153:1112-1123

Imai K, Kanno M, Kimoto H, Shigemoto K, Yamamoto S, Taniguchi M (1986) Sequence and expression of transcripts the T-cell antigen receptor α-chain gene in a functional, antigen-specific suppressor-T-cell hybridoma. Proc Natl Acad Sci USA 83:8708-8712

Katabami S, Matsuura A, Chen H, Imai K, Kikuchi K (1998) Structural organization of rat CD1 typifies evolutinarily conserved CD1D class genes. Immunogenetics 48:22-31

Kingsmore S, Watson ML, Howard TA, Seldin MF (1989) A 6000 kb segment of chromosome 1 is conserved in human and mouse. EMBO J 8:4073-4080

Knudsen E, Seierstad T, Vaage JT, Naper C, Benestad HB, Rolstad B, Maghazachi AA (1997) Cloning, functional activities and in vivo tissue distribution of rat NKR-P1[+] TCRαβ[+] cells. Int Immunol 9:1043-1051

Lantz O, Bendelac A (1994) An invariant T cell receptor α chain is used by a unique subset of major histocompatibility complex class I-specific CD4[+] and CD4'8[-] T cells in mice and humans. J Exp Med 180:1097-1106

Matsuura A, Kinebuchi M, Chen H, Yamada K, Yoshida MC, Horie M, Kikuchi K (1997) Assignment of the evolutionarily conserved CD1D class gene (Cd1d) to rat chromosome 7q32. Cytogenet Cell Genet 79:235-236

Matsuura A, Takayama S, Kinebuchi M, Hashimoto Y, Kasai K, Kozutsumi D, Ichimiya S, Honda R, Natori T, Kikuchi K (1997) RT1.P, rat class Ib genes related to mouse TL: evidence that CD1 molecules but not authentic TL antigens are expressed by rat thymus. Immunogenetics 46:293-306

Matsuura A, Chen HZ, Kinebuchi M, Hashimoto Y, Kasai K (1999) Identification of a rat invariant T-cell receptor α chain similar to mouse Vα14-Jα281 and human Vα24-JαQ. Transplant Proc 31:1577-1578

Matsuura A, Kinebuchi M, Katabami S, Chen H, Yamada K, Yoshida MC, Horie M, Watanabe KT, Kikuchi K, Sato N (1999) Correction and confirmation of the assignment of the evolutionarily conserved CD1D class gene to rat chromosome 2q34 and its relationship to human and mouse loci. Cytogenet Cell Genet: in press

Matsuura A, Kinebuchi M, Itoh Y, Kasai K, Yamada K, Yoshida MC, Horie M, Kikuchi K, Sato N (1999) Genomic organization and chromosome location of the rat CD3ζ locus (Cd3z). Cytogent Cell Genet 85:301-305

McMichael AJ, Pilch JR, Galfre G, Mason DY, Fabre JW, Milstein C (1979) A human thymocyte antigen defined by a hybrid myeloma monoclonal antibody. Eur J Immunol 9:205-210

Moseley WS, Seldin MF (1989) Definition of mouse chromosome 1 and 3 gene linkage groups that are conserved on human chromosome 1: evidence that a conserved linkage group spans the centromere of human chromosome 1. Genomics 5:899-905

Obata Y, Satta Y, Moriwaki K, Shiroishi T, Hasegawa H, Takahashi T, Takahata N (1994) Structure, function, and evolution of mouse TL genes, nonclassical class I genes of the major histocompatibility complex. Proc Natl Acad Sci USA 91:6589-6593

Old LJ, Boyse EA, Stockert E (1963) Antigenic properties of experimental leukemias. I. Serological studies in vitro with spontaneous and radiation-induced leukemias. J Natl Cancer Inst 31:977-986

Spada FM, Koezuka Y, Porcelli SA (1998) CD1d-restricted recognition of synthetic glycolipid antigens by human natural killer T cells. J Exp Med 188:1529-1534

Sugita M, Jackman RM, van Donselaar E, Behar SM, Rogers RA, Peters PJ, Brenner MB, Porcelli SA (1996) Cytoplasmic tail-dependent localization of CD1b antigen-presenting molecules to MIICs. Science 273:349-352

Porcelli SA (1995) The CD1 family: A third lineage of antigen-presenting molecules. Adv Immunol 59:1-98

Porcelli SA, Modlin RL (1999) The CD1 system:antigen-presenting molecules for T cell recognition of lipids and glycolipids. Annu Rev Immunol 17:297-329

Rhind SM, Hopkins J, Dutia BM (1999) Amino-terminal sequencing of sheep CD1 antigens and identification of a sheep CD1D gene. Immunogenetics 49:225-230

Walter L, Levan G, Guenther E (1998) Mapping of the rat Cd1 gene to chromosome 2. Rat Genome 4:8-12

Vega MA, Strominger JL (1989) Constitutive endocytosis of HLA class I antigens requires a specific portion of the intracytoplasmic tail that shares structural features with other endocytosed molecules. Proc Natl Acad Sci USA 86:2688-2695

Yu CY, Milstein C (1989) A physical map linking the five CD1 human thymocyte differentiation antigen genes. EMBO J 8:3727-3732

Are chicken *Rfp-Y* class I genes classical or non-classical?

Marielle Afanassieff[1,2], Ronald M. Goto[1], Jennifer Ha[1], Rima Zoorob[3], Charles Auffray[3], Françoise Coudert[2], W. Elwood Briles[4], and Marcia M. Miller[1]

[1]Department of Molecular Biology, Beckman Research Institute, City of Hope National Medical Center, Duarte, CA 91010, USA
[2]INRA, Station de Pathologie Aviaire et Parasitologie, Nouzilly 37380, France
[3]CNRS, UPR 420, Genetique Moleculaire et Biologie du developpement, Villejuif, France
[4]Department of Biological Sciences, Northern Illinois University, DeKalb, Il 60115, USA

Summary. *Mhc* genes in the chicken and at least some other gallinaceous birds are organized into genetically independent clusters containing both class I and class II genes. The genes within these clusters are of interest in that they may provide insights into the function and evolution of the *Mhc* not found previously in the mammalian paradigms. In the chicken the two gene clusters called *B* and *Rfp-Y* reside on the same chromosome but are separated by a region supporting highly frequent meiotic recombination such that *B* and *Rfp-Y* haplotypes are inherited independently. Genes within *Rfp-Y* are apparently transcriptionally active but whether any class I molecules encoded within *Rfp-Y* present peptides in the fashion of classical class I molecules or perform more specialized functions has not been determined. Consistent with the capacity to hybridize under highly stringent conditions with *B* system class I DNA probes, the *Rfp-Y* class I loci are about 73% identical in coding region sequence with the classical class I genes located in the *B* system. At least one *Rfp-Y* class I locus is transcribed in nearly all organs and exhibits allelic sequence diversity similar to that of classical class I genes. These similarities suggest that the *Rfp-Y* class I molecules could be functionally nearly identical to their classical *B* system counterparts. However, distinctive substitutions of residues that are highly conserved in the peptide binding groove of classical molecules occur in the *Rfp-Y* encoded molecules making it unlikely that they are capable of presenting antigen in the manner of classical class I molecules. This observation together with those made previously for *Rfp-Y* class II loci demonstrate that the *Rfp-Y* gene cluster likely represents specialization not previously encountered in the study of *Mhc* evolution.

Key words. Chicken, *Rfp-Y*, *Mhc* class I loci, Antigen binding region

Introduction

Chicken major histocompatibility genes (*Mhc*) are arranged into two genetically independent clusters. One cluster, the *B* system, has been extensively studied since its discovery 50 years ago. The *B* system was defined initially as a blood group system (Briles et al. 1950). Over 20 years ago the *B* system was shown to exert a strong influence in the capacity of chickens to survive infections with the highly virulent Marek's herpes virus causing T cell lymphomas (Briles et al. 1977; Longenecker et al. 1977). Other studies suggest a role for some *B* haplotypes in resistance to tumors induced by retroviral pathogens (Brown et al. 1984; Collins et al. 1985). Resistance to disease appears to rest in the heart of the *B* system. The association is with particular haplotypes of a compact chromosomal segment containing classical *Mhc* class I, class IIβ, TAP, at least one lectin gene, and several other genes (Briles et al. 1983; Hala et al. 1998; Kaufman et al. 1999). The adjacent cluster of highly polymorphic genes called *B-G* has more minor influences, if any, in resistance to these viral diseases (Aeed et al. 1993; Auclair et al. 1995; Briles et al. 1983; Hala et al. 1998). It has been postulated that the small number of *B* class I and class II loci limits the spectrum of antigens that can be effectively presented and results in the strong disease associations seen with particular *B* haplotypes (Kaufman et al. 1995).

More recently, in an analysis of fully pedigreed families, a second, genetically independent cluster of chicken *Mhc*-like genes was identified that contains both class I and class II loci (Briles et al. 1993). The name of the second cluster, *Rfp-Y*, derives from the restriction fragment patterns revealed in this initial study. DNA within *Rfp-Y* hybridizes under highly stringent conditions with probes originating from *B* system class I and class II cDNA clones indicating high sequence similarity over at least portions of the genes (Miller et al. 1994b). The degree of similarity is such that the *Rfp-Y* loci were placed originally among the *B* genes in the first molecular map of the chicken *Mhc* genes (Guillemot et al. 1988). Cosmid clusters III and II/IV are now known to contain the *Rfp-Y* gene cluster (Miller et al. 1994a, 1996). The two class I genes within these clusters have been renamed *Y-FV* and *Y-FVI*. The three class IIβ loci located in these cosmid clusters, which mostly lack allelic sequence variability (Zoorob et al. 1993), have been renamed *Y-LIV*, *Y-LV* and *Y-LVI* to reflect their position in *Rfp-Y* (Miller et al. 1996). Additional genes mapping to *Rfp-Y* include *17.5*, a lectin gene (Bernot et al. 1994), and two additional genes, *17.8* and *13.1*, of unknown nature (Guillemot et al. 1988). *Rfp-Y* was found to reside on chromosome 16 through analyses of the inheritance of extra copies of chromosome 16 in aneuploid lines (Miller et al. 1996) and by two-color fluorescent *in situ* hybridization (Fillon et al. 1996). Hence *Rfp-Y* resides on the same chromosome with the *B* gene cluster and the single chicken nucleolar organizer region (*NOR*) (Bloom and Bacon 1985) (Fig. 1). The chromosomal region supporting the highly frequent meiotic recombination between the two clusters remains to be identified, but may actually separate *B* from the *NOR* and *Rfp-Y*. In segregation analyses of chromosomes marked for *B*

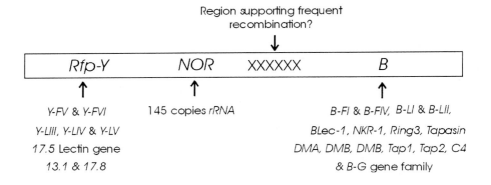

Fig. 1. Segment of chicken chromosome 16 drawn to illustrate the location of the *Rfp-Y* and *B* gene clusters with respect to the nucleolar organizer region (*NOR*).

and the *NOR*, parental and recombinant chromosomes were found to be inherited in a ratio approximating 1:1, indicating that the point of recombination lies between the *NOR* and the *B* gene cluster (Bloom et al. 1990; Miller et al. 1996).

The arrangement of highly similar class I and class II loci into two genetically independent units is so far rare and perhaps occurs only in chickens and other closely related gallinaceous species (Wittzell et al. 1995). The functional relevance of the *Rfp-Y* loci is a matter of considerable interest. It could be that the *Rfp-Y* system broadens the number of antigens that can be presented beyond those "fitting" the small number of *B* system presentation molecules.

The organization of chicken *Mhc* genes into two clusters is also of interest in that the two clusters may be the result of duplication of a chromosomal segment. Maybe after duplication of an ancestral chromosomal segment to form two *Mhc* gene clusters, the clusters evolved to become specialized such that the *B* cluster performs the classical *Mhc* functions. Perhaps the *Rfp-Y* cluster, cleanly separated genetically at a physical location not far removed, carries out more specialized functions. The presence of lectin genes in both *Rfp-Y* and *B* supports the idea that the two gene clusters share a common ancestor.

What little is known about the function of *Rfp-Y* indicates that it indeed has special properties making it distinct from the classical *B* complex. Historically, *Rfp-Y* was not detected in alloimmunizations leading to the identification of the *B* system and 11 other chicken blood group systems (Briles 1984). Even in later experiments with birds differing in known *Rfp-Y* genotypes, immunizations failed to produce Rfp-Y alloantisera (W.E. Briles, unpublished data). In addition, there is little evidence for the presence of Rfp-Y class I protein in immunoprecipitations with an anti-$\beta_2$microglobulin monoclonal antibody that otherwise efficiently precipitates *B* system proteins originating from the major *B* class I locus (Kaufman et al. 1995). However, well-controlled experiments show that *Rfp-Y* haplotypic differences are capable of stimulating the rejection of skin grafts at speeds that resemble rejection of minor histocompatibility antigens (Pharr et al. 1996). The evidence for the ability of *Rfp-Y* differences to stimulate lymphocyte proliferation in vitro is varied, leaving uncertainty as to whether *Rfp-Y* differences induce allogeneic lymphoproliferative responses (Juul-Madsen et al. 1997; Pharr et al. 1996).

Provided here is a summary of our attempts to answer questions about how the *Rfp-Y* class I loci differ from known classical and non-classical loci in structure, transcriptional expression, and sequence variability.

Structural analysis of the *Y-FV* and *Y-FVI* loci

We sequenced the two *Rfp-Y* class I loci identified in the chicken CB line cosmid cluster II/IV (Guillemot et al. 1988; Miller et al. 1994a; M. Afanassieff et al. submitted). The sequences of the two class I loci, *Y-FV* and *Y-FVI*, are highly similar to each other. *Y-FV* and *Y-FVI* exhibit the same intron/exon organization as classical class I molecules. Both loci exhibit a single conserved N-glycosylation site and both have cysteine pairs in α2 and α3 domains in positions typical of classical class I molecules. At both loci three intact exons are present for encoding a typical class I cytoplasmic tail. However, only the *Y-FV* gene appears to be intact in the *Rfp-Y* haplotype represented in Line CB. *Y-FVI* is disrupted by the insertion of 48 copies of a hexanucleotide repeat in exon 1 and it also lacks a polyadenylation signal. Apart from the presence of the hexanucleotide repeats, the two loci are 94% identical in nucleotide sequence.

As Southern hybridizations suggest, the *Y-FV* and *Y-FVI* loci are quite similar to the two *B* loci. Excluding the hexanucleotide repeats in *Y-FVI*, the *Y-F* and *B-F* sequences share about 73% identity over coding regions. As was evident through earlier Southern hybridizations with segments of a *B* class I cDNA clone, the similarity is distributed nearly uniformly over the entire coding length of *B* and *Rfp-Y* genes (Miller et al. 1994b). The percent similarity values for the first five exons are 79%, 66%, 74%, 87%, and 73%, respectively. The *Y-F* and *B-F* gene sequences are only 60% similar in exons 6-8 and show specializations in 5'- and 3'-untranslated regions. The divergence of the α1 domain sequence reflected in the comparatively lower similarity (66% compared to 74% or greater identity with *B-F* genes) is of particular interest and could reflect functional specialization of the *Y-F* genes.

Sequence variability among *Y-FV* alleles

One of the chief criteria most often applied for separating class I molecules into classical and non-classical categories is allelic polymorphism. Polymorphism is inherent in the alloimmune responses defining classical class I molecules. Most non-classical molecules are not very polymorphic. Hence, non-classical class I

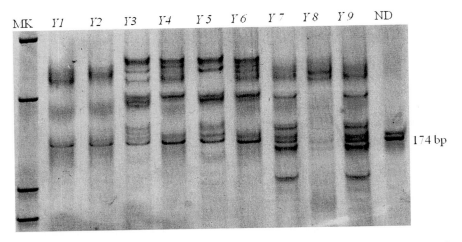

Fig. 2. SSCP patterns derived from a portion of exon 2 in the *Y-F* loci in nine different *Rfp-Y* haplotypes. The patterns are mostly distinctive reflecting sequence variability among the *Y-F* loci in nine *Rfp-Y* haplotypes.

molecules in the mouse were mostly undetected or only detected by special characteristics (Wang et al. 1991) before nucleotide sequencing of the *H-2* region. At the moment, *Rfp-Y* class I genes are paradoxical. They are apparently non-classical since alloantisera to them have not been raised, but they are also apparently highly polymorphic as defined by DNA variability making it difficult to unequivocally categorize them as non-classical.

Polymorphic restriction fragment patterns of the *Y-F* genes are inherent in the definition of the *Rfp-Y* gene cluster (Briles et al. 1993). Hence, the *Y-F* loci have either inherent sequence variability or the regions surrounding them carry sequence variability. Polymorphic patterns found in SSCP (single-stranded conformational polymorphism) analyses of *Y-F* exon 2 sequences indicate that at least a portion of the sequence variability contributing to *Y-F* polymorphism occurs within exon 2. Illustrated in Figure 2 are the SSCP patterns for exon 2 of *Y-F* genes in nine different restriction fragment-defined *Rfp-Y* haplotypes. Nearly all the haplotypes can be distinguished from one another by either quite obvious or somewhat subtle variations in SSCP pattern.

The variations in SSCP patterns reflect nucleotide sequence variability with exon 2. Nucleotide sequences of four *Y-F* genes confirm the variability seen in the SSCP patterns. The predicted amino acid sequences from these clones have been used in the illustration in Figure 3. The positions of sequence variability are mapped in black onto a model of the Y-F MHC fold (Fig. 3b). The four clones each contribute about equally to this variability. Three of the four clones were

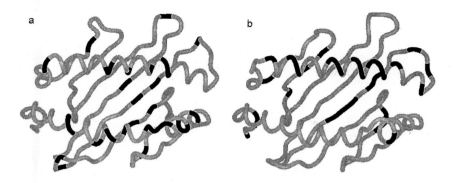

Fig. 3. Comparison of the distribution of polymorphic residues in HLA-A and Rfp-Y sequences. (**a**) Positions of sequence variability found among the four most common HLA-A alleles are provided on the HLA-A structure. (**b**) Positions of sequence variability found among four Y-F α-1 and α-2 domain sequences are displayed on a model of the Y-F MHC-fold.

obtained from different *Rfp-Y* haplotypes and based on their near identity in exon 5 (TM) sequences (single nucleotide differences) almost certainly represent alleles at the *Y-FV* locus. The fourth clone obtained from a cosmid library from a *Rfp-Y* heterozygous bird differs from the others by five nucleotides in exon 5 (TM) and may represent another *Y-FV* allele or another, highly similar *Rfp-Y* locus distinct from *Y-FV* and *Y-FVI*. For comparison, the distribution of polymorphic positions defined by the variability in the four most common *HLA-A* alleles is mapped on the HLA-A2 structure (Fig. 3a). The number of positions where variability is found in the Y-F model is not greatly different from the number of positions in the four *HLA-A* alleles mapped on the HLA-A2 structure. Whether any of the differences in the distribution of the variability seen between the two images in Figure 3, such as the conspicuous lack of variability over much of the Y-FV α2 α-helix, are significant is not yet known but may be resolved when more *Y-FV* sequences become available. At this point, it appears that, at least superficially, *Y-FV* allelic variability is similar to that found at classical class I loci.

Transcription of *Y-FV*

Since many non-classical *Mhc* class I genes are restricted in expression to particular organs, we investigated the sites of expression of the *Y-FV* locus to determine whether *Y-FV* expression might also be restricted and hence more like non-classical class I genes. A riboprobe specific for the *Y-FV* allele in CB and C

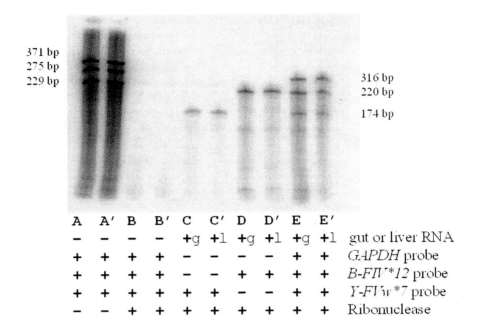

Fig. 4. Ribonuclease protection assays demonstrate transcription of the *Y-FV* locus in small intestine (gut) and liver in Line C chickens. *Y-FV* mRNA is protected by the *Y-FVw*7* riboprobe, as is classical class I mRNA by the B-FIV riboprobe, *B-FIV*12*, and glyceraldehyde-3'-phosphate dehydrogenase mRNA by the *GAPDH* riboprobe. The full-length probes are shown in lanes A and A' and protected fragments in lanes C and C' through E and E'. In the absence of target mRNA, the riboprobes are fully degraded (B and B') by ribonuclease.

line chickens was used in ribonuclease protection assays to examine a number of organs for the presence of *Y-FV* transcripts. As illustrated in Figure 4, sufficient *Y-FV* mRNA is present in RNA preparations from liver and small intestine to provide protection to the *Y-FV* riboprobe equivalent to that provided by mRNA for *B-FIV* and the *GAPDH* control. Similar results were obtained for other organs including spleen, thymus, bursa, heart, lung, and kidney. These results are corroborated by reverse transcriptase PCR assays described elsewhere (M. Afanassieff et al. submitted) where *Y-FV* transcripts were easily detected in late stage embryos and in adult tissues. No evidence has been found for mRNA originating from the structurally disrupted *Y-FVI* allele. Hence, at least one *Rfp-Y* class I locus is widely transcribed as is typical of classical class I loci.

While it remains to be determined whether *Y-FV* transcripts are transcribed in

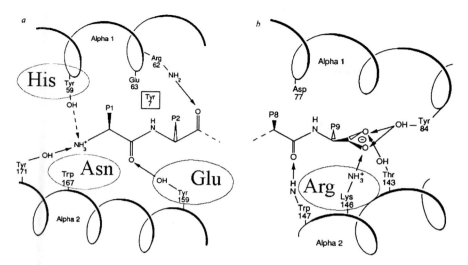

Fig. 5. Positions of substitutions occurring in the Y-F sequences at positions occupied in classical class I molecules by highly conserved residues that interact with main-chain atoms at the left (**a**) and right (**b**) ends of bound peptide antigen. Positions of substitutions occurring in the Y-F sequences are circled with substituting residues noted in larger font. Sketch of the binding environment of peptide termini in HLA-B27 is reproduced with permission (Madden et al. 1991).

vivo, transfected full-length cDNA clones in expression vectors can be transcribed in chicken cells. One such full-length cDNA clone isolated from the small intestine has been FLAG-epitope-tagged and expressed in chicken DF1 fibroblasts and in the Rous sarcoma virus-transformed B cell line RP9. Flow cytometry and immunoprecipitation data indicate that protein of the predicted size for the epitope-tagged Y-FV reaches the cell surface in conjunction with β2-microglobulin. Implicit in these results is the presence in cultured chicken fibroblasts and RP9 cells of whatever antigen that may be needed, if any, for the Y-F molecule to reach the cell surface.

Specializations in the Y-F antigen binding region

At the same time that the *Y-F* loci exhibit allelic variability and expression characteristics quite like classical class I loci, special features in the predicted amino acid sequences indicate that *Y-F* genes are not typical classical class I genes.

Table 1. The amino acid residues in the environment of peptide amino- and carboxy-termini in the antigen binding region of HLA-A are compared with the equivalent residues encoded in *Y-FV* alleles and in a number of different non-classical class I genes

HLA-A[1]	Amino-terminus					Carboxy-terminus			
	Y7	Y59	Y159	W167	Y171	Y84	T143	K146	W147
Y-FV	Y	H	E	N	Y	Y	T	R	W
HLA-E	Y	Y	Y	W	Y	Y	S	K	S
HLA-F	Y	Y	Y	L	Y	R	T	F	Y
HLA-G	Y	Y	Y	W	Y	Y	S	K	C
H2M3	Y	Y	Y	F	L	Y	T	R	L
DLA79	F	Y	F	W	H	Y	T	K	W
XNC6	Y	H	Y	H	Y	Y	M	Q	W
ZnAGP	Y	D	Y	T	Y	Y	T	K	W
Mr1	Y	H	W	W	F	Y	I	A	W
HFe	Y	M	Y	Q	L	N	T	E	W
FcRn	Y	Y	F	R	H	-	I	R	W
MICA	Y	T	A	E	Y	H	V	I	L
CD1c	H	F	L	F	L	A	L	Q	Y

[1]Residues included are conserved in most classical Mhc class I molecules represented here by HLA-A2. As described previously (Madden et al. 1991), eight of these residues are positioned to contact the charged termini and adjacent polar atoms in the main-chain of the peptide bound within the peptide binding groove. The ninth, W167, is considered to define the left margin of the peptide binding through contact with Y59 (Saper et al. 1991). Sequences were obtained from Genbank accession numbers as follows: human HLA-A (P01892), human HLA-E (P13747), human HLA-F (CAA34947), human HLA-G (P17693), mouse H2M3 (AAA60335), dog DLA79 (S35940), *Xenopus* XNC6 (S39603), human ZnAGP 9(4699584), human MR1 (NP_001522), human Hfe (NP_000401), human FcRn (P55899), human MICA (AAC28945), and human CD1c (p29017).

In classical class I molecules of many species there are nine residues, five at one end of the antigen binding groove and four at the other, that are conserved almost without exception (Madden et al. 1991; Saper et al. 1991). These residues define the margin of the binding groove and bond with atoms of the main-chain of the bound peptide largely independent of the peptide amino acid side-chains. At the left end of the groove, hydrogen bonds form between the amino terminus of the peptide and four highly conserved tyrosine residues (Fig. 5). The highly conserved tryptophan at position 167 forms a non-polar contact with Tyr59 (Saper et al.

1991). In the Y-F sequences, histidine and glutamic acid replace Tyr59 and Tyr159, respectively. Trp167 is replaced with asparagine.

At the right end of the groove in classical class I molecules, there are four additional residues that are highly conserved. Tyr84, Thr143, Lys146, and Trp147 hydrogen bond with the peptide carboxylate and the carbonyl oxygen of the penultimate peptide amino acid, again largely independent of the peptide amino acid side-chains. In the Y-F sequences three of these residues are conserved. The only change is the substitution of arginine at the position corresponding to Lys146.

The substitutions found in the Y-F sequences, especially those at the left margin of the groove, make it unlikely that peptide is bound in Y-F molecules in a manner typical of classical class I molecules. In this characteristic then, the Y-F molecules are more like non-classical class I molecules. However, as can be seen in Table 1, while a number of substitutions occur at these conserved positions in non-classical molecules, none of the substitutions observed among non-classical class I molecules are identical to those occurring in the Y-F sequences.

Concluding remarks and future directions

From these analyses, it can be seen that the two class I loci that map to *Rfp-Y* in the chicken share characteristics with both classical and non-classical class I genes. The pattern of *Y-FV* transcription in many different organs and the variability in gene sequence at the *Y-FV* locus in different *Rfp-Y* haplotypes are features associated with classical class I loci. If it were not for the unusual substitutions in the amino acid sequences encoded in *Y-FV* in the clusters of conserved amino acids typical of classical class I molecules, assignment of *Rfp-Y* to the classical category would be tenable. However because of these substitutions at positions that define the binding environment of the peptide termini in classical molecules, it is unlikely that Y-F molecules bind peptide antigen in the manner typical of the classical class I molecules. Therefore it is more appropriate to consider *Y-FV* as a newly recognized type of non-classical class I gene.

Future experiments will explore possible Y-F antigen presentation functions and the conditions under which allogeneic immune responses predicted by the Y-F sequence variability can be produced.

Acknowledgments

The authors thank Dr. Mark Sherman, Molecular Modeling Core, Dr. Yate-Ching Yuan, Bioinformatics, and Tim Chan, RCSC, at the City of Hope for excellent computer support. We thank Dr. Larry D. Bacon and Dr. Pierrick Thoraval for providing chickens. This material is based upon work supported by an USDA National Research Initiative Competitive Grant (92-37204), by a USDA federal assistance award for international collaborative research, and by the National Science Foundation Grant No. MCB-9604589.

References

Aeed PA, Briles WE, Zsigray RM, Collins WM (1993) Influence of different B-complex recombinants on the outcome of Rous sarcomas in chickens. Anim Genet 24:177-181

Auclair BW, Collins WM, Zsigray RM, Briles WE (1995) B-complex recombinants and sarcoma regression: role of B-L/B-F region genes. Poult Sci 74:434-440

Bernot A, Zoorob R, Auffray C (1994) Linkage of a new member of the lectin supergene family to chicken Mhc genes. Immunogenetics 39:221-229

Bloom SE, Bacon LD (1985) Linkage of the major histocompatibility (B) complex and the nucleolar organizer in the chicken. Assignment to a microchromosome. J Hered 76:146-154

Bloom SE, Briles WE, Briles RW (1990) Recombination between the major histocompatibility complex (MHC) and the ribosomal RNA gene cluster. Poult Sci 69:21 (Abstract)

Briles WE (1984) Early chicken blood group investigations. Immunogenetics 20:217-226

Briles WE, McGibbon WH, Irwin MR (1950) On multiple alleles effecting cellular antigens in the chicken. Genetics 35:633-652

Briles WE, Stone HA, Cole RK (1977) Marek's disease: effects of B histocompatibility alloalleles in resistant and susceptible chicken lines. Science 195:193-195

Briles WE, Briles RW, Taffs RE, Stone HA (1983) Resistance to a malignant lymphoma in chickens is mapped to subregion of major histocompatibility (B) complex. Science 219:977-979

Briles WE, Goto RM, Auffray C, Miller MM (1993) A polymorphic system related to but genetically independent of the chicken major histocompatibility complex. Immunogenetics 37:408-414

Brown DW, Collins WM, Briles WE (1984) Specificity of B genotype response to tumors induced by each of three subgroups of RSV. Immunogenetics 19:141-147

Collins WM, Zervas NP, Urban WE, Jr., Briles WE, Aeed PA (1985) Response of B complex haplotypes B22, B24, and B26 to Rous sarcomas. Poult Sci 64:2017-2019

Fillon V, Zoorob R, Yerle M, Auffray C, Vignal A (1996) Mapping of the genetically independent chicken major histocompatibility complexes *B@* and *RFP-Y@* to the same microchromosome by two-color fluorescent in situ hybridization. Cytogenet Cell Genet 75:7-9

Guillemot F, Billault A, Pourquie O, Behar G, Chausse AM, Zoorob R, Kreibich G, Auffray C (1988) A molecular map of the chicken major histocompatibility complex: the class II beta genes are closely linked to the class I genes and the nucleolar organizer. EMBO J 7:2775-2785

Hala K, Moore C, Plachy J, Kaspers B, Bock G, Hofmann A (1998) Genes of chicken MHC regulate the adherence activity of blood monocytes in Rous sarcomas progressing and regressing lines. Vet Immunol Immunopathol 66:143-157

Juul-Madsen HR, Zoorob R, Auffray C, Skjodt K, Hedemand JE (1997) New chicken Rfp-Y haplotypes on the basis of MHC class II RFLP and MLC analyses. Immunogenetics 45:345-352

Kaufman J, Volk H, Wallny HJ (1995) A "minimal essential Mhc" and an "unrecognized Mhc": two extremes in selection for polymorphism. Immunol Rev 143:63-88

Kaufman J, Jacob J, Shaw I, Walker B, Milne S, Beck S, Salomonsen J (1999) Gene organisation determines evolution of function in the chicken MHC. Immunol Rev 167:101-117

Longenecker BM, Pazderka F, Gavora JS, Spencer JL, Stephens EA, Witter RL, Ruth RF (1977) Role of the major histocompatibility complex in resistance to Marek's disease: restriction of the growth of JMV-MD tumor cells in genetically resistant birds. Adv Exp Med Biol 88:287-298

Madden DR, Gorga JC, Strominger JL, Wiley DC (1991) The structure of HLA-B27 reveals nonamer self-peptides bound in an extended conformation. Nature 353:321-325

Miller MM, Goto R, Bernot A, Zoorob R, Auffray C, Bumstead N, Briles WE (1994a) Two Mhc class I and two Mhc class II genes map to the chicken Rfp-Y system outside the B complex. Proc Natl Acad Sci USA 91:4397-4401

Miller MM, Goto R, Zoorob R, Auffray C, Briles WE (1994b) Regions of homology shared by Rfp-Y and major histocompatibility B complex genes. Immunogenetics 39:71-73

Miller MM, Goto RM, Taylor RL, Jr., Zoorob R, Auffray C, Briles RW, Briles WE, Bloom SE (1996) Assignment of Rfp-Y to the chicken major histocompatibility complex/NOR microchromosome and evidence for high-frequency recombination associated with the nucleolar organizer region. Proc Natl Acad Sci USA 93:3958-3962

Pharr GT, Gwynn AV, Bacon LD (1996) Histocompatibility antigen(s) linked to Rfp-Y (Mhc-like) genes in the chicken. Immunogenetics 45:52-58

Saper MA, Bjorkman PJ, Wiley DC (1991) Refined structure of the human histocompatibility antigen HLA-A2 at 2.6 Å resolution. J Mol Biol 219:277-319

Wang CR, Loveland BE, Lindahl KF (1991) H-2M3 encodes the MHC class I molecule presenting the maternally transmitted antigen of the mouse. Cell 66:335-345

Wittzell H, von Schantz T, Zoorob R, Auffray C (1995) Rfp-Y-like sequences assort independently of pheasant Mhc genes. Immunogenetics 42:68-71

Zoorob R, Bernot A, Renoir DM, Choukri F, Auffray C (1993) Chicken major histocompatibility complex class II B genes: analysis of interallelic and interlocus sequence variance. Eur J Immunol 23:1139-1145

Xenopus class I proteins

Martin F. Flajnik and Yuko Ohta

Department of Microbiology and Immunology, University of Maryland at Baltimore, Room 13-009, 655 West Baltimore Street, Baltimore, MD 21201, USA

Summary. Class I molecules in the amphibian *Xenopus* are encoded by a single copy MHC gene. Alleles are found in 2 highly diverse sets, defined primarily by their $\alpha 2$ domains and cytoplasmic regions, that match perfectly 2 divergent allelic sets of *TAP2* and *lmp7* genes. It is likely that these sets of processing, transport, and presenting genes have been coselected in the *Xenopus* lineage. Previous biochemical analyses of the class I proteins already suggested that two lineages existed, and there are other unusual properties of the molecules, such as associations with non-$\beta 2$-microglobulin proteins, that await future study.

Key words. Class I proteins, Allelic lineages, Processing molecules, Unusual biochemical associations

Introduction

Among ectothermic vertebrates, including the cartilaginous and bony fish, amphibians, and reptiles, the amphibian *Xenopus* has long stood out as an excellent model for studies of the immune system. The propagation of inbred strains and cloned lines, the ease of embryo manipulation, and the rapid changes of the immune system at metamorphosis have made the model very attractive (Du Pasquier et al. 1989; Rollins-Smith 1998). The *Xenopus* MHC is of special interest, both for the changes in expression of class I and class II at metamorphosis (Rollins-Smith 1998) and the silencing of MHC genes in polyploid species (reviewed in Flajnik et al. 1999a). Furthermore, the MHC gene organization of an amphibian is pivotal to our understanding of the original architecture of this gene complex (Nonaka et al. 1997).

Xenopus class I proteins were first identified, as was done in mammals much earlier, with immunoprecipitating alloantisera (Flajnik et al. 1984). The IgY (IgG-equivalent) fraction of sera from frogs immunized to MHC-mismatched skin-grafts and blood cells recognized mammalian-like class I molecules with glycosylated heavy chains of 40-44,000 daltons and non-glycosylated $\beta 2$-microglobulin ($\beta 2$-m)-like light chains of 13,000 daltons (Fig. 1). The class I

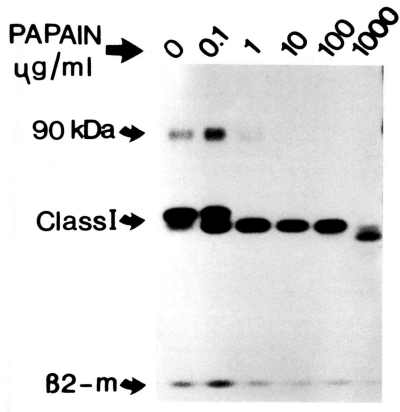

Fig. 1. Papain digestion of alloantiserum-purified class I molecules (labeled with ^{125}I) from the *f* haplotype reveals a proteolytic pattern similar to mammalian class I molecules. The 90 kDa protein that is also precipitated with the alloantiserum in this experiment does not associate with class I heavy chains and is not MHC-linked (data not shown).

molecules show a digestion pattern with the protease papain that is quite similar to digests of mammalian class I molecules: low concentrations of the enzyme (0.1-1 micrograms/ml) are likely to cleave the cytoplasmic tail while higher concentrations (1 milligram/ml) probably remove the transmembrane region (Mann et al. 1968). Such data suggest that at least a portion of the alloantiserum-precipitable *Xenopus* class I protein is in a compact form (papain degrades biosynthetically-immature class I heavy chains into more fragments), with the heavy chain bound to peptides.

From our laboratory stocks of *Xenopus* only a single class I molecule could be detected for each defined MHC haplotype, but a higher (44 kDa) or lower (40 kDa) molecular weight (mw) form was seen by one-dimensional SDS-PAGE in approximately the same number of haplotypes (Flajnik et al. 1984, 1991). In addition, mouse monoclonal antibodies (mAbs), and even xenoantisera raised to human or *Xenopus* class I molecules, recognized *Xenopus* class Ia alleles more like "allo" than "xeno," i.e. reacted with some alleles and not others. Lastly, class I proteins were shown to associate with another molecule, christened the "red blood cell (RBC) molecule" that from preliminary experiments appeared to associate with only the β2-m-free form of class I proteins on the surface of erythrocytes (Flajnik et al. 1991 and see below). In this short paper, we review these biochemical and genetic properties of *Xenopus* class Ia molecules.

Cloning of *Xenopus* class I genes

Two lineages

Class I heavy chains encoded by the *f* haplotype were mAb-purified from erythrocyte detergent lysates and the sequence of the N-terminal 22 amino acids was obtained (Shum et al. 1993). Degenerate oligonucleotide primers were designed and used to select cDNA clones. The class Ia gene was shown to be single copy in all of the defined *Xenopus* haplotypes (one band on genomic Southern blots under high stringency conditions), but the presumed exon encoding the α-3 domain cross-hybridized under low stringency conditions to a large number of non-classical class I genes that are not MHC-linked (Flajnik et al. 1993). The first major surprise in the studies of class Ia gene expression was the demonstration that RNA from animals with different haplotypes showed distinct mRNA sizes of either 3 or 1.8 kb; the mRNA size correlated perfectly with the sizes of the proteins detected by one-dimensional SDS-PAGE (44,000 daltons encoded by the 3 kb mRNA and 40,000 daltons for the 1.8 kb mRNA). Sequence analysis of class I cDNA clones from 8 haplotypes (4 of each type, determined by the size of the protein and the mRNA) showed that the lineages could be distinguished by the cytoplasmic tails and their α-2 domains (Flajnik et al. 1999b). The α-1 domains are so diverse that the lineages cannot be distinguished for individual alleles, and the α-3 domains are homogenized between the allelic sets suggesting intra-allelic recombination or gene conversion. The deduced mw of class I heavy chains in both sets is the same, so it is likely that some unknown property(ies) of the cytoplasmic tail of each allelic set accounts for the differences seen in the apparent mw of the protein on SDS-PAGE.

Perhaps the most surprising (and pleasing) aspect of the discovery of the 2 class I allelic sets was that a perfect concordance was found for the 2 allelic sets previously described for *lmp7* (Namikawa et al. 1995) and *TAP2* (Ohta et al.

1999) genes. This concordance was first demonstrated for the 8 defined laboratory haplotypes (Flajnik et al. 1999b), and now has been shown for all of the *Xenopus* species tested so far (Ohta, Nonaka, Du Pasquier, and Flajnik, unpublished). Such a "trans-species" presence of both *lmp7/TAP2*/class Ia allelic sets, in animals that have diverged from each other as long as 100 million years ago, strongly suggests that the 2 sets of alleles are under balancing selection, and there is a strong pressure to maintain linkage of the particular types of proteasome, transporter, and presenting genes. Clearly, biochemical analyses are in order to test this hypothesis.

Xenopus class I molecules

Serology

As described, we have known for a long time that antisera of all sorts recognize subsets of *Xenopus* alleles rather than all possible alleles. Two mAbs prepared to *Xenopus* class I proteins encoded by the *f* haplotype demonstrated this trait: mAbs TB1, 15 and 18 react with f and r but not g and j class I proteins while mAb TB17 reacts with f, g, and j but not r class I (Flajnik et al. 1991; note that we found TB1, 15, and 18 to be identical mAbs in other experiments). TB1 can immunoprecipitate native class I from detergent lysates or recognize denatured class I by western blotting, but cannot bind to class I on the cell surface (and Figs. 2 and 3); indeed once the sequence of the class I alleles was determined we found that TB1 reacts with a peptide (GKPDAGYTAAANRDSPPSSI, first residue is #314, HLA-A2 numbering, see Shum et al. 1993) encompassing the cytoplasmic tail of f and r class I—recall that the g and j class I cytoplasmic tail sequence is very different from f and r (Flajnik et al. 1999b). mAb TB17 recognizes class I on the cell surface as well as by western blotting and immunoprecipitation (Flajnik et al. 1991 and Figs. 2 and 3); however, TB17 will not recognize the β2-m-associated form of class I, despite its ability to bind to cell surface class I (see later). TB17 recognizes a peptide (GRDEPATATQWMKQKQKEGPEY, the first residue is #43, HLA-A2 numbering, see Shum et al. 1993 and Kaufman et al. 1992) in the region of mammalian class I molecules where the β strands of the α-1 domain turns into the α helix, a section of class I that differs in sequence between f, g, j on the one hand and r on the other (Flajnik et al. 1999b). The r class I sequence differs in an invariant proline (Pro47 in f, g, and j class I, changed to Ala in r) that presumably breaks the strand structure, initiating the turn into α helix. Note that while the mAb TB1 recognizes an epitope that distinguishes the two allelic lineages, the TB17 epitope does not fit this pattern.

It is likely that other class I-specific antisera also recognize these (and other) polymorphic regions and might explain our previous data, i.e. the great diversity

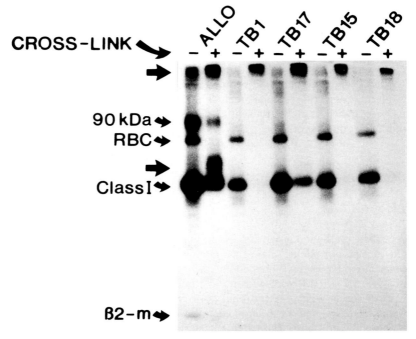

Fig. 2. Cross-linking experiments with the chemical cross-linker DSP reveals different types of class I complexes purified from ^{125}I erythrocyte proteins (*f* haplotype). Only the alloantiserum (ALLO) immunoprecipitates the class I/β2-m complex, but also recognizes the class I/RBC molecule complex. The mAbs TB1, 15, and 18 recognize exclusively the class I/RBC molecule complex while TB17 recognizes this complex as well as an unassociated, β2-m-free form of class I heavy chain. The thick arrow at the top of the gel indicates class I cross-linked to the RBC molecule while the thick arrow in the middle of the gel denotes the class I/β2-m complex. The RBC molecule under non-reducing and non-crosslinked conditions is a dimer with each monomer being approximately 40 kDa. -, no cross-linking; +, cross-linked in detergent lysates with DSP.

among the allelic sets and the alleles within each set could easily account for the unusual serological reactivities.

Class I complexes

In general terms the *Xenopus* class I molecules are quite similar to their mammalian counterparts. As shown above in Figure 1, alloantiserum-purified

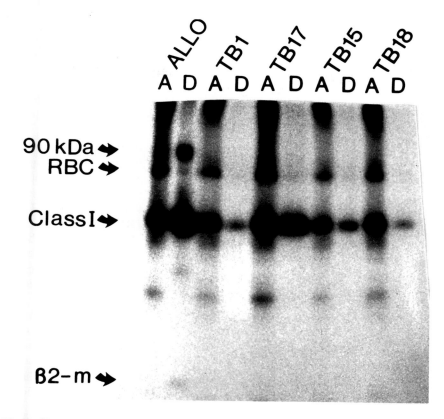

Fig. 3. β2-m-associated class I heavy chains partition into the detergent phase of Triton X (TX)-114 lysates prepared from ^{125}I-labelled erythrocyte membranes, while the RBC molecule-associated heavy chain is present in the aqueous phase. Note that a relatively larger amount of TB17-precipitable material is found in the detergent phase as compared to the TB1, 15, 18 immunoprecipitations. A, aqueous phase; D, detergent phase; ALLO, alloantiserum-immunoprecipitated. See Figure 1 for a description of the 90-kDa molecule.

class I is relatively protected from papain digestion, probably just susceptible to clipping of the cytoplasmic tail, and then at higher concentrations the TM region. However, class I molecules on the surface of frog erythrocytes are found in at least two and perhaps three forms that can be detected serologically: the typical β2-m-positive form, a second form in association with another protein, the red blood cell (RBC) molecule, and maybe a third form not associated with any molecule (other than peptides?).

The mAb TB1 (15, 18) immunoprecipitates exclusively the class I heavy chain associated with the RBC molecule (Fig. 2). This form of heavy chain partitions primarily into the aqueous phase of TX114 lysates (Fig. 3), and is entirely degraded by papain (not shown). Since TB1 is reactive with the cytoplasmic tail it is not clear why it cannot immunoprecipitate the β2-m-positive form as well; perhaps the epitope is masked because of detergent-binding to the class I transmembrane region or there is a conformational change in the cytoplasmic tail of the β2-m-positive form that precludes mAb binding (Little et al. 1995).

The mAb TB17, while capable of recognizing cell surface class I, does not immunoprecipitate the β2-m-positive form of class I in detergent lysates (Fig. 2); however, TB17 does immunoprecipitate the heavy chain in association without the RBC molecule (Fig. 2) and also recognizes another form of β2-m-free heavy chain that partitions into the detergent phase of TX-114 lysates (Fig. 3). Thus, considering the epitope that is recognized by mAb TB17 (see above), it is possible that this mAb displaces β2-m or associates with heavy chain as β2-m "falls off." Alternatively, there are three forms of class I on the cell surface of erythrocytes, β2-m-associated, RBC molecule-associated, and free but compact. Other experiments (not shown) have demonstrated conclusively that the RBC molecule- and β2-m-positive forms of class I are mutually exclusive.

There are differences in the amount of β2-m-free heavy chain for the different class I alleles (Flajnik et al. 1984). In part this relates to the alloantiserum tested, i.e. depending on the class I epitopes to which the allogeneic frog responded. However, preclear experiments have made it relatively clear that the g class I molecule on erythrocytes is more tightly bound to β2-m than are f, j, or r class I (not shown). There are no obvious sequence features of the g class I molecule that would suggest a tighter association with β2-m; there is only one residue that is specific to g class I that is predicted to interact with β2-m: residue #204 in the α3 domain (Shum et al. 1993; Flajnik et al. 1999b) is Gln in all other sequences and His in g class I (side chain-α3/main chain β2-m interaction in human class I, Saper et al. 1991).

Despite the apparently tighter association of g heavy chain with β2-m, the mAb TB17 nevertheless immunoprecipitates only β2-m-negative g heavy chain, again implying a displacement mechanism when the mAb binds its epitope. One residue in this region bound by TB17 makes direct contact with β2-m (Arg-35 in sequences from all other vertebrates; Ala in all of the *Xenopus* sequences).

Consistent with the class Ia sequence data, when the class I heavy chains of different haplotypes are treated with V8 protease, a number of polymorphic bands is detected (Fig. 4). However, digestion of the RBC molecule with V8 results in the same three bands regardless of the MHC genotype, suggesting that this molecule may be monomorphic. IEF data (not shown) gave similar results. Both the class I heavy chain and the RBC molecule apparently bear one complex Asn-linked glycan (Fig. 5).

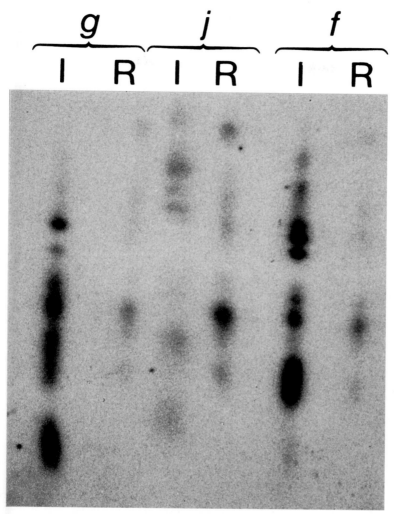

Fig. 4. Digestion with V8 protease shows that class I molecules are polymorphic while the RBC molecule is not. Consistent with the diversity of sequences of the g, j, and f class I alleles (Flajnik et al. 1999b), a polymorphic pattern is easily discerned, while the RBC molecule shows absolutely no differences. I, class I; R, RBC molecule.

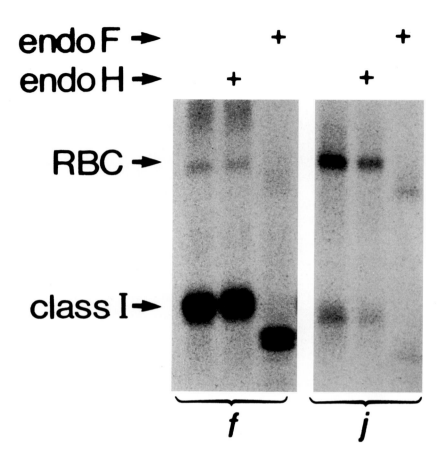

Fig. 5. The class Ia molecule and the RBC molecule both bear complex N-linked glycans, as there is a decrease in size after treatment with endoglycosidase F but not endoglycosidase H (which only cleaves high mannose glycans).

Role of the RBC molecule

The association of class I heavy chains with other molecules on the erythrocyte surface is interesting and the molecule begs for a function. First, it is not known why class I molecules should be expressed at all on the surface of cells that presumably cannot be productively infected by intracellular pathogens in the mature stages. The class I/RBC molecule complex may have a distinct function besides antigen presentation; there certainly are enough examples of class I molecules having distinct functions, and the case most similar to ours is likely the direct association of hemochromatosis protein (HFE) with the transferrin receptor (Lebron et al. 1998). Thus, an obvious goal is to molecularly clone the gene specifying RBC molecule. The facts that the heavy chain in the RBC complex is degraded by papain and partitions into the aqueous phase of TX-114 lysates suggest that the heavy chain/RBC complex is in a very different conformation compared to the heavy chain/β2-m/peptide complex. Class I heavy chain/RBC complexes are also found on erythrocyte membranes from reptiles (but not birds) suggesting that the association has been maintained in evolution for an important reason (Kaufman et al. 1991). Since the RBC molecule is on the cell surface and is glycosylated it seems likely that it associates with class I heavy chain on the surface of erythrocytes, rather than during class I biosynthesis. It will be interesting to determine whether tadpole erythrocytes, which do not bear class I molecules, nevertheless express the RBC molecule defined in adults.

Conclusions

While the RBC molecule is apparently a feature unique to *Xenopus* class I biology and is potentially very interesting, the more tangible studies of coevolution of *TAP*, *lmp*, and class I should be of general interest to the immunological community. There is good evidence in the rat (Joly et al. 1998) and chicken (Kaufman et al. 1996) for such coevolution of *TAP* and class I, but not of the proteasome genes. Furthermore, the identification of other players in the processing pathway, such as tapasin, TAP1, and other recently discovered members of the class I pathway will be important and will either add or detract from the hypothesis that the allelic 'sets' we have uncovered are propagated as such because the proteasome, transport, and presenting genes must "work as a team" to accomplish efficient immune responses. We further hope that the study of MHC in non-mammalian vertebrates may yield even more members of the class I/II biosynthetic pathway that have not been discovered in mouse/man.

Acknowledgments

This work has been supported by NIH grant AI27877. We thank Louis Du Pasquier, with whom we collaborated on the biochemical experiments; Jim Kaufman for technical and theoretical suggestions at every step; and our collaborators Masanori Kasahara, Masaru Nonaka, and Simon Powis for collegial interactions.

References

Du Pasquier L, Schwager J, Flajnik MF (1989) The immune system of *Xenopus*. Annu Rev Immunol 7:251-275

Flajnik MF, Kaufman JF, Riegert P, Du Pasquier L (1984) Identification of class I major histocompatibility complex encoded molecules in the amphibian *Xenopus*. Immunogenetics 20:433-442

Flajnik MF, Taylor, E, Canel C, Grossberger D, Du Pasquier L (1991) Reagents specific for MHC class I antigens of *Xenopus*. Amer Zool 31:580-591

Flajnik MF, Kasahara M, Shum BP, Salter-Cid L, Taylor E, Du Pasquier L (1993) A novel type of class I gene organization in vertebrates: a large family of non-MHC-linked class I genes is expressed at the RNA level in the amphibian *Xenopus*. EMBO J 12:4385-4396

Flajnik MF, Ohta Y, Namikawa-Yamada C, Nonaka M (1999a) Insight into the primordial MHC from studies in ectothermic vertebrates. Immunol Rev 167:59-67

Flajnik MF, Ohta Y, Greenberg AS, Salter-Cid L, Carrizosa A, Du Pasquier L, Kasahara M (1999b) Two ancient allelic lineages at the single classical class I locus in the *Xenopus* MHC. J Immunol 163:3826-3833

Joly E, Le Rolle A-F, Gonzalez AL, Mehling B, Stevens J, Coadwell WJ, Huenig T, Howard JC, Butcher GW (1998) Co-evolution of rat TAP transporters and MHC class I RT1-A molecules. Curr Biol 8:169-172

Kaufman J, Flajnik MF, Du Pasquier L (1991) The MHC molecules of ectothermic vertebrates. In: Warr GW, Cohen N (eds) Phylogenesis of immune function. CRC Press, Boca Raton FL, pp 125-149

Kaufman J, Andersen D, Avila D, Engberg J, Lambris J, Salomonsen J, Welinder K, Skjoedt K (1992) Different features of the MHC class I heterodimer have evolved at different rates; chicken B-F and β2-microglobulin sequences reveal invariant surface residues. J Immunol 148:1532-1546

Kaufman J, Wallny HJ (1996) Chicken MHC molecules, disease resistance, and the evolutionary origin of birds. Curr Top Microbiol Immunol 212:129-141

Lebron JA, Bennett MJ, Vaughn DE, Chirino AJ, Snow PM, Mintier GA, Feder JN, Bjorkman PJ (1998) Crystal structure of the hemochromatosis protein HFE and characterization of its interaction with transferrin receptor. Cell 93:111-123

Little A-M, Nössner E, Parham P (1995) Dissociation of β2-microglobulin from HLA class I heavy chains correlates with acquisition of epitopes in the cytoplasmic tail. J Immunol 154:5205-5215

Mann D, Rogentine GN, Fahey L, Nathenson SG (1968) Solubilization of human leucocyte membrane isoantigens. Nature 217:1180-1181

Nonaka M, Namikawa C, Kato Y, Sasaki M, Salter-Cid L, Flajnik MF (1997) Major histocompatibility complex gene mapping in the amphibian *Xenopus* implies a primordial organization. Proc Natl Acad Sci USA 94:5789-5791

Ohta Y, Powis SJ, Coadwell WJ, Haliniewski DE, Liu Y, Li H, Flajnik MF (1999) Identification and mapping of *Xenopus TAP2* genes. Immunogenetics 49:171-182

Rollins-Smith LA (1998) Metamorphosis and the amphibian immune system. Immunol Rev 166:221-230

Saper M, Bjorkman PJ, Wiley D (1991) Refined structure of the human histocompatibility antigen HLA-A2 at 2.6Å resolution. J Mol Biol 219:277-319

Shum BP, Avila, D, Du Pasquier L, Kasahara M, Flajnik MF (1993) Isolation of a classical MHC class I cDNA from an amphibian: evidence for only one class I locus in the *Xenopus* MHC. J Immunol 151:5376-5386

Two MHC class II *A* loci in the channel catfish

Ulla B. Godwin[1], Michael Flores[1], Melanie R. Wilson[2], Sylvie Quiniou[2], Norman W. Miller[2], L. William Clem[2], and Thomas J. McConnell[1]

[1]Department of Biology, N108 Howell Science Complex, East Carolina University, Greenville, NC 27858, USA
[2]Department of Microbiology and Immunology, University of Mississippi Medical Center, Jackson, MS 39216, USA

Summary. The use of a striped bass MHC class II *A* gene as a heterologous probe allowed for the isolation and identification of MHC class II *A* genes from both an outbred and a second generation gynogenetic, and presumably homozygous, channel catfish (*Ictalurus punctatus*). Four different clones were isolated from a cDNA library generated from a heterozygous catfish, suggestive of the presence of at least two functional loci. Additionally, both genomic and corresponding class II *A* cDNA clones were isolated for each of the two loci from gynogenetic cell lines. The amino acids sequences encoding the putative MHC class II α chain of the catfish possess the general features of those of the three known teleost species. However, the catfish lacks N-linked glycosylation sites. Comparison of the *Icpu-DAA1*01* and *Icpu-DBA1*01* genomic sequence with the corresponding cDNA sequences revealed that the channel catfish MHC class II A gene has 5 exons and 4 phase 1 introns. Introns 1 and 2 showed only short stretches of identity between the two class II *A* loci. All three introns differ in length between the two loci. The catfish II *A* gene has a significantly longer intron 2 versus that found in the zebrafish gene. The relationship of these two loci to the *A* locus identified in zebrafish, or to the two loci implicated to be present in striped bass, carp, and shark, remain to be elucidated. Genomic sequence determination of class II *A* genes in these and other species would help determine the evolutionary history of the II *A* gene in fishes.

Key words. Channel catfish, Major histocompatibility complex, MHC class II *A* genes, Gynogenetic catfish

Introduction

Mammalian classical major histocompatibility complex (MHC) class I and class II molecules are polygenic and highly polymorphic cell-surface glycoproteins that are known to play a central role in adaptive immunity via their ability to bind and present processed foreign peptides to CD8 and CD4 positive T cells, respectively (Klein 1986; Rothbard and Gefter 1991; Germain 1994; Germain et al. 1996). Much of what is known about MHC class II function and gene organization has come from mammalian studies. Hashimoto et al. (1990) first isolated MHC class I and class II *B* genes in carp (*Cyprinus carpio*), which led to numerous sequence studies of the MHC in fishes. Most of the research involved with MHC class II genes has focused on the class II *B* loci (for reviews, see Klein et al. 1997; McConnell et al. 1998). Little information on the class II *A* genes in nonmammalian vertebrates is available to date (Kasahara et al. 1992; Sültmann et al. 1993; Hardee et al. 1995; van Erp et al. 1996). Our understanding of the evolution of the MHC and of the functional role of class II molecules in the teleost immune response would benefit by the analyses of MHC class II *A* genes in other teleost species. The channel catfish, *Ictalurus punctatus*, is an excellent candidate for both evolutionary and functional studies of teleost MHC molecules. *In vitro* culture systems are well established (Miller et al. 1994a; Clem et al. 1991; Vallejo et al. 1992). Previous studies have demonstrated significant functional similarities among the catfish and mammalian specific immune systems (Miller et al. 1985; Vallejo et al. 1991a, b). Also, clonal cell lines are available from gynogenetic catfish, allowing extensive mapping in this homozygous model system. In this study we have isolated catfish major histocompatibility class II *A* genes from the 1B10 cell line (Miller et al. 1994b), to complement our previous findings on the *B* genes (Godwin et al. 1997), and from gynogenetic catfish cell lines. We obtained four cDNAs (*Icpu*-A-1 through *Icpu*-A-4; Accession #s: AF103006 through AF103009). We also isolated *MhcIcpu-DAA1*01* (Accession # AF103002 and AF103003) and *Icpu-DBA1*01* (Accession # AF103004 and AF103005), both as genomic sequences and cDNAs, from gynogenetic catfish cell lines. *Icpu* refers to *Ictalurus punctatus*, *D* to class II, *A* or *B* to the locus designation, and the last letter *A* to the α-chain encoding genes, as proposed by Klein et al. (1990).

Materials and methods

Gynogenetic catfish (Goudie et al. 1995; Hogan et al. 1999) were obtained from the Catfish Genetics Research Unit, USDA-ARS (Stoneville, MS). Peripheral blood leukocytes (PBLs) were isolated as previously described (Miller et al. 1994a). The clonal B cell line (1B10) was developed from an outbred channel catfish (Miller et al. 1994b), and the clonal T cell line (G14-A) was developed from a second-generation gynogenetic catfish (Hogan et al. 1999). The G14F fibroblast cell line was derived from an explanted fin snip from the same gynogenetic catfish that yielded G14-A (Antao et al. 1999).

Poly (A)+ mRNA was isolated from catfish PBLs and the 1B10 cell line (Miller et al. 1994b), reverse transcribed, and cloned into λ ZAPII (Stratagene, LaJolla, CA) as previously reported (Godwin et al. 1997). Replica NitroPure transfer membranes of the PBL library (MSI, Westboro, MA) initially screened under low stringency conditions with a full-length striped bass MHC class II *A* (*Mosa*-A-S5) cDNA probe (Hardee et al. 1995). A partial MHC classII *A* cDNA clone isolated from the PBL library was then used to screen a 1B10 cell line cDNA library. Positive phage clones were subcloned according to the manufacturer's instruction and subsequently sequenced.

Polymerase chain reaction (PCR) amplification was carried out on the cDNA library stock with sense-primer TM 345 (GCTGCTGATTGAACACACACT) corresponding to the 5'untranslated region and antisense-primer TM 302 (GCG*TCTAGA*TCTGGACTCGTCACTGAAACATCA) with an *Xba*I site and including a portion of the 3'UTR up to the stop codon. The primer sequences were designed based on the information obtained from the catfish cDNA library-derived class II *A* clones. Borriello and Krauter's (1990) PCR+1 protocol was used to prevent heteroduplex formation. Reaction conditions were as follows: 4 µl of phage stock, 0.2 mM dNTP mix, 1.5 pmole TM 345, 12 pmole TM 302, 13 µl ddH$_2$O, 3 µl Buffer A, and 7 µl of Buffer B were mixed and denatured for 1 minute at 94°C. After cooling to 80°C, 20 µl of a 1:10 ELONGase mix (Gibco BRL, Gaithersburg, MD) were added. The initial 35 cycles (94°C for 1 min, 56°C for 1 min, 68°C for 1 min) were followed by the addition of 15 pmole of primer TM 346 (identical to TM 345 except for a 9 bp sequence with a *Hind*III site at the 5'end). A 1 min cycle for denaturing and for annealing followed, with a final extension of 15 min. The PCR products were ligated into the pCRII cloning vector (InVitrogen, San Diego, CA) followed by standard transformations, plasmid DNA preparations, and *Hin*d III/*Xba* I digests.

For cDNA synthesis 1 µg of total RNA from G14-A T cells was reverse transcribed as follows: 3 µl of DEPC-H$_2$0 in the presence of 40 µM oligo dT (17) were heated for 2 min at 72°C. After cooling to 4°C, 2 µl of 5x First Strand Buffer, 10 mM DTT, 1 mM dNTP mix, and 1 µl of reverse transcriptase (Gibco BRL, Gaithersburg, MD) were added and heated to 42°C for 30 min. The mix was then heated to 70°C for 15 min, followed by the addition of RNAse H for 10 min at 55°C and then placed on ice. Thirty-five PCR cycles using reagents and conditions as recommended by the Advantage cDNA PCR kit (Clontech, Palo Alto, CA) were performed (see above). Sense primer TM 345 (5' end) together with antisense-primer TM 302 (3' end) as well as TM 345 together with antisense-primer TM 490 (see below) were used to amplify the cDNA, which was subsequently cloned using the TOPO TA Cloning System (Invitrogen, San Diego, CA).

One µg of G14-A genomic DNA was PCR amplified with sense-primer TM 345 and antisense-primer TM 490 (ACAGGTCAATAAAATGTGGAACTCA), based on a region 200 basepairs (bp) 3' of the stop codon. The sequence for this region was obtained as part of a linkage study of MHC class II A and B genes (manuscript in preparation). Amplification of the second locus was obtained with sense primer TM 345 and antisense-primer TM 491 (AGCAAT CCTACAGCCAGACCCACTC) from the conserved transmembrane-encoding region. Thirty PCR cycles using reagents and conditions as recommended by Advantage Genomic PCR kit (Clontech, Palo Alto, CA) were performed, followed by agarose gel electrophoresis, QIAquick gel extraction (QIAGEN, Chatsworth, CA), and ligation into the pGEM-T vector (Promega, Madison, WI).

Sequencing was performed using universal forward and reverse primers (DNA International, Lake Oswego, OR), followed by gene-specific primers derived from these sequences. Both strands were sequenced by the dideoxy chain termination method (Sanger et al. 1977) using the fluorescence-based PRISM AmpliTaq™ Ready Reaction FS (Applied Biosystems, Foster City, CA) according to the manufacturer's protocol. The results were analyzed on the Applied Biosystems Model 373A DNA Sequencing System. We employed the Fragment Assembly System (FAS-Version 8-UNIX) of the Genetics Computer Group (Devereux et al. 1984) as well as the PILEUP and PRETTY options to align segments of DNA sequence, obtain amino acid translations, overlay homologous sequences and determine consensus. We performed NCBI Blast (Altschul et al. 1997) searches for sequences similar to the catfish class II A genes and preliminary nucleotide alignments. To predict the cleavage site of the signal peptide, we utilized the SignalP computer program (Nielsen et al. 1997). The MEGA program (Kumar et al.1993) was used to construct the dendrograms, calculating pairwise distances for exon 3 (α2-domain encoding) with the p-distance algorithm and the neighbor-joining method of Saitou and Nei (1987).

Results and discussion

Initially, a full length striped bass (*Morone saxatilis*) class II A gene cDNA (Hardee et al. 1995) was used as a heterologous probe to screen a channel catfish PBL cDNA library. This screening yielded one positive 5' truncated cDNA clone which showed sequence relatedness to other known teleost class II A genes. This partial class II A cDNA was then used to screen a cDNA library constructed from a clonal B cell line (1B10) previously used to identify class I A (Antao et al. 1999) and class II B (Godwin et al. 1997) genes. From this screen, seven positive hybridizing cDNA clones were sequenced and found to have homology to known MHC class II A genes. Of these, two different full-length MHC class II A cDNA clones were identified. The first was 1053 bp in length and designated *Icpu*-A-1 (accession # AF103006), and the second was 1062 bp in length and was designated *Icpu*-A-2 (accession # AF103007). Based on the conserved sequences of *Icpu*-A-1 and *Icpu*-A-2, PCR primers (TM 345 and TM 302) were designed in order to determine if coding regions for other possible MHC class II A cDNAs were present in the 1B10 cDNA library. To this end, PCR+1 protocols yielded two additional unique MHC class II A gene sequences which were designated *Icpu*-A-3 (accession # AF103008) and *Icpu*-A-4 (accession # AF103009). Each of four catfish MHC class II A cDNAs have a coding region of 705 bp, which would encode 26 kDa proteins. The identification of four different sequences from a single clonal diploid B cell line strongly suggests that channel catfish possess at least two functional MHC class II *A* loci. Since A-1 to A-4 sequences are highly similar to each other at both the nucleotide (i.e., 89-98% identical, data not shown) and deduced amino acid levels (80-95% identical, see Fig. 1), it is not possible at this time to assign locus designations to these sequences. As shown in Figure 1, the majority of the deduced amino acid differences occur in the putative peptide-binding region (PBR) of the α1 domain, as expected.

Many of the general features of the three known teleost species' putative MHC class II α molecules are also present in the catfish. All six of the catfish sequences reported here include typical hydrophobic leader sequences of 18 aa residues, nearly identical to one another, and similar in length to those of other reported teleost class II α proteins (Fig. 1). The catfish α1 domains are 83 residues (except for *Icpu-DAA1*01*, at 84 residues, see below), as found in zebrafish (*Danio rerio*) (Sültmann et al. 1993), striped bass (*Morone saxatilis*) R2 and R5 clones (Hardee et al. 1995), and carp (*Cyprinus carpio*) (Van Erp et al. 1996), based on the leader/α1 cleavage site shown (Fig. 2). Of seven striped bass class II *A* sequences previously cloned, two encoded putative proteins with 84 residue α1 domains as found for *Icpu-DAA1*01*, based on the leader cleavage site shown, probably representing a separate class II *A* locus in the striped bass (Hardee et al. 1995). The nurse shark α1 domain is 82 residues long. Two cysteine residues 53 residues apart (except for *Icpu-DAA1*01*, at 54 residues apart) are positioned so that an intradomain disulfide bond may form. The other teleost α1 domains all possess cysteines in the identical positions, though the nurse shark does not (Kasahara et al. 1992). In human class II α sequences, cysteines in identical positions of the α1 domain are only found in HLA-DMA (Kelly et al. 1991). The teleost class II α molecules, therefore, are likely different from mammalian sequences in possessing an intradomain disulfide-bound α1 domain that is typically polymorphic. The α2 domain of all of the catfish sequences and the nurse shark sequence are 94 residues, whereas the other known transcribed teleost α2 sequences are 93 residues. All of the fish α2 domains have conserved cysteines 55 residues apart that may form intradomain disulfide bonds, similar to known mammalian sequences. The connecting peptide, transmembrane, and cytoplasmic tail regions in catfish are similar to other teleosts, as has been described (Sültmann et al. 1993). The catfish putative MHC class II α sequences, in contrast to other known class IIα sequences, lack any consensus glycosylation signal sequence. The other known teleost class II α sequences contain glycosylation signals in the α2 domain (Fig. 1). For both zebrafish and carp, these consensus signals align with the human glycosylation signals of HLA-DRA (Schamboeck et al. 1983), HLA-DNA (Jonsson and Rask 1989), HLA-DPA (Lawrance et al. 1985), HLA-DQA (Jonsson et al. 1989); the striped bass consensus glycosylation signal sequence is nine residues further towards the carboxy-terminus. The HLA-DMA glycosylation signal sequence is also in the α2 domain, but aligns 34 residues closer towards the carboxy-terminus (HLA sequences not shown) than the other human sequences.

```
                        LEADER PEPTIDE       α-1 DOMAIN
                         -19               1         10         20
Consensus               ----MR LFLLCFTLVC -VKDTEAQNK HHDLQLSACS DTDKEYMVG-
Icpu-DAA1*01            --    ---------- .-------i- ---ikvi--- ------vl-i
Icpu-DBA1*01            --    ---------- .--------- l-h---av-- -------i-d
Icpu-A-1               --    ---------- .--------- f-h-e----- ek------s
Icpu-A-2               --    ---------- .--------- f-h------p ek------s
Icpu-A-3               --    ---------- .m-------- f-h-e----- ek------s
Icpu-A-4               --    ---------- .--------- l-h---av-- -------i-d
Cyca-DXA1*01           mygvllm-al .ivs--t-vv nr-v-fvg-- --er-fli-f
Dare-1.3.4             -d --gfll-ftv .ilsnv--ae -r-vdffg-- --e---lq-f
Mosa-A-R2              -k mmkimmv--l .fscvs-ddl -e-friag-- -s-g-e-y-l
Mosa-A-R5              -k mmkmmmv--l .fscvs-dvl -e-iaid--- -s-g-k-y-l
Gici-DAA              mearny fsv-vlv-iq ggwagkylyd ftqvyfvqqr spe-hfd-me
```

```
                        30         40         50         60         70
Consensus               DGEEMFYADF IKKDIVNA-P PFADPIEYPE GGY-GAEAKM AIC-QNLQV-
Icpu-DAA1*01           -k--vy---- v-ql--k-l- ----tlnp-- ta-qs-g-ei d--kt--g-y
Icpu-DBA1*01           ---ve----- k-----ymf- ------r--. s-fa---gw- sl-qndie-f
Icpu-A-1               ---------- e-------l- -----g-ft. --faf--s-t -n-qa----l
Icpu-A-2               ---------- e-------f- -----g-ft. --fgv--s-t -n-qa----f
Icpu-A-3               ---k------ e-------w- -----g-ft. --faf--s-t -n-qa----f
Icpu-A-4               ---ve----- k-----ymf- ------r--. s-fa---gw- sl-qndie-f
Cyca-DXA1*01           ----lwh--- -r-eg-vtv- d-----gf-. -f-etgv-l- ev-k---aln
Dare-1.3.4             ----lyhs-- -r-vg-vta- d----ms--. -f-ensv-q- ev-k-d-atd
Mosa-A-R2              ----vw---- -n-kg-epq- s-i-hts-v- -t-eq-v-nq q--r---gla
Mosa-A-R5              ----vw---- -n-kg-epq- s-i-h-t--. ---es-v-qq q--r---k-a
Gici-DAA              --d-i--m-- nl-ke-ari- e--h..l-mq --ea-is-ni --vkn--k-v
```

```
                                  α-2 DOMAIN
                        80         90         100        110        120
Consensus               S-EFKDKPLP QD-PQSS-IY PRDDVQLGSE NTLICHVT-F FPPPVRVRWT
Icpu-DAA1*01           rt----t-t- --v--n-.-- --v--n---k ------s-r- ---ti-it--
Icpu-DBA1*01           -v-------- --a--t-.v- -s-------- ------s-r- -----h----
Icpu-A-1               -v-------- --a----.-- a-tg------ -l----asr- ----------
Icpu-A-2               -gd------- --a----.-- --------k ------air- ----------
Icpu-A-3               -v-------- --a----.-- a-tg------ -l-v--asr- ----------
Icpu-A-4               -v-------- --a--t-.v- -s-------- ------s-r- -----h----
Cyca-DXA1*01           ikvy-ptdeq lap-da-.v- seg--v--vq --------gl -----nrs--
Dare-1.3.4             ikaynspeeq l-p-vt-.-- se-e-v-der --------g- -----nrs--
Mosa-A-R2              lkav--pq.k f-p-s-pv-- -----e--ek --------g- y-a--k-y--
Mosa-A-R5              lqay-np--q l-r-s-pm-- t--n-e--ek --------g- y-a--k-y--
Gici-DAA              mnlsggt-e- kvp-ev-.v- se-l-ew-ql -----fadg- y--hitmk-r
```

```
                        130        140        150        160        170
Consensus               KNNVDVTEES SLSQYYPNED NT-NQFSHLP FTPQEGDVY- CTV-HEALQT

Icpu-DAA1*01           --g----d-- ---------- --y------- ---k-----t ---q-----
Icpu-DBA1*01           ---l---g-- ---------- --y------- ---------t ---e-----
Icpu-A-1               ---l---dk- ----------d- e-f------- ---------t ---q-----
Icpu-A-2               -------g-- ---------- e-f------- ---------t ---q-----
Icpu-A-3               -------g-- ---------- --y------- ---------t ---g-----
Icpu-A-4               -------g-- ---------- --y------- ---------t ---e-----
Cyca-DXA1*01           ---qi---dv -----rrkn- g-f-i--s-k ---a---i-s ---y-k--es
Dare-1.3.4             ---di----i -f---rr-s- g-f-m--a-k ---a---i-s ---n-rsi-g
Mosa-A-R2              --gknvt-gt -invp-l-k- --ft-t-r-e -i--l--m-s -s-k-ls-k.
Mosa-A-R5              --gknvt-gt -invp---k- gsft-t-r-e -i--l--m-s -s-k-ls-k.
Gici-DAA              r--epm-dgd nitef-ikd- f-yrr--y-s iv-sp--m-s -h-e-ss--.
```

```
                      CP            TM                           CT
                     180           190          200          210           220
Consensus        DPDTRTWEV  VDLPSVGP-A VFCGVGLAVG LLGVATGTFF L-KGNQCN-
Icpu-DAA1*01     .--------d  ---------.-  ----------  ----------  -v------*
Icpu-DBA1*01     .--------n  ---------.-  ----------  ----------  -v------*
Icpu-A-1         .--------n  ---------.-  ------v---  ----------  -v------*
Icpu-A-2         .--------n  ---------.-  ------v---  ----------  -v------*
Icpu-A-3         .--------d  ---------.-  ------v---  ----------  -v------*
Icpu-A-4         .--------n  ---------.-  ----------  ----------  -v------*
Cyca-DXA1*01     rfi-k----d  -av-g---.-  -------sl-  -----a----  -i-l-n--*
Dare-1.3.4       q-n-k----d  -e------.-  --------vl-  -----a----  i---n--*
Mosa-A-R2        --l--f-d-e  kpe--i--.-  ----l--t--  -----a----  i---e-s*
Mosa-A-R5        --l--f-d-e  kpe--i--.-  ----l--t--  -----a----  i---e-s*
Gici-DAA         --v-vf-dqg  -peeks--gt ii-al--tl-  iisavv-iil- i-er-rlqa
```

Fig. 1. Comparison of the deduced amino acid sequences of the channel catfish MHC class II α chains with other teleost and elasmobranch: Cyca-DXA1*01 (Van Erp et al. 1996), Dare-1.3.4 (Sültmann et al. 1993), Mosa-A-R2 and A-R5 (Hardee et al. 1995), and Gici-DAA (Kasahara et al. 1992). The domains are indicated above the sequences, and the numbering is that of the putative mature protein of the channel catfish α chain. Dashes "-" indicate with the consensus (simple majority) and dots "..." were introduced for better alignment. Asterisks "*" denotes the STOP codon, the symbol "~" where data was not added, and the N-glycosylation sites are bold.

The nurse shark is unusual in possessing a glycosylation signal sequence in the α1 domain. The significance of the lack of any glycosylation site in the channelcatfish is difficult to determine. Less is known about the role of chaperonins for MHC class II assembly in the endoplasmic reticulum (ER) than for class I assembly. Calnexin, known to associate with several membrane glycoproteins based on interactions with N-linked glycans, associates rapidly with newly synthesized trimers of class II-invariant chain complexes in the ER (Anderson and Cresswell 1994). Calnexin dissociates from this complex before egress of the class II αβ heterodimer from the ER. Calnexin may not, however, act simply as a lectin in associating with the class II-invariant chain complex, although association of calnexin with invariant chain may help retain invariant chain in the ER for association with newly synthesized class II heterodimers (Romagnoli and Germain 1995). Whatever the role of calnexin in class II assembly, glycosylation of the class II β chain in catfish (Godwin et al. 1997) may provide a sufficient target site for chaperonins guiding assembly of MHC class II αβ molecules in the cell. To address questions concerning the number of MHC class II *A* loci present in the channel catfish, as well as the exon-intron structure, class II *A* sequences were identified by PCR using genomic DNA and mRNA from cell lines derived from a homozygous line of gynogenetic channel catfish (Hogan et al. 1999). The rationale for using gynogenetic catfish is that these fish are homozygous at a given locus and therefore different sequences would indicate different loci rather than different alleles. When genomic DNA from a gynogenetic fibroblast line was subjected to PCR pairing sense primer TM345 (5'-

UTR) with anti-sense primer TM490 (3'-UTR) a 2324 bp fragment was obtained (data not shown). Sequence analysis revealed this DNA fragment to contain a full-length MHC class II *A* gene, designated *MhcIcpu-DAA1*01* (accession# AF103002). In a different PCR reaction pairing TM345 with anti-sense TM491 (a primer within the transmembrane region) two fragments of approximately 2.0 and 1.7 kb in size were obtained. Sequencing revealed the larger fragment to be identical to *MhcIcpu-DAA1*01*, and the shorter fragment (due to shorter introns) to represent a different class II *A* locus which was designated *MhcIcpu-DBA1*01* (accession# AF103004). RT-PCR of mRNA isolated from a clonal T cell line derived from a gynogenetic catfish resulted in two cDNA sequences (accession# AF103003 and AF103005), one identical to *MhcIcpu-DAA1*01* and the other identical to *MhcIcpu-DBA1*01* (data not shown). It should be noted that RT-PCR protocols using mRNA from the fibroblast line did not result in the amplification of MHC class II *A* cDNA, implying that such genes are not expressed in fibroblasts (data not shown). These two different MHC class II *A* cDNA sequences not only indicate that the two gene loci are probably functional, but also that they are expressed in a *bona fide* catfish T cell line as they are in a clonal B cell line (1B10). The two sequences demonstrate 64% nucleotide identity in exon 2, and only 44% identity for the respective encoded amino acids. More telling for distinguishing loci, the two sequences share only 86% identity in the usually highly conserved $\alpha2$ domain (82% aa identity). These are the first fish class II *A* sequences for which the genomic and the exact corresponding cDNA sequences have been isolated. In zebrafish, both genomic and cDNA sequences were isolated for a class II *A* gene, but they did not represent the identical sequence, and only one expressed class II *A* locus has been identified (Sültmann et al. 1993). The identification of the cDNAs enabled the exact intron-exon structure of the two catfish alleles, representative of the two MHC class II *A* loci, to be determined.

The catfish MHC class II *A* gene structure is similar to that seen with other known MHC class II *A* genes and consists of 5 exons and 4 introns (see Fig. 2). Like other MHC class II genes, the first exon encodes the 5' UTR (untranslated region), leader peptide, and the first amino acid of the $\alpha2$ domain; the second and third exons encode the $\alpha1$ and $\alpha2$ domains, respectively. As shown with *MhcIcpu-DAA1*01*, the fourth exon encodes for the TM (transmembrane), Cy (cytoplasmic tail), and part of the 3'UTR, with the rest of the 5' UTR encoded by the fifth exon (Fig. 2). We suspect that the same intron-exon structure is also present in the *MhcIcpu-DBA1*01* locus, but isolation of a full-length genomic clone is required for verification. This exon-intron organization is in contrast to the only other known complete teleost class II *A* gene, zebrafish, wherein the fourth exon encodes the TM, Cy, and all of the 3'UTR (Sültmann et al. 1993). A gene organization similar to that of the catfish II *A* gene, although with significantly larger introns, has been reported in the human *DQA*, *DRA*, and mouse *EA* genes (Trowsdale 1995). The carp MHC class I *A* gene has a similar pattern at the 3' end of the gene, with a 92 bp intron starting 4 bp after the stop codon (Van Erp et al. 1996). All of the introns for the catfish MHC loci have canonical GT/AG splice

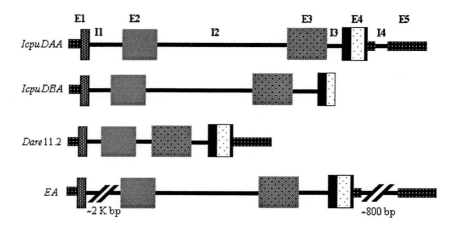

Fig. 2. Comparison of MHC class II *A* gene organization of channel catfish (*Icpu-DAA*, *Icpu-DBA*) with zebrafish (*Dare* 11.2) and mouse (*EA*). 'E' and 'I' denote 'exons' and 'introns'.

signals (Senepathy et al. 1990), and, like those of other MHC genes (class I and II), are phase one introns which divide codons between the first and second nucleotides. Comparison of the first three introns of *DAA1* with those of *DBA1* shows that they are quite different in both length (intron 1, 330 vs. 260 bp; intron 2, 887 vs. 638 bp; and intron 3, 156 vs. 195 bp; respectively) and sequence (Fig. 2). However, for both loci the 3' ends of intron 1 and 3 are CT/GT rich. Tracy et al. (1997) reported that both the prokaryotic RecA gene and its eukaryotic homologue RAD51 utilize a GT-rich segment for genetic recombination. For this reason they proposed that GT-rich regions might be areas that allow for increased genetic exchange. It is tempting to speculate that the presence of such GT/CT-rich intronic regions bordering the α1 and α2 encoding domains may in part be responsible for maintaining allelic exchange and diversity in MHC molecules.

To study the phylogenetic relationships of fish class II *A* genes, a dendrogram was constructed with the neighbor-joining method of Saitou and Nei (1987) based on p-distances as calculated by MEGA (Kumar et al. 1993). The resulting phylogenetic tree (Fig. 3), based on exon 3 DNA sequences, clearly shows *Icpu-DAA1* to be distinct from *DBA1*. The A1 through A4 sequences and the *DBA1*01* sequences are all more related to one another than to *DAA1*01*. Among the members of this cluster, A-2 emerges just before the other four class II *A* sequences. The A-1 and A-3 sequences cluster together. Also, *DBA1*01* and A-4 group together, differing by a single nucleotide in exon 3, with an additional 3 nucleotide differences in the CP/TM/Cy coding region (exon 4), strongly

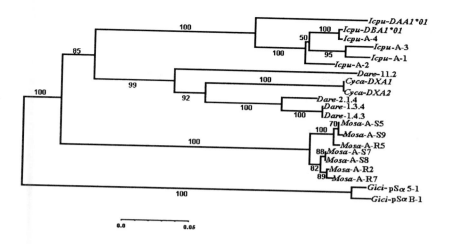

Fig. 3. A phylogenetic tree based on the genetic distances between exon 3 nucleotide sequences (α-2 domain encoding). The amino acid comparison of the different species was used to align gaps of this domain. The tree was constructed as described under Materials and methods. Numbers on the nodes indicate frequency with which this node was recovered per 100 bootstrap replications in a total of 500. References: see Figure 1.

suggestive that A-4 and *DBA1*01* are alleles of the same locus, *DBA1*. The topology of the dendrogram makes it tenuous to assign definitive locus assignments for the *Icpu* A1 through A4 sequences, and may even suggest the possibility of three MHC class II A loci in channel catfish. The unique one codon indel in exon 2 of *DAA1*01*, the extensive exon-intron differences between *DAA1*01* and *DBA1*01*, and the extensive differences in coding regions for both the α1 and α12 domains all provide additional support for the presence of two distinct class II A loci.

The carp and the zebrafish cluster together in Figure 3 as expected, both are members of the order Cypriniformes. The catfish sequences are related to those of the zebrafish and carp sequences; all three species are members of the superorder Ostariophysi, though channel catfish belong to the order Siluriformes. A transcript has not been detected for the zebrafish clone *Dare*-11.2. The unusual position of this zebrafish sequence, emerging before any of the other zebrafish or carp sequences, could be accounted for if it is eventually identified as a pseudogene. In fact, to date only one zebrafish MHC class II A locus has been identified as transcribed (Sültmann et al. 1993; Bingulac-Popovic et al. 1997; Graser et al. 1998). The two carp sequences shown in Figure 3, *Cyca-DXA1* and *-DXA2* have been identified as alleles of closely related but distinct loci (Van Erp et al. 1996). The presence of two distinct class II *A* loci has been implicated in nurse shark (Kasahara et al. 1993) and striped bass (Hardee et al. 1995).

In the current study we provide evidence for the presence of two MHC class II *A* loci, both transcribed, in the channel catfish. Of the six allelic sequences identified, none have a consensus glycosylation signal sequence. With the identification of two MHC class II *B* loci in the same gynogenetic catfish (manuscript in preparation), it is apparent that the MHC is polygenic and polymorphic in a number of fish species. This MHC diversity is likely the result of selective pressures on fishes and may be a reflection of parallel mechanisms operating in higher vertebrates.

References

Altschul SF, Madden TL, Schaffer AA, Zhang J, Zhang Z, Miller W, Lipman DJ (1997) Gapped BLAST and PSI-BLAST: a new generation of protein database search Programs. Nucleic Acids Res 25:3389-3402

Anato AB, Chinchar VG, McConnell TJ, Miller NW, Clem LW, Wilson MR (1999) MHC class I genes of the channel catfish: sequence analysis and expression. Immunogenetics 49:303-311

Anderson KS, Cresswell P (1994) A role for calnexin (IP90) in the assembly of class II MHC molecules. EMBO J 13:675-682

Bingulac-Popovic J, Figueroa F, Sato A, Talbot WS, Johnson SL, Gates M, Postlethwait JH, Klein J (1997) Mapping of Mhc class I and II regions to different linkage groups in the zebrafish, *Danio rerio*. Immunogenetics 46:129-134

Boriello F, Krauter KS (1990) Reactive site polymorphism in the murine protease inhibitor gene family is delineated using a modification of the PCR reaction (PCR+1) Nucleic Acids Res 18:5481-5487

Clem LM, Miller NW, Bly JE (1991) Evolution of lymphocyte subpopulations, their interaction and temperature sensitivities. In: Warr GW, Cohen N (eds) The Phylogenesis of Immune Functions. CRC Press, Boca Raton FL, pp 191-212

Devereux J, Haeberli P, Smithies O (1984) A comprehensive set of sequence analysis programs for the VAX. Nucleic Acid Res 12:387-395

Germain RN (1994) MHC-dependent antigen processing and peptide presentation: Providing ligands for T lymphocyte activation. Cell 76:287-299

Germain RN, Castellino F, Han R, Sousa CRE, Romangnoli P, Sadegh-Nasseri S, Zhong GM (1996) Processing and presentation of endocytically acquired protein antigens by MHC class I and class II molecules. Immunol Rev 151:5-30

Godwin UB, Antao A, Wilson MR, Chinchar VG, Miller NW, Clem LW, McConnell TJ (1997) Mhc class II *B* genes in the channel catfish *(Ictalurus punctatus)*. Dev Comp Immunol 18:13-23

Goudie CA, Simco BA, Dais KB, Liu Q (1995) Production of gynogenetic and polyploid catfish by pressure-induced chromosome set manipulation. Aquaculture 133:185-198

Graser R, Vincek V, Takami K, Klein J (1998) Analysis of zebrafish Mhc using BAC clones. Immunogenetics 47:318-325

Hardee JJ, Godwin UB, Benedetto R, McConnell TJ (1995) Major Histocompatibility complex class II *A* gene polymorphism in the Striped Bass *(Morone saxatilis)*. Immunogenetics 41:229-238

Hashimoto K, Nakanishi T, Kurosawa Y (1990) Isolation of carp genes encoding major histocompatibility complex antigens. Proc Natl Acad Sci USA 87:6863-6867

Hogan RJ, Waldbieser GC, Goudie CA, Antao A, Godwin UB, Wilson MR, Miller NW, Clem LW, McConnell TJ, Wolters WR, Chinchar VG (1999) Molecular and immunologic characterization of gynogenetic channel catfish (*Ictalurus punctatus*). Mar Biotechnol: in press

Jonsson AK, Rask L (1989) Human class II *DNA* and *DOB* genes display low sequence variability. Immunogenetics 29:411-413

Jonsson AK, Andersson L, Rask L (1989) Complete sequences of DQA and DQB1 cDNA clones corresponding to the DQw4 specificity. Immunogenetics 30:232-234

Kasahara M, Vazquez M, Sato K, McKinney EC, Flajnik MF (1992) Evolution of the major histocompatibility complex: Isolation of class II *A* cDNA clones from the cartilaginous fish. Proc Natl Acad Sci USA 89:6688-6692

Kasahara M, McKinney EC, Flajnik MF, Ishibashi T (1993) The evolutionary origin of the major histocompatibility complex: polymorphism of class II α chain genes in the cartilaginous fish. Eur J Immunol 89:2160-2165

Kelly AP, Monaco JJ, Cho S, Trowsdale J (1991) A new human HLA class II-related locus, *DM*. Nature 353:571-573

Klein J (1986) Natural History of the Major Histocompatibility Complex. John Wiley, New York

Klein J, Bontrop RE, Dawkins RL, Ehrlich HA, Gyllensten UB, Heise ER, Jones PP, Parham P, Wakeland EK, Watkins DI (1990) Nomenclature for the major histocompatibility complexes for different species: a proposal. Immunogenetics 31:217-219

Klein J, Klein D, Figueroa F, O'hUigin, C (1997) Major histocompatibility genes in the study of fish phylogeny. In: Kocher TD, Stepien CA (eds) Molecular Systematics of Fishes. Academic Press, San Diego CA, pp 271-283

Kumar S, Tamura K, Nei M (1993) MEGA: Molecular Evolutionary Genetics Analysis, Version 1.02. The Pennsylvania State University, University Park PA

Lawrance SK, Das HK, Pan J, Weissman SM (1985) The genomic organization and nucleotide sequence of the HLA-SB (DP) alpha gene. Nucleic Acids Res 13:7515-7528

McConnell TJ, Godwin UB, Cuthbertson BJ (1998) Expressed major histocompatibility complex class II loci in fishes. Immunol Rev 166:294-300

Miller NW, Deuer A, Clem LW (1986) Phylogeny of lymphocyte heterogeneity: the cellular requirements for the MLR in channel catfish. Immunology 59:123-128

Miller NW, Chinchar VG, Clem LW (1994a) Development of leukocyte cell lines from the channel catfish (*Ictalurus punctatus*). J Tiss Cult Meth 16:117-123

Miller NW, Rycyzyn MA, Wilson MR, Warr GW, Naftel JP, Clem LW (1994b) Development and characterization of channel catfish long term B cell lines. J Immunol 152:2180-2189

Miller NW, Sizemore RC, Clem LW (1985) Phylogeny of lymphocyte heterogeneity: The cellular requirements for in vitro antibody responses of channel catfish Leukocytes. J Immunol 134:2884-2888

Nielsen H, Engelbrecht J, Brunak S, Von Heijne G (1997) Identification of prokaryotic and eukaryotic signal peptides and prediction of the cleavage sites. Protein Eng 10:1-6

Romagnoli P, Germain R (1995) Inhibition of invariant chain (Ii)-calnexin interaction results in enhanced degradation of Ii but does not prevent the assembly of alpha beta Ii complexes. J Exp Med 182:2027-2036

Rothbard JB, Gefter ML (1991) Interactions between immunogenic peptides and MHC proteins. Annu Rev Immunol 9:527-565

Saitou N, Nei M (1987) The neighbor-joining method: a new method in reconstructing phylogenetic trees. Mol Biol Evol 4:406-425

Sanger F, Nicklen S, Coulson AR (1977) DNA sequencing with chain-terminating inhibitors. Proc Natl Acad Sci USA 74:5463-5467

Senepathy P, Shapiro MB, Harris NL (1990) Splice junctions, branch point sites, and exons: Sequence statistics, identification, and applications to genome projects. Meth Enzymol 183:252-278

Schamboeck A, Korman AJ, Kamb A, Strominger JL (1983) Organization of the transcriptional unit of a human class II histocompatibility antigen: HLA-DR heavy chain. Nucleic Acids Res 11:8663-8675

Sültmann H, Mayer WE, Figueroa F, O'hUigin C, Klein J (1993) Zebrafish MHC class II α chain-encoding genes: polymorphism, expression, and function. Immunogenetics 38:408-420

Tracy RB, Baumohl JK, Kowalczykowski SC (1997) The preference for GT-rich DNA by the yeast Rad51 protein defines a set of universal pairing sequences. Genes Dev 11:3423-3431

Trowsdale J (1995) "Both man & bird & beast": comparative organization of MHC genes. Immunogenetics 41:1-17

Vallejo A, Miller NW, Clem LW (1992) Antigen processing and presentation in teleost immune responses. Annu Rev Fish Dis 2:73-89

Vallejo A, Ellsaesser CF, Miller NW, Clem LW (1991a) Spontaneous development of functionally active long term monocyte-like cell lines from the channel catfish. In Vitro Cell Dev Biol 27A:279-286

Vallejo A, Miller NW, Clem LW (1991b) Phylogeny of immune recognition: role of alloantigens in antigen presentation in channel catfish immune responses. Immunology 74:165-168

Van Erp SHM, Egberts E, Stet RJM (1996) Characterization of major histocompatibility complex class II A and B genes in a gynogenetic carp clone. Immunogenetics 44:192-202

The evolution of MHC class I genes in cattle

Shirley A. Ellis[1], Edward C. Holmes[2], Karen A. Staines[1], and W. Ivan Morrison[1]

[1]Institute for Animal Health, Compton, Newbury, RG20 7NN, UK
[2]The Wellcome Trust Centre for the Epidemiology of Infectious Disease, Department of Zoology, University of Oxford, South Parks Road, Oxford OX1 3PS, UK

Summary. Detailed analysis of cattle MHC class I haplotypes has been carried out using inbred homozygous animals. Transcribed class I genes have been identified and characterised by cDNA cloning, sequence analysis, and transfection/expression studies. The analysis demonstrated that multiple genes are transcribed, and also revealed substantial variation between haplotypes. Phylogenetic analysis of all available BoLA class I cDNA sequences revealed complex evolutionary relationships. While some groupings of sequences may demonstrate a correlation with individual loci, overall it was not possible to define the origin of alleles using this approach. These data indicate that there may be four or more classical class I genes in cattle, with striking differences in the array of these genes expressed in different haplotypes. This situation may reflect the complex origins of modern domestic cattle breeds.

Various approaches are being used to determine the origin and possible functional significance of the observed haplotype variation in cattle. These include mapping of the class I region, assessment of expression levels of different class I gene products, and phylogenetic analysis of class I sequences from related species.

Key words. Cattle, MHC class I, Evolution

Introduction

Cattle MHC class I polymorphisms have, in common with other species, traditionally been detected by serological typing (Bernoco et al. 1991). These early studies suggested that cattle expressed products of a single class I locus (termed 'A'), and while serological specificities are still prefixed -A, more recent biochemical and molecular studies have shown that at least some haplotypes express the products of more than one gene (Joosten et al. 1992; Garber et al. 1994; Ellis et al. 1998).

The aim of this study was to determine how many classical class I genes are in the cattle genome, and expressed, and to establish their relative functional significance. Cattle production is of great economic importance worldwide, and MHC studies are an important component in development of disease control strategies. In addition, such studies contribute to our understanding of MHC evolution and diversity, and their role in the immune response in outbred species.

273

Haplotype analysis

Transcribed and expressed class I genes

A number of haplotypes have been studied in detail, using MHC homozygous cattle. In the case of 5 haplotypes (expressing class I specificities A10/KN104, A18, A11, A31, A14) all transcribed class I genes were identified, cloned and sequenced, and expression studies were carried out. Transcribed class I genes have been sequenced in several additional haplotypes, but in these cases no data have been generated to conirm expression. Table 1 details the 5 fully characterised haplotypes.

Two genes are transcribed in 3 of the haplotypes (A31, A11 and A10/KN104), the A18 haplotype has a single transcribed gene, and the A14 haplotype has 3 transcribed genes. Previous studies of cattle MHC suggested that variation in transmembrane domain (TM) length might correlate with class I locus (Ellis et al. 1992; Garber et al. 1994). Our data show that there are variable combinations of genes with different TM domain lengths in the different haplotypes. Haplotypes have been identified which encode a single gene (encoding either a 35 or a 37 amino acid TM domain), 2 genes (with 2 of 35, one each of 35 and 37, or one each of 36 and 37 amino acid TM domains), or 3 genes (one of 37 and 2 of 35, or one of 36 and 2 of 37 amino acid TM domains) (Ellis et al. 1999; and unpublished data). No haplotype has been identified

Table 1. MHC homozygous animals used in haplotype analysis (from Ellis et al. 1999, with permission)

Animal	Breed	Expressed class I specificity (serologically defined)	Transcribed genes	Transmembrane domain (aa)
E98	Boran	A10, KN104	*A10*	37
			KN104	36
4229	H/F	A18	*HD6*	35
1197	H/F	A31	*HD1*	35
			HD7	35
4222	H/F	A14	*D18.1*	37
			D18.4	35
			D18.5	35
3225	H/F	A11	*D18.2*	37
			D18.3	35

H/F: Holstein / Friesian

with more than 3 transcribed, polymorphic class I genes. Expression studies using a combination of monoclonal antibodies (mAbs) demonstrated that all transcribed genes generated products, and in most cases, it was possible to confirm cell surface expression. Quantitative FACS analysis showed that while most of the genes were expressed at comparable levels, 2 (HD1 and D18.2) were present at significantly lower levels (~10%) (Ellis et al. 1999; Smith and Ellis 1999, unpublished data). These data indicate that there are at least 3 polymorphic, expressed class I genes in cattle, but do not facilitate assignment of sequences to loci.

Phylogenetic analysis

Phylogenetic analyses have been performed on 22 cattle class I cDNA sequences (Ennis et al. 1989; Bensaid et al. 1991; Garber et al. 1994; Ellis et al. 1999). The tree shown in Figure 1 was constructed using sequence data which excluded the peptide binding region in each case, as this is most likely to be subject to selection and is therefore not appropriate for an analysis designed to investigate evolutionary relationships (Cadavid et al. 1997; Holmes and Ellis 1999; Ellis et al. 1999).

Such analysis of human (and other primate) class I sequences generally results in the generation of clear groups which correlate with loci (Knapp et al. 1998). In contrast, phylogenetic analysis of cattle class I sequences revealed complex evolutionary relationships, which may in part result from interlocus recombination (Ellis et al. 1999). Although it is not possible to delineate firm phylogenetic groups, some of the sequences do cluster together, suggesting that they might represent alleles of the same locus e.g. HD1, BSX, MAN3, and HD6, MAN2, 3349.1. However, the relationships between such clusters is not clear, and some sequences do not appear to be closely related to any others e.g. D18.1, KN104.

These data do fit well with the haplotype data shown in Table 1, in that sequences from a single haplotype always fall into separate groups, or appear unrelated (as in A14, with D18.1, D18.4 and D18.5). The 2 sets of data together suggest that there are at least 4 expressed loci, and that different haplotypes do not consistently express the product of any particular class I locus.

In other investigations, it has proved useful to include sequences from related species in order to determine evolutionary relationships between MHC genes (Holmes and Ellis 1999). Class I sequences from species closely related to cattle e.g. yak, buffalo, bison, gaur have been examined, and are found to cluster closely with the cattle sequences. These data are preliminary, and it has not yet been possible to demonstrate that all loci present in cattle are present in these species (data not shown). In contrast, sequences from much more distantly related members of the order Artiodactyla do not cluster with the cattle sequences. Nine pilot whale class I sequences clustered in a species-specific manner, and preliminary analyses indicated a possible distant relationship with a putative non-classical cattle class I sequence, HD15. Further work in this area may reveal information about the origins and functional significance of such genes.

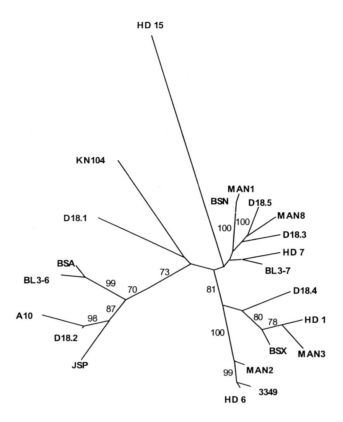

Fig. 1. Neighbour-joining tree of 22 bovine class I sequences. Numbers on the branches refer to bootstrap values (percentages) recovered after 1000 replications.

Mapping the cattle MHC class I region

The cattle MHC is located on chromosome 23, and while the overall arrangement has been well studied (Lewin et al. 1999), the precise arrangement of genes within each region has yet to be elucidated. Southern blotting and analysis of genomic libraries suggests that there are in excess of 10 class I (or related) genes in the cattle MHC (Lindberg and Andersson 1988; Barker et al. 1997). It is at present unknown whether the observed variation in number of expressed class I genes is due to genetic rearrangement or transcriptional control, and interpretation of limited published mapping data is complicated by problems of assignment of sequences to loci (Bensaid et al. 1991; Skow et al. 1996).

A BAC library has been constructed using DNA from an MHC homozygous animal carrying the A14 specificity, and expressing 3 class I genes (Table 1).

Preliminary analysis has shown that all 3 expressed class I genes (D18.1, D18.4 and D18.5), 2 putative non-classical class I genes, and a further 6 class I-like genes or gene fragments are present on a contig spanning ~400kb of DNA (diPalma and Ellis, manuscript in preparation). These results (in common with previously published data) suggest that the cattle MHC class I genes may be clustered relatively closely together. However, analysis of this region in additional haplotypes is needed to determine the source of variations observed in MHC gene expression.

Origins of modern cattle

It is not known why cattle show such variation in class I expression, and may have a larger number of potentially expressed class I genes than some other species. A possible contributory factor is that modern cattle (both *Bos taurus* and *Bos indicus*) are believed to be the result of multiple domestication events involving potentially diverse progenitor populations of wild oxen (*Bos primigenius*). Existing breeds have been established relatively recently (in the last 200 years), but mitochondrial DNA (mt DNA) analysis has demonstrated that sequence variations correlate with geographical distribution rather than breed, and reflect these complex origins (Bailey et al. 1996). Thus although many cattle may now be subjected to selective breeding, they still maintain a high level of genetic diversity. There exist, in addition, populations/breeds that are adapted to their environment, and tolerant to some extent to endemic pathogens, for example the N'dama cattle of West Africa are tolerant to trypanosomiasis (Murray and Trail 1984).

Conclusions

Our data show that cattle exhibit a broad range of MHC class I expression patterns. It is likely that the haplotype configurations reflect the origin of particular breeds/populations, although in some cases selection by pathogens may also play an important role. Cattle are not unique in demonstrating such variations in MHC expression. Horses (Holmes and Ellis 1999), rats (Joly et al. 1997), pigs (Chardon et al. 1999) and several primate species (Cadavid et al. 1997) all show some degree of variation between haplotypes. However, cattle may be unusual because of their close relationship with man over thousands of years. Genetically distinct populations and subspecies have been brought together artificially, introduced into potentially hostile environments, and subjected to various degrees of intensive inbreeding. The effect of this manipulation and the resulting genetic variation on population fitness and response to vaccination is yet to be revealed.

References

Bailey JF, Richards MB, Macauley VA, Colson IB, James IT, Bradley DG, Hedges REM, Sykes BC (1996) Ancient DNA suggests a recent expansion of European cattle from a diverse wild progenitor species. Proc R Soc Lond B Biol Sci 263:1467-1473

Barker N, Young JR, Morrison WI, Ellis SA (1997) Sequence diversity present within the 5' upstream regions of BoLA class I genes. Immunogenetics 46:352-354

Bensaid A, Kaushal A, Baldwin CL, Clevers H, Young JR, Kemp SJ, MacHugh ND, Toye PG, Teale AJ (1991) Identification of expressed bovine class I MHC genes at 2 loci and demonstration of physical linkage. Immunogenetics 33:247-254

Bernoco D, Lewin HA, Andersson L, Arriens MA, Byrns G, Cwik S, Davies CJ, Hines HC, Leibold W, Lie O, Meggiolaro D, Oliver RA, Ostergaard H, Spooner RL, Stewart-Haynes JA, Teale AJ, Templeton JW, Zanotti M (1991) Joint Report of the Fourth International BoLA Workshop, 1990. Anim Genet 22:477-495

Cadavid LF, Shufflebotham C, Ruiz FJ, Yeager M, Hughes AL, Watkins DI (1997) Evolutionary instability of the MHC class I loci in New World primates. Proc Natl Acad Sci USA 94:14536-14541

Chardon P, Renard C, Vaiman M (1999) The MHC in swine. Immunol Rev 167:179-193

Ellis SA, Ballingall KT (1999) Cattle MHC: evolution in action? Immunol Rev 167:159-168

Ellis SA, Braem KA, Morrison WI (1992) Transmembrane and cytoplasmic domain sequences demonstrate at least two expressed bovine MHC class I loci. Immunogenetics 37:49-56

Ellis SA, Staines KA, Stear MJ, Hensen EJ, Morrison WI (1998) DNA typing for BoLA class I using sequence-specific primers. Eur J Immunogenet 25:365-370

Ellis SA, Holmes EC, Staines KA, Smith KB, Stear MJ, McKeever DJ, MacHugh ND, Morrison WI (1999) Immunogenetics, in press

Ennis P, Jackson AP, Parham P (1989) Molecular cloning of bovine class I MHC cDNA. J Immunol 141:642-651

Garber TL, Hughes AL, Watkins DI, Templeton JW (1994) Evidence for at least three transcribed BoLA class I loci. Immunogenetics 39:257-265

Holmes EC, Ellis SA (1999) Evolutionary history of MHC class I genes in the mammalian order Perissodactyla. J Mol Evol: in press

Joly E, Graham M, Coadwell E, Deverson E, Leong L, Le Rolle A, Gonzalez A, Butcher GW (1997) Novel expressed MHC class I genes of the rat: RT1-U, RT1-V, RT1-Y and RT1-Z. Rat Genome 3:133-137

Joosten I, Teale AJ, van der Poel JJ, Hensen EJ (1992) Biochemical evidence of the expression of 2 MHC class I genes on bovine PBMC. Anim Genet 23:113-123

Knapp LA, Cadavid LF, Watkins DI (1998) The MHC-E locus is the most well conserved of all known primate class I histocompatibility genes. J Immunol 160:189-196

Lewin HA, Russell GC, Glass EJ (1999) Comparative organization and function of the MHC of domesticated cattle. Immunol Rev 167:145-158

Lindberg PG, Andersson L (1988) Close association between DNA polymorphism of bovine MHC class I genes and serological BoLA-A specificities. Anim Genet 19:245-250

Murray M, Trail J (1984) Genetic resistance to animal trypanosomiasis in Africa. Prev Vet Med 2:541-551

Skow LC, Snaples SN, Davis SK, Taylor JF, Huang B, Gallagher DH (1996) Localisation of BoLA DYA and class I loci to different regions of chromosome 23. Mamm Genome 7:388-389

Smith KB, Ellis SA (1999) Standardisation of a procedure for quantifying surface antigens by indirect immunofluorescence. J Immunol Methods: in press

Evidence for four functional *DQA* loci in cattle with distinct distributions amongst European and African populations

Keith T. Ballingall, Anthony Luyai, Bernard Marasa, and Declan J. McKeever

International Livestock Research Institute (ILRI), Box 30709, Nairobi, Kenya

Summary. Full length cDNAs derived from the bovine MHC class II *DQA1*, *DQA2*, *DQA3* and putative *DQA4* gene loci have been cloned by RT-PCR from three MHC-homozygous Kenya Boran (*Bos indicus*) cattle and transiently expressed in COS-7 cells. Immunocytochemical analysis of these transfectants with a panel of monoclonal antibodies specific for bovine surface DQ revealed differential staining, which is consistent with the cDNAs encoding mature surface-expressed products. This suggests that each cDNA encodes a functional molecule capable of presenting antigens to CD4⁺ T cells. Analysis of genomic DNA derived from MHC-homozygous cattle (9 African *B. indicus* and 14 European *B. taurus*) revealed distinct differences in the distribution of the *DQA* loci. European *B. taurus* cattle are characterised exclusively by haplotypes containing either a single *DQA1* gene or duplicated *DQA1* and *DQA2* genes. Although these haplotypes also occur in African *B. indicus* cattle, these animals also feature haplotypes that are characterised by *DQA2/DQA3* or *DQA2/DQA4*. The apparent absence of *DQA3* and *DQA4* in this cohort of European *B. taurus* cattle may reflect the diverse evolutionary origins and environmental pressures acting on the two populations.

Key words. Bovine, *DQA*, Expression, Populations

Introduction

T cells recognise antigens in the form of processed peptides associated with products of the major histocompatibility complex (MHC). The capacity of an individual's immune system to respond to foreign peptides is therefore influenced by the number of MHC genes expressed by each haplotype. The human MHC region is characterised by 4 expressed *A/B* class II gene pairs (*HLA DM, DP, DQ* and *DR*), with some haplotype polymorphism at the *HLA DRB* loci (Beck and Trowsdale 1999). It has been assumed that a similar rigid organisational situation exists in other mammalian species. However, analysis of class II regions of the bovine MHC (*BoLA*) at both genomic and transcriptional levels has revealed considerable variation between haplotypes, both in the number and types of transcribed *DQ* genes (Andersson and Rask 1988; Sigurdardottir et al. 1992; Ballingall et al. 1997, 1998). We now describe an analysis of the distribution of known *DQ* subtypes in European and African cattle and present evidence that they encode surface-expressed products.

Methods

Full length *BoLA DQA* cDNAs were amplified by RT-PCR from *T. parva*-transformed lymphoblasts derived from MHC homozygous cattle as described (Ballingall et al. 1997, 1998). For expression studies, the genes were subcloned into the mammalian expression vector pcDNA3.1 (Invitrogen) and transfected in COS-7 cells by DEAE-dextran-mediated transfer. After 48 hrs, transfected cells were fixed in acetone/ethanol and analysed for *DQA* gene expression by immunoperoxidase staining using a panel of MAbs (Table 1) raised against sheep and cattle MHC class II. Genomic DNA samples from 14 MHC-homozygous *B. taurus* cattle were kindly provided by Dr. S. Ellis (IAH, Compton, UK) and Dr. S. Miko (Uppsala, Sweden), while samples from 9 MHC homozygous African *B. indicus* cattle were prepared at ILRI. The *BoLA DQA* genes in these samples were characterised by PCR-RFLP analysis as described by Ballingall et al. (1997).

Results

Nucleic acid sequences of four full-length *BoLA DQA* cDNAs amplified from MHC-homozygous Kenya Boran cattle are aligned in Figure 1. *DQA* haplotype and phylogenetic analysis (Figs. 2 and 3) of these sequences suggests that each represents an allele at a unique locus. Transfection of individual *DQA1*, *DQA2*, *DQA3* and *DQA4* genes into COS-7 cells resulted in intracellular expression of a product recognised by the panel of antibodies generated against mature surface-expressed cattle and sheep MHC class II (Table 1). Representation of known *BoLA DQA* genes in 14 European *B. taurus* and 9 African *B. indicus* haplotypes cattle was determined by PCR-RFLP of genomic DNA from MHC-homozygous cattle. This analysis (Fig. 3) revealed a total absence of the *DQA3* and *DQA4* genes in the European samples, while 7 of 9 *Bos indicus* haplotypes contained these genes. This is consistent with a considerably higher frequency of *DQA3/A4*-containing haplotypes in *B. indicus* cattle than in European taurine animals. Conversely, haplotypes containing a single *DQA1* gene or duplicated *DQA1* and *DQA2* genes were found with much greater frequency in the European *B. taurus* cattle (14/14) than in the African samples (2/9).

Table 1. MAb reactivity to *DQA* transfected COS-7 cells

Mab	DQA1	DQA2	DQA3	DQA4
VC9	-	-	-	-
RI	-	-	-	-
ILA21	-	-	-	-
VPM36[*]	-	+++	+++	+++
B3.107[+]	+++	-	-	-
J11	-	-	-	-

Monoclonal antibodies VC9, R1, ILA21 and J11 are published bovine class II MHC-specific reagents raised against the native protein at ILRI, Kenya. B3.107 is an unpublished reagent generated at ILRI by Jan Naessens. VPM 36 was raised against sheep class II MHC by John Hopkins (University of Edinburgh, Scotland, UK).

```
            10         20         30         40         50         60         70         80         90        100
QA1   1 ATGATCCTGA ACAGAGCCCT GATTTTGGGG GCCCTCGCCC TGACCACCAT GATGGGTCCC TCTGGGAGTG AAGACATTGT GGCTGACCAC ATTGGCGCCT
QA2   1 ...G...... .........T ....C..... .......... .......... ...A.CT..AG...AG... .......... .......... G....T..
QA3   1 ...G...A.. .........T ....C..... .......A.. .......... .CAA.C.T..G..AG... .......... ..G....... G....A.T.
QA4   1 ...G...A.. .........T ....C..... .......A.. .......... .CAA.C.T..G..AG... .......... .......... ....A...

           110        120        130        140        150        160        170        180        190        200
QA1 101 ATGGCATAAA CGTCTACCAC TCATATGGTC CCTCTGGCTA CTATACCCAT GAATTTGATG GAGATGAAGA GTTCTACGTG GACCTGGAAA AGAGGGAGAC
QA2 101 ......C.G. GA......A ..TC...... .......... .C. G..C....G .......... ...C..GAT ...T..T... .......GG ...A......
QA3 101 ...C..G. .T......A ..TC...... .......T ......C. G..C.T...C C......... ...C..G.. ...T..T.. .......G. .GAA...G.
QA4 101 .C...GC.G. .T......A ..TC...... .......... .C. G..C.T...C .......... ...C..G.. ...T..T.. .......G. .GAA...G.

           210        220        230        240        250        260        270        280        290        300
QA1 201 TGTCTGGCGT CTGCCTGTGT TTAGTAAATT TGCAAGTTTT GACCCTCAGG GTGCGCTGAG AAACATAGCT GTGGGGAAAC GGACTTTGGA GGTCATGATT
QA2 201 ......A.G .......A.. ...CC.G... .......G... .......A... C...A..... TG.A...... ....ACATCA ....AC.AC... T..C...C.
QA3 201 ......A.G .......A.. ...GA..... A---C..... C...G..A. .......... .......A... A.A.C..... AC.AC.... T..TC...CA
QA4 201 ......AG .......A.. ..G..G.... AA........ ...AG.A..A A.........A TG.A...... AAA.CA.... AC..C.... T..C...C.

           310        320        330        340        350        360        370        380        390        400
QA1 301 CGAAGGTCCA ACTCTACTGC TGCTACCAAC AAGGTTCCTG AGATGACTGT GTTTCCCAAG TCTCCTGTAA TGCTGGGCCA GCCCAACACC CTCATCTGTC
QA2 301 AA.C.C.... .......T..CC. ..T..T...T G.......A. ..G....... .T........ ...C..G... .......T.. .......... ..........
QA3 301 AA.CTC.A.. ...T...CC. ..T..T...T G.......A. ..G....... .T........ .......C..G ....A.T.. .......T... ..........
QA4 301 AA.C.C.... ...T...CC. ..T..T...T G.......... ..G....... .T........ .......C..G ....A.T... .......... ..........

           410        420        430        440        450        460        470        480        490        500
QA1 401 ACGTGGACAA CATCTTTCCT CCTGTGATCA ACATTACATG GCTGAGGAAC GGGCACTTGG TCATAGAGGG TATTTCTGAG ACCAGCTTCC TCTCCAAGGA
QA2 401 .......... ...T.....C .......... .......... .....A...T ......TGCA...C...... .G........ .......C..T...
QA3 401 .......... .........C .......... ...C...... .T..A... .......CT. ..C...CA .G........ .......CT..GAAG
QA4 401 .......... .........C .......... ...C...... .T........T ......TGCA...C... .G........ .......C...AAG

           510        520        530        540        550        560        570        580        590        600
QA1 501 TGATCATTCC TTCGCCAAGA TCAGTTACCT CACCTTCCTT CCTTCTGATG ATGATGTTTA TGACTGCAAA GTGGAGCACT GGGGCCTGGA TAAGCCACTG
QA2 501 ......T ...CT...... TG...T.... .......... .......A ...CA..... .......... .......... ....T...G....T
QA3 501 ...T.... A.CT...... A...T..... .C....GC .......... .......... C......... .......... .G.......
QA4 501 ...T.... ...CT...... TG...T. T.......C .......... .......... C......... .......... .G.......T

           610        620        630        640        650        660        670        680        690        700
QA1 601 CTAAAACACT GGGAACCTGA GATTCCAGCC CCTATGTCAG AGCTGACAGA GACTGTGGTC TGTGCCTTGG GGTTGACCGT GGGCCTTGTG GGCATCGTGG
QA2 601 ..G....... .......G.. ..G....... .......... .......... .......... ...C...... .......... .......... ..T......
QA3 601 ..G....... .......G.. ..A....... .......... .......... .......... ...C...... .......T.. ...C...... .........
QA4 601 ..G....... .......G.. .......... .......... .......... .......... ...C...... .......T.. ...C...... .........

           710        720        730        740        750        760        770        780        790        800
QA1 701 TGGGCACTGT CCTCATCATC CGAGGTCTGC GCTCAGGCGG CCCCTCCAGA CACCAGGGGC CGTTGTGAGT CACACTCCAG -AAGGAAGGT GCAATGTTCA
QA2 701 ......CA. .T........ ..A..C.... .......T. GG........ .......T ..C....... .G..C.T..A .......... ..TC..CCG.
QA3 701 .......... .......C.. .......... ...A.T. GG........ ...A...... .C........ .......... T.G.-..... .TC..C...
QA4 701 .......... .......... .......... ...A.T. GG........ ...A..T ..C....... .......... T.G.-..... ..C...C..

           810        820        830        840        850        860
QA1 801 TCTTCGAGAA CAGAAAAAAT GGA------ --TAACCTAG AATTATTTT CTGACCAAG.
QA2 801 ...AT...G ...G.G-. ...CGTGCTA GACG...... ..C..G... ...G.A...
QA3 801 ...A.A.... ........-. ...CATACTA GA.G...G. ..C...... ...G.....
QA4 801 ...A.A.... ........-. ...CATACTA GA.G...G. ..C...... ...G.....
```

Fig. 1. Alignment of BoLA *DQA1 DQA2, DQA3* and *DQA4* nucleotide sequences (Ballingall et al. 1998).

Animal	DQA1 PCR-RFLP patterns	DQA2 PCR-RFLP patterns	DQA3/4 PCR-RFLP patterns	DRB3 allele	Class I serotype	
E54	CCA			35	A25	Kenyan *Bos Indicus*
E55	AAB	BAB		27	KN8	
F187		BAB	ECD	31	KN103	
F188		BAB	ABC	06	A10/KN104	
E223		BAB	ECD	02	KN12	
G277		BAB	CBA	31	A1	
G307		BAB	CBA	31	A7	
4229	AAA			15	A18	European *Bos Taurus*
UPS 007	AAB			27	A1	
UPS 059	AAA			24	A9	
UPS 075	AAA			24	A2	
UPS 102	BCA			07	A7	
UPS 340	BCA			09	A8	
4198	AAA	BAB		28	A31	
5350	ADA	BAB		08	A20	

Fig. 2. BoLA DQA genes represented in 14 European *B. taurus* and 9 African Zebu (*B. indicus*) MHC homozygous cattle. Repeated haplotypes are not shown.

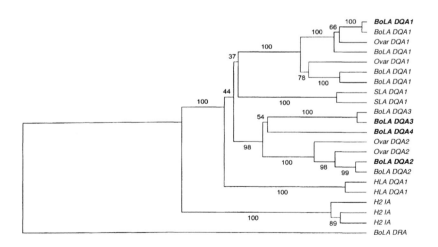

Fig. 3. UPGMA tree showing the relationships between the sequences in Figure 1 (in bold) and other class II MHC *DQA*-like genes.

Discussion

Previous analyses of haplotype and allelic polymorphism in bovine *DQA* genes are consistent with three or possibly four distinct loci (Andersson and Rask 1988; Ballingall et al. 1997, 1998) expressed in different combinations on individual haplotypes. We now provide evidence that each of these loci gives rise to an expressed product, based on their recognition by a panel of mAbs raised against surface-expressed cattle and sheep MHC class II antigens. This is consistent with each being capable of surface expression after association with the analogous *BoLA DQB* gene and highlights the potential for substantial variation in the number of MHC class II antigens expressed by individual cattle.

To examine the evolutionary basis of bovine *DQA* haplotype polymorphism we analysed MHC-homozygous European *B. taurus* and African *B. indicus* cattle. Only haplotypes with a single *DQA1* gene or duplicated *DQA1* and *DQA2* genes were represented among the European haplotypes. Conversely, haplotypes incorporating *DQA2* in combination with either *DQA3* or *DQA4* were present in 7 of the 9 haplotypes derived from African *B. indicus* cattle. This suggests that the *DQA3/A4* haplotypes occur with much greater frequency in *B. indicus* cattle when compared with *B. taurus* animals. This could reflect a number of possibilities: a) reduction in frequency of *DQA3/A4* haplotypes in European *B. taurus* cattle as a result of modern AI-based breeding, b) *DQA3/A4* haplotypes are of African origin and are therefore still common in African cattle, c) *DQA3/A4* are of Asian origin and were imported into Africa over the last 2000 years. Alternatively, it is possible that these differences in *DQA* gene expression reflect distinct environmental and pathogen selection pressures in Europe and Africa over the last 10,000 years (Ellis and Ballingall 1999; Miko et al. 1999; Parham 1999).

Acknowledgements

This work is supported by the Consultative Group on International Agricultural Research (CGIAR). This represents ILRI publication number 990125.

References

Andersson L, Rask L (1988) Characterization of the MHC class II region in cattle. The number of *DQ* genes varies between haplotypes. Immunogenetics 27:110-120

Ballingall KT, Luyai A, McKeever DJ (1997) Analysis of genetic diversity at the *DQA* loci in African cattle: evidence for a *BoLA DQA3* locus. Immunogenetics 46:237-247

Ballingall KT, Marasa B, Luyai A, McKeever DJ (1998) Identification of diverse BoLA *DQA3* genes consistent with non-allelic sequences. Anim Genet 29:123-129

Beck S, Trowsdasle J (1999) Sequence organisation of the class II region of the human MHC. Immunol Rev 167:201-210

Ellis S, Ballingall KT (1999) Cattle MHC: Evolution in action. Immunol Rev 167:159-168

Sigurdardottir S, Borsch C, Gustafsson K, Andersson L (1992) Gene duplications and sequence polymorphism of bovine class II *DQB* genes. Immunogenetics 35:205-213

Miko S, Roed K, Schmutz S, Andersson L (1999) Monomorphism and polymorphism at the MHC *DRB* loci in domestic and wild ruminants. Immunol Rev 167:169-178

Parham P (1999) Virtual reality within the MHC. Immunol Rev 167:5-16

4. Natural killer gene complex

The NKC and regulation of natural killer cell-mediated immunity

Michael G. Brown[1], Anthony A. Scalzo[2], and Wayne M. Yokoyama[1]

[1]Washington University School of Medicine, St. Louis, MO 63110, USA
[2]University of Western Australia, Nedlands, Western Australia 6907, Australia

Summary. The natural killer gene complex (NKC) contains many linked genes that encode type II protein members of the C-type lectin superfamily. Most of these genes are discretely clustered in families within the gene complex. NKC gene order and organization is conserved among humans and rodents suggesting common ancestry and conservation of function. Most known NKC encoded molecules are expressed at the plasma membrane of natural killer cells as disulfide-linked, dimeric receptors that may either activate or inhibit NK cell activity, dependent upon the presence or absence of cytoplasmic domain regulatory sequences. In addition, both classical and non-classical MHC molecules may modulate natural killing through engagement of these receptors. Notably, several NKC loci control various immunologic phenotypes, including the *Cmv1* locus that regulates NK cell-mediated immunity to murine cytomegalovirus. As a means to understand NKC regulation of viral immunity, we have cloned and mapped the NKC. Moreover, by genetic analysis of intra-NKC recombinant mice we have physically defined a *Cmv1* critical region (<300 kb) between the NKC-linked *Ly49* and *Prp* gene clusters. Thus, this detailed evaluation of the NKC will lead to enhanced understanding of the role of NK cells in innate immunity.

Key words. *Cmv1*, MCMV, Mouse genetics, NKC, NK cells

Features of the NK gene complex

The NKC was initially defined after genetic analysis of recombinant inbred (RI) mouse strains demonstrated that the *Ly49a* and *Nkrp1c* genes were linked to *Prp*, an anchor locus on distal mouse chromosome 6 (Yokoyama et al. 1990; Yokoyama et al. 1991). Both genes are members of related but different gene families that encode structurally similar molecules. Ly-49A and NKR-P1C (NK1.1) molecules are selectively expressed on the surface of natural killer cells and each may modulate natural killing. Hence, the *Ly49* and *Nkrp1* gene families established the backbone of the NK gene complex (Yokoyama and Seaman 1993). Herein, we have briefly reviewed the salient features of the NKC and the molecules that are encoded by this large genomic complex, and report on our recent genetic mapping analysis of the *Cmv1* locus.

NKC encoded NK cell receptors

A striking feature of many of the NKC encoded molecules is the overall structural relatedness of each unique molecule to another (reviewed in Brown et al. 1997a). NKR-P1, and NKG2 are each representative of different NKC encoded polypeptide families. Unique family members are most related to other family members (i.e. NKR-P1 polypeptides share 73-85% amino acid identity; Ly-49 polypeptides share 48-91% amino acid identity and NKG2 polypeptides share 36-70% amino acid identity), but are also similar to members of different NKC encoded polypeptide families and to NKC members whose genes are not clustered with other NKC genes (e.g. CD69). Indeed, NKR-P1, Ly-49, CD69, CD94 and NKG2 polypeptides are all type II integral membrane proteins and members of the C-type lectin superfamily. Inclusion in this superfamily suggests that the NKC encoded NK receptors should bind carbohydrates in a calcium-dependent manner. Consistent with this prediction, Ly-49 molecules do interact with carbohydrate ligands (Daniels et al. 1994a; Brennan et al. 1995). On the other hand, Ly-49A may bind preferentially to a protein ligand and to a somewhat lesser extent to a carbohydrate ligand (Matsumoto et al. 1998). Interestingly, CD94 structural analysis revealed a disrupted putative Ca^{2+}-binding pocket suggesting that ligand binding is not dependent upon Ca^{2+} (Boyington et al. 1999). Inasmuch as the NKC type II proteins are more closely related to one another than to other C-type lectins, some members of this family may bind protein ligands in a Ca^{2+}-independent manner.

In addition to the structural relationships discussed above, many of the NKC type II proteins are selectively expressed on the plasma membrane of natural killer cells as disulfide-linked dimers. Prototypical NKR-P1C (NK1.1 antigen) homodimers are expressed on the surface of most C57BL/6 (B6) mouse NK cells, but it is not known whether these (all or a subset) coexpress NKR-P1A and/or NKR-P1B homodimers. Expression of NKR-P1B homodimers has been demonstrated for NK cells from SJL/J inbred mice (Kung et al. 1999). CD94/NKG2 heterodimers also are expressed on most NK cells, including Ly-49+ and Ly-49- NK cells (Ho et al. 1998). In contrast, Ly-49A homodimers are expressed only on 15-20% of NKR-P1+ cells and these may overlap with other distinct Ly-49 subsets. Hence, NK cells may be finely divided by virtue of their distinctive pattern of NKC encoded type II receptor expression profile.

Such a profile may be complex since the *Nkrp1* (3 or more genes), *Ly49* (14 genes) and *Nkg2* (4 or more genes) gene families each include multiple gene members. Additionally, multiple alleles may exist for each unique gene member, similar to *Nkrp1c* (4 alleles) and *Ly49a* (5 alleles) polymorphism among different inbred mouse strains (Yokoyama et al. 1990; Yokoyama et al. 1991). Using polymorphic microsatellite NKC markers we have found similar levels of polymorphism for *Cd69*, *Cd94*, *Nkg2* and distal NKC microsatellite sequences suggesting that haplotypes may likely extend throughout the NKC ((Scalzo et al. 1999); A.A.S. and M.G.B., unpublished data). Thus, NK cell repertoire diversity may be expanded at several different levels; each level is determined by the NKC type II receptor expression profiles. A single NK1.1+ cell may coexpress gene products encoded by the *Nkrp1* gene family, *Cd94/Nkg2* genes and more than one *Ly49* gene. NKC gene polymorphism could thus augment receptor diversity within a given population. NKC receptor variability may be relevant for innate immunity since allelic differences could alter ligand specificity, and thus natural killer cell effector function.

The NKC encoded NK receptors are also functionally related since they can modulate NK cell-mediated natural killing. For example, Ly-49A and NKR-P1B homodimers and CD94/NKG2A heterodimers contain intracellular domain consensus inhibitory motifs and initiate negative signalling in NK cells that will inhibit cytolysis and/or cytokine release (Brooks et al. 1997; Kim and Yokoyama 1998; Carlyle et al. 1999; Kung et al. 1999). For Ly-49A, this response is regulated by engagement of the $\alpha 1/\alpha 2$ domains of H-2 D^d class I molecules which globally inhibits natural killing and can diminish Ly-49A surface expression (Karlhofer et al. 1992; Karlhofer et al. 1994). Support for this interaction has been obtained using various cell adhesion assays (Kane 1994; Daniels et al. 1994b; Chang et al. 1996). Moreover, functional assays with Ly-49A+ NK cells and target cells expressing peptide-substituted D^d molecules, peptide stabilized "empty" D^d or recombinant K^d-D^d class I molecules have confirmed Ly-49A association with a peptide-induced D^d peptide-binding domain, although association was not peptide- or glycosylation-specific (Orihuela et al. 1996; Matsumoto et al. 1998). Ly-49C and Ly-49G interactions with MHC class I molecules may also determine natural killing since these molecules contain consensus regulatory sequences within their cytoplasmic domains, similar to Ly-49A and NKR-P1B (Brennan et al. 1994; Smith et al. 1994; Mason et al. 1995; Brennan et al. 1996). Correspondingly, cytolytic regulation by CD94/NKG2 heterodimers appears to be dependent upon target cell H-2 Qa-1^b expression in mice or HLA E expression in humans (Borrego et al. 1998; Braud et al. 1998; Lee et al. 1998; Vance et al. 1998). Interestingly, the CD94 cytoplasmic domain contains only several amino acids and thus, appears to lack any intrinsic signalling capacity (Chang et al. 1995). However, CD94 may form heterodimers with either NKG2A or NKG2C molecules and each heterodimer pair appears competent to transmit negative or positive signals, respectively (Lazetic et al. 1996; Brooks et al. 1997; Carretero et al. 1997; Houchins et al. 1997).

Signal transduction by the NKC encoded inhibitory-type NK receptors is likely dependent upon the immunoreceptor tyrosine based inhibitory motif (ITIM) located within their cytoplasmic domains. Tyrosine phosphorylation of these amino acid sequences promotes interaction with the intracellular tyrosine phosphatase, SHP-1 (Olcese et al. 1996; Brooks et al. 1997; Carretero et al. 1998; Kung et al. 1999).

However, some of the NKC encoded NK receptor sequences do not contain consensus ITIM sequences within their cytoplasmic domains. Rather, Ly-49D homodimers which may also engage MHC class I molecules and NKR-P1C homodimers and CD94/NKG2C heterodimers each include a charged amino acid within their transmembrane domains which may facilitate association with the membrane coreceptor DAP-12 (Mason et al. 1996; Lanier et al. 1998; Smith et al. 1998; Nakamura et al. 1999). DAP-12 contains an immunoreceptor tyrosine based activation motif in its cytoplasmic domain that can transduce activation signals. Hence, individual NKC gene clusters include both inhibitory (ITIM-containing) and activation NK receptors (no ITIM but associate with ITAM-containing co-receptors) that may modulate NK cell effector function.

A physical map of the NKC

To further our understanding of natural killer cell-mediated immunity, we have chosen to characterize the NKC encoded NK cell expressed genes at the molecular level. Initially, we constructed a 2.1 megabase (Mb) yeast artificial chromosome (YAC) contig containing *Nkrp1*, *Cd69*, *Cd94*, *Nkg2*, and *Ly49* genes and these were mapped within the NKC (Brown et al. 1997b; Ho et al. 1998). By this analysis, it was apparent that the *Nkrp1*, *Nkg2* and *Ly49* genes are separately clustered in different NKC regions, but all reside within an ~2.0 Mb interval of the NKC. Although most of the known *Ly49* genes have been mapped within the *Ly49* gene cluster (Brown et al. 1997b; McQueen et al. 1998), *Ly49b* does not reside in this cluster of genes, nor does it reside on the previously established 2.1 Mb YAC contig. Furthermore, since the B6 mouse resistance locus (*Cmv1*; see below) for murine cytomegalovirus (MCMV) resides between *Ly49* and *Prp*, it was apparent that *Cmv1*-containing genomic clones had not been identified.

Recently, we have identified and characterized YAC clones containing *Ly49b* and different members of the NKC-linked *Prp* gene cluster (Brown et al. 1999). Thus, those genomic clones were exploited to seed construction of a YAC contig over the *Cmv1* critical region between *Ly49* and *Prp* and this YAC contig was aligned with the previously established 2.1 Mb YAC contig. The current overall NKC YAC contig and physical map now spans ~4.7 Mb of mouse chromosome 6 DNA and contains the proximal NKC gene clusters, *Ly49b*, *Prp* and the *Cmv1* locus (Fig. 1).

NKC gene conservation

Similar to the mouse, the human genome includes a complex of numerous genes that encode type II integral membrane proteins that have been included in the C-type lectin superfamily. Human NKG2 transcripts were initially cloned from natural killer cells by subtractive hybridization techniques (Houchins et al. 1990). The NKG2 genes were mapped to human chromosome 12p13, a region that is syntenic with distal mouse chromosome 6 (Yabe et al. 1993). Subsequently, human *CD69, NKRP1A, CD94* and *LY49L* genes also have been mapped to human chromosome 12 by FISH and/or PCR analysis of a YAC contig constructed by the human genome project (Lopez-Cabrera et al. 1993; Lanier et al. 1994; Chang et al. 1995; Krauter et al. 1995; Plougastel et al. 1996; Renedo et al. 1997; Suto et al. 1997; Westgaard et al. 1998). Notably, each of these molecules is expressed by human NK cells. Although several additional human NKC-linked genes that encode type II integral membrane proteins have been identified (i.e. *LOX-1, AICL* and *MAFA-L*), some members may not be expressed in the NK cell lineage (Rong and Pecht 1996; Hamann et al. 1997; Yamanaka et al. 1998).

Using mouse, rat and human sequence probes, rat *Ly49, Nkrp1, Cd94,* and *Nkg2* sequences have been cloned and genetically mapped to a region of rat chromosome 4 that also is syntenic with mouse chromosome 6 (Dissen et al. 1996). Indeed, the syntenic regions of the mouse, rat and human chromosomes, extend over an ~17 cm interval, including many orthologous genes identified in each species (Ansari-Lari et al. 1997; Brown et al. 1997a; and see Fig. 2). Notably, *Nkrp1, Cd69, Cd94, Nkg2* and *Ly49* genes that are selectively expressed in natural killer cells have been identified for each organism and these each map to similar NKC regions (Lanier et al. 1994; Dissen et al. 1996; Brown et al. 1997b; Renedo et al. 1997; Suto et al. 1997; Westgaard et al. 1998). Comparison of NKC type II receptors from different species reveals that these molecules have been conserved.

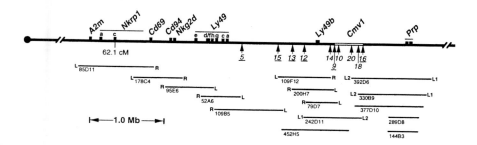

Fig. 1. Alignment of the C57BL/6 mouse NKC proximal and distal YAC contigs spanning the *Cmv1* locus. At top, mouse chromosome 6 and positions of the proximal NKC gene clusters, *Ly49b* and the *Prp* gene cluster are shown (centromeric-telomeric, respectively). An approximate position for *D6Wum15* is shown between the pYAC4L (L) end of YAC 109F12 and the 109F12 adjacent distal *Mlu* I site. Proximal (*D6Wum9*) and distal (*D6Wum16*) NKC boundaries for the *Cmv1* critical region (open bar) on the physical map are shown. Novel NKC STS markers (*D6Wum*) are shown below the chromosome map. NKC STS markers that distinguish different alleles in inbred mouse strains are underlined. At bottom, an overall NKC YAC contig is shown from which the physical map was established (from Brown et al. 1999, *Copyright 1999, The American Association of Immunologists*).

For example, rat and mouse sequences are quite similar for NKR-P1 (73% amino acid identity), CD94 (77% amino acid identity), NKG2D (81% amino acid identity) and Ly-49 (70-94% amino acid identity) sequences. Although rodent and human molecules retain somewhat less homology, it is apparent that these interspecies similarities for members of a particular NKC cluster family (i.e. NKR-P1 ~50%; CD94 56%; NKG2D ~60% and Ly-49 40-45%) are greater than intraspecies similarities among members of different NKC cluster families (20-25% amino acid indentities). Likewise, rodent and human *Nkg2* orthologs have been identified and similarly positioned within the *Nkg2* gene clusters of these different species (Plougastel et al. 1996; Plougastel and Trowsdale 1997; Vance et al. 1998). A similar comparison of *Ly49* gene cluster conservation has not been performed since individual gene orthologs have not been designated. Although it is not known whether humans contain functional *Ly49* genes, human LY49L is most closely related to mouse Ly-49B, the least conserved of the mouse LY49 sequences (Wong et al. 1991; Westgaard et al. 1998). Perhaps other human *LY49* gene sequences will be identified that are more related (structurally and functionally) to members of the mouse *Ly49* gene cluster.

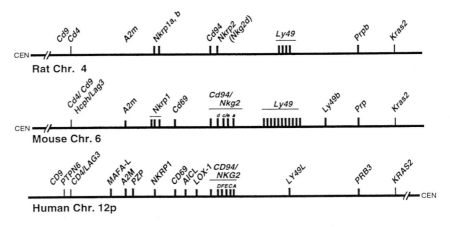

Fig. 2. Alignment of syntenic regions of rodent and human chromosomes containing the NKC. The rodent and human chromosomes (horizontal lines) are designated and the relative positions for each centromere (CEN) is indicated. Relative gene positions defined by physical and/or genetic analyses are designated by chromosomal tics. For the rat A2m gene, an inferred position is indicated based on comparison with the mouse and human NK gene complexes. This map is based on data obtained from multiple reports (Yokoyama et al. 1990; Yokoyama et al. 1991; Lopez-Cabrera et al. 1993; Yabe et al. 1993; Lanier et al. 1995; Chang et al. 1995; Dissen et al. 1996; Plougastel et al. 1996; Brown et al. 1997; Ho et al. 1998; Vance et al. 1998; Westgaard et al. 1998; Rong and Pecht 1996; Hamann et al. 1997; Yamanaka et al. 1998 and Brown et al. 1999) (adapted from Brown et al. 1997a).

Phenotypic traits controlled by the NKC

In addition to the known structurally and/or functionally characterized NKC genes, this genomic region also contains several loci that determine important immunologic phenotypes. In each case, NKC regulation of the immune response appears to be controlled by natural killer cells.

One of the best studied of these NKC-linked loci, *Cmv1*, regulates mouse immune responsiveness to MCMV differently in unique inbred mouse strains (Scalzo et al. 1990). Although overall mouse resistance to MCMV is controlled by more than one gene, acute splenic viral replication is determined solely by this autosomal, dominant locus following low titer viral infection. Moreover, this correlates with host mortality following high titer MCMV challenge among inbred mouse strains. Furthermore, NK cells are required for this *Cmv1*-dependent protective effect (Scalzo et al. 1992). Hence, B6 mouse NK cells possess the *Cmv1[r]* resistance allele which controls MCMV replication within the spleen and presumably promote host survival. On the other hand, BALB/c mouse NK cells (otherwise functionally competent) do not limit MCMV replication, and these MCMV-susceptible (*Cmv1[s]*) mice do not survive high titer MCMV challenge.

Similar to *Cmv1*-mediated resistance to MCMV, B6 mice also are resistant to ectromelia virus-induced necrosis and disease (mouse pox). As noted for MCMV resistance, ectromelia virus resistance is multigenic in nature and thus far chromosomal linkages have been demonstrated for four mouse resistance loci (Brownstein et al. 1991; Delano and Brownstein 1995). Notably, the NKC-linked *Rmp1* locus regulates necrotizing hepatitis and possibly splenomegaly following ectromelia infection (Brownstein and Gras 1997). *Rmp1*-dependent protection appears to occur through NK cell-mediated immunity since depletion of NK cells from mice prior to viral challenge abrogates resistance (Delano and Brownstein 1995). Although *Rmp1* is NKC-linked, it is not known whether this locus and *Cmv1* are the same or distinct loci.

Differences in natural killing profiles among inbred mouse strains have also been elegantly demonstrated by studies of the *Chok* locus (Idris et al. 1998). B6 and BALB/c mice differentially lyse chinese hamster ovary (CHO) cells but not YAC-1 cells. Genetic analysis of CHO killing in RI strains and NKC congenic mice demonstrated *Chok* linkage to *Ly49* within the NKC. Recently, Idris and coworkers demonstrated that Ly-49D can determine CHO killing by B6 NK cells and that transduction of Ly-49D[B6] using recombinant vaccinia vectors was necessary and sufficient to convert BALB/c NK cells into B6-like NK cells that are able to mediate CHO killing (Idris et al. 1999).

Regulation of natural killing also has been well documented by analysis of NK cell-mediated alloreactivity to MHC-incompatible lymphocytes from different inbred rat strains. Rat NK cell-mediated alloreactivity is controlled by the autosomal dominant locus, *Nka*, that is closely linked to the rat *Ly49* gene cluster on chromosome 4 (Dissen et al. 1996). Interestingly, alloreactive rat natural killer cells may recognize both classical and non-classical MHC molecule allodeterminants (Vaage et al. 1994; Naper et al. 1999).

Localization of *Cmv1* within the NKC

In our previous backcross analysis in mice, *Cmv1* had been mapped to a 0.5 cM genetic interval between the *Ly49* and *Prp* gene clusters (Forbes et al. 1997). However, a physical location for *Cmv1* not been determined since genomic clones spanning the genetic interval had not been identified or characterized in our initial physical analysis of the mouse NKC. From the current physical map, the *Cmv1* critical region extends beyond 2 Mb, spanning several YAC clone inserts. Recently, we utilized several important tools to position *Cmv1* on the NKC physical map, including *Cmv1*-aligned genomic clones, intra-NKC recombinant mice that are recombinant in this genetic interval, and novel NKC sequence tagged site (STS) markers.

Genetic analysis of backcross and intra-NKC recombinant congenic mice

From our backcross analysis, six backcross mice (of eight total mice that contained recombination breakpoints within the NKC) were recombinant in the distal NKC between *Ly49a* and *Prp* (Forbes et al. 1997; and see Fig. 3). Likewise, ten of our eighteen intra-NKC recombinant congenic inbred strains of mice are recombinant in the distal NKC (Scalzo et al. 1999). Thus, as new NKC genetic markers were developed from the genomic clones shown in our NKC physical map, these were used to genotype the informative backcross and intra-NKC recombinant congenic mouse strains.

Importantly, we demonstrated that *D6Wum9* could distinguish a restriction fragment length polymorphism (RFLP) for B6 and BALB/c alleles (Brown et al. 1999). *D6Wum9* analysis of our backcross and intra-NKC recombinant congenic mice demonstrated tight genetic linkage of *D6Wum9* and *Ly49a*. Although no backcross mice were identified that were recombinant between *Ly49a* and *D6Wum9*, two mice were identified that were recombinant between *D6Wum9* and *Cmv1* (Brown et al. 1999; and see Fig. 3). Hence, we concluded that the *Cmv1*-flanking marker *D6Wum9* would serve as a novel centromeric boundary for the critical region. Likewise, we demonstrated that *D6Wum16* (an NKC microsatellite marker) could distinguish B6 and BALB/c alleles. *D6Wum16* analysis of our backcross and intra-NKC recombinant congenic mice revealed tight genetic linkage with *Prp* (Brown et al. 1999). Indeed, four of the informative backcross mice and all of the informative intra-NKC recombinant congenic mice containing distal NKC recombination breakpoints were recombinant between *Cmv1* and *D6Wum16*.

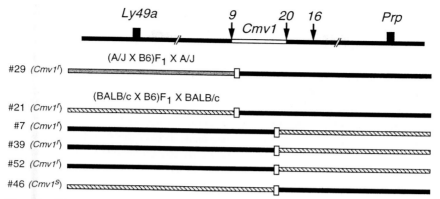

Fig. 3. Diagrammatic representation of informative backcross mouse chromosomes that delineate the non-recombinant interval for *Cmv1*. At top, mouse chromosome 6 between *Ly49a* and *Prp* is depicted. *D6Wum* NKC markers that flank *Cmv1* are positioned above the chromosome map. Below, recombinant chromosomes of individual backcross mice that are recombinant between *D6Wum9* and *D6Wum20* are depicted. B6 (black), BALB/c (hatched) and A/J (shaded) chromosomes are represented by the horizontal boxes below the chromosome map. Recombination breakpoints for the recombinant chromosomes are indicated (unfilled boxes).

Recently, we have identified an allelism for the NKC-linked *D6Wum20* microsatellite marker. Using radiolabeled *D6Wum20*-specific oligonucleotides in PCR, the dominant B6 allele, but not the BALB/c or A/J alleles, could be amplified from genomic DNA templates (Fig. 4). Amplification of a secondary product (146 nt) from each genomic template confirmed the assignment of BALB/c or A/J alleles. Thus, *D6Wum20* was also utilized to genotype the backcross and intra-NKC recombinant congenic mice discussed above (Fig. 4). Notably, *D6Wum20* analysis of these informative mice demonstrated tight genetic linkage with *D6Wum16* and *Prp*. Indeed, the recombination breakpoints in four of the six distal NKC recombinant backcross mice and all ten of the distal NKC recombinant intra-NKC recombinant congenic mice that reside between *Cmv1* and *D6Wum16*, also reside between *Cmv1* and *D6Wum20*. We conclude that *D6Wum20* resides distal to the *Cmv1* locus and is a novel telomeric boundary for the *Cmv1* locus, such that the non-recombinant interval must reside between *D6Wum9* and *D6Wum20*.

Assignment of the *Cmv1* critical region on the NKC physical map

Previously, we demonstrated that the *Cmv1* locus resides on a single YAC clone fragment (~390 kb) between the NKC markers *D6Wum9* and *D6Wum16*, an interval that may be recombinogenic in mice (Brown et al. 1999). Moreover,

D6Wum20 which also resides on YAC 242D11 was positioned ~90 kb centromeric to *D6Wum18* between *D6Wum9* and *D6Wum18*. Since *D6Wum18* is ≤65 kb centromeric to *D6Wum16*, *D6Wum20* should reside 90-155 kb centromeric to *D6Wum16*. Hence, the *Cmv1* critical region now corresponds to a maximum physical distance of 300 kb on this YAC insert. Furthermore, since all of the informative distal NKC recombinant animals contain recombination breakpoints within these narrow genetic/physical boundaries, the data remain consistent with the possibility of hotspots for recombination within the NKC.

To enhance our physical and genetic map analysis over the *Cmv1* locus, we are now aligning bacterial artificial chromosome (BAC) clone inserts to the critical region. Novel NKC STS markers will be developed from the cloned insert ends of BACs and BAC sequences that are generated in our laboratory. Those novel NKC markers that are useful in genetic analysis will be utilized to aid in determining the location, physical size and ultimately the sequence of the non-recombinant interval.

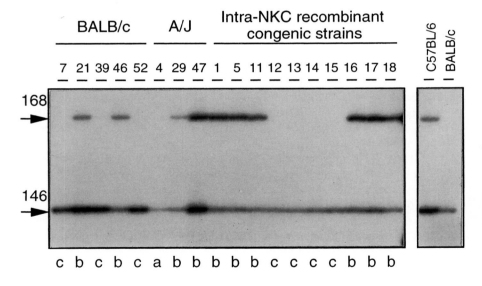

Fig. 4. *D6Wum20* genotype analysis of backcross and intra-NKC recombinant congenic mice. ^{32}P-radiolabeled *D6Wum20*-specific oligonucleotides were used in the PCR to amplify (BALB/c × C57BL/6)F$_1$ × BALB/c (designated BALB/c), (A/J × C57BL/6)F$_1$ × A/J (designated A/J), and intra-NKC recombinant congenic mouse DNA samples. Denatured PCR products were separated by 6% denaturing polyacrylamide gel electrophoresis and visualized by autoradiography. The expected, NKC-linked C57BL/6 (B6) allele product size (168 nt) has been designated. Below, individual allele designations [B6 (b), BALB/c (c) or A/J (a)] for the recombinant chromosome of the indicated mice have been listed and these are identical for *D6Wum20*, *D6Wum16* and *Prp*.

Acknowledgments

MGB is supported by an individual National Research Service Award from the National Institute of Allergy and Infectious Diseases. This research was supported in part by the Barnes-Jewish Hospital Research Foundation, by grants from the National Institutes of Health, and by the National Health and Medical Research Council of Australia. WMY is an investigator of the Howard Hughes Medical Institute.

References

Ansari-Lari MA, Shen Y, Muzny DM, Lee W, Gibbs RA (1997) Large-scale sequencing in human chromosome 12p13: experimental and computational gene structure determination. Genome Res 7:268-280

Borrego F, Ulbrecht M, Weiss EH, Coligan JE, Brooks AG (1998) Recognition of human histocompatibility leukocyte antigen (HLA)-E complexed with HLA class I signal sequence-derived peptides by CD94/NKG2 confers protection from natural killer cell-mediated lysis. J Exp Med 187:813-818

Boyington JC, Riaz AN, Patamawenu A, Coligan JE, Brooks AG, Sun PD (1999) Structure of CD94 reveals a novel C-type lectin fold: implications for the NK cell-associated CD94/NKG2 receptors. Immunity 10:75-82

Braud VM, Allen DSJ, O'Callaghan CA, Söderström K, D'Andrea A, Ogg GS, Lazetic S, Young NT, Bell JI, Phillips JH, McMichael AJ (1998) HLA-E binds to natural-killer-cell receptors CD94/NKG2A, B and C. Nature 391:795-799

Brennan J, Mager D, Jefferies W, Takei F (1994) Expression of different members of the Ly-49 gene family defines distinct natural killer cell subsets and cell adhesion properties. J Exp Med 180:2287-2295

Brennan J, Takei F, Wong S, Mager DL (1995) Carbohydrate recognition by a natural killer cell receptor, Ly-49C. J Biol Chem 270:9691-9694

Brennan J, Mahon G, Mager DL, Jefferies WA, Takei F (1996) Recognition of class I major histocompatibility complex molecules by Ly-49: specificities and domain interactions. J Exp Med 183:1553-1559

Brooks AG, Posch PE, Scorzelli CJ, Borrego F, Coligan JE (1997) NKG2A complexed with CD94 defines a novel inhibitory natural killer cell receptor. J Exp Med 185:795-800

Brown MG, Scalzo AA, Matsumoto K, Yokoyama WM (1997a) The natural killer gene complex - a genetic basis for understanding natural killer cell function and innate immunity. Immunol Rev 155:53-65

Brown MG, Fulmek S, Matsumoto K, Cho R, Lyons PA, Levy ER, Scalzo AA, Yokoyama WM (1997b) A 2-Mb YAC contig and physical map of the natural killer gene complex on mouse chromosome 6. Genomics 42:16-25

Brown MG, Zhang J, Du Y, Stoll J, Yokoyama WM, Scalzo AA (1999) Localization on a physical map of the NKC-linked *Cmv1* locus between *Ly49b* and the *Prp* gene cluster on mouse chromosome 6. J Immunol 163:1991-1999

Brownstein DG, Bhatt PN, Gras L, Jacoby RO (1991) Chromosomal locations and gonadal dependence of genes that mediate resistance to ectromelia (mousepox) virus-induced mortality. J Virol 65:1946-1951

Brownstein DG, Gras L (1997) Differential pathogenesis of lethal mousepox in congenic DBA/2 mice implicates natural killer cell receptor NKR-P1 in necrotizing hepatitis and the fifth component of complement in recruitment of circulating leukocytes to spleen. Amer J Pathol 150:1407-1420

Carlyle JR, Martin A, Mehra A, Attisano L, Tsui FW, Zuniga-Pflucker JC (1999) Mouse NKR-P1B, a novel NK1.1 antigen with inhibitory function. J Immunol 162:5917-5923

Carretero M, Cantoni C, Bellon T, Bottino C, Biassoni R, Rodriguez A, Perezvillar JJ, Moretta L, Moretta A, Lopez-Botet M (1997) The CD94 and NKG2-A C-type lectins covalently assemble to form a natural killer cell inhibitory receptor for HLA class I molecules. Eur J Immunol 27:563-567

Carretero M, Palmieri G, Llano M, Tullio V, Santoni A, Geraghty DE, López-Botet M (1998) Specific engagement of the CD94/NKG2-A killer inhibitory receptor by the HLA-E class Ib molecule induces SHP-1 phosphatase recruitment to tyrosine-phosphorylated NKG2-A: evidence for receptor function in heterologous transfectants. Eur J Immunol 28:1280-1291

Chang C, Rodriguez A, Carretero M, Lopez-Botet M, Phillips JH, Lanier LL (1995) Molecular characterization of human CD94: a type II membrane glycoprotein related to the C-type lectin superfamily. Eur J Immunol 25:2433-2437

Chang CS, Shen LJ, Gong DE, Kane KP (1996) Major histocompatibility complex class I-dependent cell binding to isolated LY-49a - evidence for high-avidity interaction. Eur J Immunol 26:3219-3223

Daniels BF, Nakamura MC, Rosen SD, Yokoyama WM, Seaman WE (1994a) Ly-49A, a receptor for H-2D^d, has a functional carbohydrate recognition domain. Immunity 1:785-792

Daniels BF, Karlhofer FM, Seaman WE, Yokoyama WM (1994b) A natural killer cell receptor specific for a major histocompatibility complex class I molecule. J Exp Med 180:687-692

Delano ML, Brownstein DG (1995) Innate resistance to lethal mousepox is genetically linked to the NK gene complex on chromosome 6 and correlates with early restriction of virus replication by cells with an NK phenotype. J Virol 69:5875-5877

Dissen E, Ryan JC, Seaman WE, Fossum S (1996) An autosomal dominant locus, Nka, mapping to the Ly-49 region of a rat natural killer (NK) gene complex, controls NK cell lysis of allogeneic lymphocytes. J Exp Med 183:2197-2207

Forbes CA, Brown MG, Cho R, Shellam GR, Yokoyama WM, Scalzo AA (1997) The Cmv1 host resistance locus is closely linked to the Ly49 multigene family within the natural killer cell gene complex on mouse chromosome 6. Genomics 41:406-413

Hamann J, Montgomery KT, Lau S, Kucherlapati R, van Lier RAW (1997) AICL: a new activation-induced antigen encoded by the human NK gene complex. Immunogenetics 45:295-300

Ho EL, Heusel JW, Brown MG, Matsumoto K, Scalzo AA, Yokoyama WM (1998) Murine Nkg2d and Cd94 are clustered within the natural killer complex and are expressed independently in natural killer cells. Proc Natl Acad Sci USA 95:6320-6325

Houchins JP, Yabe T, McSherry C, Miyokawa N, Bach FH (1990) Isolation and characterization of NK cell or NK/T cell-specific cDNA clones. J Mol Cell Immunol 4:295-304

Houchins JP, Lanier LL, Niemi EC, Phillips JH, Ryan JC (1997) Natural killer cell cytolytic activity is inhibited by NKG2-A and activated by NKG2-C. J Immunol 158:3603-3609

Idris AH, Iizuka K, Scalzo AA, Yokoyama WM (1998) Genetic control of natural killing and in vivo tumor elimination by the Chok locus. J Exp Med 188:2243-2256

Idris AH, Smith HR, Mason LH, Ortaldo JR, Scalzo AA, Yokoyama WM (1999) The natural killer gene complex genetic locus Chok encodes Ly-49D, a target recognition receptor that activates natural killing. Proc Natl Acad Sci USA 96:6330-6335

Kane KP (1994) Ly-49 mediates EL4 lymphoma adhesion to isolated class I major histocompatibility complex molecules. J Exp Med 179:1011-1015

Karlhofer FM, Ribaudo RK, Yokoyama WM (1992) MHC class I alloantigen specificity of Ly-49+ IL-2-activated natural killer cells. Nature 358:66-70

Karlhofer FM, Hunziker R, Reichlin A, Margulies DH, Yokoyama WM (1994) Host MHC class I molecules modulate in vivo expression of a NK cell receptor. J Immunol 153:2407-2416

Kim S, Yokoyama WM (1998) NK cell granule exocytosis and cytokine production inhibited by Ly-49A engagement. Cell Immunol 183:106-112

Krauter K, Montgomery K, Yoon SJ, LeBlanc-Straceski J, Renault B, Marondel I, Herdman V, Cupelli L, Banks A, Lieman J, et al. (1995) A second-generation YAC contig map of human chromosome 12. Nature 377:321-333

Kung SK, Su RC, Shannon J, Miller RG (1999) The NKR-P1B gene product is an inhibitory receptor on SJL/J NK cells. J Immunol 162:5876-5887

Lanier LL, Chang C, Phillips JH (1994) Human NKR-P1A. A disulfide-linked homodimer of the C-type lectin superfamily expressed by a subset of NK and T lymphocytes. J Immunol 153:2417-2428

Lanier LL, Corliss B, Wu J, Phillips JH (1998) Association of DAP12 with activating CD94/NKG2C NK cell receptors. Immunity 8:693-701

Lazetic S, Chang C, Houchins JP, Lanier LL, Phillips JH (1996) Human natural killer cell receptors involved in MHC class I recognition are disulfide-linked heterodimers of CD94 and NKG2 subunits. J Immunol 157:4741-4745

Lee N, Llano M, Carretero M, Ishitani A, Navarro F, López-Botet M, Geraghty DE (1998) HLA-E is a major ligand for the NK inhibitory receptor CD94/NKG2A. Proc Natl Acad Sci USA 95:4791-4794

Lopez-Cabrera M, Santis AG, Fernandez-Ruiz E, Blacher R, Esch F, Sanchez-Mateos P, Sanchez-Madrid F (1993) Molecular cloning, expression, and chromosomal localization of the human earliest lymphocyte activation antigen AIM/CD69, a new member of the C-type animal lectin superfamily of signal-transmitting receptors. J Exp Med 178:537-547

Mason LH, Ortaldo JR, Young HA, Kumar V, Bennett M, Anderson SK (1995) Cloning and functional characteristics of murine large granular lymphocyte-1: a member of the Ly-49 gene family (Ly-49G2). J Exp Med 182:293-303

Mason LH, Anderson SK, Yokoyama WM, Smith HRC, Winklerpickett R, Ortaldo JR (1996) The Ly-49D receptor activates murine natural killer cells. J Exp Med 184:2119-2128

Matsumoto N, Ribaudo RK, Abastado J-P, Margulies DH, Yokoyama WM (1998) The lectin-like NK cell receptor Ly-49A recognizes a carbohydrate-independent epitope on its MHC class I ligand. Immunity 8:245-254

McQueen KL, Freeman JD, Takei F, Mager DL (1998) Localization of five new *Ly49* genes, including three closely related to *Ly49c*. Immunogenetics 48:174-183

Nakamura MC, Linnemeyer PA, Niemi EC, Mason LH, Ortaldo JR, Ryan JC, Seaman WE (1999) Mouse Ly-49D recognizes H-2D^d and activates natural killer cell cytotoxicity. J Exp Med 189:493-500

Naper C, Ryan JC, Kirsch R, Butcher GW, Rolstad B, Vaage JT (1999) Genes in two major histocompatibility complex class I regions control selection, phenotype, and function of a rat Ly-49 natural killer cell subset. Eur J Immunol 29:2046-2053

Olcese L, Lang P, Vely F, Cambiaggi A, Marguet D, Blery M, Hippen KL, Biassoni R, Moretta A, Moretta L, Cambier JC, Vivier E (1996) Human and mouse killer-cell inhibitory receptors recruit PTP1C and PTP1D protein tyrosine phosphatases. J Immunol 156:4531-4534

Orihuela M, Margulies DH, Yokoyama WM (1996) The natural killer cell receptor Ly-49A recognizes a peptide-induced conformational determinant on its major histocompatibility complex class I ligand. Proc Natl Acad Sci USA 93:11792-11797

Plougastel B, Jones T, Trowsdale J (1996) Genomic structure, chromosome location, and alternative splicing of the human NKG2A gene. Immunogenetics 44:286-291

Plougastel B, Trowsdale J (1997) Cloning of NKG2-F, a new member of the NKG2 family of human natural killer cell receptor genes. Eur J Immunol 27:2835-2839

Renedo M, Arce I, Rodriguez A, Carretero M, Lanier LL, Lopez-Botet M, Fernandez-Ruiz E (1997) The human natural killer gene complex is located on chromosome 12p12-p13. Immunogenetics 46:307-311

Rong X, Pecht I (1996) Clustering the mast cell function-associated antigen (MAFA) induces tyrosyl phosphorylation of the Fc epsilonRI-beta subunit. Immunol Lett 54:105-108

Scalzo AA, Fitzgerald NA, Simmons A, La Vista AB, Shellam GR (1990) *Cmv-1*, a genetic locus that controls murine cytomegalovirus replication in the spleen. J Exp Med 171:1469-1483

Scalzo AA, Fitzgerald NA, Wallace CR, Gibbons AE, Smart YC, Burton RC, Shellam GR (1992) The effect of the *Cmv-1* resistance gene, which is linked to the natural killer cell gene complex, is mediated by natural killer cells. J Immunol 149:581-589

Scalzo AA, Brown MG, Chu DT, Heusel JW, Yokoyama WM, Forbes CA (1999) Development of intra-natural killer complex (NKC) recombinant and congenic mouse strains for mapping and functional analysis of NK cell regulatory loci. Immunogenetics 49:238-241

Smith HRC, Karlhofer FM, Yokoyama WM (1994) Ly-49 multigene family expressed by IL-2-activated NK cells. J Immunol 153:1068-1079

Smith KM, Wu J, Bakker AB, Phillips JH, Lanier LL (1998) Cutting edge: Ly-49D and Ly-49H associate with mouse DAP12 and form activating receptors. J Immunol 161:7-10

Suto Y, Yabe T, Maenaka K, Tokunaga K, Tadokoro K, Juji T (1997) The human natural killer gene complex (NKC) is located on chromosome 12p13.1-p13.2. Immunogenetics 46:159-162

Vaage JT, Naper C, Lovik G, Lambracht D, Rehm A, Hedrich HJ, Wonigeit K, Rolstad B (1994) Control of rat natural killer cell-mediated allorecognition by a major histocompatibility complex region encoding nonclassical class I antigens. J Exp Med 180:6416-6451

Vance RE, Kraft JR, Altman JD, Jensen PE, Raulet DH (1998) Mouse CD94/NKG2A is a natural killer cell receptor for the nonclassical major histocompatibility complex (MHC) class I molecule Qa-1. J Exp Med 188:1841-1848

Westgaard IH, Berg SF, Orstavik S, Fossum S, Dissen E (1998) Identification of a human member of the Ly-49 multigene family. Eur J Immunol 28:1839-1846

Wong S, Freeman JD, Kelleher C, Mager D, Takei F (1991) Ly-49 multigene family. New members of a superfamily of type II membrane proteins with lectin-like domains. J Immunol 147:1417-1423

Yabe T, McSherry C, Bach FH, Fisch P, Schall RP, Sondel PM, Houchins JP (1993) A multigene family on human chromosome 12 encodes natural killer-cell lectins. Immunogenetics 37:455-460

Yamanaka S, Zhang XY, Miura K, Kim S, Iwao H (1998) The human gene encoding the lectin-type oxidized LDL receptor (OLR1) is a novel member of the natural killer gene complex with a unique expression profile. Genomics 54:191-199

Yokoyama WM, Kehn PJ, Cohen DI, Shevach EM (1990) Chromosomal location of the Ly-49 (A1, YE1/48) multigene family. Genetic association with the NK 1.1 antigen. J Immunol 145:2353-2358

Yokoyama WM, Ryan JC, Hunter JJ, Smith HR, Stark M, Seaman WE (1991) cDNA cloning of mouse NKR-P1 and genetic linkage with Ly-49. Identification of a natural killer cell gene complex on mouse chromosome 6. J Immunol 147:3229-3236

Yokoyama WM, Seaman WE (1993) The Ly-49 and NKR-P1 gene families encoding lectin-like receptors on natural killer cells: the NK gene complex. Ann Rev Immunol 11:613-635

5. MHC-pathogen coevolution

Manipulation of MHC-encoded proteins by cytomegaloviruses

Dagmar Bauer[1], Frank Momburg[2], and Hartmut Hengel[1]

[1]Max von Pettenkofer-Institut, Lehrstuhl Virologie, Ludwig-Maximilians-Universität München, Pettenkoferstr. 9a, 80336 München, Germany
[2]Abteilung für Molekulare Immunologie, Deutsches Krebsforschungszentrum, Im Neuenheimer Feld 280, 69120 Heidelberg, Germany

Summary. Cytomegaloviruses (CMVs) represent species-specific herpesviruses which persist for life in the host despite of vigorous CMV-specific immune responses. In *vivo*, CMVs are under stringent CD8[+] T cell control. This selective pressure has led to the evolution of immune evasion functions which target the MHC class I pathway of antigen presentation. So far seven CMV genes have been identified which subvert MHC I molecules. Among viruses, this is the most extensive genetic repertoire generated to evade MHC class I restricted T cell responses. The viral genes code for type I transmembrane glycoproteins and are members of virus-specific gene families located at the genome termini. The common phenotype is the loss of MHC class I molecules on the cell surface. However, the target proteins and the underlying molecular mechanisms differ. While the *US2-* and *US11*-encoded glycoproteins of human CMV (HCMV) attack free MHC class I heavy chains, the HCMV US3 glycoprotein and the mouse CMV (MCMV) glycoproteins gp40, gp34 and gp48 all affect heterotrimeric MHC class I complexes. The earliest inhibition with respect to the MHC class I pathway is mediated by the endoplasmic reticulum (ER) resident, 21 kDa US6 glycoprotein, gpUS6. This factor blocks peptide import into the ER by the MHC-encoded transporter associated with antigen processing (TAP). Unlike the herpes simplex virus (HSV) inhibitor ICP47, gpUS6 does not affect peptide binding to TAP but inhibits peptide translocation into the ER. gpUS6 is found in complexes with the transient TAP1/2-MHC class I heavy chain-$\beta_2$microglobulin-calreticulin-tapasin assembly complex and calnexin. The former interaction is most likely directed to TAP1/2 and does not require MHC I, $\beta_2$microglobulin, tapasin nor calnexin. Binding and functional inactivation of TAP transporters by gpUS6 is species-restricted. The cytoplasmic and transmembrane part of the viral protein are dispensable for its function, suggesting that gpUS6 regulates TAP via its luminal face and prevents pore opening and peptide release into the ER. Most recent findings indicate that CMVs have also found multiple means to cope with CD4[+] T cell immunity. This is accomplished by interfering with cytokine responses and by selective proteolysis of MHC class II molecules.

Key words. Cytomegaloviruses, MHC class I pathway, Transporter associated with antigen processing TAP1/2, Jak/Stat pathway, MHC class II

Cytomegaloviruses — their biology and immune control

Cytomegaloviruses (CMVs) represent a prototypic β-subgroup of herpesviruses widely distributed among mammals (Mocarski 1996). CMVs are enveloped viruses and characterized by their slow replication in a limited number of cell types and their typical cytopathology. CMVs possess double-stranded DNA genomes of about 240 kbp in length encompassing more than 200 separate open reading frames which represent the highest herpesviral potential coding capacity (Chee et al. 1990; Rawlinson et al. 1996). Gene blocks located in the central 100 kbp of the CMV genomes are closely related to each other, whereas the sequences near both ends of the genomes are divergent and harbour gene families which are specific for each of the CMV's. Most of the gene families are clustered in tandem arrays and encode glycoproteins (gps). Like other herpesviruses, CMVs are ancient viruses closely linked to their natural host. Human cytomegalovirus (HCMV) is prevalent in probably all human populations, and many genetically unique strains of the virus continually circulate in the general population throughout the world. As with other herpesviruses, primary CMV infection is followed by a lifelong persistent infection. After clearance of productive infection, the CMV genome persists in a latent form associated with only minimal viral gene expression. Viral DNA replication is frequently reactivated from latency and results in recurrent infection and virus shedding. The peculiar symbiotic relationship between CMVs and their host usually results in asymptomatic infections in immunocompetent individuals. In the immunologically immature or immunocompromised, however, CMV replication results in severe or even fatal disease manifestations (Britt and Alford 1996).

The antiviral defence against CMVs is critically dependent on hierarchical but also redundant effector functions of T lymphocytes (Polic et al. 1998). Natural killer (NK) cells and antibodies can significantly contribute to virus control, but fail to protect against lethal CMV infection on its own (Shellam et al. 1981; Welsh et al. 1991; Jonjic et al. 1994). In vivo depletion experiments and adoptive transfer studies in mouse CMV (MCMV)-infected mice identified the CD8[+] T cell subset as essential and sufficient to clear acute infection in visceral organs and to mediate protection from otherwise lethal disease (Reddehase et al. 1985; Reddehase et al. 1987; Jonjic et al. 1988). The efficacy of adoptively transferred CMV-specific CD8[+] T cells for controlling CMV infection could be confirmed in humans (Riddell et al. 1992). In addition, an important role for CD4[+] T lymphocytes was demonstrated for controlling CMV infection in the salivary gland (Jonjic et al.

1989). A strong antiviral effect operating in *vivo* was demonstrated for certain cytokines. In particular, depletion of endogenous IFNγ and TNFα resulted in an increase in MCMV replication (Lucin et al. 1992; Pavic et al. 1993), CD8[+] T cell effector function (Hengel et al. 1994) and viral peptide processing (Geginat et al. 1997).

To achieve permanent coexistence with their hosts, CMVs avoid immune recognition by multiple and highly sophisticated strategies (Hengel et al. 1998). During latency, viral gene expression is quantitatively and qualitatively reduced, thereby limiting the generation and exposure of epitopes. Furthermore, CMV replicates efficiently at sanctuaries which are kept under a less stringent immune control like epithelial cells of the salivary glands which lack MHC class I expression. Virus production at this site allows virus shedding into the saliva and thus subsequent transmission to naive individuals. Finally, CMVs exploit immune responses of the host for their own purposes and manipulate immune receptors. The elucidation of the multigenetic basis for the subversion of MHC class I molecule function by CMVs established a paradigm for immune receptor modulation by pathogens which is one of the best characterized evasion strategies known so far.

The MHC class I pathway of antigen presentation

The evolution and diversification of the MHC class I antigen processing and presentation system in mammals may have created a selective pressure for CMVs to acquire functions which put this pathway under viral control in order to avoid attack and eradication by CD8[+] cytotoxic T lymphocytes. In this pathway, viral proteins are cleaved into peptides by the ubiquitin/proteasome system in the cytosol (Heemels and Ploegh 1995). Due to their cytosolic origin, peptides have to be translocated across the membrane of the endoplasmic reticulum (ER) to encounter the peptide binding site of MHC class I molecules in the ER lumen. Vectorial transport of peptides into the ER is mediated by a specific peptide pump, transporter associated with antigen processing (TAP), which comprises two MHC-encoded subunits, TAP1 and TAP2 (reviewed by Momburg and Hämmerling 1998). Acquisition of peptides by MHC I molecules in the ER requires the concerted and sequential action of a number of auxiliary proteins, including calnexin, calreticulin, thiol-dependent reductase ER-60 and a MHC-encoded protein, tapasin (Degen et al. 1991; Rajagopalan and Brenner 1994; Sadasivan et al. 1996; Ortmann et al. 1994; Lindquist et al. 1998). Tapasin mediates the association of MHC I/$\beta_2$microglobulin/calreticulin/ER-60 complexes with the TAP1/2 transporter resulting in a transient TAP1/2 - MHC class I heavy chain - $\beta_2$microglobulin - calreticulin - ER-60 - tapasin assembly complex (Ortmann et al. 1997). Following peptide binding, stably formed heterotrimeric MHC I/$\beta_2$microglobulin/peptide complexes dissociate from TAP1/2 and exit from the

ER for their transport via the constitutive secretory route to the plasma membrane to present the peptide to the T cell receptor.

Cytomegaloviral interference with the MHC class I pathway of antigen presentation

CMVs have generated the largest genetic repertoire to subvert MHC class I functions among viruses (Hengel and Koszinowski 1997). The responsible CMV genes are dispensable for virus replication in *vitro* and belong to virus-specific gene families without homologous sequences in the database. To date, four HCMV glycoproteins and three MCMV glycoproteins have been identified which either downregulate MHC class I surface expression, bind to MHC I molecules, or both (listed in Table 1). The HCMV *US2/US6* gene family (Fig. 1) represents a paradigm of a CMV gene family. Typically, the genes are tandemly arranged in the outer regions of the CMV genomes (Chee et al. 1990; Rawlinson et al. 1996). The *US2* family is a clustered pair of two homologous genes, *US2* and *US3*,

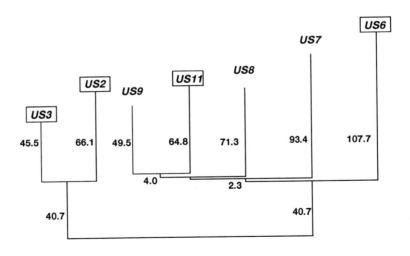

Fig. 1. Dendrogram illustrating the amino acid relatedness of the HCMV *US2* and *US6* gene families. Horizontal distances are proportional to the relative sequence deviations between individual amino acid sequences and indicated as arbitrary values. Genes coding for MHC I subversive gps are framed.

coding for small transmembrane glycoproteins. Sequence comparison reveals an identity of 23% between both gps, and a similarity of 56%. Their sequences are further related to the members of the *US6* gene family (see dendrogram depicted in Fig.1). Two members of the US6 family, *US11* and *US6*, downregulate MHC I expression as well, while all others do not (Jones et al. 1995; Hengel et al. 1997; Ahn et al. 1997). Some have been implicated in different virus-host interactions (Maidji et al. 1996). It is tempting to speculate that the modern members of the *US2/US6* gene family have evolved by gene duplications from a primordial gene and diverged over time to fulfill now different tasks.

Although redundant regarding the resulting MHC I phenotype, all MHC I reactive CMV glycoproteins appear unique and distinct in their molecular function. Inhibition of MHC I molecules occurs at different stages of MHC I biogenesis, indicating that, in principle, every single step along the MHC I pathway is fair game for CMVs.

The common outcome of the isolated expression of the viral genes is the loss of plasma membrane expressed MHC class I molecules. In most cases, the viral factors are designed to interact directly with MHC I, but binding to MHC I molecules occurs at different checkpoints of assembly and maturation. Some HCMV proteins attack MHC I heavy chains already prior to their loading with peptides, while other CMV gps recognize only fully assembled ternary MHC I complexes. Furthermore, MHC I downregulation can be also achieved by reducing

Table 1. Cytomegaloviral glycoproteins interfering with the MHC class I pathway of antigen presentation

Gene	Protein	Mechanism employed	References
HCMV *US3*	gp22/23	retention of MHC I in the ER	Ahn et al. 1996a Jones et al. 1996
HCMV *US2*	gp24	back transport of MHC I heavy chains from the ER via Sec 61 to the cytosol for proteolytic degradation	Wiertz et al. 1996b Jones and Sun 1997
HCMV *US11*	gp33	dislocation of MHC I into the cytosol for proteolytic degradation	Wiertz et al. 1996a
HCMV *US6*	gp21	inhibition of the TAP transporter	Hengel et al. 1997 Ahn et al. 1997 Lehner et al. 1997
MCMV *m152*	gp37/40	retention of MHC I in the ERGIC/ *cis*-Golgi compartment	Ziegler et al. 1997
MCMV *m04*	gp34	binding to MHC I complexes exposed on the cell surface	Kleijnen et al. 1997
MCMV *m06*	gp48	directing MHC I complexes into endolysosomes for proteolysis	Reusch et al. 1999

the amount of peptides supplied for MHC class I loading, making TAP an attractive target for distortion.

CMV proteins targeting MHC I molecules

In MCMV-infected fibroblasts, viral early genes abrogate CD8+ T cell recognition and downregulate MHC I surface expression (Del Val et al 1989, 1992). Two of the three MHC I reactive gps are implicated in the reduction of MHC class I surface expression on infected cells, gp37/40 encoded by *m152* (Ziegler et al. 1997) and gp48 encoded by *m06* (Reusch et al. 1999). *m152* belongs to the first set of early genes transcribed in MCMV-infected cells reaching a maximum after 4 hours p.i. before declining at later times of the replication cycle. The *m06* gene is expressed later but is still active during the late phase of infection. While gp48 associates tightly with MHC I complexes, direct binding of gp37/40 could not be demonstrated (Ziegler et al. 1997). The fact that gp37/40 mediated retention of mouse MHC I complexes in the ER-Golgi-Intermediate compartment (ERGIC) (Fig. 2) is highly selective for mouse MHC I alleles but ignores human MHC I molecules supports the notion that MHC I molecules are indeed recognized by gp37/40. This interaction occurs transiently since gp37/40 molecules reach the endo/lysosome while MHC I complexes get stuck in the ERGIC (H. Ziegler and U. Koszinowksi, personal communication).

Upon binding of gp48 to MHC I, the complex exits from the ER and passes the Golgi. Due to two di-leucine motifs (Bakke and Dobberstein 1990; Letourner and Klausner 1992) present in the cytoplasmic tail of gp48 the complex is delivered to the endolysosome where the proteins are proteolytically destructed (Fig. 2). Belonging to the same *m02* gene family of MCMV, *m04*/gp34 shares the MHC I binding properties of *m06*/gp48. After complexing with β_2m-associated MHC class I molecules gp34 exits from the ER and reaches the cell surface (Kleijnen et al. 1997). It was speculated that gp34/MHC I complexes displayed on the cell surface may serve as a decoy receptor for natural killer (NK) cells or prevent peptide recognition by CD8[+] cytotoxic T lymphocytes (CTL) (Kleijnen et al. 1997).

As in MCMV-infected cells presentation of antigenic peptides to CD8[+] CTL is abrogated during HCMV infection (Warren et al. 1994; Hengel et al. 1995). The analysis of defined HCMV deletion mutants guided the identification of four genes in the short component of the HCMV genome which independently downregulate MHC I surface expression (Jones et al. 1995; Hengel et al. 1996) (listed in Table 1). The genes belong to different kinetic classes and are expressed in a strictly ordered fashion, thus covering the complete replication cycle of HCMV which lasts at least 72 hours. Already during the immediate early phase of infection, *US3* is transcribed expressing a 22/23 kDa glycoprotein which retains peptide loaded MHC I complexes in the ER (Ahn et al. 1996a; Jones et al. 1996). Transcription of *US3* is followed by simultaneous expression of the neighbouring

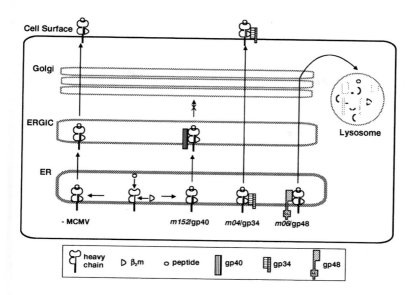

Fig. 2. The MHC class I pathway of antigen presentation and interfering MCMV glycoproteins. The *m152*-encoded gp37/40 protein disrupts MHC I transport by retaining MHC I complexes in the ER-Golgi Intermediate compartment (ERGIC). *m06*/gp48 binds to MHC I in the ER and targets the complex to the endo/lysosome for proteolytic degradation. *m04*/gp34 associates with MHC I in the ER. The complex is exposed on the cell surface for a function not yet known.

gene, *US2*, encoding a glycoprotein of 24 kDa, and *US11*, both following early expression kinetics. The function of gpUS2 is similar to that of the 33 kDa protein encoded by *US11*. In *US11*-expressing cells, the half-life of peptide-free ("empty") MHC class I heavy chains is shortened to 1-2 minutes due to the retrograde transport of glycosylated and therefore membrane inserted MHC I heavy chains from the ER to the cytosol (Wiertz et al. 1996a). In the cytosol, MHC I is rapidly destructed by the proteasome. Misfolded MHC I heavy chains have been shown to follow this route physiologically ending up in the cytosol for proteasome-mediated degradation (Hughes et al. 1997). Co-immunoprecipitation of a deglycosylated MHC class I heavy chain intermediate and a deglycosylated 20 kDa product of *US2* together with the heterotrimeric Sec61 complex suggests that the export of MHC class I molecules is mediated through the translocon (Wiertz et al. 1996b). This finding implies that the translocon also allows back transport from the ER, which is supported by genetic evidence from yeast (Plemper et al. 1997). This link suggests that the Sec61 complex is part of a more generally used retrograde transport pathway of ER degradation. Remarkably,

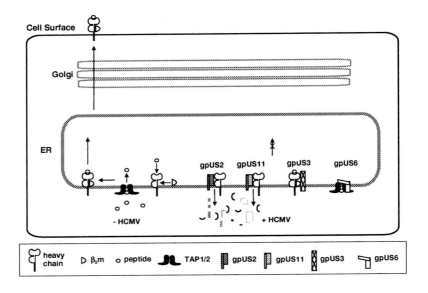

Fig. 3. Interference of HCMV gps with the MHC I pathway of antigen presentation. gpUS3 inhibits export of MHC I complexes from the ER and anterograde transport to the cell surface. The *US2*/gp24 and *US11*/gp33 proteins attack MHC class I heavy chains and induce their retrograde transport from the ER back into the cytosol. The *US6*-encoded glycoprotein prevents the import of peptides by the TAP1/2 transporter.

gpUS2 and gpUS11 do not affect HLA-G and HLA-C molecules which bind to inhibitory receptors of NK cells (Schust et al. 1998). This selective function may avoid otherwise enhanced NK susceptibility of HCMV-infected cells resulting from MHC I downregulation (Kärre 1995).

HCMV gpUS6 targets the MHC-encoded TAP transporter

Assembly of ternary MHC I complexes requires the import of peptides which are produced in the cytosol into the lumen of the ER. This is achieved by the MHC-encoded peptide transporter associated with antigen processing (TAP). This transporter is composed of two homologous subunits of approximately 70 kDa, TAP1 and TAP2, both encoded in the MHC (reviewed by Momburg and Hämmerling 1998). The translocation of peptides by TAP1/2 requires two coupled but independent events: binding to TAP which occurs in an ATP-independent fashion, and subsequent translocation of the peptide ligand which requires the hydrolysis of ATP (see Fig. 4). Prior to the binding of peptides, MHC class I heavy chains are assisted in their proper folding by the transient and sequential interaction with ER-resident chaperones, specifically calnexin, calreticulin, ER-60 and tapasin (Degen et al. 1991; Sadasivan et al. 1996; Ortmann et al. 1997; Lindquist et al. 1998). The late MHC class I assembly complex is formed by MHC I, β_2 microglobulin, calreticulin, and ER-60 which are linked to TAP1/2 by the MHC-encoded ER glycoprotein tapasin (see Fig. 4).

To date, two herpesviral proteins have been identified which shut off the peptide transporter. The herpes simples virus (HSV) protein ICP 47 is a cytosolic protein of 9 kDa expressed as an immediate early gene product and was demonstrated to inhibit recognition of HSV-infected cells (York et al. 1994). ICP 47 binds with high affinity directly to the cytosolic face of TAP1/2, thereby competitively inhibiting the binding of peptide ligands and subsequent peptide translocation (Früh et al. 1995; Hill et al. 1995). ATP binding to TAP1/2 is not affected by ICP47. In competition experiments, ICP47 was shown to bind to human TAP with a 100 fold higher affinity compared to murine TAP, indicating a species-specific function of the protein (Tomazin et al. 1996; Ahn et al. 1996b). Based on chemical cross-linking experiments it was suggested that ICP 47 destabilizes the TAP1/2 heterodimer, while peptide substrates stimulated the formation of TAP1/2 heterodimers (Lacaille and Androlewicz 1998). The significance of this finding awaits further experimentation.

Phenotypically similar to HSV, HCMV shuts down the translocation of cytosolic peptides into the ER lumen during permissive infection despite the fact that TAP1/2 synthesis is significantly enhanced (Hengel et al. 1996). This effect is mediated by the 21 kDa HCMV US6 glycoprotein the expression of which reaches peak levels in the late phase of HCMV replication (Hengel et al. 1997; Ahn et al 1997; Lehner et al 1997). The subcellular distribution of gpUS6 is ER restricted. The viral protein is found in complexes with calnexin and the late assembly complex comprising MHC I, $\beta_2$microglobulin, calreticulin, ER-60, tapasin and TAP1/2 (see Fig. 4). The blockade of TAP by gpUS6 is independent of MHC I, tapasin and calnexin and does not involve peptide binding to TAP1/2 (Hengel et al. 1997; Ahn et al. 1997). A gpUS6 mutant lacking the transmembrane and cytoplasmic domains of the molecule still downregulates MHC I surface expression, confirming the interaction with TAP1/2 via its luminal face (Ahn et al. 1997). Although controlling TAP by an entirely different molecular mechanism than HSV ICP 47, most recent data indicate that gpUS6, like ICP47, directly binds to TAP1 and TAP2 in a species-restricted fashion leaving out murine TAPs (D. Bauer and H. Hengel, unpublished observation). Taken together, the available data are compatible with the hypothesis that gpUS6 binding to TAP prevents pore opening and peptide release into the ER.

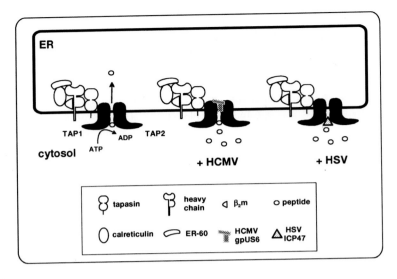

Fig. 4. Inactivation of the peptide transport mediated by TAP1/2. HCMV gpUS6 binds to the luminal face of TAP1/2, thus inhibiting peptide import into the ER. In contrast, HSV-ICP47 binds to the cytoplasmic surface of TAP1/2, thereby preventing the association of peptides to TAP1/2.

Cytomegaloviral mechanisms interfering with MHC class II molecule expression

CD4+ T cells recognize peptide antigens presented by MHC II molecules. The highly polymorphic MHC class II αβ heterodimer assembles together with the invariant chain (Ii) in the ER (for review see Pieters et al. 1997; Watt 1997). The MHC II-Ii complex is targeted to the endocytic pathway due to sorting signals in the Ii cytoplasmic tail. At endolysosomal sites, antigenic peptides are loaded onto MHC II molecules, a process which is facilitated by HLA-DM molecules, before MHC class II-peptide complexes are transported to the plasma membrane. In contrast to MHC I molecules, MHC II molecules are constitutively expressed on a limited number of cell types like B lymphocytes and dendritic cells. On other cell types, MHC class II molecule expression requires induction by interferon-γ (IFNγ). MHC class II gene transcription is activated by the Jak/Stat signal transduction pathway (Darnell et al. 1994; O'Shea 1997). After IFNγ binding to its receptor, a cascade of events is initiated, starting with the phosphorylation of the receptor associated Janus kinases (Jaks) Jak1 and Jak2 as well as the cytoplasmic tail of the IFNγ-receptor itself. Upon phosphorylation, a member of the family of signal transducers and activators of transcription (Stats), Stat1α, can be bound, which is in turn phosphorylated by the Jaks. This allows dimerization of Stats and migration to the nucleus where it binds to specific sites present in promoters of IFNγ-inducible genes.

HCMV and MCMV affect constitutive as well as IFNγ-inducible MHC class II expression. Both viruses disrupt the Jak/Stat signal transduction pathway, but the underlying viral mechanisms appear clearly different. In HCMV-infected cells levels of Jak1 are significantly decreased due to HCMV-associated enhancement of Jak1 protein degradation (Miller et al. 1998). In contrast, MCMV infection interferes with the induction of MHC II genes at a stage downstream of Stat1α activation and nuclear translocation (Heise et al. 1998). Constitutively expressed MHC class II molecules on macrophages from MCMV-infected cultures show significant lower densities compared to controls (Redpath et al. 1999). This effect was selective and observed irrespective whether individual cells were MCMV-infected or not, pointing to a soluble factor mediating this general reduction. This effect is caused by viral induction of host cell Interleukin-10, demonstrating an autocrine pathway exploited by the virus (Redpath et al. 1999). In HCMV infection, gpUS2 dislocates not only MHC I heavy chains into the cytosol (Wiertz et al. 1996a; Jones et al. 1997), but affects also MHC II β chains as well as HLA-DM (Tomazin et al. 1999), indicating that CMVs are even able to kill two birds with one stone.

Acknowledgements

Our work is supported by the Deutsche Forschungsgemeinschaft through the Graduiertenkolleg "Infection and Immunity" and grant He 2526/3-1.

References

Ahn K, Angulo A, Ghazal P, Peterson PA, Yang Y, Früh K (1996a) Human cytomegalovirus inhibits antigen presentation by a sequential multistep process. Proc Natl Acad Sci USA 93:10990-10995

Ahn K, Gruhler A, Galocha B, Jones TR, Wiertz EJ, Ploegh HL, Peterson PA, Yang Y, Früh K (1997) The ER luminal domain of the HCMV glycoprotein US6 inhibits peptide translocation by TAP. Immunity 6:613-621

Ahn K, Meyer TH, Uebel S, Sempe P, Djaballah H, Yang Y, Peterson PA, Früh K, Tampe R (1996) Molecular mechanism and species specificity of TAP inhibition by herpes simplex virus protein ICP47. EMBO J 15:3247-3255

Bakke O, Dobberstein B (1990) MHC class II-associated invariant chain contains a sorting signal for endosomal compartments. Cell 63:707-716

Britt WJ, Alford CA (1996) Cytomegaloviruses. In: Fields BN, Knipe DM, Howley PM (eds) Virology, Volume 2 Lippincott-Raven, Philadelphia, pp 2493-2523

Chee MS, Bankier AT, Beck S, Bohni R, Brown CM, Cerny R, Horsnell T, Hutchison III CA, Kouzarides T, Martignetti JA, Preddie E, Satchwell SC, Tomlinson P, Weston KM,

Barrell BG (1990) Analysis of the protein-coding content of the sequence of human cytomegalovirus strain AD169. Curr Top Microbiol Immunol 54:125-169

Darnell JE, Kerr IM, Stark GR (1994) Jak-STAT pathway and transcriptional activation in response to IFNs and other extracellular signaling proteins. Science 264:1415-1421.

Del Val M, Hengel H, Häcker H, Hartlaub U, Ruppert T, Lucin P, Koszinowski UH (1992) Cytomegalovirus prevents antigen presentation by blocking the transport of peptide-loaded major histocompatibility complex class I molecules into the medial-Golgi compartment. J Exp Med 172:729-738

Del Val M, Münch K, Reddehase MJ, Koszinowski UH (1989) Presentation of CMV immediate-early antigen to cytolytic T lymphocytes is selectively prevented by viral genes expressed in the early phase. Cell 58:305-315

Früh K, Ahn K, Djaballah H, Sampe P, van Endert PM, Tampe R, Peterson PA, Yang Y (1995) A viral inhibitor of peptide transporters for antigen presentation. Nature 375:415-418

Geginat G, Ruppert T, Hengel H, Holtappels R, Koszinowski UH (1997) Interferon-γ is a prerequisite for efficient antigen processing of viral peptides in vivo. J Immunol 158:3303-3310

Heemels MT, Ploegh HL (1995) Generation, translocation, and presentation of MHC class I-restricted peptides. Annu Rev Biochem 64:463-491

Heise M, Connick M, Virgin HW (1998) Murine cytomegalovirus inhibits interferon γ-induced antigen presentation to CD4 T cells by macrophages via regulation of expression of major histocompatibility complex class II-associated genes. J Exp Med 187:1037-1046

Hengel H, Brune W, Koszinowski UH (1998) Immune evasion by cytomegalovirus - survival strategies of a highly adapted opportunist. Trends Microbiol 6:190-197

Hengel H, Eßlinger C, Pool J, Goulmy E, Koszinowski UH (1995) Cytokines restore MHC class I complex formation and control antigen presentation in human cytomegalovirus-infected cells. J Gen Virol 76:2987-2997

Hengel H, Flohr T, Hämmerling GJ, Koszinowski UH, Momburg F (1996) Human cytomegalovirus inhibits peptide translocation into the endoplasmic reticulum for MHC class I assembly. J Gen Virol 77:2287-2296

Hengel H, Lucin P, Jonjic S, Ruppert T, Koszinowski UH (1994) Restoration of cytomegalovirus antigen presentation by gamma interferon combats viral escape. J Virol 68:289-297

Hengel H , Koszinowski UH (1997) Interference with antigen processing by viruses. Curr Opin Immunol 9:470-476

Hengel H, Koopmann JO, Flohr T, Muranyi W, Goulmy E, Hämmerling GJ, Koszinowski UH, Momburg F (1997) A viral ER resident glycoprotein inactivates the MHC encoded peptide transporter. Immunity 6:623-632

Hill A, Jugovic P, York I, Russ G, Bennink J, Yewdell J, Ploegh H, Johnson D (1995) Herpes simplex virus turns off the TAP to evade host immunity. Nature 375:411-415

Hughes EA, Hammond C, Cresswell P (1997) Misfolded major histocompatibility complex class I heavy chains are translocated into the cytoplasm and degraded by the proteasome. Proc Natl Acad Sci USA 94:1896-1901

Jones TR, Hanson LK, Sun L, Slater JS, Stenberg RS, Campbell AE (1995) Multiple independent loci within the human cytomegalovirus unique short region down-regulate expression of major histocompatibility complex class I heavy chains. J Virol 69:4830-4841

Jones TR, Wiertz EJ, Sun L, Fish KN, Nelson JA, Ploegh HL (1996) Human cytomegalovirus US3 impairs transport and maturation of major histocompatibility complex class I heavy chains. Proc Natl Acad Sci USA 93:11327-11333

Jones T, Sun L (1997) Human cytomegalovirus US2 destabilizes major histocompatibility complex class I heavy chains. J Virol 71:2970-2979

Jonjic S, Del Val M, Keil GM, Reddehase MJ, Koszinowski UH (1988) A nonstructural viral protein expressed by a recombinant vaccinia virus protects against lethal cytomegalovirus infection. J Virol 62:1653-1658

Jonjic S, Mutter W, Weiland F, Reddehase MJ, Koszinowski UH (1989) Site-directed persistent cytomegalovirus infection after selective long-term depletion of CD4$^+$ T lymphocytes. J Exp Med 169:1199-1212

Jonjic S, Pavic I, Polic B, Crnkovic I, Lucin P, Koszinowski UH (1994) Antibodies are not essential for the resolution of primary cytomegalovirus infection but limit dissemination of recurrent virus. J Exp Med 179:1713-1717

Kärre K (1995) Express yourself or die: peptides, MHC molecules, and NK cells. Science 267: 978-979

Kleijnen M, Huppa JB, Lucin P, Mukherjee S, Farrell H, Campbell A, Koszinowski UH, Hill AB, Ploegh HL (1997) A mouse cytomegalovirus glycoprotein, gp34, forms a complex with folded class I MHC molecules in the ER which is not retained but is transported to the cell surface. EMBO J 16:685-694

Lacaille VG, Androlewicz MJ (1998) Herpes simplex virus inhibitor ICP47 destabilizes the transporter associated with antigen processing (TAP) heterodimer. J Biol Chem 28:17386-17390

Lehner PJ, Karttunen JT, Wilkinson GW, Cresswell P (1997) The human cytomegalovirus US6 glycoprotein inhibits transporter associated with antigen processing-dependent peptide translocation. Proc Natl Acad Sci USA 94:6904-6909

Letourner F, Klausner RD (1992) A novel di-leucine motif and a tyrosine-based motif independently mediate lysosomal targeting and endocytosis of CD3 chains. Cell 69: 1143-1157

Lindquist JA, Jensen ON, Mann M, Hämmerling GJ (1998) ER-60, a chaperone with thiol-dependent reductase activity involved in MHC class I assembly. EMBO J 17:2186-2195

Lucin P, Pavic I, Polic B, Jonjic S, Koszinowski UH (1992) Gamma-interferon-dependent clearance of cytomegalovirus infection in salivary glands. J Virol 66:1977-1984

Maidji E, Tugizov S, Jones T, Zheng Z and Pereira L (1996) Accessory human cytomegalovirus glycoprotein US9 in the unique short component of the viral genome promotes cell-to-cell transmission of virus in polarized epithelial cells. J Virol 70:8402-8410

Mocarski ES (1996) Cytomegaloviruses and their replication. In: Fields BN, Knipe DM, Howley PM (eds) Virology, Volume 2 Lippincott-Raven, Philadelphia, pp 2447-2492

Momburg F, Hämmerling GJ (1998) Generation and TAP-mediated transport of peptides for major histocompatibility complex class I molecules. Adv Immunol 68:191-256

Miller DM, Rahill BM, Boss J M, Lairmore MD, Durbin JE, Waldman WJ, Sedmak DD (1998) Human cytomegalovirus inhibits major histocompatibility complex class II expression by disruption of the Jak/Stat pathway. J Exp Med 187:675-683

Ortmann B, Androlewicz M, Cresswell P (1994) MHC class I / β_2-microglobulin complexes associate with TAP transporters before peptide binding. Nature 368:864-867

Ortmann B, Copeman J, Lehner PJ, Sadasivan B, Herberg JA, Grandea AG, Riddell SR, Tampe R, Spies T, Trowsdale J, Cresswell P (1997) A critical role for tapasin in the assembly and function of multimeric MHC class I-TAP complexes. Science 277:1306-1309

Polic B, Hengel H, Krmpotic A, Trgovcich J, Pavic I, Luccaronin P, Jonjic S, Koszinowski UH (1998) Hierarchical and redundant lymphocyte subset control precludes cytomegalovirus replication during latent infection. J Exp Med 188:1047-1054

Pavic I, Polic B, Crnkovic I, Luccaronin P, Jonjic S, Koszinowski UH (1993) Participation of endogenous tumor necrosis factor α in host resistance to cytomegalovirus infection. J Gen Virol 74:2215-2223

Pieters J (1997) MHC class II restricted antigen presentation. Curr Opin Immunol 9:89-96

Plemper R, Bömler K, Bordallo J, Sommer T, Wolf DH (1997) Mutant analysis links the translocon and BiP to retrograde protein transport for ER degradation. Nature 388:891-895

Rajagopalan S, Brenner MB (1994) Calnexin retains unassembled major histocompatibility complex class I free heavy chains in the endoplasmic reticulum. J Exp Med 180:407-412

Rawlinson WD, Farrell HE, Barrell BG (1996) Analysis of the complete DNA sequence of murine cytomegalovirus. J Virol 70:8833-8849

Reddehase MJ, Mutter W, Münch K, Bühring HJ, Koszinowski UH (1987) CD8 positive T lymphocytes specific for murine cytomegalovirus immediate-early antigens mediate protective immunity. J Virol 61:3102-3108

Reddehase MJ, Weiland F, Münch K, Jonjic S, Lüske A, Koszinowki UH (1985) Interstitial murine cytomegalovirus pneumonia after irradiation: characterization of cells that limit viral replication during established infection of the lungs. J Virol 55:264-273

Redpath S, Angulo A, Gascoigne NR, Ghazal P (1999) Murine cytomegalovirus infection down-regulates MHC II expression on macrophages by induction of IL-10. J Immunol 162:6701-6707

Reusch U, Muranyi W, Lucin P, Burgert HG, Hengel H, Koszinowski UH (1999) A cytomegalovirus glycoprotein re-routes MHC class I complexes to lysosomes for degradation. EMBO J 18:1081-1091

Riddell SR, Watanabe KS, Goodrich JM, Li CR, Agha ME, Greenberg PD (1992) Restoration of viral immunity in immunodeficient humans by the adoptive transfer of T cell clones. Science 257:238-241

Sadasivan B, Lehner PJ, Ortmann B, Spies T, Cresswell P (1996) Roles for calreticulin and a novel glycoprotein, tapasin, in the interaction of MHC class I molecules with TAP. Immunity 5:103-114

Schust DJ, Tortorella D, Seebach J, Phan C, Ploegh HL (1998) Trophoblast class I major histocompatibility complex (MHC) products are resistant to rapid degradation imposed by the human cytomegalovirus (HCMV) gene products US2 and US11. J Exp Med 188:497-503

Shellam GR, Allan JE, Papadimitriou JM, Bancroft GJ (1981) Increased susceptibility to cytomegalovirus infection in beige mutant mice. Proc Natl Acad Sci USA 78:5104-5108

Tomazin R, Boname J, Hegde NR, Lewinsohn DM, Jones TR, Cresswell P, Nelson JA, Riddell SR, Johnson DC (1999) Cytomegalovirus US2 destroys two components of the MHC class II pathway preventing recognition by CD4+ T cells. Abstract 9001. 24th International Herpesvirus Workshop, July 17-23, Boston, USA

Tomazin R, Hill AB, Jugovic P, York I, van Endert P, Ploegh HL, Andrews DW, Johnson DC (1996) Stable binding of the herpes simplex virus ICP47 protein to the peptide binding site of TAP. EMBO J 15: 3256-3266

Warren AP, Ducroq DH, Lehner PJ, Borysiewicz LK (1994) Human-cytomegalovirus-infected cells have unstable assembly of major histocompatibility complex class I complexes and are resistant to lysis by cytotoxic T lymphocytes. J Virol 68:2822-2829

Watts C (1997) Capture and processing of exogenous antigens for presentation on MHC molecules. Annu Rev Immunol 15:821-850

Welsh RM, Brubaker JO, Vargas-Cortes M, O'Donnell CLO (1991) Natural killer (NK) cell response to virus infections in mice with severe combined immunodeficiency. The stimulation of NK cells and the NK cell-dependent control of virus infections occur independently of T and B cell function. J Exp Med 173:1053-1063

Wiertz EJ, Jones TR, Sun L, Bogyo M, Geuze HJ, Ploegh HL (1996a) The human cytomegalovirus US11 gene product dislocates MHC class I heavy chains from the endoplasmic reticulum to the cytosol. Cell 84:769-779

Wiertz EJ, Tortorella D, Bogyo M, Yu J, Mothes W, Jones TR, Rapoport TA, Ploegh HL (1996b) Sec61-mediated transfer of a membrane protein from the endoplasmic reticulum to the proteasome for destruction. Nature 384:432-438

York IA, Roop C, Andrews DW, Riddell SR, Graham FL, Johnson DC (1994) A cytosolic herpes simplex virus protein inhibits antigen presentation to CD8+ T lymphocytes. Cell 77:525-535

Ziegler H, Thäle R, Lucin P, Muranyi W, H, Flohr T, Hengel H, Farrell H, Rawlinson W, Koszinowski UH (1997) A mouse cytomegalovirus glycoprotein retains MHC class I complexes in the ERGIC/cis-Golgi compartments. Immunity 6:57-66

An animal model for understanding the immunogenetics of AIDS virus infection

Carol M. Kiekhaefer[1], David T. Evans[2], David H. O'Connor[1], and David I. Watkins[1]

[1]Wisconsin Regional Primate Research Center, 1220 Capitol Court, Madison, WI 53715-1299, USA
[2]New England Regional Primate Research Center, One Pine Hill Drive, Southborough, MA 01772-9102, USA

Summary. Heterogeneity in disease progression after HIV-1 infection can be partly attributed to genetic variance among HIV strains and host genetic differences. Host factors such as the CKR5Δ32 deletion and the beneficial effects of certain MHC class I and class II alleles may serve to enhance immune resistance in the setting of HIV infection. Population studies of HIV-infected humans aimed at detecting associations between host genotype and influence on AIDS/disease progression have been limited by a number of confounding variables. For these reasons we are proposing the use of MHC-defined, related rhesus macaques infected with cloned SIV in order to clarify the complex immunogenetic factors involved in disease progression after AIDS-virus infection.

Key words. CD8-positive, MHC class I, Genetic predisposition to disease, HIV infections, Major histocompatibility complex

Introduction

The AIDS epidemic has been characterized by heterogeneity in the clinical course after HIV-1 infection. Among HIV-infected individuals, the rate of disease progression and the clinical outcome vary greatly. Approximately 10% of HIV infected individuals progress to AIDS within two to three years of HIV infection and these are termed rapid progressors. Most HIV infected subjects develop AIDS within approximately ten years and these are called typical progressors. At the other end of the spectrum, 10-17% of HIV infected individuals will likely remain AIDS-free for more than ten years (Phair et al. 1992; Sewell et al. 1999). These differences can be partly attributed to genetic variance among HIV strains and host genetic differences (Hill 1998). We are attempting to use the SIV-infected rhesus macaque as an animal model to understand the immunogenetic basis of differential survival after AIDS-virus infection.

The relationship of HLA-genotype to disease progression after HIV infection

The steps involved in HIV infection and the subsequent progression to disease are multifactorial. One factor that influences disease progression is the genetic makeup of the host. The genetic makeup, in turn, influences such disease parameters as the susceptibility of cells to HIV infection and the efficacy of the antiviral immune response.

For example, the chemokine receptor gene CCR5 and its 32 bp deletion allele (Δ 32 CCR5) may play a role in AIDS pathogenesis. CCR5 serves as a secondary receptor on CD4+ T lymphocytes for certain strains of HIV. Survival analyses in a number of AIDS cohorts have demonstrated that disease progression is slower in CKR5 deletion heterozygotes than in individuals homozygous for the wild type CKR5 gene. Thus the CKR5Δ32 deletion may serve as a resistance gene in the setting of HIV infection, delaying progression to AIDS among infected individuals (Buchbinder et al. 1994; Dean et al. 1996; Phair et al. 1992).

There are several mechanisms by which MHC-encoded molecules may influence rate of progression to AIDS in HIV-infected individuals. The HLA class I and class II loci located within the MHC complex comprise the most polymorphic set of genes known in humans (Dupont 1995). Evolutionary and population studies have lead to the general idea that the diversity and distribution of allelic frequencies observed in class I and class II genes in the MHC in human are maintained through selective forces such as infectious disease morbidity (Parham and Ohta 1996). The hypothesis of overdominant selection (heterozygote advantage) at the MHC proposes that individuals heterozygous at their HLA-loci are able to present a greater variety of antigen-derived peptides than homozygotes, resulting in a more productive immune response to a diverse array of pathogens (Doherty and Zinkernagel 1975; Zinkernagel and Hengartner 1996). If maximum diversity in the repertoire of antigen-presenting molecules is advantageous in prolonging the onset of AIDS after HIV-1 infection, then homozygosity at the class I genes should associate with more rapid progression to AIDS. Carrington et al. (1999) recently explored the relationship between post-HIV survival and HLA expression by examining the genotypes of 498 seroconverters. The HLA class I alleles HLA-B35 and HLA-Cw04 were consistently associated with rapid development of AIDS defining conditions in Caucasians. The extended survival of 28-40% of HIV-1-infected Caucasian patients who avoided AIDS for ten years was attributed to heterozygosity at all HLA class I loci or to lacking the class I alleles HLA-B35 and HLA-Cw04. Additional studies have documented a constant association between the DQ2-DR3-B8-Cw7-A1 and DQ1-DR1-B35-Cw4-A2 haplotypes and disease progression (Haynes et al. 1996; Itescu et al.

1991; Just 1995; Klein et al. 1994; Rowland-Jones et al. 1997; Sahmoud et al. 1992; Shiga et al. 1996).

Confounding variables of HIV infection in human populations

Despite the previously mentioned studies correlating particular HLA haplotypes with HIV disease progression, determining the relevance of these findings has been complicated by variables intrinsic to population studies of HIV-infected humans. Issues that complicate cross-study comparisons can include viral heterogeneity and biased participation by particular races or risk groups.

Exposure to cofactors and routes of infection vary by risk group and may contribute to inconsistent findings across studies. Cofactors can include exposure to other infectious agents or chemicals which have a direct influence on the immune system. These cofactors could therefore interact with or confound the detection of associations between host HLA genotype and AIDS. Infectious agents may influence AIDS progression by interacting with HIV or by inducing an immune response that facilitates the pathogenicity of HIV. Drug and alcohol use may also influence AIDS progression due to their immunosuppressive properties.

Furthermore, advances in the therapeutic modalities used to treat patients has dramatically prolonged asymptomatic infection. However, an untoward consequence of the increased survivorship is a decreased ability to study the natural course of HIV-infection. Similarly, the long monitoring required to track and evaluate disease progression complicates the assessment of disease progression. The introduction of highly active antiretroviral therapy (HAART) has exacerbated this issue.

An animal model for understanding the genetic basis for differential disease progression in HIV infection

Unraveling the contribution of HLA-associated factors in HIV-infection has been difficult in outbred human populations due to high polymorphism among MHC genes and infection with heterogenous HIV strains by a number of different routes.

For these reasons we have used SIV-infected rhesus macaques in an attempt to understand the immunogenetics of disease progression. These animals are housed in controlled, homogeneous environments which decrease the number of confounding variables associated with long-term HIV-infection. Infection of related, MHC-defined rhesus macaques with a well-characterized viral stock should facilitate our ability to understand the immunogenetics of disease progression. The opportunity to control matings and carry out *in vitro* fertilization (IVF) facilitates production of pedigreed macaques that will be useful for studies of immunogenetic susceptibility that could not be readily accomplished in humans. The use of the macaque as an animal model for HIV pathogenesis is facilitated by the reduced length of disease course compared to HIV-infected humans. The absence of HAART also facilitates our ability to understand the role of the MHC in disease progression.

We have conducted an extensive immunogenetic study of a family of rhesus macaques (Fig. 1) The class I and class II alleles expressed by all members of this family have been determined. Interestingly, all of the siblings in this family are Mamu-DRβ-identical. We have infected 5 individuals in this family with SIV and mapped 5 CTL epitopes. Two of the individuals (Fig. 1) died soon after SIV infection. In contrast, their full sibling mounted CTL responses against two Nef epitopes and an Env epitope before succumbing 511 days post-infection (Fig. 1). This animal differs by only a single MHC class I haplotype from the pair of rapidly progressing animals. We also infected this animal's half-sibling who recognized all five CTL epitopes and died 889 days post-infection.

MHC Class I Haplotypes and Survival in a Family of DRB1-Identical Macaques

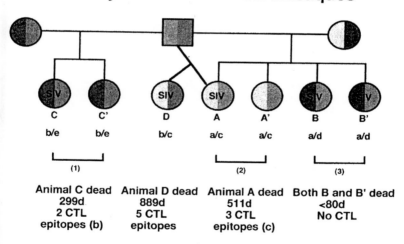

Fig. 1. The MHC Class I haplotypes were determined for all members of a Class II-DRB identical family of rhesus macaques. This family was unique as two sibling pairs shared identical haplotypes, while other sibling pairs differed from each other by only a single haplotype. Infection of these macaques with a well-defined stock of SIV facilitated a study of viral CTL epitopes recognized by each of these animals. Somewhat surprisingly, the number of epitopes presented in each animal was predictive of the length of disease course. Remarkably, two rapid progressors who died within a day of each other possessed an identical MHC-class I haplotype and did not effectively elicit a detectable SIV-specific CTL response.

Given the association between the number of detectable CTL responses and disease outcome (Fig 1.), we hypothesize that individuals that mount a broad and effective cellular immune response (Dalod et al. 1999) will live longer than those that make a modest cellular immune response to a more narrow repetoire of epitopes. An additional prediction of this hypothesis is that homozygosity will correlate with poor prognosis, since the reduced MHC diversity of a homozygote will narrow the diversity of the immune response to HIV. To test this hypothesis we will measure CTL responses in additional MHC-defined family members, all infected with the same dose of the same virus. We will create these MHC-defined homozygotes and heterozygotes from this family through the use of IVF technologies (Fig. 2). These animals will be challenged with SIV and the disease progression will be monitored.

Conclusions

It is possible that not all MHC class I haplotypes contain alleles that encode cell-surface expressed MHC class I molecules able to present epitopes from a particular pathogen to CTL. The inability of certain MHC class I molecules to

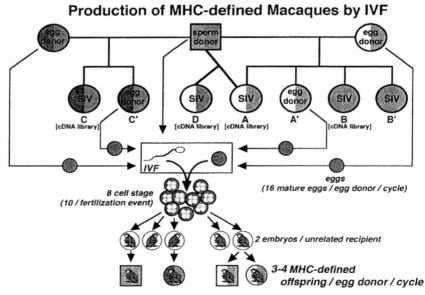

Fig. 2. Production of MHC-defined macaques by IVF. The family shown in Figure 1 can be expanded rapidly by IVF with eggs from both "mothers" and from offspring C' and A'. Currently, each female can undergo 4-5 cycles of egg collection and 4 offspring can be produced. This yield should improve with advances in macaque reproductive biology.

present epitopes from an emerging pathogen such as HIV may constitute a recent evolutionary disadvantage with regards to survival. Recent data from several laboratories indicates that potent CTL responses are frequently directed against a variety of HIV epitopes. However, the MHC class I molecules of a heterozygous individual are more likely to present a broad range of viral peptides relative to a homozygous individual. If AIDS remains a threat throughout many generations, selection for protective genotypes may occur. The use of MHC-defined, related rhesus macaques infected with SIV should help to clarify the complex immunogenetic factors involved in disease progression after AIDS-virus infection.

Acknowledgements

This work was supported by grants from the National Institutes of Health (RR00167 and AI41913). Dr. Watkins is an Elizabeth Glaser Scientist.

References

Buchbinder SP, Katz MH, Hessol NA, O'Malley PM, Holmberg SD (1994) Long-term HIV-1 infection without immunologic progression. AIDS 8:1123-1128

Carrington M, Nelson GW, Martin MP, Kissner T, Vlahov D, Goedert JJ, Kaslow R, Buchbinder S, Hoots K, O'Brien SJ (1999) HLA and HIV-1: Heterozygote advantage and B*35-Cw*04 disadvantage. Science 283:1748-1752

Dalod M, Dupuis M, Deschemin JC, Sicard D, Salmon D, Delfraisy JF, Venet A, Sinet M, Guillet JG (1999) Broad, intense anti-human immunodeficiency virus (HIV) ex vivo CD8+ responses in HIV type 1-infected patients: comparison with anti-Epstein-Barr virus responses and changes during antiretroviral therapy. J Virol 733:7108-7116

Dean M, Carrington M, Winkler C, Huttley GA, Smith MW, Allikmets R, Goedert JJ, Buchbinder SP, Vittinghoff E, Gomperts E, Donfield S, Vlahov D, Kaslow R, Saah A, Rinaldo C, Detels R, O'Brien SJ (1996) Genetic restriction of HIV-1 infection and progression to AIDS by a deletion allele of the CKR5 structural gene; Hemophilia Growth and Development Study, Multicenter AIDS Cohort Study, Multicenter Hemophilia Cohort Study, San Francisco City Cohort, ALIVE Study. Science 273:1856-1862

Doherty PC, Zinkernagel RM (1975) Enhanced immunological surveillance in mice heterozygous at the H-2 gene complex. Nature 256:50-52

Dupont B (1995) Immunobiology of HLA. Springer-Verlag, New York

Haynes BF, Pantaleo G, Fauci AS (1996) Toward an understanding of the correlates of protective immunity to HIV infection. Science 271:324-328

Hill AV (1998) The immunogenetics of human infectious diseases. Annu Rev Immunol 16:593-617

Itescu S, Mathur-Wagh U, Skovron ML, Brancato LJ, Marmor M, Zeleniuch-Jacquotte A, Winchchester R (1991) HLA-B35 is associated with accelerated progression to AIDS. J Acquir Immune Defic Syndr 5:37-45

Just JJ (1995) Genetic predisposition to HIV-1 infection and acquired immune deficiency virus syndrome: a review of the literature examining associations with HLA. Hum Immunol 44:156-169

Klein MR, Keet IPM, D'Amaro J, Bende RJ, Hekman A, Mesman B, Koot M, de Waal LP, Coutinho RA, Miedema F (1994) Associations between HLA frequencies and pathogenic features of human immunodeficiency virus type 1 infection in seroconverters from the Amsterdam cohort of homosexual men. J Infect Dis 169:1244-1249

Parham P, Ohta T (1996) Population biology of antigen presentation by MHC class I molecules. Science 272:67-74

Phair J, Jacobson L, Detels R, Rinaldo C, Saah A, Schrager L, Munoz A (1992) Acquired immune deficiency syndrome occurring within 5 years of infection with human immunodeficiency virus type-1: the Multicenter AIDS Cohort Study. J AIDS 5:490-496

Rowland-Jones S, Tan R, McMichael A (1997) Role of cellular immunity in protection against HIV infection. Adv Immunol 65:277-346

Sahmoud T, Laurian Y, Gazengel C, Sultan Y, Gautreau C, Costagliola D (1992) Progression to AIDS in French haemophiliacs: association with HLA-B35. AIDS 7:497-500

Sewell AK, Price DA, Teisserenc H, Booth BL Jr, Gileadi U, Flavin FM, Trowsdale J, Phillips RE, Cerundolo V (1999) IFN-gamma exposes a cryptic cytotoxic T lymphocyte epitope in HIV-1 reverse transcriptase. J Immunol 162:7075-7079

Shiga H, Shioda T, Tomiyama H, Yakamiya Y, Oka S, Kimura S, Yamaguchi Y, Gojobori T, Rammensee H-G, Miwa K, Takiguchi M (1996) Identification of multiple HIV-1 cytotoxic T-cell epitopes presented by human leukocyte antigen B35 molecules. AIDS 10:1075-1083

Zinkernagel RM, Hengartner H (1996) Correlates of protective viruses damaging to HIV infection. Science 272:1362

6. MHC polymorphism

Evolution of HLA-DRB loci, DRB1 lineages, and alleles: analyses of intron-1 and -2 sequences

Tomas F. Bergström[1], Steven J. Mack[2,4], Ulf Gyllensten[3], and Henry A. Erlich[4,2]

[1]University of Washington Genome Center, University of Washington, Seattle, WA, USA
[2]Children's Hospital Oakland Research Institute, Oakland, CA, USA
[3]Department of Genetics and Pathology, Rudbeck Laboratory, University of Uppsala, Uppsala, Sweden
[4]Department of Human Genetics, Roche Molecular Systems, Alameda, CA, USA

Summary. Phylogenetic analyses of DRB intron-1 and intron-2 sequences derived from a variety of primate species suggest that a sequence of duplication events has given rise to the DRB1, B3, B4, B5, and B6 loci. Allelic diversification of DRB1 appears to have preceded the origin of some of these loci. The DRB3 sequences cluster with alleles from the DR51 lineage (DR1 and DR2) and DRB4 is also more closely related to the DR51 than to the DR52 and DR53 lineages. DRB5 and DRB6 sequences diverged before separation of the DRB1 allelic lineages. For DRB1, the divergence of most of the lineages occurred prior to the separation of the hominoids (5-7myr ago), while alleles within a lineage, based on the paucity of intron sequence variation, appear to have diverged recently (within the last 200-300,000 years). Within lineage sequence variation can be detected, however, by analysis of a complex microsatellite in intron-2. We postulate that the structure of this microsatellite has evolved by point mutations from a putative ancestral $(GT)_x (GA)_y$ complex dinucleotide repeat. In all contemporary human DRB1 allelic lineages, with the exception of the DR4 lineage, the $(GA)_y$ repeat is interrupted, often by a G to C transversion, giving rise to a variety of more complex structures. In general, the length of the 3' (GA) repeat correlates with a specific allelic lineage and thus evolves more slowly than the middle (GA) repeat, whose length correlates with a specific allele. The longer 5' (GT) repeat evolves more rapidly than both the middle (GA) repeat and the 3' (GA) repeat, consistent with a higher mutation rate for longer tracts. The length variation in this complex microsatellite repeat was used to trace the origin of new DRB1 alleles, such as the novel DRB1*08 alleles found in South America.

Key words. Phylogenetic, HLA class II, Microsatellite, Polymorphism evolution, Populations

Introduction

The HLA class I and class II loci are the most polymorphic coding sequences in the human genome, with several loci, such as HLA-B and HLA-DRB1, having well over 200 alleles. The class II loci encode an alpha and a beta chain that form a heterodimer on the surface of antigen presenting cells, such as B cells, macrophages, and dendritic cells. These heterodimeric class II molecules, HLA-DR, DQ, and DP, bind processed peptides derived predominantly from membrane and extracellular proteins, and present them to CD4+ T cells. Both the HLA-DQ and DP regions contain one functional gene for each of their alpha (DQA1 and DPA1) and beta (DQB1 and DPB1) chains as well as the pseudogenes DQA2, DQB2, DPA2, and DPB2. The HLA-DRB region can contain either one or two functional DRB loci (DRB1 alone or with DRB3, with DRB4, or with DRB5); some DR haplotypes also contain several pseudogenes, such as the DRB6 locus, which is found on DR1 and DR2 (DRB1*15 and DRB1*16) haplotypes. All DR haplotypes contain a single, monomorphic DRA locus.

Virtually all of the extensive class II sequence diversity is localized to the second exon, which encodes the peptide binding groove with its characteristic beta-sheet "floor" and two alpha-helical "walls". All the class II loci encoding beta-chains (e.g. DRB1, DPB1, DQB1) are highly polymorphic, but among the alpha chain loci only the DQA1 locus shows significant second exon polymorphism. For the class II beta-chain loci, the first part of the second exon encodes the beta-sheet floor and the second part encodes the alpha-helix wall of the peptide binding groove.

X-ray crystallographic studies have revealed the structure of class I and class II molecules and the interaction of the peptide binding groove with the bound peptide (Stern et al. 1994; Jardetzky et al. 1996). The peptides bound in the cleft of class II molecules are usually 12-14-mers, with the ends protruding from the groove, unlike the peptides bound to the class I molecule, where the ends of the peptide are buried within pockets of the cleft. Based on the sequences of peptides eluted from purified HLA molecules, specific peptide-binding motifs have been inferred for different class I and class II molecules allowing modeling of the interaction between these residues and the polymorphic residues within the pockets of the binding groove. Some polymorphic HLA residues within the binding groove bind to the peptide while others bind to the T cell receptor (Garboczi et al. 1996). It is the complex of specific peptide bound within a given HLA class II molecule that is recognized by the TCR of CD4+ cells.

Evolution of HLA polymorphism

The evolution of this extensive allelic sequence diversity at the HLA loci has been a topic of considerable interest and controversy. The genetic mechanisms that have generated the HLA sequence polymorphism include point mutation, recombination and gene conversion-like events (segmental exchanges between different alleles and loci). The sharing of particular polymorphic sequence motifs among alleles within a species or between species has been attributed to 1) common ancestry, 2) convergent evolution, and 3) gene conversion (interallelic or interlocus segmental exchange). (Erlich and Gyllensten 1991). The relative importance of these explanatory models remains a subject of continued investigation and discussion. In particular, the rate of sequence diversification and the "age" of the HLA polymorphisms have been topics of intense debate. Some investigators have noted that because the coding sequence polymorphism is subject to selective pressures and appears to participate in recombinational/gene conversion events, it may be more appropriate to make evolutionary inferences from phylogenetic analyses of intron sequences rather than of exon sequences (Erlich et al. 1996; Bergström et al. 1998).

Results and discussion

Evolutionary diversification of loci and lineages, based on phylogenetic analysis of intron sequences

Because of these considerations about selection and recombination in the polymorphic second exon, we have focused our attention on the intron sequences surrounding exon 2 of various DRB loci; these sequences were generated by using sequence-specific PCR primers for polymorphic motifs within the second exon which permit the determination of flanking intron sequences for a given allele (Bergström et al. 1998). Phylogenetic analyses of DRB intron-1 and intron-2 sequences derived from a variety of primate species suggest that a series of gene duplication events has given rise to the DRB1, B3, B4, B5, and B6 loci (Bergström et al. 1999b) Allelic diversification of the DRB1 locus appears to have preceded the duplication events which gave rise to some of these loci (see below).

Based on coding sequence comparisons and serological specificities, the alleles at the HLA-DRB1 locus have been divided into 13 allelic lineages, denoted *01, *03, *04, *07, *08, *09, *10, *11, *12, *13, *14, *15, and *16. The sharing of allelic lineages and of loci among primate species provides some information about the timing of allelic diversification and of gene duplication. Representatives of the DRB2, DRB3, DRB4, DRB5, DRB6, DRB7, DRB8 and DRB9 loci have been identified in both chimpanzees and humans, indicating that the gene duplications leading to these DRB loci occurred prior to the hominoid divergence (> 5-7 mya ago). (In our studies of intron sequences, we have focused on the DRB1, the DRB3, the DR4, and the DRB5 loci (all functional) and the DRB6 locus, a pseudogene in humans). The DRB1 allelic lineages *01/*10 (denoted *10 in chimpanzees), *03/*11/*13/*14 (denoted *03 in chimpanzees), *15/*16 (denoted *02 in chimpanzees), and *07 have also been found in both species (Fan et al. 1989; Gyllensten et al. 1991; Corell et al. 1992; Kenter et al. 1992; Mayer et al. 1992; Vincek et al. 1992; Bontrop et al. 1993; Gongora et al. 1996; Bontrop 1997). In addition, most of the DRB1 allelic lineages and DRB loci shared between humans and chimpanzees are present in gorillas, although gorillas appear to lack the DRB4 and DRB7 loci and the DRB1*07 allelic lineage found in humans and chimpanzees. Conversely, the DRB1*08 allelic lineage is found among humans and gorillas, but appears to be absent in chimpanzees (Figueroa et al. 1991; Vincek et al. 1992; Gyllensten et al. 1991; Gongora et al. 1996; Bontrop 1997; Klein et al. 1992; Kasahara et al. 1992; Kupfermann et al. 1992; Gyllensten and Erlich 1993; Kenter et al. 1993; Slierendregt et al. 1993).

Human DRB1

Our analyses have compared the human DRB1 sequence diversity calculated for 1) intron sequences vs. exon 2 sequences, and for 2) alleles within a lineage vs. alleles in different lineages (Bergström et al. 1998). As is shown in the phylogenetic trees in Figure 1, analyses of human intron and exon sequences indicate that most of the DRB1 allelic lineages diverged from one another prior to hominoid speciation. The estimated "age" of alleles within a lineage, however, varies dramatically, depending on whether intron or exon 2 sequences are used for the analyses. These calculations involve making certain "molecular clock" assumptions and using specific estimates for the rate of evolution for both introns sequences and exon 2 sequences (Bergström et al. 1999). Given that the exon 2 sequences are subject to selection and appear to participate in gene conversion events, estimates based on intron sequences are more likely, in our view, to reflect the evolutionary history of these alleles.

While some lineages are estimated to have diverged as early as 40 million years ago, analyses of intron sequences seem to indicate that the diversification of alleles within lineages has occurred much more recently (Bergström et al. 1998). With the exception of the *03, *11, *13, and *14 lineages (known as the DR52 group), and of the *15 and *16 lineages (the two representatives of the DR2 group), all DRB1 allelic lineages appear to predate the separation of Homo and Pan (Fig. 1). However, the intron sequences of all alleles within a lineage are, with the exception of a complex microsatellite (see below), identical or highly similar, implying that DRB1 alleles within lineages have been generated recently. The mean sequence difference for intron 1 and 2 sequences among alleles within a lineage corresponds to an average age of 180,000 to 320,000 years, implying that the vast majority of the contemporary HLA-DRB1 alleles have a very recent origin.

Alternative explanations of the similarity of the intron sequences have been proposed (e.g., homogenization by gene conversion and genetic draft within introns) (Yeager and Hughes 1999). However, this model does not account for the significant differences in intron sequences between lineages, nor does it explain why gene conversion would, by this alternative hypothesis, generate diversity in exon 2 while it results in homogeneity in the flanking introns.

Since diversifying selection is assumed to be acting on exon 2, it is likely that hitchhiking effects will counterbalance, to some extent, genetic drift in the flanking introns. The relative contribution of these two factors, drift and selection, will naturally vary in different populations and be dependent on selection pressure

(a)

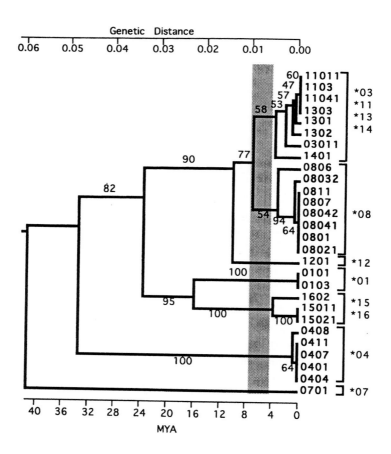

(b)

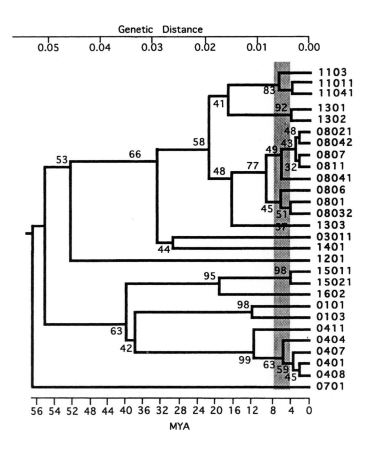

Fig. 1. Phylogenetic trees for 28 DRB1 alleles. The trees were constructed by the UPGMA method (Sneath et al. 1973) from the pairwise genetic distances calculated by using the Jukes-Cantor correction for multiple hits and rooted by the midpoint method. The numbers on the branches represent the proportion of replicates in a bootstrap analysis (500 replications) that were consistent with respect to a particular branching point. The solid bar between 4 and 7 million years ago (MYA) represents the estimate for the hominoid divergence. (a) Intron −1 and −2 sequences. Tree based on the combined sequences for intron 1 and 2, but excluding the microsatellite. The divergence time in tree was estimated using the substitution rate for introns of 1.4 x 10^{-9}/site/year (Li et al. 1996). (b) Exon 2 sequences. Tree based on all positions in the second exon. The divergence time was estimated using the substitution rate for DRB1 exon of 0.97 x 10^{9}/site/year (Satta et al. 1991).

and effective population size. Also, since the same motifs in the α-helical region are found frequently in different allelic lineages with very different intron-2 sequences it seems more plausible that gene-conversion is primarily confined to the α-helical encoding part of exon 2 and does not frequently extend out to intron-2. If that is the case, then we would expect gene conversion to have limited or no homogenization effect on the intron sequence. On the contrary, since selection may act on a new allele with a new combination of amino acid motifs, we would expect mutations in the introns to have a greater probability of being maintained as a result of hitchhiking effects. Therefore, the simplest interpretation of the data is that the limited sequence variation in intron sequences within the same lineage is a result of recent ancestry rather than homogenization by gene conversion in the introns.

Given the relative homogeneity of intron sequences of alleles within a lineage, we infer that the gene conversion-like events that have generated the exon 2 diversity do not extend, in general, to the introns. (Diversification by point mutation, of course, would not affect the flanking introns). There are a few exceptions to this pattern, such as the DRB1*0806 allele (Bergström et al. 1998) and a few other DRB1 alleles (see the chapter by K. Kotsch and R. Blasczyk, this volume), where the putative segmental exchange in exon 2 appears to have involved part of the flanking intron sequence.

The hypothesis that gene conversion and selection may be responsible for generating much of the DRB1 allelic sequence diversity is consistent with the observed pattern of synonymous substitutions within exon 2. In general, the rate of synonymous substitutions, in the absence of selection and recombination, is presumed to be similar to the rate of intron sequence evolution. However, for the DRB1 locus, we found that the level of synonymous changes in the antigen recognition site (ARS) codons (ds, A) is higher than for the non-ARS codons (ds, n), as well as higher than the rate of intron changes (Bergström et al. 1998). One interpretation of this pattern is that the ARS codons are "older" than those in the non-ARS codons and that the ARS codons have been transferred from other alleles or loci by gene conversion-like events (Ohta 1995). In this model, the silent substitutions are carried along with the replacement changes in the putative sequence blocks that are exchanged. In contrast, between lineages, the number of synonymous changes in the non-ARS codons is similar to the divergence in introns 1 and 2, indicating that the non-ARS codons are evolving mainly by the accumulation of point mutations.

In general, a ratio of dn/ds greater than 1 has been interpreted as evidence for balancing selection in the ARS codons of the HLA class I and class II loci (Hughes and Nei 1988, 1989). A ratio less than 1 is consistent with purifying selection. However, in the analysis of DRB1 alleles within a lineage, the value of dn/ds within the ARS codons is less than 1 (Bergström et al. 1998). The dn, a value for within lineage comparison is, as expected, lower than for between lineage comparisons. The ds, a value for <u>within</u> lineage comparisons, is, however, unexpectedly high and may account for this surprisingly low ratio. Presumably, balancing selection is operating for diversification of these ARS codons and this paradoxical result may reflect the failure to incorporate gene conversion into the model upon which the algorithms that calculate dn/ds are based.

As noted above, both the intron and exon trees in Figure 1 indicate a close relationship among the DRB1*03, *11, *12, *13, and *14 allelic lineages (collectively known as the DR52 lineage family, see below), as well as between these lineages and the DRB1*08 lineage. A common origin for the DR52 lineages is further supported by the high degree of similarity of the complex microsatellite located 50 bp downstream of exon 2 (see following section).

Evolution of the DRB loci

The number of human DRB loci found on a given chromosome can vary, and five general arrangements, known as the HLA-DR1, DR8, DR51, DR52 and DR53 allelic lineage families, have been identified. While all of these arrangements carry a DRB1 locus, those carrying alleles of the DRB1*03, *11, *12, *13 and *14 lineages also carry a DRB3 locus, those carrying alleles of the DRB1*04, *07, and *09 lineages also carry a DRB4 locus, and those carrying alleles in the DRB1*15 and *16 lineages also carry the DRB5 locus. The DRB1 locus is the only DRB locus present in the DRB1*01, *08 and *10 lineages.

Phylogenetic analyses of intron 1 and 2 sequences for the DRB1, DRB3, DRB4 and DRB5 loci, as well as the DRB6 pseudogene, have also been carried out recently (Bergström et al. 1999), and suggest a history of gene duplication interspersed with allelic diversification. The phylogenetic tree shown in Figure 2 (intron 1 and 2 tree for DRB loci) is based on the analysis of these intron sequences from humans, chimpanzees and gorillas, and suggests the following scenario for the generation of DRB locus diversity. Based on the tree branching pattern, DRB6 is the first locus to separate from the primordial DRB locus via gene duplication, suggesting that this locus may be one of the most ancient DRB loci known. While the DRB2, DRB6, DRB7, DRB8, and DRB9 loci have all been identified as pseudogenes, DRB6 is the only pseudogene locus included in this analysis; all the other DRB loci shown are expressed. The DRB5 locus separates next, and this event may represent an early duplication of DRB6 or another early DRB locus. The next division in the tree contains DRB1 allelic lineages in the DR53 lineage family (DRB1*04, and *07). While the intron sequences of these two DRB1 allelic lineages are quite different from each other, indicating that they diverged a very long time ago, the analysis, nevertheless, indicates that they have a common origin.

The major cluster in the tree contains all other DRB1 lineages, as well as the DRB3 and DRB4 loci, indicating a common origin for these loci and the alleles in the DR51 and DR52 lineage families (the *01, *10, *15 and *16 alleles, and the*03, *08, *11, *12, *13 and *14 alleles, respectively). Interestingly, the DRB1 allelic lineages found on the DR51 family cluster with the DRB3 locus, suggesting

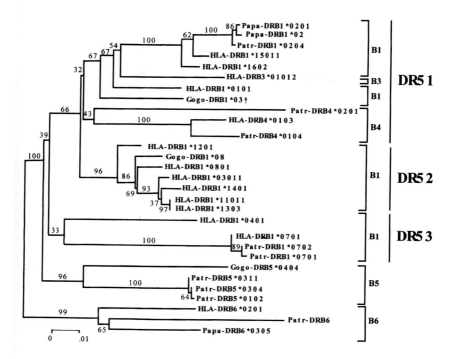

Fig. 2. Phylogenetic trees for the DRB1 loci and DRB1 allelic lineages. The trees were constructed by the neighbor-joining (NJ) (Saitou and Nei 1987) method from the distance values corrected by the method of Jukes and Cantor (1969). The robustness of the branching order was tested by 500 bootstrap replications. The percentage of replications showing a consistent branching pattern is given on each branch. The analyses were based on intron 1 and 2 sequences.

that DRB3 is the result of the duplication of a DRB1 allele related to the ancestral sequences for DRB1*01, *15 and *16. The presence of DRB1 allelic lineages and other DRB loci in the same cluster supports the notion that loci such as DRB3 and DRB4 have been generated through duplications of ancestral DRB1 allelic lineages, as opposed to duplications of a primordial DRB locus. In principle, allelic variants found on DR53 haplotypes predate the DRB3 and DRB4 loci.

The conclusion drawn from these observations is that the DRB1 alleles do not form a monophyletic cluster, separate from the other DRB loci; some of the DRB1 lineages appear to have diverged as much from each other as they have from some of the other DRB loci, consistent with a history in which gene duplication has been interspersed with allelic diversification.

Evolutionary inferences based on the complex microsatellite in DRB1 intron 2

Although the intron -1 and -2 sequence of alleles within a DRB1 lineage are virtually identical (Bergström et al. 1998), sequence variation within a complex microsatellite in intron 2 of the DRB1 gene provides valuable clues to the evolution of HLA sequence diversity. This microsatellite, which consists of a series of dinucleotide repeat tracts, is located approximately 50 bp downstream of the polymorphic second exon of the DRB1 locus (Epplen et al. 1997; Bergström et al. 1999a). The polymorphism at this microsatellite locus displays a complex pattern of both sequence and length variation between alleles. As shown in Table 1, the simplest microsatellite structure observed is denoted as $(GT)_x(GA)_y$, and is present in alleles from the HLA-DRB1*04 lineage, as well as in a gorilla DRB1*08 microsatellite sequence (Bergström et al. 1999a). Because all other microsatellite structures can be derived from this pair of dinucleotide repeats via a small number of nucleotide substitutions, we assume that $(GT)_x(GA)_y$ is the ancestral microsatellite structure. The microsatellite structure in all other allelic lineages is more complex and can be accounted for by a small number of point mutations interrupting the di-nucleotide repeat tract. In general, this structure can be subdivided into three parts: a 5' $(GT)_x$ repeat, a 3' $(GA)_y$ repeat, and a central $(GA)_z$ region containing at least one dinucleotide repeat. While all alleles from a given allelic lineage show the same microsatellite structure, the structure differs between each of the individual allelic lineages (Bergström et al. 1999a). However, significant length polymorphism of these dinucleotide repeats is observed for alleles within a lineage.

When the microsatellite structure is "superimposed", using parsimony considerations, onto the topology of the phylogenetic tree constructed using intron 1 and intron 2 sequences shown in Figure 1A, the minimum number of point mutations required to generate the different microsatellite structures may be inferred (Bergström et al. 1999a). A minimum of 11 nucleotide substitutions are required to generate the eight different human microsatellite structures observed among the DRB1*01, *03, *04, *07, *08, *11, *12, *13, *14, *15 and *16 lineages, where individual lineages have undergone zero to three substitutions. In this model, all substitutions distinguishing allelic lineages, except in the *07 lineage, have occurred in the 3' (GA)y repeat. The microsatellite sequence of the *0701 allele has been interrupted by a total of three point mutations, located in both the (GT)x and the (GA)y repeats. The topology of the tree indicates that the DRB1*07 allelic lineage is the most ancient. On the basis of this topology, it might be argued that the microsatellite structure of the DRB1*07 allelic lineage is the most ancestral, but the *07 microsatellite structure is considerably more complex than that of most other lineages. Since the direction of change of the microsatellite structure in other parts of the tree tends toward complex rather than simple patterns, it seems likely that the ancestral microsatellite structure was simple and of greater similarity to that seen in the contemporary *04 lineage than

Table 1

Microsatellite Variation among *DRB1* Alleles

Allele	Origin	No of Alleles Examined	5'-GT Repeat	Middle Repeat	3'-GA Repeat
0101	Eur	1	$(GT)_{16}$	AA GA AA	$(GA)_6$
0101	Eur	1	$(GT)_{18}$	AA GA AA	$(GA)_6$
0103	Eur	1	$(GT)_{16}$	AA GA AA	$(GA)_6$
03011	Eur	1	$(GT)_{17}$	$(GA)_6$ CA $(GA)_3$ CA	$(GA)_6$
GogoB1*03OR287	Gorilla	1	$(GT)_{14}$	$(GA)_3$ GG $(GA)_6$ CA	$(GA)_6$
PatrB1*03662	Chimpanzee	1	$(GT)_{17}$	$(GA)_6$ AA	(GA)-
0401	Eur	2	$(GT)_{20}$	...	$(GA)_{16}$
0402	Eur	1	$(GT)_{22}$	...	$(GA)_{20}$
0404	Eur	1	$(GT)_{22}$	...	$(GA)_{19}$
0404	Sai	1	$(GT)_{20}$	...	$(GA)_{19}$
0407	Eur	1	$(GT)_{22}$	...	$(GA)_{16}$
0407	Sai	1	$(GT)_{22}$	...	$(GA)_{21}$
0408	Eur	2	$(GT)_{20}$	...	$(GA)_{16}$
0411	Sai	1	$(GT)_{21}$	...	$(GA)_{20}$
0411	Asi	1	$(GT)_{22}$	...	$(GA)_{19}$
0701	Eur	1	$(GT)_2$	GG TT $(GT)_7$ $(GA)_8$ GC	$(GA)_2$
PatrB1*0701/Debbie	Chimpanzee	1	$(GT)_2$	GG TT $(GT)_{14}$ $(GA)_8$ GC	$(GA)_2$
PatrB1*0701/Brigitte	Chimpanzee	1	$(GT)_2$	GG TT $(GT)_{10}$ (GA)- GC	$(GA)_2$
0801	Eur	5	$(GT)_{18}$	(GA)- CA	$(GA)_6$
0801	Sai	1	$(GT)_{18}$	(GA)- CA	$(GA)_6$
0801	Eur	1	$(GT)_{19}$	(GA)- CA	$(GA)_6$
08021*	Sai	3	$(GT)_{17}$	$(GA)_6$ CA	$(GA)_6$
08021	Nai	1	$(GT)_{15}$	$(GA)_6$ CA	$(GA)_5$
08021*	Eur	2	$(GT)_{17}$	$(GA)_6$ CA	$(GA)_5$
08032*	Eur	1	$(GT)_{21}$	(GA)- CA	$(GA)_5$
08032	Eur	1	$(GT)_{20}$	(GA)- CA	$(GA)_5$
08032*	Asi	2	$(GT)_{17}$	$(GA)_6$ CA	$(GA)_5$
0803	Aus?	1	$(GT)_{16}$	$(GA)_8$ CA	$(GA)_5$
08041	Afr	1	$(GT)_{15}$	$(GA)_6$ CA	$(GA)_5$
08041	Afr	2	$(GT)_{16}$	$(GA)_6$ CA	$(GA)_5$
08041	Afr	1	$(GT)_{17}$	$(GA)_6$ CA	$(GA)_5$
08042	Sai	1	$(GT)_{17}$	$(GA)_6$ CA	$(GA)_5$
0806	Eur	1	$(GT)_{21}$	$(GA)_6$ CA	$(GA)_5$
0806	Afr	4	$(GT)_{21}$	$(GA)_6$ CA	$(GA)_5$
0807	Sai	1	$(GT)_{17}$	$(GA)_6$ CA	$(GA)_5$
0811	Nai	1	$(GT)_{17}$	$(GA)_9$ CA	$(GA)_5$
GogoB1*0812/OR759	Gorilla	1	$(GT)_{14}$	...	$(GA)_{21}$
11011	Eur	1	$(GT)_{18}$	$(GA)_4$ CA $(GA)_3$ CA	$(GA)_6$
11011	Asi	1	$(GT)_{21}$	$(GA)_4$ CA $(GA)_3$ CA	$(GA)_6$
1102	Sai	1	$(GT)_{22}$	$(GA)_3$ CA $(GA)_3$ CA	$(GA)_6$
11041	Eur	1	$(GT)_{23}$	$(GA)_4$ CA $(GA)_3$ CA	$(GA)_6$
11041	Eur	1	$(GT)_{24}$	$(GA)_4$ CA $(GA)_3$ CA	$(GA)_6$
1103	Eur	1	$(GT)_{21}$	$(GA)_{10}$ CA $(GA)_3$ CA	$(GA)_6$
1201	Eur	1	$(GT)_{27}$	$(GA)_{10}$ $(CA)_2$	$(GA)_{10}$
1201	Eur	1	$(GT)_{27}$	$(GA)_{11}$ $(CA)_2$	$(GA)_{10}$
1301	Afr	1	$(GT)_{23}$	$(GA)_{10}$ CA $(GA)_3$ CA	$(GA)_6$
1301	Eur	1	$(GT)_{21}$	$(GA)_{10}$ CA $(GA)_3$ CA	$(GA)_6$
1302	Eur	1	$(GT)_{18}$	$(GA)_{11}$ CA $(GA)_3$ CA	$(GA)_6$
1303	Eur	1	$(GT)_{25}$	(GA)- CA $(GA)_3$ CA	$(GA)_6$
1401	Eur	1	$(GT)_{26}$	$(GA)_{11}$ CA	$(GA)_6$
1401	Asi	1	$(GT)_{24}$	$(GA)_{12}$ CA	$(GA)_6$
15011	Eur	1	$(GT)_{18}$	$(GA)_3$ CA $(GA)_4$ CA $(GA)_3$ GG AA	$(GA)_6$
15011	Eur	1	$(GT)_{19}$	$(GA)_3$ CA $(GA)_4$ CA $(GA)_3$ GG AA	$(GA)_6$
15021	Eur	1	$(GT)_{27}$	$(GA)_2$ CA $(GA)_4$ CA $(GA)_3$ GG AA	$(GA)_6$
15021	Asi	1	$(GT)_{18}$	$(GA)_3$ CA $(GA)_4$ CA $(GA)_3$ GG AA	$(GA)_6$
PatrB1*02/Wodka	Chimpanzee	1	$(GT)_{18}$	$(GA)_{14}$ CA $(GA)_4$ CA $(GA)_3$ GG AA	$(GA)_6$
1602	Eur	1	$(GT)_{17}$	$(GA)_8$ GG AA	$(GA)_6$
1602	Sai	4	$(GT)_{18}$	$(GA)_8$ GG AA	$(GA)_6$

NOTE—.Abbreviations: Eur = European, Nai = North American Indian, Sai = South American Indian, Afri = African, Aus = Australian, and Asi = Asian.

[a] No intron sequence available.

[b] Until recently, the *DRB1*0811* allele has been found only in North American Indians. This sample was collected from an African American of American Indian ancestry, and is indicative of recent admixture.

to the structure seen in the *07 lineage.

Based on the microsatellite sequences of the DRB1*02 lineages (*15 and *16), the *16 lineage appears to be ancestral to the *15 lineage, because the microsatellite structure of the *15 lineage can be derived from the structure of the *16 lineage by two additional GA->CA changes. While the possibility cannot be excluded that the *16 lineage was derived from the *15 lineage by the deletion of a CA $(GA)_n$CA tract, a specific deletion seems less likely than the two-mutation model. In addition, the tree topology indicates that the microsatellite structure of the *03, *11, and *13 lineages is derived from the *14 lineage, and that *11 and *13 appear to be the most recently diverged allelic lineages.

The microsatellite sequences from chimpanzees and gorillas support these observations. The gorilla *08 microsatellite has the simplest structure, $(GT)_x(GA)_y$, and differs from the human *08 microsatellite structure by a single point mutation that splits the 3' $(GA)y$ dinucleotide repeat. The chimpanzee *02 microsatellite has a structure identical to that of the HLA-DRB1*15 lineage, indicating that this structure is at least as old as the 5-7 million year divergence between humans and chimpanzees. In addition, while the structure of the *patr*B1*03 allele differs from that of the HLA-DRB1*03 alleles by two substitutions, the gorilla *03 structure differs from the HLA-DRB1*03 lineage by a single substitution. Finally, the two *patr*B1*07 alleles have a microsatellite structure that is virtually identical to that of the HLA-DRB1*07 lineage, and differs only in the length of the dinucleotide repeats. This structure appears to predate the separation of the two species; the phylogenetic tree indicates that it is present on one of the most ancient allelic lineages.

Relationships between DRB1 alleles in the DR8 lineage inferred from microsatellite sequences

Variation in the dinucleotide-repeat lengths can be used to infer phylogenetic relationships among alleles, including those within the same lineage. For example, all the *08 alleles have the 3' $(GA)_5$ repeat. Among all other alleles tested, only DRB1*0301 has this repeat length. However, the overall repeat structure for the DRB1*0301 allele is distinct from the *08 structure in that the *03 microsatellite has a second G->C mutation interrupting the middle $(GA)_z$ repeat. Thus, the 3' $(GA)_5$ appears to be a lineage-specific repeat, and presumably has not changed in length since the divergence of *08 from other lineages. However, the middle $(GA)_z$ repeat varies among different *08 alleles, and the length of this repeat constitutes an allele-specific marker. All the *0801 alleles tested (n = 7) have a middle $(GA)_7$ repeat, *08021 (n = 6) has a $(GA)_9$ repeat, *08041 (n = 5) has a $(GA)_8$ repeat, and *0806 (n = 5) has a $(GA)_6$ repeat. The *08032 allele is the only exception to this allele-specific pattern; an *08032 allele of European origin has a middle $(GA)_7$ repeat, while a *08032 allele of Asian origin has a $(GA)_8$ repeat. Generally, the middle $(GA)_z$ repeat does not appear to have been altered since the time when these DRB1 alleles diverged from each other. The 5' $(GT)_x$ repeat evolves faster than the middle and 3' $(GA)_y$ repeats, and may be useful in tracing the origin of populations, rather than of alleles or of lineages.

The association of the middle $(GA)_z$ repeat length with specific *08 alleles can be used to test hypotheses for the origin of recently discovered *08 alleles in Native American populations. These *08 alleles could have been generated recently, *in situ*, from a putative parental allele (Titus-Trachtenberg et al. 1994; Mack and Erlich 1998), could represent an allele from an ancestral population, or could represent a more recent admixture.

For example, the *08042 allele is uniquely found (f = 0.05) among the Cayapa of Ecuador. This allele has been postulated to have arisen from a point mutation in codon 86, from the pan-American *08021 allele, subsequent to the colonization of South America (Titus-Trachtenberg et al. 1994). This inference was made on the basis of the distribution of DRB1 alleles and the pattern of DR-DQ linkage-disequilibrium in Native American and other populations. The microsatellite of the *08042 allele is consistent with this hypothesis; like the *08021 allele, the *08042 allele has a middle $(GA)_9$ repeat. Similarly, the newly discovered *0807 allele, which differs from *08021 by a single point substitution in codon 57, and has been observed at very high frequency (f = .23) among the Ticuna of Brazil (Mack and Erlich 1998), as well as in other native South American groups, has been postulated to have been generated by a point mutation in the *08021 allele. The microsatellite of the *0807 allele also has a middle $(GA)_9$ repeat. In addition, the *0811 allele, which has been identified in the Na-Dene-speaking Native American populations, also differs from the *08021 allele by a single substitution at codon 57, and has a middle $(GA)_9$ repeat. This pattern is consistent with the hypothesis that the DRB1 *08021 allele is the ancestral Native American *08 allele. The *0807 and *0811 alleles were derived recently (i.e., since humans colonized the Americas), from the *08021 allele, by point mutations or gene conversion events involving codon 57, and the *08042 allele has also been generated recently by a point mutation in codon 86. As noted previously, the observation that most of the new alleles are position 57 or 86 variants of the putative parental sequence suggests selection at these sites, which are known to affect peptide binding.

The middle $(GA)_z$ repeat can also be used to distinguish between an *in situ* origin of an allele and a recent admixture. The microsatellite of the single DRB1*0801 allele found in the Ticuna, has a middle $(GA)_7$ repeat, which is identical to Caucasian and African *0801 alleles, but is distinct from the repeat length of the *08021 allele. The origin of this allele was tentatively attributed to admixture, because of its absence from other American Indian groups, rather than to generation from *08021 (Mack and Erlich 1998). The microsatellite sequence suggests admixture rather than local generation of the *0801 allele from *08021.

The length of the 5' $(GT)_x$ repeat in the *08 lineage varies among samples with the same allele and may serve as a population marker. For instance, all Native

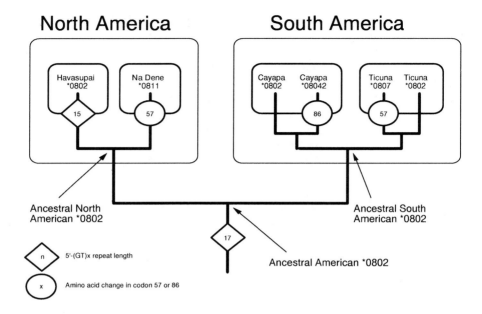

Fig. 3. Variation at the 5' (GT) repeat tract and at codons 57 and 86 of DRB1*08 alleles in the Americas.

American *08 alleles (*08042, *0807, and *0811) of putatively recent origin, have the 5' $(GT)_{17}$ repeat length, as do the proposed parental Cayapa and Ticuna *08021 alleles. However, the Havasupai (Arizona, USA) *08021 allele has a 5' $(GT)_{15}$ repeat length, suggesting that the length of this repeat has changed since the Havásupai population became isolated from other Native American groups (Fig. 3).

Evolution of length polymorphism in the DRB1 complex microsatellite

The putative rate for point mutations altering the microsatellite structure was estimated to be slightly, but not significantly, lower than the rate estimated for the surrounding intron sequences. The length of the dinucleotide repeat tracts, however, evolves more rapidly and the different tracts within the microsatellite appear to have different rates of evolution. The 5' $(GT)_x$ repeat showed greater variability among alleles than the 3' $(GA)_z$ repeat. Comparison of multiple copies of the same allele revealed a 3-5-fold higher variation in the 5' $(GT)_x$ repeat length than in the 3' $(GA)_z$ repeat. This difference in variability could reflect a mutational bias that is related to repeat length. The 5' $(GT)_x$ repeat is generally the longest of the dinucleotide repeats, which suggests that the difference in stability reflects the length of uninterrupted dinucleotide repeats. Similar results have been obtained from sequence analysis of orthologous microsatellite sequences between human and chimpanzee. In yeast, it has been demonstrated that microsatellite variability is dependent on the length of uninterrupted repeat units (Wierdl et al. 1997; Petes et al. 1997). The length of the dinucleotide repeat tracts cannot explain, however, all the differences in variability within the DRB1 microsatellite. For instance, the increased variability in the middle $(GA)_y$ repeat, relative to the 3' $(GA)_z$ repeat, cannot be attributed to repeat length because these dinucleotide repeat tracts are similar in length.

Multiple vs. single origins of DRB1 alleles

In examining multiple copies of the same DRB1 allele from different global populations, it was observed that all such alleles had similar intron sequences, consistent with the notion that all the copies were derived from the same lineage (Bergström et al. 1999a). In addition, the same exon 2 sequences were never found in combination with different microsatellite structures. There are, however, a few rare examples where the different copies of the same DRB1 allele have different dinucleotide repeat lengths (see below). These observations are consistent with the idea that alleles with identical exon 2 sequences could, in principle, be generated repeatedly within a lineage via point mutations or gene conversion events within the second exon. Comparisons of alleles within lineages indicate that some of the *04 alleles represent such cases of potential multiple origins; a *0404 allele from Europe has the $(GT)_{22}(GA)_{19}$ repeat sequence, whereas the *0404 allele from South America has the $(GT)_{20}(GA)_{19}$ sequence. This microsatellite difference may reflect independent origins of the *0404 allele, which has been suggested by (Mack and Erlich 1998) to explain the diversity of *04 alleles in the Americas. Alternatively, the microsatellite sequence could have diverged after the colonization of the Americas from Asia from a single ancestral *0404 allele. In addition, two pairs of different *04 alleles have the same microsatellite sequence; *0401 (Europe) and *0408 (Europe) are both $(GT)_{20}(GA)_{16}$. Perhaps *0408 was derived from *0401, by an exchange at codon 70. Alternatively, the similarity could be coincidental. Similarly, *0404 (Europe) and *0411 (Asia) are both $(GT)_{22}(GA)_{19}$.

The European *08032 and the Asian and Australian *08032 alleles differ in the middle GA repeat; $(GA)_7$ is observed in Europe, and $(GA)_8$ in Asia and Australia. This observation could reflect independent origins for the exon 2 sequence or a change in the length of the middle GA repeat in the microsatellite subsequent to migration or expansion. In the first scenario, the $(GA)_8$ *08032 might have been generated from another *08 allele with $(GA)_8$. DRB1 *08041, the only other $(GA)_8$ allele, differs at multiple exon 2 positions from *08032, and does not seem a likely parental allele. In the second, and perhaps more likely, scenario, the $(GA)_8$ *08032 allele is generated from the $(GA)_7$ *08032.

These analyses of the polymorphic complex microsatellite 50 bp 3' of the second exon of DRB1 have proved useful for the examination of the evolution of microsatellite sequences and DRB1 allelic lineage diversity, as well as for tracing the origins of and relationships between individual DRB1 alleles. The phylogenetic analyses of other sequences in introns-1 and −2 have also been highly informative for inferences about the evolution and allelic diversification of the DRB loci.

References

Bergström TF, Engkvist H, Erlandsson R, Josefsson A, Mack SJ, Erlich HA, Gyllensten U (1999a) Tracing the origin of HLA-DRB1 alleles by microsatellite polymorphism. Am J Hum Genet 64:1709-1718

Bergström TF, Erlandsson R, Engkvist H, Josefsson A, Erlich HA, Gyllensten U (1999b) Phylogenetic history of hominoid DRB loci and alleles inferred from intron sequences. Immunol Rev 167:351-365

Bergström TF, Josefsson A, Erlich HA, Gyllensten U (1998) Recent origin of HLA-DRB1 alleles and implications for human evolution. Nat Genet 18:237-242

Bontrop RE, Kenter M, Otting N, Jonker M (1993) Major histocompatibility complex polymorphism in humans and chimpanzees. J Med Primatol 22:50-59

Bontrop RE (1997) MHC class II genes of nonhuman primates. In: Blancher A, Klein J, Socha WW (eds) Molecular Biology and Evolution of Blood Group and MHC Antigens in Primates. Springer-Verlag, Berlin-Heidelberg, pp 358-371

Chimge N-O, Tanaka H, Kashiwase K, Ayush D, Tokunaga K, Saji H, Akaza T, Batsuuri J, Juji T (1996) The HLA system in the population of Mongolia. Tissue Antigens 49:477-483

Corell A, Morales P, Varela P, Paz-Artal E, Martin-Villa JM, Martinez-Laso J, Arnaiz-Villena A (1992) Allelic diversity at the primate major histocompatibility complex DRB6 locus. Immunogenetics 36:33-38 [published erratum appears in Immunogenetics (1992) 36:404-405]

Erlich HA, Gyllensten U (1991) Shared epitopes among HLA Class II alleles: Gene conversion, common ancestry, and balancing selection. Immunol Today 11:411-414

Erlich HA, Bergström TF, Stoneking M, Gyllensten U (1996) HLA sequence polymorphism and the origin of humans. Science 274:1552-1554

Epplen C, Santos EJM, Guerreiro JF, Helden P, Epplen JT (1997) Coding versus intron variability: extremely polymorphic HLA-DRB1 exons are flanked by specific composite microsatellites even in distant populations. Hum Genet 99:399-406

Fan W, Kasahara M, Gutknecht J, Klein D, Mayer WE, Jonker M, Klein J (1989) Shared class II MHC polymorphisms between humans and chimpanzees. Hum Immunol 26:107-121

Figueroa F, O'hUigin C, Inoko H, Klein J (1991) Primate DRB6 pseudogenes: clue to the evolutionary origin of the HLA-DR2 haplotype. Immunogenetics 34:324-337

Gao X, Bhatia K, Trent RJ, Serjeantson SW (1992) HLA-DR, DQ nucleotide sequence polymorphisms in five Melanesian populations. Tissue Antigens 40:31-37

Gao X, Zimmet P, Serjeantson SW (1992) HLA-DR, DQ sequence polymorphisms in Polynesians, Micronesians and Japanese. Hum Immunol 34:153-161

Garboczi DN, Ghosh P, Utz U, Fan QR, Biddison WE, Wiley DC (1996) Structure of the complex between human T-cell receptor, viral peptide and HLA-A2. Nature 384:134-141

Gongora R, Figueroa F, Klein J (1996) The HLA-DRB9 gene and the origin of HLA-DR haplotypes. Hum Immunol 51:23-31

Gyllensten U, Sundvall M, Ezcurra I, Erlich HA (1991) Genetic diversity at class II DRB loci of the primate MHC. J Immunol 146:4368-4376

Gyllensten U, Erlich HA (1993) MHC class II haplotypes and linkage disequilibrium in primates. Hum Immunol 36:1-10

Hughes AL, Nei M (1988) Pattern of nucleotide substitution at major histocompatibility complex class I loci reveals overdominant selection. Nature 335:167-170

Hughes AL, Nei M (1989) Pattern of nucleotide substitution at major histocompatibility complex class II loci: evidence for overdominant selection. Proc Natl Acad Sci USA 86:958-962

Jardetzky TS, Brown JH, Gorga JC, Stern LJ, Urban RG, Strominger JL, Wiley DC (1996) Crystallographic analysis of endogenous peptide associated with HLA-DR1 suggests a common, polyproline II-like conformation for bound peptides. Proc Natl Acad Sci USA 93:734-738

Jukes TH, Cantor CR (1969) (eds) Evolution of Protein Molecules. Academic Press, New York

Kageshita T, Naruse T, Hirai S, Ono T, Horikoshi T, Nakagawa H, Tamaki K, Hayashibe K, Ichihashi M, Nakayama J, Hori Y, Ozawa A, Miyahara M, Ohkido M, Inoko H (1997) Molecular genetic analysis of HLA class II alleles in Japanese patients with melanoma. Tissue Antigens 49:466-470

Kasahara M, Klein D, Vincek V, Sarapata DE, Klein J (1992) Comparative anatomy of the primate histocompatibility complex DR subregion: evidence for combination of DRB genes conserved across species. Genomics 14:340-349

Kenter M, Otting N, Anholts J, Jonker M, Schipper R, Bontrop RE (1992) MHC-DRB diversity of the chimpanzee (Pan troglodytes). Immunogenetics 37:1-11

Kenter M, Otting N, de weers M, Anholts J, Reiter C, Jonker M, Bontrop RE (1993) MHC-DRB and -DQA1 nucleotide sequences of three lowland gorillas: implications for the evolution of primate MHC class II haplotypes. Hum Immunol 36:205-218

Klein D, Vincek V, Kasahara M, Schönbach C, O'hUigin C, Klein J (1991) Gorilla major histocompatibility complex-DRB pseudogenes orthologous to HLA-DRB8. Hum Immunol 32:211-220

Kupfermann H, Mayer WE, O'hUigin C, Klein D, Klein J (1992) Shared polymorphism between gorilla and human major histocompatibility complex DRB loci. Hum Immunol 34:267-278

Li WH, Ellsworth DL, Krushkall J, Chang BHJ, Emmett DH (1996) Rates of nucleotide substitution in primates and rodents and the generation-time effect hypothesis. Mol Phyl Evol 5:182-187

Mack SJ, Erlich HA (1998) HLA class II polymorphism in the Ticuna of Brazil: Evolutionary implications of the DRB1*0807 allele. Tissue Antigens 51:41-50

Mayer WE, O'hUigin C, Zaleska-Ruiczynska Z, Klein J (1992) Trans-species origin of MHC-DRB polymorphism in the chimpanzee. Immunogenetics 37:12-23

Ohta (1995) Gene conversion vs. point mutations in generating variability at the antigen recognition sites of major histocompatibility complex loci. J Mol Evol 41:115-119

Petes TD, Greenwell PW, Dominska M (1997) Stabilization of microsatellite sequences by variant repeats in the yeast *Saccharomyces cerevisiae*. Genetics 146:491-498

Saitou N, Nei M (1987) The neighbor-joining method: a new method for reconstructing phylogenetic trees. Mol Biol Evol 4:406-425

Satta Y, Takahata N, Schönbach C, Gutknecht J, Klein J (1991) Calibrating evolutionary rates at major histocompatibility complex loci. In: Klein J, Klein D (eds) Molecular Evolution of the Major Histocompatibility Complex. Springer-Verlag, Berlin-Heidelberg, pp 51-62

Slierendregt BL, Kenter M, Orring N, Anholts J, Jonker M, Bontrop R (1993) Major histocompatibility complex class II haplotypes in a breeding colony of chimpanzees (*pan troglodytes*). Tissue Antigens 42:55-61

Stern LJ, Brown JH, Jardetzky TS, Gorga JC, Urban RG, Strominger JL, Wiley DC (1994) Crystal structure of the human class II MHC protein HLA-DR1 complexed with an influenza virus peptide. Nature 368:215-221

Sneath PHA, Sokal RR (1973) Numerical Taxonomy. Freeman, San Francisco

Titus-Trachtenberg EA, Rickards O, De Stefano GF, Erlich HA (1994) Analysis of HLA class II haplotypes in the Cayapa Indians of Ecuador: a novel DRB1 allele reveals evidence for convergent evolution and balancing selection at position 86. Am J Hum Genet 55:160-167

Vincek V, Klein D, Figueroa F, Hauptfeld V, Kasahara M, O'hUigin C, Mach B, Klein J (1992) The evolutionary origin of the HLA-DR3 haplotype. Immunogenetics 35:263-271

Wierdl M, Dominska M, Petes TD (1997) Microsatellite instability in yeast: dependence on the length of the microsatellite. Genetics 146:769-779

Yeager M, Hughes AL (1999) Evolution of the mammalian MHC: natural selection, recombination, and convergent evolution. Immunol Rev 167:45-58

The non-coding regions of HLA-DRB uncover inter-lineage recombinations as a mechanism of HLA diversification

Katja Kotsch and Rainer Blasczyk

Department of Transfusion Medicine, Hannover Medical School, Carl-Neuberg-Str. 1, 30625 Hannover, Germany

Summary. The mechanisms generating new alleles at the MHC loci are still unknown in detail and several proposals have been made in order to explain the extent of polymorphism. The patchwork pattern of polymorphism in the second exon of HLA-DRB1 recommends this locus as a model for the study of the potential of inter-allelic gene conversion. In general, the direct detection of gene-conversion like events based exclusively on exon sequence comparisons is always unreliable, since the identity of the putative donor allele remains unknown. In this study, we describe 4 alleles of the HLA-DRB1 gene whose intron regions give evidence for inter-lineage recombination events either strictly located at the second exon or involving the adjacent introns. Furthermore, we show that the non-coding regions provide important clues to the mechanisms of the generation of new alleles and our results indicate that inter-lineage recombinations may be hidden and are perhaps more frequent than currently expected.

Key words. MHC, Generation of diversity, Evolution

Introduction

The MHC class I and class II loci are the most highly polymorphic coding regions in the human genome and the extent and mechanisms of this genetic variability have always been a subject of great interest to evolutionists. Many proposals have been made concerning these mechanisms that may have originated this polymorphism. Some alleles have clearly arisen by point mutations but in general, the patchwork pattern of polymorphism at the HLA-DRB1 locus has been proposed to have been generated by intra-locus short sequence exchanges within the coding regions (Gorski and March 1986; Erlich and Gyllensten 1991; Gyllensten et al. 1991; Belich et al. 1992; Andersson and Mikko 1995; Ohta 1995; Parham et al. 1995; Hickson and Cann 1997). For instance, the high sequence

347

similarity between alleles of the DR52 haplotype group indicates that the presence of shared sequence motifs could be due to sequence exchanges between members of different lineages.

No direct evidence for inter-allelic gene conversion events have been reported, since Zangenberg et al. (1995) described a model for estimating the frequency of variant DPB1 sequences that have been created by inter-allelic gene conversions in the germline. Nevertheless, the comparison of exon sequences is not sufficient for the direct detection of gene-conversion events, since the identity of the putative donor allele remains unknown.

It is generally assumed that introns do not seem to be involved in recombination events (Bergström et al. 1998). The extensive sharing of coding motifs among DRB alleles suggests that recombination events take place between different lineages, and that the allelic variation is generated mainly by sequence exchanges in that part of exon 2 encoding the α-helix (cds 52-92) (Gyllensten et al. 1991). This observation as well as the limited 5' conversions indicate that the exon 2 diversity must have been generated by a mechanism not extending into the introns. We have recently uncovered the remarkably conserved diversity of the HLA class I as well as the HLA-DRB introns, which is lineage specific rather than allele specific (Kotsch et al. 1997; Blasczyk et al. 1998). The finding that all alleles within a lineage share the same flanking intron sequences indicates that single crossing-over events do not constitute the main mechanism for the origin of allelic variation and we have not found any evidence that recombination events, diversifying the exons, involve the intron regions.

Now, in this article we present four HLA-DRB1 alleles (DRB1*1313, DRB1*1402, DRB1*1406, DRB1*1410) which have been generated through a recombination event between two different lineages identified by a new combination of intron 1 and intron 2 motifs.

Inter-lineage recombination events

We have analyzed the nucleotide sequences of the 3' 480 bp intron 1 fragment and the 5' 380 bp intron 2 fragment from five HLA-DRB1 (DRB1*1122, DRB1*1313, DRB1*1402, DRB1*1406, DRB1*1410) alleles. (The nucleotide sequences reported in this article have been submitted to the European Molecular Biology Laboratory Data Library.) The PCR primers, the PCR protocol as well as the cycle sequencing protocol were previously described in detail (Blasczyk et al. 1998). Each allele was sequenced from three different PCR products for the regions of interest in both directions.

DRB1*1122

One of the most informative candidates that could probably indicate whether recombination events extend into the introns or not was DRB1*1122, one of the

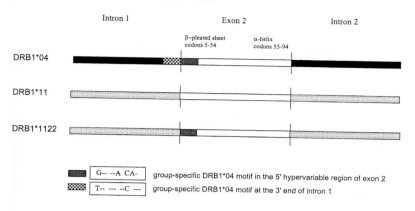

Fig. 1. Schematic exon-intron organization of DRB1. The different shaded boxes demonstrate the characteristic intron regions of a DR4 and of a DR11. The second exon region is symbolized by an open box. The allele DRB1*1122 displays the typical DR4 motif at the 5' hypervariable region of exon 2 but lacks the group-specific DR4 motif at the 3' end of intron 1.

alleles which first hypervariable region bears the DRB1*04 group-specific motif (Fig. 1). The DR4 group displays also individual motifs at the 3' end of intron 1, only 4 bp upstream from the start of exon 2, which should be present in DRB1 *1122 in the case of a recombination event involving the adjacent intron 1 region. However, sequence analysis of the directly adjacent intron regions yielded no DRB1*04 characteristics neither in intron 1 nor in intron 2 proving an inter-lineage gene conversion completely restricted to the coding region.

DRB1*1410

In contrast to DRB1*1122, in DRB1*1410, which carries the same first hypervariable DRB1*04 motif, two characteristic non-coding DRB1*04 instead of DRB1*14 motifs could be found in the adjacent part of the first intron, 4 and 12 bp upstream of the second exon (Fig. 2). In the remaining parts of the investigated intron regions, this allele displays the characteristic DR14 motifs

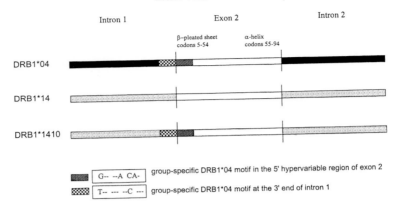

Fig. 2. In contrast to DRB1*1122, in DRB1*1410 the group-specific DR4 motif at the 3' end of intron 1 could be found indicating a recombination event involving the adjacent intron region.

and no other DR4 motif could be identified. Despite this inter-lineage recombinations, DRB1*1122 and DRB1*1410 have kept their lineage- specificity which is identical in the exon and intron sequences.

DRB1*1402 and DRB1*1406

Compared to the exon sequences of these alleles obviously created by an inter-lineage recombination event, the coding regions of DRB1*1402 and DRB1*1406 appear to be the result of a normal intra-lineage exchange. Both alleles do not show a characteristic motif of any other lineage at the 5' hypervariable region. However, sequence analysis of the first and second introns revealed exclusively DRB1*03 characteristics as shown in Figure 3.

This finding unequivocally indicates that both alleles are in reality the result of an inter-lineage recombination involving a DRB1*03 as the acceptor allele and any DRB allele bearing the consensus from codon 71-77 as the donor allele. This observation is consistent with the detection of novel alleles, which seemed to be generated by segmental exchanges in the codons of the second exon encoding theα-helix (Gyllensten et al. 1991). In both cases, the alleles have kept the intron

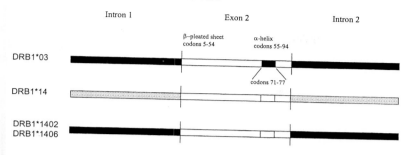

Fig. 3. Both alleles, DRB1*1402 and DRB1*1406, display in intron 1 as well as in intron 2 exclusively DR3 characteristics. This finding gives evidence for a recombination between a DRB1*03 as acceptor and any other allele bearing the consensus from codons 71-77 as donor.

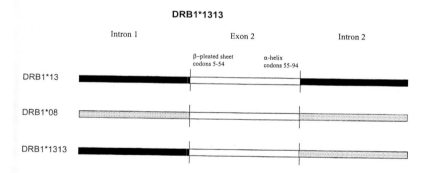

Fig. 4. The allele DRB1*1313 displays characteristic sequence motifs in intron 1 in contrast to intron 2, where this allele shows exclusively DR8 characteristics. Consequently, this allele seems to be the result of a DR13/DR8 recombination.

sequences of the acceptor allele, but belong - based on exon sequences- to the group of the donor allele.

In particular, the strong intron and exon sequence homology between the DR3, DR13 and DR14 groups indicates that alleles of these groups may have arisen by inter-lineage recombination events. Thus, in the case of DRB1*1402 and DRB1*1406, the definition of the lineage-specificity based exclusively on exon-sequences is incorrect.

DRB1*1313

Similar to the DRB1*1402 and DRB1*1406 alleles, the DRB1*1313 displays the typical DR52 associated group-specific motif at the 5' hypervariable region in the second exon, but the analysis of both adjacent introns has shown that intron 1 carries DR13 characteristics whereas intron 2 demonstrates typical DR8 motifs (Fig. 4).

Interestingly, this sample, displaying DR15 on the other haplotype, lacks the DRB3 gene. The lack of DRB3 indicates that the mechanism generating this allele is either a double crossing over event with DR8 as the acceptor or a single crossing over event with DR13 as the acceptor and DR8 as the telomeric donor. In conclusion, this allele seems to be the result of a single crossing over event between DRB1*13 and DRB1*08 at their coding regions. This recombination event could give an explanation for the absence of the DRB3 gene.

Conclusion

The intron sequences provide important clues to the mechanisms for the generation of HLA polymorphism and their analysis may deliver further insight into the genetic relationship between different alleles. It is expected, that alleles from a given allelic lineage, which encode the same amino acid positions 5-13 of exon 2, always carry the same or a closely related intron framework (Bergström et al. 1998). This region encodes the first part of the β-pleated sheet of the peptide-binding groove (codons 5-54) (Gyllensten et al. 1991) and does not seem to be generally involved in exchanges between allelic lineages. However, sequence analysis of the non-coding regions of DRB*1122 and DRB1*1410 has shown that alleles belonging to the same allelic lineage do not necessarily share an identical intron framework and the latter demonstrates that gene conversion does not need to be restricted to the region of the second exon encoding the α-helix. In the case of the DRB1*1402 and DRB1*1406, we have shown that inter-lineage recombinations may be hidden and are perhaps more frequent than currently expected. This should be considered when lineage-specificity or names of alleles are defined. In summary, we believe that intron analysis provides important information elucidating the origin and lineage specificity of alleles as well as the mechanisms generating the exon 2 diversity. The sequences of introns should be considered when looking at the molecular evolution of HLA specificities.

References

Andersson L, Mikko S (1995) Generation of MHC class II diversity by intra- and intergenic recombination. Immunol Rev 143:5-12

Belich MP, Madrigal JA, Hildebrand WH, Zemmour J, Williams RC, Luz R, Petzl-Erler ML, Parham P (1992) Unusual HLA-B alleles in two tribes of Brasilian Indians. Nature 357:326-329

Bergström TF, Josefsson A, Erlich HA, Gyllensten UB (1998) Recent origin of HLA-DRB1 alleles and implications for human evolution. Nat Genet 18:237-242

Blasczyk R, Kotsch K, Wehling J (1998) The nature of polymorphism of the HLA-DRB intron sequences is lineage specific. Tissue Antigens 52:19-26

Erlich HA, Gyllensten UB (1991) Shared epitopes among HLA class II alleles: gene conversion, common ancestry, and balancing selection. Immunol Today 11:411-414

Gorski J, Mach B (1986) Polymorphism of human Ia antigens: gene conversion between two DRβ loci results in a new HLA-D/DR specificity. Nature 322:67-70

Gyllensten UB, Sundvall M, Ezcurra I, Erlich HA (1991) Genetic diversity at class II DRB loci of the primate MHC. J Immunol 146:4368-4376

Gyllensten UB, Sundvall M, Erlich HA (1991) Allelic diversity is generated by intraexon sequence exchanges at the DRB1 locus of primates. Proc Natl Acad Sci USA 88:3686-3690

Hickson RE, Cann RL (1997) MHC allelic diversity and modern human origins. J Mol Evol 45:589-598

Kotsch K, Wehling J, Blasczyk R (1997) Sequencing of HLA class I genes based on the conserved diversity of the non-coding regions: sequencing based typing of the HLA-A gene. Tissue Antigens 50:178-191

Ohta T (1995) Gene conversion vs. point mutation in generating variability at the antigen recognition site of major histocompatibility complex loci. J Mol Evol 41:115-119

Parham P, Adams EJ, Arnett KL (1995) The origins of HLA-A, B, C polymorphism. Immunol Rev 143:141-180

Zangenberg G, Huang MM, Arnheim N, Erlich HA (1995) New HLA-DPB1 alleles generated by interallelic gene conversion detected by analysis of sperm. Nat Genet 10:407-414

Conversion or convergence? Introns of primate *DRB* genes tell the true story

Karin Kriener, Colm O'hUigin, and Jan Klein

Max-Planck-Institut für Biologie, Abteilung Immungenetik, Corrensstrasse 42,
D-72076 Tübingen, Germany

Summary. The phylogeny of *DRB* exon 2 sequences obtained from Old World monkeys, New World monkeys, and prosimians is incongruent with the phylogeny of other parts of the genes (introns 1, 2, 4, and 5, as well as exons 3 and 6). While the exon 2 sequences of the three primate groups are intermingled and seem to indicate the existence of deeply divergent trans-specific allelic lineages, the sequences of the other parts reveal a monophyletic relationship of the *DRB* genes in each of the three groups. The incongruence of the exon 2 sequences is largely ascribable to short, characteristic sequence stretches, the motifs, specifying the peptide-binding region of the DRB protein. Several of these motifs are widely shared among exon 2 sequences of the three primate groups. Statistical and phylogenetic analysis of the motifs reveals that they were assembled independently in the prosimians, New World, and Old World monkeys. The assembly occurred in a stepwise fashion by the accumulation of point mutations favored by natural selection. The motif provides a striking example of convergent evolution at the molecular level. The involvement of gene conversion in the process could be ruled out.

Key words. Convergent evolution, Gene conversion, *Mhc-DRB* genes, Primates

Major histocompatibility complex (*Mhc*) sequence alignments commonly display two conspicuous features. First, the α1 and α2 domain (and the corresponding exon 2 and exon 3) sequences of class I molecules, as well as the β1 domain (exon 2) and sometimes also the α1 domain (exon 2) sequences of the class II molecules contain characteristic motifs — short stretches of multiple substitutions differentiating allelic products (allomorphs) and the alleles themselves (Klein and O'hUigin 1995). Second, the motifs often show a mosaic distribution pattern among the various alleles (McDevitt et al. 1984; Gyllensten et al. 1991). These patterns of sequence differences are generally interpreted as being the result of motif shuffling by a process referred to as "gene conversion" (Pease et al. 1983; Weiss et al. 1983). The nature of the process is rarely specified and, indeed, behind the process is a rather nebulous concept rife with numerous *ad hoc* assumptions (Klein and O'hUigin 1995). Gene conversion is postulated to involve short stretches or long segments, alleles at the same locus or genes at different loci, very similar or very different genes, neighboring loci or loci long distances apart — whatever explanation a particular data set seems to demand (e.g., see Pease et al. 1983; Weiss et al. 1983; Gorski and Mach 1986; Wu et al. 1986; Gyllensten et al. 1991; She et al. 1991; Högstrand and Böhme 1994). At the population level, gene conversion has been made responsible for the divergence of genes, their homogenization, or both (e.g., see Andersson et al. 1987; Parham et al. 1989; Gyllensten et al. 1991; She et al. 1991). Not infrequently, gene conversion is even invoked to explain why two alleles differ by a single nucleotide substitution without considering the possibility that such a difference could be the result of point mutation (Schwaiger et al. 1993).

In most situations, it is in fact not possible to decide whether an observed pattern is the result of some form of nonreciprocal segmental exchange ("gene conversion") or of a mutational process. Here, however, we describe a situation in which such a distinction *is* possible and show that in this case the data clearly favor the alternative to gene conversion. The result of the study is relevant not only for the understanding of the process generating *Mhc* polymorphism, but also for the interpretation of the persistence of allelic lineages at *Mhc* loci of different species.

*HLA-DRB1*03* is one of 13 allelic lineages at the human *DRB1* locus; only three allelic lineages have been classified so far at the human *DRB3* locus (Bodmer et al. 1999). *HLA-DRB1* and *HLA-DRB3* are two of nine human *DRB* (class II *B*) loci, of which, however, at least four are occupied by pseudogenes or gene fragments (Klein et al. 1991). The *DRB1*03* lineage is represented in the human population by at least 16 variants (alleles); altogether, more than 220 alleles are known at the *DRB1* locus (Bodmer et al. 1999). The proteins encoded in the *HLA-DRB1*03* alleles usually differ from other *HLA-DRB*-encoded proteins at positions 10, 11, 12, 32, 67, 73, and 77, at which they have the amino acid residues, Y, S, T, H, L, G, and N, respectively (see Marsh 1999). The first three of these residues are particularly characteristic of the DRB1*03 proteins; they can be referred to as the YST motif (Klein and O'hUigin 1995). The 16 alleles at the *HLA-DRB1*03* locus differ from one another by a small number of substitutions outside these diagnostic positions. The proteins encoded in the *HLA-DRB3* genes are characterized by the amino acid residues L, L or R, L, G, Q or R, and N at positions 10, 11, 67, 73, 74, and 77, respectively, and thus by the LL/LR and GQ/GR motif at positions 10, 11, and 73, 74, respectively.

Analysis of *DRB* genes from apes (Gyllensten et al. 1991; Kenter et al. 1992; Kupfermann et al. 1992), Old World monkeys (OWM: Zhu et al. 1991; Slierendregt et al. 1992, 1994), New World monkeys (NWM; Trtková et al. 1993), and prosimians (Figueroa et al. 1994) has revealed the presence in all these taxa of exon 2 sequences resembling the *HLA-DRB1*03* sequences. This resemblance lies in the sharing of diagnostic amino acid residues in the translated protein, in particular the YST motif, and in the intermingling with the *HLA-DRB1*03* sequences on phylogenetic trees. Accordingly, all these nonhuman-primate sequences have been interpreted as representing members of the *DRB1*03* allelic lineage. The YST motif has, in addition, been found in DRB homologs of artiodactyls (Andersson et al. 1991), perissodactyls (Gustafsson and Andersson 1994), rodents (Mengle-Gaw and McDevitt 1985; Edwards et al. 1997), and marsupials (Stone et al. 1999). Here, too, the observation has been interpreted as evidence for the persistence of the motif throughout the evolution of all these taxa from a common ancestor. Similarly, based on exon 2 sequences, the presence of the *DRB3* locus has been reported in apes (Gyllensten et al. 1991; Kenter et al. 1992; Kupfermann et al. 1992), OWM (Slierendregt et al. 1992, 1994), NWM (Trtková et al. 1993; Bidwell et al. 1994; Gyllensten et al. 1994), and prosimians (Figueroa et al. 1994).

In all these cases, the sequences were taken at their face value under the assumption that the observed similarities reflected a direct descent from a common ancestor. The first indication that the evolution of the *DRB* genes might be more compounded than it originally appeared came from the study of intron 2 and exon 3, as well as exon 6 of NWM genes. Trtková and her coworkers (1995) observed that all these sequences yielded different phylogenies to the exon 2 sequences. While the NWM exon 2 sequences mingled on dendrograms with catarrhine exon 2 sequences, the intron 2, exon 3, and exon 6 sequences separated cleanly into platyrrhine and catarrhine clades. Similar results were obtained by Kupfermann and her colleagues (1999) with catarrhine, platyrrhine, and prosimian intron 4 and 5 *DRB* sequences. In this case, too, no mingling of sequences from different primate sub- and infraorders occurred in the constructed dendrograms. However, in neither of these two studies could the exon 2 sequences be related directly to the intron or exon 3 and exon 6 sequences. The possibility remained, therefore, that the two sets of sequences (exon 2 versus the rest) were derived from different genes, one set representing old allelic lineages shared by different primate sub- and infraorders, and the other corresponding to paralogous loci.

To examine this possibility, we used the polymerase chain reaction (PCR) to obtain catarrhine, platyrrhine, and prosimian exon 2 sequences together with the flanking intron sequences. Some of these sequences are described in another report (K. Kriener, C. O'hUigin, H. Tichy, J. Klein, in preparation), others are described here for the first time. A complete list of the sequences appears in Table 1. Figure 1 illustrates the motifs on which the original assignment of the exon 2 sequences to the allelic lineages and loci was based. The amino acid sequences are shown for those NWM *DRB* genes for which the entire intron 1 was obtained in the PCRs. Selected human exon 2 sequences were added to the alignment. Figure 2 shows the corresponding *DRB* intron 1 sequences from both the 5' part (Fig. 2A) and the 3' part (Fig. 2B). The intron 2 sequences are described in Kriener et al. (in preparation). In Figure 2A, sequences obtained from prosimian genes are included; these sequences are not included in Figure 2B, because they are largely unalignable in this part of the intron with the anthropoid sequences. The results of the

Table 1. Primate *DRB* genes for which intron sequences are available

Gene designation	Species	Species abbreviation	Source[a]	5'end of intron 1 (bp)[b]	3'end of intron 1 (bp)[b]	Intron 2 (bp)[b]
Old World monkeys (OWM)						
Mamu-DRB6*0106	*Macaca mulatta*	*Mamu*	1	-	1470	180
Mamu-DRB*W402			1	-	1470	180
Mafa-DRB*W3301[c]	*Macaca fascicularis*	*Mafa*	1	-	250	180
Mafa-DRB*W301[c]			1	-	-	180
Maar-DRB1*0301[c]	*Macaca arctoides*	*Maar*	1	-	250	180
Maar-DRB1*0302[c]			1	-	250	180
Maar-DRB5*0301[c]			1	-	850	180
Maar-DRB*W601[c]			1	-	810	180
Maar-DRB1*0701[c]			1	-	-	180
New World monkeys (NWM)						
Saoe-DRB1*0303	*Saguinus oedipus*	*Saoe*	1,2	600	1470[d]	180
Saoe-DRB3*0501			1	-	1470[d]	180
Saoe-DRB11*0102			1	-	1470[d]	180
Saoe-DRB11*0105			1	-	1470[d]	180
Saoe-DRB*W2209			1,2	600	1470[d]	180
Saoe-DRB*W2210[c]			1	-	1470[d]	-
Caja-DRB1*0301	*Callithrix jacchus*	*Caja*	2	600	1470[d]	-
Caja-DRB1*0303			1	-	1470[d]	-
Caja-DRB1*0304			1	-	1470[d]	180
Caja-DRB*W1201			1	-	-	180
Caja-DRB*W1612			2	600	1470[d]	-
Camo-DRB1*0301	*Callicebus moloch*	*Camo*	1	-	-	180
Camo-DRB1*0302			1,2	600	1470	180
Camo-DRB3*0501			2	600	1470[d]	-
Camo-DRB3*0503[c]			1	-	1470[d]	180
Camo-DRB3*0504[c]			1	-	1470[d]	180
Camo-DRB3*0701			1,2	600	1470	-
Camo-DRB*W1401			2	600	450	-
Camo-DRB11*0101			2	600	760	-
Ceap-DRB*W1301	*Cebus apella*	*Ceap*	1,2	600	1470[d]	180
Ceap-DRB*W1502			1,2	600	1470[d]	-
Ceap-DRB*W3201[c]			1,2	600	1470	-
Haplorrhini						
Tasy-DRB*W101[c]	*Tarsius syrichta*	*Tasy*	2	600	not alignable	-
Taba-DRB*W201[c]	*Tarsius bancanus*	*Taba*	2	600	not alignable	-
Strepsirrhini						
Gamo-DRB*W301[c]	*Galago moholi*	*Gamo*	2	600	not alignable	-
Gamo-DRB*W401[c]			2	600	not alignable	-
Gamo-DRB*W501[c]			2	600	not alignable	-

[a]Source 1 - Kriener et al. in preparation; source 2 - unpublished data.
[b]A dash (-) indicates unavailability of data.
[c]Newly identified genes; nonsimian genes are named independently of the simian sequences in the nonhuman primate database (R. Bontrop, personal communication).
[d]Including an 870 bp deletion.

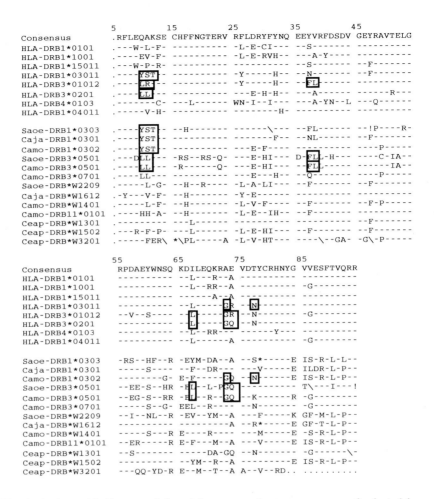

Fig. 1. Amino acid alignment inferred from second exon sequences of selected human and New World monkey (NWM) *DRB* genes. The amino acid residues are given in the IUPAC-IUB single-letter code. The simple majority consensus sequence is given at the top; identity with the consensus is indicated by a dash (-) and absence of sequence information by a dot (.). An asterisk (*), a backslash (\), and an exclamation mark (!) indicate a residue deletion, a frame shifting deletion, and a stop codon, respectively. The sites are numbered from the N-terminal residue of the β1 domain. The NWM sequences are derived from those genes for which the complete intron 1 was amplified. The species abbreviations are explained in Table 1. The amino acid motifs shared between human and NWM DRB peptides are highlighted.

phylogenetic analysis are summarized in the form of dendrograms in Figures 3A (exon 2), 3B (5' part of intron 1), 3C (3' part of intron 1), and 3D (5' part of intron 1, including prosimian sequences). The discussion that follows is based on both the intron 1 data presented here and the intron 2 data described in Kriener et al. (in preparation).

The combined data of the present and the related studies (Trtková et al. 1995; Kupfermann et al. 1999; Kriener et al. in preparation) reveal full congruence of phylogenies based on different parts of the primate *DRB* genes with the exception of the exon 2-based phylogeny. In a direct test based on contiguous or overlapping sequences, intron 1 phylogeny was shown to be congruent with intron 2 phylogeny, but incongruent with exon 2 phylogeny (Kriener et al. in preparation and this communication). In indirect tests based on noncontiguous and nonoverlapping sequences, the intron 1 /intron 2 phylogenies were congruent with intron 4 and 5, as well as with exon 3 and 6 phylogenies in terms of monophyly (Trtková et al. 1995; Kupfermann et al. 1999). These observations raise two principal questions. First, *why* does exon 2 evolve discordantly from the rest of the *DRB* gene; and second, what mechanism underlies the incongruent evolution? We deal with both of these questions in turn.

Regarding the first question, we may begin by asking: what is special about exon 2 in terms of *DRB* gene evolution? The answer is clear: it is the only part of the *DRB* gene that evolves under balancing selection (Hughes and Nei 1989). Since the selection is presumably related to the function of the exon 2-encoded β1 domain and since this function is to bind peptides for presentation to helper T lymphocytes (Klein and Horejsí 1998), the incongruence must somehow be connected to peptide presentation. This conclusion is consistent with and supported by the observation that, in fact, only some parts of exon 2, namely those specifying the motifs, evolve incongruently with the rest of the gene. When the motif-encoding parts are removed from the sequence and the remainder of the exon is used to draw a tree, the resulting phylogeny is more congruent with that of the rest of the gene (not shown). The discrepancies that remain between the phylogenies can be attributed to the shortness of the truncated exon 2 sequences. The discordance between the motif-specifying parts and the rest of exon 2 is underscored further by the fact that most NWM *DRB* sequences share a characteristic stretch in the 3' part of exon 2, which groups them together, separate from catarrhine exon 2 sequences (Trtková et al. 1993). Since the motifs are critically involved in peptide binding by DRB molecules (Brown et al. 1988), their discordant evolution is apparently somehow connected to their function. While the rest of the *DRB* genes are allowed to diverge, the evolution of the motifs is constrained by their function. Why there is a need for the presence of similar peptide-binding motifs in diverse taxa is not clear. Perhaps, there are certain parasites or certain species of the normal microbial flora (Klein and O'hUigin 1994) that are common to most primates (eutherian mammals?) or perhaps there are certain peptides shared by common primate parasites, and the motifs are necessary to ensure their binding to DRB molecules.

Fig. 2A.

```
                 1          11         21         31         41         51         61         71         81         91         101        111        121
CONSENSUS ==>    GTAAGTGCAC ATTGTGGGTG CTGAGCTACT ATGGGGTGGG G*AAAATAGG GAGTTTTGTT *AACATTGTG CCCAGGCCAT GTCCCTTAAG AAATTGTGAC GTTTTCTTCA GAGATTGC** ****CCATCT
HLA-DRB1*0101    ---------- ----C----- ---------- ---------- -*-------- ---------- *--------- ---------- ---------- ---------- ---------- --------C* *****-----
HLA-DRB1*1001    ---------- ---------- ---------- ---------- -*-------- ---------- *--------- ---------- --A------- ---------- --A--T---- ---------* *****-----
HLA-DRB1*15011   ---------- ----C----- ---------- ---------- -*-------- ---------- *--------- ---------- ---------- ---------- ---------- ---------* *****-----
HLA-DRB1*03011   ---------- ---------- ---------- ---------- -*-------- ---------- *--------- ---------- --A------- ---------- --A--T---- ---------* *****-----
HLA-DRB3*01012   -----A---- ----A----- ---------- ---------- -G--T----- -----G---- *--------- ---------- ---------- ---------G ---------- --------T* *****-----
HLA-DRB4*0103    ---------- ------AC-- -A-------- ---------- -C-------- A--------- *--------- -------T-- ---T----T- ---------- AC-------- ---------* *****-----
HLA-DRB1*04011   ---------- ---------- ----C----- ---------- **----A--- -----G---- *--------- ---------- ---------- ---G------ A--------- -G-------* *****-----
Saoe-DRB1*0303   ---------- ---------- ---------- ---------- -*-------- ---------- *--------- ---------- ---------C --T------T ---------- --------** ****------
Saoe-DRB*W2209   ---------- -C-------- ---------- ---------- -*------T- ---------- *--------- ---------- --------CC --T------T ---------- -------A** ****------
Caja-DRB1*0301   ---------- ---------- ---------- ---------- -*-------- --T------- *--------- ---------- ---------C --T------T ---------- --------** ****------
Caja-DRB*W1612   ---------- ---------- ---------- ---------- -*-------- ---------- *--------- T--------- ---------C --T------T --------T- ----C---** ****------
Camo-DRB1*0302   ---------- ---------- -------T-- ---------- -*-------- ---------- *--------- ---------- ---------- ---------T ---------- ------C-** ****------
Camo-DRB3*0701   ---------- ---------- -------T-- -----*---- -*-------- ---------- *--------- ---------- ----G----- --------G- A----A--G- ------C-** ****------
Camo-DRB3*0501   ---------- ---------- ---------- -------A-- -*-------- ---------- *--------A ---------- ---------C G-T------T ---------- --------** ****------
Camo-DRB*W1401   ---------- ---------- -------T-- -----*---- -*-------- ---------- *--------A ------G--- ---------- ---------- A----A---- ------C-** ****------
Camo-DRB11*0101  ---------- ---------- -------T-- ---------- -*-------- ---------- *--------- ---------- ----G----- --------G- A----A--G- ------C-** ****------
Ceap-DRB*W1301   ---------- ---------A ---------- ---------- -*-------- ---------- *--------- ---------- ---------C --T------T ---------- ---------* ****------
Ceap-DRB*W3201   -A-------- ---------A ---------- ---A------ T*-------- ---C-AAA-- TGTT-C---- ---------- A--------- ---------- ---------- -G-----T-C GTCT------
Ceap-DRB*W1502   ---------- ---------- ---------- ---------- -*-------- A--------- *--------- ---T---TG- ---------C --T------- A--------- --------** ****------
Tasy-DRB*W101    -------T-- -C-T------ --------TG ---------- ----G-A--- --A--AAC-- -GT--C---A T--T---T-- A--T------ --G------- A----A---- --T-GGA-** ****TT----
Taba-DRB*W201    -------T-- -C-T------ --------TG ---------- ----G-A--- --A--AAC-- -GT--C---A T--T---T-- A--T------ --G------- A----A---- --T-GCA-** ****T-----
Gamo-DRB*W301    --------C- -CAT---A-- ---G--A-T- GG---C-ACA --T-G-A-A- -----AACC- -G****---- TT----TT-- --T------- ---C--C--- A--------- -G--CCA-** ****-T----
Gamo-DRB*W401    A-------C- -CAT---A-A ---G----T- GG---T---- ----G-A--- -----AAC-- -GG-TC---- ------TT-- --T------A ---------G A--------- -G--CCA-** ****-T----
Gamo-DRB*W501    --------C- TCAT---A-- -G-G--G-T- GG---T---A -*--G-A--- ----GAAC-- -GT-TC---- T-----TT-- --T----T-A ---------T ---------- -G--CCA-** ****-T----
Roae             ------A--- -CAT-AT-G- T-*----*T- G-------T- ---GG-A--- --A--AGC-- -T*T-C-A-- T--------- --T----TGT ---C------ A--------- AT--CCA-** ****-T-CA-

                 131        141        151        161        171        181        191        201        211        221        231        241        251
CONSENSUS ==>    TTATCAT*TG GATCCCAAAT TATTT**CCA CCACAAAAGG AGCTTGGCTA CTTGCCCTCT CCATGAGACT TGTGTAAGGG GCCTCCATAC AGGCCA**TT TCTTCTCAAA TCTCCACCAA TAAAACCTT*
HLA-DRB1*0101    ---C---*-- ---------- -----**--T ---T------ -------G-- ---------- T--------- ---------- ----TTTC-- -A-T--**-- ----T----- ---------- ---------*
HLA-DRB1*1001    -------A-- ---------- A----CC--- ---------- ---------- -------A-- ---------- ---------- ---------- ---T--**-- ---A------ ---------- ---------*
HLA-DRB1*15011   -------*-- ---------- -----**--T ---T------ -------G-- ---------- T--------- ---------- ----TTGC-- -A-T--**-- ----T----- ---------- ---------*
HLA-DRB1*03011   ---------- ---------- A--------- ---------- ---------- -------A-- ---------- ---------- ---------- ---T--**-- ---A------ ---------- ---------*
HLA-DRB3*01012   ------*-G- ----G----- -----**--T ---------- -A-------- ---------- ---------- ---------- ----TTG--- ------**-- ---------- ---------- ---------*
HLA-DRB4*0103    ------G--- ---------- -----**--- ---------- C--------- ----T---G- --------T- -------A-- ----A--C-- ---T--**-- ---------- ---------- --------A*
HLA-DRB1*04011   ---------- ---------- -----**--- -------T-- -A-------- --------A- T--------- *------A-- ----TTG--- ------**-G -T-A--T--- ----T----- ---------*
Saoe-DRB1*0303   -------*-- ---------- -----**--- ---------- ---------- ----T----- ---------- ---------- ---A-A-C-- ------**-- ---------- -T--A--A-- G--------*
Saoe-DRB*W2209   ----A--*-- ---------- -----**--- ---------- ------C--- ---------- ---------- ---------- ---AGA---- -----G**-- ---------- -T-------G G--------*
Caja-DRB1*0301   -------*-- ---------- -----**--- -----G---- ---------- ------G--- ---------- ---------- ---A-G---- ------**-- ---------- -T-------- A--------*
Caja-DRB*W1612   -------*-- ---------- -----**--- -----G---- ---------- ------G--- ---------- ---------- ---A-G---- ------**-- ---------- -T-------G G--------*
Camo-DRB1*0302   -------*-- ---------- G----**--- ------G-A- ---C----C- --------T- -----G---- ---------- ----T-G--- ------**-- ---------- --A------- G--------*
Camo-DRB3*0701   -------*-- ---------- G----**--- ------G-A- ---C----C- ------G--- -----G---- ---------- ----T----- ------**-- ---------- --A------- G--------*
Camo-DRB3*0501   -*-----*-- ---------- -----**--C ---------- ---------- -----TG--- -----GA--- ---------- ---GTG---- ------**-- ---------- -T----TG-- G--------*
Camo-DRB*W1401   -------*-- ---------- G----**--- ------G-A- ---------- ------G--- -----G---- ---------- ----T-G--- ------**-- ---------- --A------- G--------*
Camo-DRB11*0101  -------*-- ---------- G----**--- ------G-A- ---C----C- ------G--- -----G---- ---------- ----T-G--- ------**-- ---------- --A------- G--------*
Ceap-DRB*W1301   -------*-- ---------- -----*--TC -ACA-T-*-- --T-AA---- ------T--- ---------- *------A-- ---A-A---- ------**-- ------T--- -T---*---- G----TA--*
Ceap-DRB*W3201   ---------- ---A------ -----*--AC -ACAG----- ------A-A- ---------- T--------- -------T-- -T-A-A---- ------**-- ----T--C-G -T----A--- -----G---*
Ceap-DRB*W1502   -------*-- ---------- -----*--AC -AC----*-- ---------- ---------- --G------- ----G----- ---A-G---- --C---**-- ---G------ -T-------- G----G---*
Tasy-DRB*W101    ---------- AT----T-*- -----**--- -A-GC----- ---C--*--- -----T--A- A--------C --CA---A-- ---------T -A-TA-**-- ----T----- -T---T---- --------C*
Taba-DRB*W201    C--------- AG----T-*- -----**--- -A-GC----- ---C--*--- -----T--A- A--------C --AA---A-- ---------T -A-TAG**-G ---------- -T---T---G -----A--C*
Gamo-DRB*W301    --TCT----- -C------G- -----**--- -A-T------ ---C------ ------T-T- -********* --******** ********** ********-G ---------- -----TG--G ------A-CT
Gamo-DRB*W401    ---------- -T------G- -----T**-- -A-------- -C-C-A---- ------T-TC -T------TG --******** ********** ********-G ---------- -A---TT--G **********
Gamo-DRB*W501    ------A-A- -CG--*--GG G-A--**--- -A-------- ---C----GG -G----T-TC ---------C C-******** ********** ********-G ----**---- ----*TT--G G----AAAC-
Roae             -------T-- ----TTC-T- -C---**-T- GA-------- G--C-A---G T--------* T*G-----AC ---AC---C- A-*------T -TAGGTCA-- ---------C -----TT--G -----T--C-
```

Fig. 2A. (continued)

362

Fig. 2A. (continued)

```
              521        531        541        551        561        571        581
CONSENSUS ==>  ***TACCCAG TGGCCTCCCC *ATTATCTCC TTTCTTTTCT TTCTGAACTC CAATGTTTAT AAAGCCTGT
HLA-DRB1*0101  ***G-----  --------   *-----     ---------  --G------  ---------  --G----
HLA-DRB1*1001  ...G.....  ........   .........  .........  ---------  ---------  -------
HLA-DRB1*15011 ***G-----  --------   *-----     .........  --G------  ---------  --G----
HLA-DRB1*03011 ***G-----  --------   *--A--     ---------  ---------  ---------  -------
HLA-DRB3*01012 ***G-----  ------T-   *-G---     ---------  ---------  ---------  -------
HLA-DRB4*0103  ***G-T---  --------   *-----     ---------  ---------  ---------  -------
HLA-DRB1*04011 ***----    --------   *--G-T-    ---------  ---------  --G------  -------
Saoe-DRB1*0303 ***----    -G-----    *--G---    ---------  --T------  ---------  --T----
Saoe-DRB*W2209 ***----    -G-----T-  *-G--G--   ---------  --T------  ---------  --T-T--
Caja-DRB1*0301 ***----    --A-----   *--G---    ---------  ----A----  --G------  -T-T---
Caja-DRB*W1612 ***----    ---TG----  *-----     ---------  ----A----  ---------  -A-T-T-
Camo-DRB1*0302 ***GG---   -----TT-   *-G---     ---------  ----C--T   ------A--  --T---T-
Camo-DRB3*0701 ***GG---   ----TTT-   *-G---     ---------  ----C--T   ---------  --T----
Camo-DRB3*0501 ***----    -G-----    *--G---    ---------  ---------  ---------  -------
Camo-DRB*W1401 ***GG---   ----TT-T   *-G---     ----C----  ----C--T   ---------  --A-T--
Camo-DRB11*0101 ***GG--*  ----TT--   *-G---     ---------  ----C--T   ---------  --T----
Ceap-DRB*W1301 ***----    -----G-A-  *--G---    ---------  ------G--  ---------  --T----
Ceap-DRB*W3201 ***----    -G-----    *----C--   ---------  ---------  ---------  -------
Ceap-DRB*W1502 ***----    -G-----    *--G---    ---------  ---------  --G------  -G-T---
Tasy-DRB*W101  ***-G-GA-  A-T-T--T   *-T----    ----A----  ---CCG---  TG---A---  --G-T--
Taba-DRB*W201  ***-GAGA-A A-T-T--T   *---T-     ----A----  ---CCG---  TGC---A-   -----TC
Gamo-DRB*W301  ATA--TTA-A A--A---T-- C---T*-C-  ----C--T   CC-A---GGG  CC-A---GGG  -----T-
Gamo-DRB*W401  G-G--T-A-A *------T-  -----C--   ----T----  CC-AA--TGG  ---------  .......
Gamo-DRB*W501  -AG--TAA-A A-------   -C-CC----  ----AT---  CC-GA--TGT G--------  -----CC
Roae           ***--T-A-  --------   ------T--  ----C--G-  CCT---C--  ----A----  --G----
```

Fig. 2B.

```
                 1          11         21         31         41         51         61         71         81         91         101        111
CONSENSUS ==>    GTTGTTGGTC ATCACTGTCT CTATTAACTT TTAACAACTT G*CCC*TTCT AAGTCACAAA CAGTGATGCT GCTGCAATTA CTTTTCACTA AGATGCCTCA GATTTCTGTA GTGATTCTAC
HLA-DRB1*0101    .......... .......... .......... .......... .......... .......... .......... .......... .......... .......... .......... ..........
HLA-DRB1*1001    -----C---- ------A--- ---------- -C-------- -*---*--G- G--------* T--------- ----A----- T--C------ -C--A----- ---------- ----------
HLA-DRB1*15011   .......... .......... .......... .......... .......... .......... .......... .......... .......... .......... .......... ..........
HLA-DRB1*03011   -----C---- ------A--- ---------- -C-------- -*---*--G- ---------- T--------- ----A----- T--C------ -C-------- ---------- ----------
HLA-DRB3*01012   ---------- -----CA--- ---------- CC-------- -*-T-*--G- ---------- T--------- ----A----- T--C---G-- -C-------- ---------- ----------
HLA-DRB4*0103    ---------- ------A--- ---------- -C-------- -*-T-*--G- ---------- T--------- ----A--C-G T--C------ -C-G------ -----G---- ----------
HLA-DRB1*04011   .......... .......... .......... .......... .......... .......... .......... .......... .......... .......... .......... ..........
Saoe-DRB1*0303   ---------- ---------- ---------- --C------- -T---C---- ---------- ---------- ---------- ---------- ---------- ----A-C--- ----------
Saoe-DRB*W2209   ---------- ---------- ---------- --C------- ---------- ---------- ---------- ---------- ---------- ---------- ----A-C--- ----------
Caja-DRB1*0301   ---------- T-A------- ---------- --T------- -C-------- ---------- ---------- ---------- ---------- ---------- ------C--- ----------
Caja-DRB*W1612   ---------- T--------- ---------- --C--G---- -----*---- ---------- ---------- ---------- ---------- ---------- ------C--- -------C--
Camo-DRB1*0302   ---------- G----****- ---------- CC-------- -*---*---- ---------- T--------- ---------- T--------- -CC------- -------A-- C--------T
Camo-DRB3*0701   ---------- -----****- ---C---A-- C--------- -*---*---- ---------- T--------- ---------- T--------- -CC------- -------A-- C--------T
Camo-DRB3*0501   ---------- G--------- ---------- --C-----C- -C-------- ---------- ---------- ---------- ---------- ---------- ---------- --A-------
Camo-DRB*W1401   .......... .......... .......... .......... .......... .......... .......... .......... .......... .......... .......... ..........
Camo-DRB11*0101  .......... .......... .......... .......... .......... .......... .......... .......... .......... .......... .......... ..........
Ceap-DRB*W1301   ---------- ---------- ---------- --C------- -C-------- ---------- ---C------ ---------- ---------- ---------- ------C--- ----------
Ceap-DRB*W3201   .......... .......... .......... .......... .......... .--------- ---------- ---------* **-------- -CC------C ---------- CC-------*
Ceap-DRB*W1502   ---------- ---------- ---------- --C------- -C-------- --------** **-------- ---------- ---------- ---------- ------C--- -------C--

                 121        131        141        151        161        171        181        191        201        211        221        231
CONSENSUS ==>    ATTTGATATT ATTCACAATG TAAAATGCTT CCATGTT*T* *TTTC*CTTT TA*CCA**GG ATTATTTT** AG****TTT* *TC**TTTCA CACTT****C AAA*A*AA*G A*AAAAACAT
HLA-DRB1*0101    .......... .......... .......... .......... .......... .......... .......... .......... .......... .......... .......... ..........
HLA-DRB1*1001    ---------- ---------- ---------- -T--T-AT-C A----G---- --C---A*-- --------TA --TATT---T G--AT----- -----****- ---C-T--A- -C--------
HLA-DRB1*15011   .......... .......... .......... .......... .......... .......... .......... .......... .......... .......... .......... ..........
HLA-DRB1*03011   ---------- ---------- ---------- -T--T-A--- ---------- -------*-- ---------- ---TA----- ---------- -----****- ---------- ----------
HLA-DRB3*01012   ---------- ---------- ---------- -T----AC-- -----A---- -------*-- ---------- ---TA----- ---G------ -----****- ---------- ----------
HLA-DRB4*0103    ----T----- ---------- ---------A -T--T-A--- -----A---- ------CA-- --------*- ---TA----- ---------- -----****- ---------- ----------
HLA-DRB1*04011   .......... .......... .......... .......... .......... .......... .......... .......... .......... .......... .......... ..........
Saoe-DRB1*0303   --------C- G--G------ ---***-TC- --T----*** ********** ********** ********** ********** ********** ********** ********** **********
Saoe-DRB*W2209   T-------C- ---G------ ---***-TC- --T----*** ********** ********** ********** ********** ********** ********** ********** **********
Caja-DRB1*0301   -C-----G-- ---------- ---***-TC- --T----*** ********** ********** ********** ********** ********** ********** ********** **********
Caja-DRB*W1612   --------C- ---G------ ---***-TC- --TC---*** ********** ********** ********** ********** ********** ********** ********** **********
Camo-DRB1*0302   -------G-- T-*------- ---------- -T-****--A -----A---- --A---C--- ---------G ----AC---- ----A----- -----****- ---G-A--*- -G--------
Camo-DRB3*0701   -------G-- T-*----G-- ---------- -T--C-A--A -----A---- --A---C--- ---------- ----AC---- ----A----- -----****- ---G-A--*- -G--------
Camo-DRB3*0501   --------C- ---G------ ---***-TC- --T----*** ********** ********** ********** ********** ********** ********** ********** **********
Camo-DRB*W1401   .......... .......... .......... .......... .......... .......... .......... .......... .......... .......... .......... ..........
Camo-DRB11*0101  .......... .......... .......... .......... .......... .......... .......... .......... .......... .......... .......... ..........
Ceap-DRB*W1301   ---------- -G-------- ---***-TC- --T----*** ********** ********** ********** ********** ********** ********** ********** **********
Ceap-DRB*W3201   *--------- ---------- ---------- -T-GC-A--A G---A----- --A----T-- --------A- --C-A----C A---A----- -----****- ---G-A--*- -G--------
Ceap-DRB*W1502   ---------- ---------- ---***-TG- --T----*** ********** ********** ********** ********** ********** ********** ********** **********
```

Fig. 2B. (continued)

```
                  241        251        261        271        281        291        301        311        321        331        341        351
CONSENSUS ==>  CAAAA*TATA *******T*T TTTTA*TATA TG**TATTT* *ACA*ATAT* T**AT*T*T* TTTATATGTA T**AA**TAC AGAAGCAC*T GTCACCA*TA AGAGCTCTGA GAC**CTTTG
HLA-DRB1*0101  ---------- *******G-G ------CA--- ---------- ------C--- -GT--G-A-A ---------- --TA--AC-- ---------- ------A--- ---------- ----------
HLA-DRB1*1001  ----A----- .......... ---------- ---------- ---------- ---------- ---------- ---------- ---------- ------A--- ---------- ----ACT---
HLA-DRB1*15011 .......... *******--- ---------- ---------- ---------- ---------- ---------- ---------- ---------- ---------- ---------- ----------
HLA-DRB1*03011 .......... *TACT*---- ---------- --TC----T ---------- ---------G ---------- --G--G---- ---------- ---------- ---------- ------C---
HLA-DRB3*01012 ---------- --A*------ ---------- --T----T- ------T--C ---------G ---------- --G------- --G------- ---------- ---------- ----------
HLA-DRB4*0103  ---------- ---------- ---------- ---------- ---------- ---------- ---------- ---------- ---------- ---------- ---------- ----------
HLA-DRB1*04011 ********** ********** ********** ********** ********** ********** ********** ********** ********** ********** ********** **********
Saoe-DRB1*0303 ********** ********** ********** ********** ********** ********** ********** ********** ********** ********** ********** **********
Saoe-DRB*W2209 ********** ********** ********** ********** ********** ********** ********** ********** ********** ********** ********** **********
Caja-DRB1*0301 ........... ---------- ---------- ---------- ---------- ---------- ---------- ---------- ---------- ---------- ---------- ----------
Caja-DRB*W1612 ********** --G-ACA-A- --TG----- --T------A ---------A ---------- --C---G-G ---G------ ------C--- ---------- ---------- ----CA----
Camo-DRB1*0302 ********** -G-A----- --TG----- --T------A ---------A ---------- --C---G-G ---G-T---- ------C--- ---------- ---------- ----CA----
Camo-DRB3*0701 ********** ********** ********** ********** ********** ********** ********** ********** ********** ********** ********** **********
Camo-DRB3*0501 ********** ********** ********** ********** ********** ********** ********** ********** ********** ********** ********** **********
Camo-DRB*W1401 .......... .......... .......... .......... .......... .......... .......... .......... .......... .......... .......... ..........
Camo-DRB11*0101 ********** ********** ********** ********** ********** ********** ********** ********** ********** ********** ********** **********
Ceap-DRB*W1301 ---G----- ---G-C--- ---TG----- --T------A ---------- --AC--T--G ---G------ ---------- ---------- ---------- ---------- ----------
Ceap-DRB*W3201 ********** ********** ********** ********** ********** ********** ********** ********** ********** ********** ********** ----CA----
Ceap-DRB*W1502 ********** ********** ********** ********** ********** ********** ********** ********** ********** ********** ********** **********

                  361        371        381        391        401        411        421        431        441        451        461        471
CONSENSUS ==>  ACC*CTTACC CTTATCAGAT ********** ********** *GAG*T* TGG*AACAA* TT**TTT*AA CTGAATTTCT GAGCTTTGTG GATTTAG*AA TGCAAAGGA* *GTTTG*GGA CATT*AC*GG
HLA-DRB1*0101  ---A------ ---------- ---------- ********** ---A----G ---------- ---------- ---------- ---------- ------A--- ------A-G- ----T----- --C--A----
HLA-DRB1*1001  .......... ---A------ ---------- ********** --T-G ---------- --TT----* ---------- ---------- ---------- ---------- ----------
HLA-DRB1*15011 .......... .......... ***------- ***------ ---------- ---------- ---------- ---------- ---------- ---------- ---------- ----------
HLA-DRB1*03011 .......... .......... GCGGATTTGCC AAAT ---------- ---------- ---------- ---------- ---------- ---------- ---------- ----T--C--
HLA-DRB3*01012 ---------- ---------- --AT------ ---------- ---------A ---A------ --A------- --A------- ---------- ---------- ---------- ----T--GA-
HLA-DRB4*0103  ---------- ---------- --A------- ---------- ---T----G ---------- ---C--T-- ----A----- --T----A-- ---------- ---A------ ----T--GA-
HLA-DRB1*04011 ********** ********** ********** ********** ********** ********** ********** ********** ---CT----- ---------- --------** **********
Saoe-DRB1*0303 ********** ********** ********** ********** ---G----A ---------- ---C--T-- ---------- ---------- ---------- ---------- **********
Saoe-DRB*W2209 ********** ********** ********** ********** ********** ********** ********** ********** ********** ********** ********** **********
Caja-DRB1*0301 .......... .......... ********** ********** ********** ********** ********** ********** ********** ********** ********** **********
Caja-DRB*W1612 --C------ ---------- --A------- ---------- ---G----C-T ---------- ---G----A ---CC---T-- ---------- ------G--- ------T--- ----T-----
Camo-DRB1*0302 --C------ ---------- --A------- ---------- --GG----C-T ---------- ---G----A ---CC---T-- ---------- ------G--- ---G-C---- --G-------
Camo-DRB3*0701 ********** ********** ********** ********** ********** ********** ********** ********** ********** ********** ********** ----------
Camo-DRB3*0501 ********** ********** ********** ********** ********** ********** ********** ********** ********** ********** ********** **********
Camo-DRB*W1401 .......... .......... .......... .......... .......... .......... .......... .......... .......... .......... .......... ..........
Camo-DRB11*0101 ********** ********** ********** ********** ********** ********** ********** ********** ********** ********** ********** **********
Ceap-DRB*W1301 --C----A- ---------- --A------- ---------- ---T----- ---G----A ---C--T-- ---------- ---------- ---------- --C------- ----T-----
Ceap-DRB*W3201 ********** ********** ********** ********** ********** ********** ********** ********** ********** ********** ********** **********
Ceap-DRB*W1502 ********** ********** ********** ********** ********** ********** ********** ********** ********** ********** ********** **********
```

Fig. 2B. (continued)

```
                481        491        501        511        521        531        541        551        561        571        581        591
CONSENSUS ==>   GA*TCATGG* TTTATTCTCC TT*AAACTCT TC*ATACTTT *CCATTGTC* TTAG**TAAA TCCAAAATCC TAACACCAC* CA*GAGGCTT T**AATACCT G*CT*C*TGT GATTTCTCC*
HLA-DRB1*0101   .......... .......... .......... .......... .......... .......... .......... .......... .......... .......... .......... ..........
HLA-DRB1*1001   --*------T ---------- --A------- --T------- C--------C ----TA---- ---------- ---------C --A------- -TC------- -A--C-*--- ---------G
HLA-DRB1*15011  .......... .......... .......... .......... .......... .......... .......... .......... .......... .......... .......... ..........
HLA-DRB1*03011  --*-----A- ---------- ---------- ---G------ --A------- ---------- ---------- ---------- ---------- ---------- -G----*--- ----------
HLA-DRB3*01012  --A------- ------G--- --T------- --G------- A--------* ---C------ ---------- ----G----- --C------- -*-------- -G--T-T--- ---------A
HLA-DRB4*0103   --*------- ---------- ---------- --A------- A--------T -----*---- ---------- ---------- --C------- ---------- -C--T----- ---------A
HLA-DRB1*04011  --*----A-- ---------- ---------- --A------- ---------T ----**---- ---------- ---------T --C------- -----C---- -G--T----- ---------A
Saoe-DRB1*0303  ********** ********** ********** ********** ********** ********** ********** ********** ********** ********** ********** **********
Saoe-DRB*W2209  ********** ********** ********** ********** ********** ********** ********** ********** ********** ********** ********** **********
Caja-DRB1*0301  ********** ********** ********** ********** ********** ********** ********** ********** ********** ********** ********** **********
Caja-DRB*W1612  ********** ********** ********** ********** ********** ********** ********** ********** ********** ********** ********** **********
Camo-DRB1*0302  --*------C ---------- --G------- --C-G----- ---------T ---------- ---------- ---------G --C------- --A------- -G--T----- ---------A
Camo-DRB3*0701  --*------C ---------- --G------- --C------- ---------A ---------- ---------- ---------G --C------- -CA------- -G--T----- ---------A
Camo-DRB3*0501  ********** ********** ********** ********** ********** ********** ********** ********** ********** ********** ********** **********
Camo-DRB*W1401  .......... .......... .......... .......... .......... .......... .......... .......... .......... .......... .......... ..........
Camo-DRB11*0101 .......... .......... .......... .......... .......... .......... .......... .......... .......... .......... .......... ..........
Ceap-DRB*W1301  ********** ********** ********** ********** ********** ********** ********** ********** ********** ********** ********** **********
Ceap-DRB*W3201  --*------- --C------- --G-----T- --C------- ---------T --G------T ---------- -----T--TG -CT--A---- --A------- -G--T----- ---------A
Ceap-DRB*W1502  ********** ********** ********** ********** ********** ********** ********** ********** ********** ********** ********** **********

                601        611        621        631        641        651        661        671        681        691        701        711
CONSENSUS ==>   G*CTAA*CTT TTACC*TCCT TCCCCTCAGC CTCTCTGCTT TAGTGAACTT T***CTAGT* T*TTGAAG** *TTCATCAAT TCAAGCTTTT GTAC*TGG*A TTTCCTA*A* **********
HLA-DRB1*0101   .......... .......... .......... .......... .......... .......... .......... .......... .......... .......... .......... ..........
HLA-DRB1*1001   -G----T--- -----C---- ---------- ---------- ---------- -GTT-----T -C------AA G--------- ---------- ----A---G- -------A-C CTGAAA*TGT
HLA-DRB1*15011  .......... .......... .......... .......... .......... .......... .......... .......... .......... .......... .......... ..........
HLA-DRB1*03011  ---------G ---------- ---------- ---------- ---------- -C-C------ -T-------- ---------- ---------- ---------- ---------- ------*---
HLA-DRB3*01012  ------C--- ---------- ---------- ---------- ---------- -C-C------ -T-------- ---------- ---------- ---------- ---------- ------*---
HLA-DRB4*0103   -C-------- -----T---- ---------- ---------- ---------- ---------- -T-------- -C-------- ----C----- ---------- ---------- ------A---
HLA-DRB1*04011  AT----C--- ---------- ---------- ---------- ---------- ---------- -T---G--TT C--------- ---------- ---------- ---------- ------*---
Saoe-DRB1*0303  ********** ********** ********** ********** ********** ********** ********** ********** ********** ********** ********** **********
Saoe-DRB*W2209  ********** ********** ********** ********** ********** ********** ********** ********** ********** ********** ********** **********
Caja-DRB1*0301  ********** ********** ********** ********** ********** ********** ********** ********** ********** ********** ********** **********
Caja-DRB*W1612  ********** ********** ********** ********** ********** ********** ********** ********** ********** ********** ********** **********
Camo-DRB1*0302  -T-------- -----G---- ---------- ---------- ---------- --C-----C- --------** *-A------- ---------- ----C-TTT- -------C-* **********
Camo-DRB3*0701  -T-------- -----G---- ---------- -A-------- ---------- --C-----C- --------** *--------- ---G-T---- ----C---T- -------C-* **********
Camo-DRB3*0501  ********** ********** ********** ********** ********** ********** ********** ********** ********** ********** ********** **********
Camo-DRB*W1401  .......... .......... .......... .......... .......... .......... .......... .......... .......... .......... .......... ..........
Camo-DRB11*0101 .......... .......... .......... .......... .......... .......... .......... .......... .......... .......... .....--CC- TGA--TTCC-
Ceap-DRB*W1301  ********** ********** ********** ********** ********** ********** ********** ********** ********** ********** ********** **********
Ceap-DRB*W3201  -T-------- ---T---TT- ---------- ---------- ---------G -A--G----- -T--A----- ---------- ---------- --------T- ---GT----* -CTG------
Ceap-DRB*W1502  ********** ********** ********** ********** ********** ********** ********** ********** ********** ********** ********** **********
```

Fig. 2B. (continued)

```
                  721        731        741        751        761        771        781        791        801        811        821        831
 CONSENSUS ==>    *CCTCC***G TTTTGT*CA* ACAGAC*C** *GG*TCC**T CTGCCCC**G GCTCAC*CCT GCTTAAC*TG T*AAGTCAC* TCTGA*C*GT CACTCTT*AG AGGGTCCTTC TCTGGCACCC
HLA-DRB1*0101     .......... .......... .......... .......... .......... .......... .......... .......... .......... .......... .......... ..........
HLA-DRB1*1001     G-----C**- ------C--A ------A-*G G-C*---AC- -------CT- ------A--- -------C-- -C-------A -----A-C-- ----A--A-- ---------- -----AC---
HLA-DRB1*15011    .......... .......... .......... .......... .......... .......... .......... .......... .......... .......... .......... ..........
HLA-DRB1*03011    -------**- ---------- --------** ---C------ ---------- ---------- ---------- ---------- ---------- ----A----- ---------- ----------
HLA-DRB3*01012    -------**- ---*------ --------** ---------- ---------- ---------- ---------- ---------- -------T-- -------C-- ---------- ----------
HLA-DRB4*0103     -------**- ---------G --------AC A--------- ---------- ----T----- ---------- ---------- -------T-- -------C-- ---------- ----------
HLA-DRB1*04011    ------**G- ---------- ---------C A--------- ---------- ---------- -------T-- -T-------- -------T-- -------*** *********- ----------
Saoe-DRB1*0303    ********** ********** ********** ********** ********** ********** ********** ********** ********** ********** ********** **********
Saoe-DRB*W2209    ********** ********** ********** ********** ********** ********** ********** ********** ********** ********** ********** **********
Caja-DRB1*0301    ********** ********** ********** ********** ********** ********** ********** ********** ********** ********** ********** **********
Caja-DRB*W1612    ********** ********** ********** ********** ********** ********** ********** ********** ********** ********** ********** **********
Camo-DRB1*0302    *-----T**- ------T--G ------C--C A------C*- -------**- ------T--- -------*-- -G-------C ----CG---- -------C-- ---------- ----------
Camo-DRB3*0701    *-----T**- C-----T--G ------C--C A------C-- -------**- ------T--- -------*-- -G-------C -----G---- -------C-- ---------- ----------
Camo-DRB3*0501    ********** ********** ********** ********** ********** ********** ********** ********** ********** ********** ********** **********
Camo-DRB*W1401    .......... .......... .......... .......... .......... .......... .......... .......... .......... .......... .......... ..........
Camo-DRB11*0101   A-AC-T-CT- ------T--G ------C--C A------C*- -------**- ---A--T--- -------*-- -G-------C ----CG---- -------C-- ---------- ----------
Ceap-DRB*W1301    ********** ********** ********** ********** ********** ********** ********** ********** ********** ********** ********** **********
Ceap-DRB*W3201    -------ATT --A-**---- ---------C A--A---C-- -------A-- ---A------ ----G----- -G-------- -----G-T-- -------C-- ---------- -T--------
Ceap-DRB*W1502    ********** ********** ********** ********** ********** ********** ********** ********** ********** ********** ********** **********

                  801        811        821        831        841        851        861        871        881        891        901        911
 CONSENSUS ==>    TCTGA*C*GT CACTCTT*AG AGGGTCCTTC TCTGGCACCC T*A*G*AA*T GAGATCATCC TATTATTCTC TG*TCTAGAA *TCCACACTT CTG****CAT TTCTCATTCC TGTCTAAGCT
HLA-DRB1*0101     .......... .......... .......... .......... .......... .........- ---------- --T------- C--------- --A***A--- ---------- ----------
HLA-DRB1*1001     -----A-C-- ----A--A-- ---------- -----AC--- -A-T-T--T- ---------- ---------- ---------- ---------- -C-***---- ---------- ----------
HLA-DRB1*15011    .......... .......... .......... .......... .......... ..A---T--- ---------- --C------- -G-------- ---***-T-- --------G- ----------
HLA-DRB1*03011    ---------- ----A----- ---------- ---------- ---------- ---------- ---A------ ---------- ---------- -C-***---- ---------- ----------
HLA-DRB3*01012    -------T-- -------C-- ---------- ---------- ---------- ---------- ---------- --C------- --T-C----- -C-***GG-- ---------- -------C--
HLA-DRB4*0103     -------T-- -------C-- ---------- ---------- ---------- ---------- -T-------- -----C---- ---------- ---***--G- ---------- ----------
HLA-DRB1*04011    -------T-- -------*** *********- ---------- ---A-G---- ---------- ---------- ---------- ---------- ---***-A-- ---------- ----------
Saoe-DRB1*0303    ********** ********** ********** ********** ********** ********** ********** ********** ********** ********** ********** **********
Saoe-DRB*W2209    ********** ********** ********** ********** ********** ********** ********** ********** ********** ********** ********** **********
Caja-DRB1*0301    ********** ********** ********** ********** ********** ********** ********** ********** ********** ********** ********** **********
Caja-DRB*W1612    ********** ********** ********** ********** ********** ********** ********** ********** ********** ********** ********** **********
Camo-DRB1*0302    ----CG---- -------C-- ---------- ---------- -G------C- ---G------ ----G----- --C--CG--- *--------- ---****--- ---C-G---- ---G------
Camo-DRB3*0701    -----G---- -------C-- ---------- ---------- -G------C- ---G------ ----G----- --C--CG--- *--------- ---****--- ---C-G---- ---G------
Camo-DRB3*0501    ********** ********** ********** ********** ********** ********** ********** ********** ********** ********** ********** **********
Camo-DRB*W1401    .......... .......... .......... .......... .......... .......... .......... .......... .......... .......... .......... ..........
Camo-DRB11*0101   ----CG---- -------C-- ---------- ---------- -G------C- ---G------ ----G----- --C--CG--- *--------- ---****--- ---C-G---- ---G------
Ceap-DRB*W1301    ********** ********** ********** ********** ********** ********** ********** ********** ********** ********** ********** **********
Ceap-DRB*W3201    -----G-T-- -------C-- ---------- -T-------- GG-*****-G ---------- ---------- --C------- *--------- ---A***--- ---------- ---G------
Ceap-DRB*W1502    ********** ********** ********** ********** ********** ********** ********** ********** ********** ********** ********** **********
```

Fig. 2B. (continued)

```
                     921        931        941        951        961        971        981        991        1001       1011       1021       1031
CONSENSUS ==>        CTTGTGT*TT T**TTTTTGG CCATCACTT* CACTGCTCTT TAAGCTCCCC CAGCGGAGTG GAGAGGTCTG TTT*CCCTTG TTTGGATTCC TAGAGGCAGC GCAGGCCTGG CACAAGGTCG
HLA-DRB1*0101        -------G-- -GG------- -----G---T ---------- ---------- ---TA----- ---------- ---T----CA ---------- --C------- ---------- ---------A
HLA-DRB1*1001        ---------- ----G--G-- ---------- ---------- ---------- ---------- ---------- --------C- ---------- ---G------ ---------- ---------A
HLA-DRB1*15011       ---------- ---------- -----G---- ---------- ---------- ---A------ ---------- --------C- ---------- ---------- ---------- ---------A
HLA-DRB1*03011       ---------- ----G--G-- ---------- ---------- ---------- ---------- ---------- --------C- ---------- ---------- ----A--A-- ---------A
HLA-DRB3*01012       ---------C ---------- -----G---- ---------- ---------- ---------- ---------- ---------- ---------- ---------- ---------- ---------A
HLA-DRB4*0103        ---------- ---------- -------G-- ---------- ---------- ---------- ---------- ------T--- ---------- ---------- ----A--A-- ---------A
HLA-DRB1*04011       ---------- ---------- ---------- ---------- A--------- ---------- ---------- -------G-- ---------- ---------- ---------- ---------A
Saoe-DRB1*0303       ********** ********** ********** ********** ********** ********** ********** ********** ********** ********** *******A-- ----------
Saoe-DRB*W2209       ********** ********** ********** ********** ********** ********** ********** ********** ********** ********** *******--- ----------
Caja-DRB1*0301       ********** ********** ********** ********** ********** ********** ********** ********** ********** ********** *******--- --------G-
Caja-DRB*W1612       ********** ********** ********** ********** ********** ********** ********** ********** ********** ********** *******--- ----------
Camo-DRB1*0302       -C-----*-G -**G-G---- --G------* ------C--- ---------- ---------- ------C--- ---*------ --C------- C--------- ---------- -------C--
Camo-DRB3*0701       -C-----*-G -**G-G---- ---------* ------C--- ---------- ---------- ------C--- ---*------ --C------- CC-------- ---******* **********
Camo-DRB3*0501       ********** ********** ********** ********** ********** ********** ********** ********** ********** ********** *******--- --G-------
Camo-DRB*W1401       .......... .......... .......... .......... .......... .......... .......... .......... .......... ....------ ---------- ----------
Camo-DRB11*0101      -C-----*-G -**G-G---- ---------* ------C--- ---------- ---------- ------C--- ---*------ --C------- C--------- ---------- ----------
Ceap-DRB*W1301       ********** ********** ********** ********** ********** ********** ********** ********** ********** ********** *******--- --------G-
Ceap-DRB*W3201       -C-------- G*AG------ ------A-** ---------- ---------T ---------- -T-------- ---*------ ---------- ---------- ----TA---- --------TA
Ceap-DRB*W1502       ********** ********** ********** ********** ********** ********** ********** ********** ********** ********** *******--- ----------

                     1041       1051       1061       1071       1081       1091       1101       1111       1121       1131       1141       1151
CONSENSUS ==>        TCACTAAGGA AGTGTTCACA GGATGAACGC GGTGGGGGCT GTTGAAGGAA CCGGTAAAGC CTGTGGGATG AGAGAAGGGG CAGAGAGTGT CTTTGGG*TG GAGGCTCCCA GGAGGAGGC*
HLA-DRB1*0101        ---------- ---------- --G----G-- ------T--- ---------- ---------- ---------- --------A- ---------- T------G-- ---------- ---------*
HLA-DRB1*1001        G--------- ---------- -A-------- ------T--- ---------- ------G--- G--------- --------A- ---------- ---------- ---A------ ---------*
HLA-DRB1*15011       ---------- ---------- --G------- ------T--- --------T- ---------- G--------- --------A- ---------- -------*** ***------- ---------*
HLA-DRB1*03011       G--------- --G-----G- ---------- ------T--- ---T------ --------*- A--------- --------A- --**------ ---------- ---------- ---------*
HLA-DRB3*01012       --------A- ---------- ---------- ------T--- ---T------ ---------- A--------- --------A- ---**----- ---------- --------T- ---------*
HLA-DRB4*0103        G--------- --G------G --------A- ------T--- ---------- ---------- G--------- ---***--A- ---------- ----CA---- A--A------ -A-------*
HLA-DRB1*04011       ---------- ---------- -------A-- ----C-T--- ---T----C- AA-------- --*----T-- --------A- --A**----C ---------- ---------- ---------*
Saoe-DRB1*0303       ---G------ -A-A------ ---------- ---------- ---------- ---------- ---------- ------T*C- ---------- -------*-- ---------- ---------*
Saoe-DRB*W2209       ---------- ----G----- ---------- T--------- ---------- ---------- ---------- ------T*C- ---G------ -------*-- ---------- ---------*
Caja-DRB1*0301       -----****- -A-------- -A-------- ---------- ---------- TA---G-C-- -G-------- ---------- ---------- -------*-- ---------- ---------*
Caja-DRB*W1612       ---------- -A-------- -------T-- ---------- --G------- --A-----A- ---------- ---------- ---------- -------*-- ---------- ---------*
Camo-DRB1*0302       ----C----- -----G---- ------G--- ---------- --C-C----- -G---G---- ---------- ----C--*-- **---G---- --**---*-- ---------- --G------*
Camo-DRB3*0701       ********** ********** ********** **-------- --C-C----- -G---G---- ---------- ----C-***- GG---G---- --**---*-- ---------- ---------*
Camo-DRB3*0501       -------A-- -A-------- --G--C---- ---------- ---------- -G-*-G---- ---------- ---------- ---------C -------*-- -------TG- ------A-T*
Camo-DRB*W1401       ----C----- -----C---- ------G--- ---------- --C-C----- -G---G---- ---------- ----C----- *-**-G---- --**---*-- -G-------- -----C---*
Camo-DRB11*0101      ----C----- -----C---- ------G--- ---------- -CC-C----- -G---G---- ---------- ----C----- -----G---- --**---*-- ---------- ---------*
Ceap-DRB*W1301       ---------- ---------- --G----TAT ----T----- ---------- -T---G---- ---------- ---------- -----G---- -------*-- ---------- --------T*
Ceap-DRB*W3201       ----C----- ---------- --G------- CA-------- -------A-- TG-------- ---------- ---------- --------T- ---------- ---------- ----AG--T*
Ceap-DRB*W1502       -------A-- -A-------- ---------- ---------- A--------- -G---G---- -C---A---- ---------- ---------- G--------- ---------- ---A-----*
```

Fig. 2B. (continued)

```
                   1161       1171       1181       1191       1201       1211       1221       1231       1241       1251       1261       1271
CONSENSUS ==>      **GGCGCGGG CTGCGGTGCG GGGCGGATCC TCCTCCAGCT CCTGCTTGGA GGTCTCCAGA ACAGGCTGGA GGCAGGGAGG GGG*TCCCAA AAGCCTGGGG ATCAGAGGGG GTTTTCCCGC
HLA-DRB1*0101      **--GA---- ---------T ---------- ---------- ---------- ---------- ---------- ---------- ---*------ ---------- --G---A--- ----------
HLA-DRB1*1001      **C---A--- ---------T ---------- -------A-- ---------- ---------- ---------- ------A--- ---*------ ---------A ------A--- ----------
HLA-DRB1*15011     **-------- ---------T ---------- ---------- ---------- ---------- ---------- ---------- ---*------ ---------- --G---A--- ----------
HLA-DRB1*03011     **-------- ---------T ---------- ---------- ---------- ---------- ---------- --T------- ---*------ ---------- ------C-T- ----------
HLA-DRB3*01012     **--G----- ---------- ---------- ---------- ---------- ---------- ---A------ ---------- ---*------ ---------- --G---A--- ----------
HLA-DRB4*0103      **C---T--- -C-------T ---------- ---------- ---------- -T-------- ---------- ---------- ---*------ -------A-- ------A--- ----------
HLA-DRB1*04011     **-------- ---------T --A------- ---------- -----C---- ---------- ---------- ---------- ---*------ ------T--- --------TA -----T--A-
Saoe-DRB1*0303     **-------- ---------T ---------- ---------- G------C-- ---------- ------G--- ---------C ---G*----- --TA--T--- --------C- --A-------
Saoe-DRB*W2209     **-------- ---------- ---------- ---------- ---------- ---------- ------G--- ---C------ ----*----- ---A--T--- G--------- --A-----TT
Caja-DRB1*0301     **-----A-- ---------- ----A---T- ---------- -----G---- ---------- ---------- ---C---G-- ----G----- ---A--T--- --------A- --------T-
Caja-DRB*W1612     **-----A-- ---------- ----*----T ---------- -------C-- ---------- G--------- ---C------ ----*----- ---A--T-A- ---------- --------T-
Camo-DRB1*0302     **-------- ---------- ---------- ---------- ---------- --C------- --C------- ---------- ---**---T- ---------- G------A-- *-------T-
Camo-DRB3*0701     **-------- ---T------ ---------- ---------- ---------- --C------- --C------- ---------- ---**---T- ---------- G------A-- -*------T-
Camo-DRB3*0501     **-----C-- ---------- ---------- ---------- -T-------- ---------- ---------- ---------- ---**----- ---A--T--- ----A--A-- A---GT--A-
Camo-DRB*W1401     **-------- ---T-----C ---------- ---------- ---------- --C------- --C------- ---------- ---**---T- ---------- T------A-- ----*---T-
Camo-DRB11*0101    **-------- ------G--- ---------- ---------- ---------- --C------- --C------- ---------- ---**---T- ---------- T------A-- ----*---T-
Ceap-DRB*W1301     **-------- --------T- ------G--- ---------- ---------- ---------- ---------- A-----*--- C---*G---- ---A--T--- --------C- --A-------
Ceap-DRB*W3201     **--T-T--- -----A---- ---------- -------C-- ---------- ---------- ---------- --T------A -A-*-----C ---A--T--- G-ATC-CA-* *---------
Ceap-DRB*W1502     **-------- ---------T ---------- ---------- -------A-- ---------- ---------- ---G-A---- -----GG--- -TTA--T--- --------AT --------T-

                   1281       1291       1301       1311       1321       1331       1341       1351       1361       1371       1381       1391
CONSENSUS ==>      CTGGTCCCCC AGCCCCC*GT CCGCCTCAGG AAGACGGAGG ATGAGCCCCT GGGCTGCGGG TGGTGGGGGT TGCGGGTGGG GCCGGTCAAG GTTCCCAGAG CCCGCACCCC GCC*AGGGAG
HLA-DRB1*0101      A--------- --G----C-- T--------- ---------- ------T--- -------T-- -------C-- ---------- ------T--- --------T- ---------T ---C------
HLA-DRB1*1001      ---------- --T------- T--------- ---------- ---------- -------T-- -C-----C-- ---------- *-G---T--- --------T- ---------- A---------
HLA-DRB1*15011     A--------- --G------- T--------- ---------- ------T--- -------A-- -------C-- ---------- ------T--- -----*--C- ---------- --T-------
HLA-DRB1*03011     ---------- --G-----T- T--------- -----A---- -G-------- -------T-- -------C-- -----CG--- ------T--- --------T- ---------- A---------
HLA-DRB3*01012     A-A------T --G------- T-----G--- --A------- ------T--- -------A-- -------C-- ---------- ------T--- --------T- ---------- A---------
HLA-DRB4*0103      ---------- --A----G-- T--------- -----T---- -C-------- -------T-- -------C-- ---------- ------T--- --------C- ---------- ----------
HLA-DRB1*04011     ---------- --A------- ---------A -----A---- ---------- --------T- -T--C----- ---------- ---A-***** ********** ********** **********
Saoe-DRB1*0303     ----C----- -A-----*-- ---------- ----G----- -----G---- -------ACT ---------- ---------- --G------- ---------- ---------- ----------
Saoe-DRB*W2209     *****----- --T----*A- ---------- ----TA--A- -A-------- -------*-A *--CC----- ---------- ---------- ----T----- ---------* ----------
Caja-DRB1*0301     -A-------- -TA---**-- ---------- -----AC-A- -G-------- --C---T--- G**CC---T- ---C------ ---------- ----T----- -A-------G C--*------
Caja-DRB*W1612     -A-------- -------*-- ---------- ----TA--A- -G-------- --*------- G**CC----- ---C------ ---------A ----T----- ---------G C--*------
Camo-DRB1*0302     ----C-T-G- -------*-- --C-----A- ---------- ---------A --*-----C- --------*- ---------- ---------- ---------- ---------- ---*------
Camo-DRB3*0701     ----C-T-G- -------*-- ----T--G-- ---------- ---------A --*-----C- --------*- ---------- -------G-- ---------- ---------- ---*------
Camo-DRB3*0501     ---*C----- --GT---*-- ---------- -----C---- --C------- --C-----C- --------CG --A------- --T------- ---------- ------T--- ---*------
Camo-DRB*W1401     ----C-T-G- -------*-- ----T--G-- ---------- ---------A --*-----C- --------*- ---------- -------G-- ---------- ---------- ---*------
Camo-DRB11*0101    ----C-T-G- -------*-- ----T--G-- ---------- ---------A --*-----C- --------*- ---------- ---------- -G-------- ---------- ---*------
Ceap-DRB*W1301     ----C----- -T-----*-- ---------A ---T-C---- ---------- --------C- -A--C----- ---------- --T------- A--------- ---------- ----------
Ceap-DRB*W3201     ----C----- -----G-*-G ---------- -G---A---- ---------- -------AC- ----C----- -C-------- --G------- -C-------- ---------- ---*------
Ceap-DRB*W1502     -C-------- -------*-- -----C---- -----A---- ---------- --C---GC-- G**C------ ------C--- ---------- ---------- ---------G C--*------
```

As for the second question, there are two principal mechanisms that could explain the sharing of motifs among taxonomically diverse primate (eutherian mammal) taxa, and their independent evolution from the rest of the *DRB* genes — exchange of gene segments (recombination) and independent, de novo origin of motifs (convergent evolution). Recombination can, theoretically, shuffle the entire exon 2 (by the classical crossing over mechanism) or the motifs alone (by gene conversion-like events). Classical crossing over does take place within *DRB* genes, but is relatively rare (Bergström et al. 1998; Blasczyk et al. 1998). It is excluded as an explanation for the sharing of motifs by the fact that the sharing does not extend beyond the motifs. If reciprocal recombination were responsible for the motif sharing, the flanking intron on one or the other side of exon 2 should also have been exchanged, but there is no sign of such swapping. To save the crossing over hypothesis, one would have to postulate multiple exchanges, either simultaneous or sequential, but to dissect the motifs out cleanly from the rest of the gene by recombination would require prohibitive recombinational promiscuity.

The gene conversion hypothesis avoids the high frequency requirement because a single event can introduce one or more motifs into a gene nonreciprocally. The problem with this hypothesis is, however, that it is defined so nebulously that almost any observation can be made to fit into it. Nevertheless, the hypothesis does make certain predictions which can be used to confront it with an alternative explanation, the hypothesis of convergent evolution. Before we come to these predictions, however, it may be useful to point out the circumstances under which gene conversion might explain motif sharing.

```
                1401      1411       1421       1431       1441        1451       1461        1471
   CONSENSUS ==> CCCC*GATGG CGGCGTCGCT GTCGGTGTCT TCCCCGGAGG CCGCCCCTGT GACCGGATCG TTCGTGTCCC CACAG
HLA-DRB1*0101   ----G----- T--------- ---A------ --T-A----- -T----G--- ---------C ---------- -----
HLA-DRB1*1001   ---------- T--------- ---A------ --T-A----- ------G--- ---------- ---------- .....
HLA-DRB1*15011  ---------- T--------- ---C------ --T-A----- ------*--- ---------G- ---------- -----
HLA-DRB1*03011  ---------- -------A--- ---A------ --T-A----- -----TG--- ---T------ ---------- -----
HLA-DRB3*01012  ---------- ---------- ---A------ --T-A----- ------G--- ---------C -------A-- -G---
HLA-DRB4*0103   T--G--T--- T--------- ---A------ --T-A----- ------G--- ---------- ---------- -----
HLA-DRB1*04011  ********** ********** *ATA------ ---------- ----TT---- A--------- ---T------ -C---
Saoe-DRB1*0303  ----*----- ---------- --T--C---- -T---C---- ---------- ---------- ---------- -----
Saoe-DRB*W2209  ---A*----- ---------- -----C---- -T-------- ---------- ---------- --------T-- -----
Caja-DRB1*0301  ----*----- T--------- ---A-C---- -T---T---- ---------- ---------- ---------- -----
Caja-DRB*W1612  ----*----- ---------- -----C---- -T---T---- ---------- ---------C ---------- -----
Camo-DRB1*0302  ----*----- G-*A------ ---------- ---------- ------*--- ---------- ---------- -----
Camo-DRB3*0701  ---**----- G-*A------ ---------- -T--T----- ---------* ---------- ---------- -----
Camo-DRB3*0501  ----*----- G-*A------ ---------- -T--T----- ---------- ---------- --T------- -----
Camo-DRB*W1401  ----*----- G-*A------ ---------- ---------- ------*--- ---------- ---------- -----
Camo-DRB11*0101 ----*----- G-*A------ ---------- ---------- ------*--- ---------- ---------- -----
Ceap-DRB*W1301  ----*----- ---T------ ---------- -T-------- ---------- ---------- ------T-- -----
Ceap-DRB*W3201  ---------- -A-------- -G-A------ -*--A--G-- ---------- -----CG--A --T------- -----
Ceap-DRB*W1502  ---------A ---------- ---------- -T-------- ---------- A--------- ------T-- -----
```

Fig. 2. Alignment of intron 1 sequences of primate *DRB* genes. NWM, tarsier, and prosimian sequences were obtained from amplified PCR products which included the entire intron 1 and most of exon 2. They were aligned to human sequences and to one sequence of *Rousettus aegyptiacus* (*Roae*). In **A**, the 5' part of intron 1 is shown. The consensus sequence was determined by simple majority, A dash (-) indicates identity with the consensus, an asterisk (*) an indel, and a dot (.) unavailability of sequence information. Numbering above the sequences starts with the first nucleotide of the alignment. In **B**, the 3' part of intron 1 is shown. Only human and NWM sequences are alignable in this region.

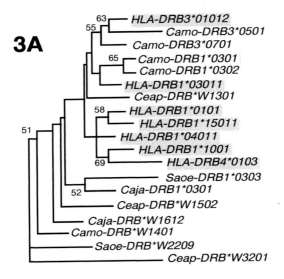

3A

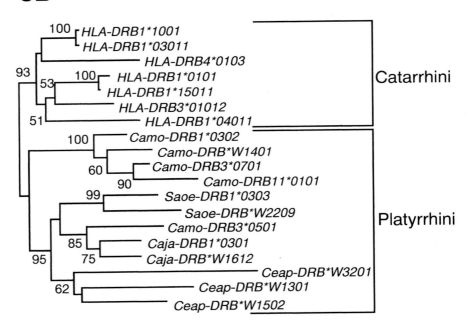

3B

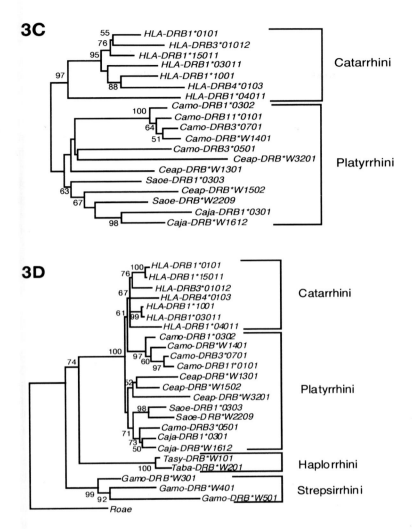

Fig. 3. Phylogenetic trees of the primate *DRB* exon and intron sequences shown in Figures 1 and 2. The trees were obtained by using the neighbor-joining method (Saitou and Nei 1987) in the version specified by the program MEGA (Kumar et al. 1993). Genetic distances were calculated by the two-parameter method (Kimura 1980). The figure shows trees of exon 2 (**A**), the 5' part of intron 1 (**B**), and the 3' part of intron 1 (**C**) for human and NWM genes. Prosimian sequences were included in a second dendrogram of the 5' part of intron 1 (**D**). Numbers at nodes indicate the percentage of recovery of that node in 500 bootstrap replications. In **A**, human sequences are highlighted.

The separate clustering of catarrhine, platyrrhine, and prosimian *DRB* genes implies that each of the three clusters evolved from a single ancestral gene. This gene could have therefore possessed only one of the several variants of motif present in a population at corresponding positions. Consequently, numerous donor genes must have existed alongside the ancestral gene in the population for long enough to transfer all the variants of the motifs into the different copies of the ancestral gene. In such a situation, however, it would have been much simpler to keep the various *DRB* alleles and allelic lineages in the evolving species, instead of generating new ones from the ancestral gene. It is difficult to come up with an appropriate selection pressure favoring the generation of new genes with old motifs, while maintaining the existing genes already possessing these motifs. Since the selection pressure is on the motifs rather than on the rest of the gene, there does not seem to be any advantage in gene replacement compared to the retention of allelic lineages. One could, of course, argue that the gene replacement took place by random genetic drift during a bottleneck phase, but if this were the case, one would expect the reduction in the population size to have led to the loss of many allelic lineages and so also of the motifs. As a result, there would not have been enough donor genes in the bottleneck population to transfer the various motifs onto the new genes spreading through the population.

The alternative to gene conversion is convergent evolution. In a certain sense, it is the simpler of the two explanations because it avoids all the vagaries associated with motif transfers from one gene to another (e.g., the probability of donor and recipient genes coming together in a single individual, the probability of gene conversion taking place in the right part of the gene, and so on). It also avoids the problem of explaining the necessity for the transfer of motifs onto new genes when they are already on hand in the old genes. The assumption here is that, for whatever reason (e.g., because of a drastic reduction in population size), existing motifs are lost, but as there is a continuing need for them (e.g., in a population expanding after a bottleneck), there is a selection pressure to generate them anew. (In the case of the gene conversion hypothesis, once motifs have been lost, they obviously cannot be passed onto new genes by this mechanism in any species of an evolutionary lineage originating after the loss.) On the other hand, convergent evolution at the molecular level is generally considered a highly improbable event if it is expected to involve multiple clustered changes (Takahata 1995). It has, however, been argued that the *Mhc* genes might be an exception and suggested that "*however improbable* statistically in a usual model of nucleotide substitutions, convergence has taken place by natural selection so as to reproduce characteristic motifs in the PBR [peptide binding region]" (Takahata 1995, p. 236). We concur with this conclusion and submit that the incongruence of exon 2 phylogeny with the phylogeny of the rest of each of the *DRB* genes provides strong evidence for convergent evolution at the molecular level.

We base this proposition on three considerations. First, discounting for a moment gene conversion as an explanation for the sharing of motifs among the catarrhine, platyrrhine, and prosimian *DRB* genes, which are clearly of independent origin, the only remaining alternatives are either chance or convergent evolution. We therefore assessed the probability that the observed sequence similarities are the result of chance rather than of selection-driven convergence in nucleotide substitutions. To this end we used the ANCESTOR program (Zhang and Nei 1997) to infer the ancestral sequences in our collection of extant sequences and then estimated the probabilities of homoplasies being due to chance by using CONVERG2 (Zhang and Kumar 1997). The assessment clearly showed the observed convergences and parallelisms to be significantly above chance expectations (Kriener et al. in preparation). The observed similarities are therefore not caused by chance.

Second, alignments of *DRB* exon 2 sequences from a variety of species reveal the existence of intermediate forms between motifs. As an example, some of the intermediates of the YST motifs are shown in Figure 4, arranged so as to show the possible transition pathways. The existence of incompletely assembled motifs in populations is more consistent with their de novo evolution than with their being handed over from one gene to another in a fully assembled form. It also suggests that the partially assembled motifs are favored by natural selection and hence that selection acts along the entire pathway of motif assembly.

	9	10	11	12	13
Consensus sequence	E	Q	A	K	S
Mus domesticus (Mudo Eb18)	E	Y	V **GUN**	T	S
*Bos taurus (BoLADRB3*0801)*	E	Y	A **GCN**	T	S
*Equus caballus (ELA-DRB*4)*	E	Y	T **ACN**	T	S
Platyrrhine *DRB1*03* genes	E	Y	S **AGY**	T	S
Mus domesticus (Mudo Eb13)	E	Y	Y **UAY**	T	S
Syncerus caffer (DRBSyn2e)	E	Y	C **UGY**	T	S
Catarrhine *DRB1*03* genes	E	Y	S **UCN**	T	S
*Macaca mulatta (Mamu-DRB1*0309)*	E	Y	L **UUR**	T	S
Catarrhine *DRB1*03* genes	E	Y	S **UCN**	T	S

Fig. 4. Intermediate forms of the YST motif in eutherians. The amino acid residues are given in the IUPAC-IUB single-letter code. Numbers above the residues refer to the position in the amino acid alignment shown in Figure 1. Possible transition pathways to the YST motif in platyrrhine and catarrhine primates are highlighted by showing the codons corresponding to amino acid residue 11. The codons are given in the IUPAC-IUB standard code for nucleic acids.

Third, and most significant, the gene conversion and the convergent evolution hypotheses of motif origin differ in their predictions regarding the codons specifying amino acid motifs. A motif transferred from one gene to another by gene conversion can be expected to be identical not only in its amino acid but also in the nonsynonymous positions of the corresponding codons. If it has arisen only once and selection has acted on it, the motif should retain its nucleotide identity throughout evolution. By contrast, two motifs arisen independently may have used different codons as starting points of their evolution. To take the example of the YST motif again, its serine residue is specified by the AGY codon in all platyrrhine *DRB* genes bearing this motif, but by the UCN codon in all catarrhine genes. These codons cannot be interconverted by synonymous substitutions only. Any nonsynonymous change leads to the intermediate forms shown in Figure 4. The YST motifs specified by the platyrrhine and catarrhine *DRB* genes therefore are of independent origin and their sharing by the two primate taxa is not accountable by the gene conversion hypothesis.

Acknowledgments

We thank Ms. Jane Kraushaar for editorial assistance. We also thank Dr. Herbert Tichy, MPI für Biologie, Tübingen, Germany for the NWM and prosimian DNA samples and Dr. Philippe Dijan, CNRS, Paris for providing us with tarsier DNA.

References

Andersson G, Larhammer D, Widmark E, Servenius B, Peterson PA, Rask L (1987) Class II genes of the human major histocompatibility complex. Organization and evolutionary relationships of the DRß genes. J Biol Chem 262:8748-8758

Andersson L, Sigurdardóttir S, Borsch C, Gustafsson K (1991) Evolution of MHC polymorphism: extensive sharing of polymorphic sequence motifs between human and bovine DRB alleles. Immunogenetics 33:188-193

Bergström TF, Josefsson A, Erlich HA, Gyllensten U (1998) Recent origin of HLA-DRB1 alleles and implications for human evolution. Nat Genet 18:237-242

Bidwell JL, Lu P, Wang Y, Zhou K, Clay TM, Bontrop RE (1994) *DRB, DQA, DQB*, and *DPB* nucleotide sequences of *Sanguinus oedipus* B95-8. Eur J Immunogenet 21:67-77

Blasczyk R, Kotsch K, Wehling J (1998) The nature of polymorphism of the HLA-DRB intron sequences is lineage specific. Tissue Antigens 52:19-26

Bodmer JG, Marsh SG, Albert ED, Bodmer WF, Bontrop RE, Dupont B, Erlich HA, Hansen JA, Mach B, Mayr WR, Parham P, Petersdorf EW, Sasazuki T, Schreuder GM, Strominger JL, Svejgaard A, Terasaki PI (1999) Nomenclature for factors of the HLA system, 1998. Hum Immunol 60:361-395

Brown JH, Jardetzky T, Saper MA, Samraoui B, Bjorkman PJ, Wiley DC (1988) A hypothetical model of the foreign antigen binding site of class II histocompatibility molecules. Nature 332:845-850

Edwards SV, Chesnut K, Satta Y, Wakeland EK (1997) Ancestral polymorphism of Mhc class II genes in mice: implications for balancing selection and the mammalian molecular clock. Genetics 146:655-668

Figueroa F, O'hUigin C, Tichy H, Klein J (1994) The origin of primate *Mhc-DRB* genes and allelic lineages as deduced from the study of prosimians. J Immunol 152:4455-4465

Gorski J, Mach B (1986) Polymorphism of human Ia antigens: gene conversion between two DR b loci results in a new HLA-D/DR specificity. Nature 322:67-70

Gustafsson K, Andersson L (1994) Structure and polymorphism of horse MHC class II *DRB* genes: Convergent evolution in the antigen binding site. Immunogenetics 39:355-358

Gyllensten U, Sundvall M, Erlich HA (1991) Allelic diversity is generated by intraexon sequence exchange at the *DRB1* locus of primates. Proc Natl Acad Sci USA 88:3686-3690

Gyllensten UB, Bergström T, Josefsson A, Sundvall M, Savage A, Blumer ES, Humberto Giraldo L, Soto LH, Watkins DI (1994) The cotton-top tamarin revisited: *Mhc* class I polymorphism of wild tamarins, and polymorphism and allelic diversity of the class II *DQA1, DQB1*, and *DRB* loci. Immunogenetics 40:167-176

Högstrand K, Böhme J (1994) A determination of the frequency of gene conversion in unmanipulated mouse sperm. Proc Natl Acad Sci USA 91:9921-9925

Hughes AL, Nei M (1989) Nucleotide substitution at major histocompatibility complex class II loci: Evidence for overdominant selection. Proc Natl Acad Sci USA 86:958-962

Kenter M, Otting N, Anholts J, Jonker M, Schipper R, Bontrop RE (1992) Mhc-DRB diversity of the chimpanzee. Immunogenetics 37:1-11

Kimura M (1980) A simple method for estimating evolutionary rates of base substitutions through comparative studies of nucleotide sequences. J Mol Evol 16:111-120

Klein J, Horejsi V (1998) Immunology. 2nd ed. Blackwell Scientific Publishers, Oxford

Klein J, O'hUigin C (1994) MHC polymorphism and parasites. Philos Trans R Soc Lond B Biol Sci 346:351-357

Klein J, O'hUigin C (1995) Class II B *Mhc* motifs in an evolutionary perspective. Immunol Rev 143:89-111

Klein J, O'hUigin C, Kasahara M, Vincek V, Klein D, Figueroa F (1991) Frozen haplotypes in Mhc evolution. In: Klein J, Klein D (eds) Molecular Evolution of the Major Histocompatibility Complex. Springer-Verlag, Heidelberg, pp 261-286

Kumar S, Tamura K, Nei M (1993) MEGA: Molecular Evolutionary Genetic Analysis, version 1.01. The Pennsylvania State University, University Park, PA 16802

Kupfermann H, Mayer WE, O'hUigin C, Klein D, Klein J (1992) Shared polymorphism between gorilla and human major histocompatibility complex *DRB* loci. Hum Immunol 34:267-278

Kupfermann H, Satta Y, Takahata N, Tichy H, Klein J (1999) Evolution of *Mhc-DRB* introns: Implications for the origin of primates. J Mol Evol 48:663-674

Marsh SGE (1999) HLA sequences. http://www.anthonynolan.com/HIG/index.html

McDevitt HO, Mathis DJ, Benoist C, Kanter MR, Williams VEd (1984) Structure, regulatory polymorphisms, and allelic hypervariability regions in murine I-A alpha. Fed Proc 43:3021-3024

Mengle-Gaw L, McDevitt HO (1985) Predicted protein sequence of the murine I-E-S-beta polypeptide chain from cDNA and genomic clones. Proc Natl Acad Sci USA 82:2910-2914

Parham P, Lawlor DA, Lomen CE, Ennis PD (1989) Diversity and diversification of HLA class I alleles. J Immunol 142:3937-3950

Pease LR, Schulze DH, Pfaffenbach GM, Nathenson SG (1983) Spontaneous H-2 mutants provide evidence that a copy mechanism analogous to gene conversion generates polymorphism in the major histocompatibility complex. Proc Natl Acad Sci USA 80:242-246

Saitou N, Nei M (1987) The neighbor-joining method: A new method for reconstructing phylogenetic trees. Mol Biol Evol 4:406-425

Schwaiger F-W, Weyers E, Epplen C, Brün J, Ruff G, Crawford A, Epplen JT (1993) The paradox of MHC-DRB exon/intron evolution: a-helix and ß-sheet encoding regions diverge while hypervariable intronic simple repeats coevolve with ß-sheet codons. J Mol Evol 37:260-272

She JX, Boehme SA, Wang TW, Bonhomme F, Wakeland EK (1991) Amplification of major histocompatibility complex class II gene diversity by intraexonic recombination. Proc Natl Acad Sci USA 88:453-457

Slierendregt BL, van Noort J, Bakas RM, Otting N, Jonker M, Bontrop RE (1992) Evolutionary stability of trans-species major histocompatibility complex class II DRB lineages in man and rhesus monkey. Hum Immunol 35:29-39

Slierendregt BL, Otting N, Besouw N, Jonker M, Bojntrop R (1994) Expansion and contraction of rhesus macaque DRB regions by duplication and deletion. J Immunol 152:2298-2306

Stone WH, Bruun DA, Fuqua C, Glass LC, Reeves A, Holste S, Figueroa F (1999) Identification and sequence analysis of an Mhc class II B gene in a marsupial (Monodelphis domestica). Immunogenetics 49:461-463

Takahata N (1995) Mhc diversities and selection. Immunol Rev 143:225-247

Trtková K, Kupfermann H, Grahovac B, Mayer WE, O'hUigin C, Tichy H, Bontrop RE, Klein J (1993) Mhc-DRB genes of platyrrhine primates. Immunogenetics 38:210-222

Trtková K, Satta Y, Mayer WE, O'hUigin C, Klein J (1995) Mhc-DRB genes and the origin of New World monkeys. Mol Phylogenet Evol 4:408-419

Weiss EH, Mellor A, Golden L, Fahrner K, Simpson E, Hurst J, Flavell RA (1983) The structure of a mutant H-2 gene suggests that the generation of polymorphism in H-2 genes may occur by gene conversion-like events. Nature 301:671-674

Wu S, Sanders T, Bach F (1986) Polymorphism of human Ia antigens generated by reciprocal intergenic exchange between two DR beta loci. Nature 324:676-679

Zhang J, Kumar S (1997) Detection of convergent and parallel evolution at the amino acid sequence level. Mol Biol Evol 14:527-536

Zhang J, Nei M (1997) Accuracies of ancestral amino acid sequences inferred by the parsimony, likelihood, and distance methods. J Mol Evol 44:S139-S146

Zhu Z, Vincek V, Figueroa F, Schönbach C, Klein J (1991) Mhc-DRB genes of the pigtail macaque (Macaca nemestrina): Implications for the evolution of human DRB genes. Mol Biol Evol 8:563-578

Intron 1 sequence analysis of the MHC-DRB1, 3, 4, 5, and 6 genes in five non-human primate species

Katja Kotsch and Rainer Blasczyk

Department of Transfusion Medicine, Hannover Medical School, Carl-Neuberg-Str. 1, 30625 Hannover, Germany

Summary. The common chimpanzee is regarded as one of the closest relatives of humans and its MHC has been extensively studied so far. The analysis of HLA-DRB exon 2 sequences demonstrated that humans and chimpanzees share related alleles which group into lineages predating their speciation. Nevertheless, little is known about the non-coding regions of the different DRB alleles of non-human primates. In order to provide an evolutionary framework to interpret the polymorphism of the HLA-DRB non-coding regions found in humans, we have analyzed the 500 bp 3' intron 1 fragment of the chimpanzee (Pan troglodytes), bonobo (Pan paniscus), gorilla (Gorilla gorilla), orangutan (Pongo pygmaeus) and rhesus monkey (Macaca mulatta). For that purpose, DNA samples from 32 individuals were obtained to amplify 48 alleles of the expressed loci DRB1, -DRB3, -DRB4, -DRB5, and of the pseudogene DRB6. The intron sequences turned out to be highly polymorphic. The phylogenetic tree based on these intron data and closely related human alleles displayed a high sequence similarity for some lineages, for example PatrDRB1*02, GogoDRB1*02 and human DRB1*15/16 showing a high sequence stability of the introns. This finding is in concordance with the observation made for exon 2 sequences, indicating that these lineages have evolved in a trans-species manner. Therefore, the new DRB intron sequence data of non-human primates provide insight into the mechanisms by which polymorphism is generated at the MHC loci.

Key words. DRB, Non-human primates, Introns

Introduction

The observed high sequence similarity between several human HLA-DRB and non-human primate MHC-DRB alleles indicates that most of the alleles now present in the human population were founded before the divergence of humans and their nearest living relatives from their last common ancestor and that the

377

polymorphism has been developed along the evolutionary history of the higher primates (Klein 1987). The MHC-DRB6 lineage for example has been detected in humans, chimpanzees, gorillas, orangutans and rhesus macaques indicating that this lineage is at least 36 million yr old (Slierendregt et al. 1994). Consequently, further understanding of the evolution of the HLA-DRB lineages can be obtained by the study of MHC-DRB sequences of related species. Although recent studies of human, chimpanzees, gorilla and rhesus macaques MHC-DRB genes indicate that at least some of the HLA-DRB variation is shared by these species (Fan et al. 1989; Gyllensten and Erlich 1993; Kenter et al. 1992; Kupfermann et al. 1992; Slierendregt et al. 1992, 1994), little is known about the MHC-DRB non-coding sequences of non-human primates.

The analysis of HLA-DRB introns 1 and 2 (Blasczyk et al. 1998; Bergström et al. 1998; Kotsch et al. 1999) has shown a limited sequence variability within the different lineages which was group-specific rather than allele-specific. This low amount of variation was interpreted as a result of recent origin of the alleles (Bergström et al. 1998).

Therefore the aim of this study was to examine the evolutionary relationships between different DRB lineages based on non-coding sequences. We have analyzed 48 alleles of the 500 bp 3' intron 1 fragment of the chimpanzee (Pan troglodytes), bonobo (Pan paniscus), gorilla (Gorilla gorilla), orangutan (Pongo pygmaeus) and rhesus monkey (Macaca mulatta) for the DRB1, -DRB3, -DRB4, -DRB5 and DRB6 loci. For that purpose genomic DNA was prepared from peripheral blood leukocytes by a salting out procedure for the samples of the gorilla (4 samples), bonobo (2 samples), orangutan (3 samples) and rhesus macaque (13 samples). The 10 DNA samples of the chimpanzees were kindly provided by Ronald Bontrop (Rijswijk, The Netherlands).

The samples were typed for the second exon and several new alleles were detected. The designation of the unknown subtypes is according to the published rules of MHC genes (Klein et al. 1990).

For the sequencing of the non-coding regions, haplotype-specific PCR was carried out using generic primers located in intron 1 and group- or allele-specific primers in the second exon. The PCR protocol as well as the cycle sequencing protocol for sequence analysis of the second exon and the intron sequences were previously described in detail (Blasczyk et al. 1998; Kotsch et al. 1999). Each allele was sequenced up to three times from different PCR products. The human exon 2 sequences were obtained from the EMBL Databank and the non-human primate sequences were obtained from the publication of O'hUigin (1997).

Phylogenetic analysis of the non-human primate intron sequences as well as some human sequences was performed by using the neighbour-joining algorithm (Saitou and Nei 1987) with the Jukes-Cantor method (Jukes and Cantor 1969) as implemented in MEGA (Kumar et al. 1993).

Intron variability

The alignment of the first 100 bp 3' end of intron 1 of the non-human primate MHC-DRB1-5 loci is shown in Figure 1. The intron sequences turned out to be highly polymorphic and within the different lineages of some non-human primates group-specific motifs could be identified. This is especially the case for the PatrDRB1*02, PatrDRB5 and PatrDRB6 group, displaying unique sequence motifs. With a few exceptions, the variability for these lineages is arrested on the level of the serological diversity. The same feature was observed for the human HLA-DRB intron 1 and 2 sequences (Blasczyk et al. 1998; Bergström et al. 1998; Kotsch et al. 1999).

The observation made for human HLA-DRB sequences that all alleles within one lineage show almost no sequence variation is not the case for all non-human primate MHC-DRB lineages. In contrast, some alleles display great subtype differences within several lineages. Particularly for some alleles of the MamuDRB1*03 and MamuDRB1*04 lineages, a characteristic nucleotide deletion of 240 bp was detected (Fig. 1).

Within the GogoDRB1*03 lineage, the GogoDRB1*0308 and GogoDRB1*03new1 show identical sequences for the investigated region in contrast to the GogoDRB1*0307 allele, which displays unique sequence motifs. Also within the PatrDRB3*01 and PatrDRB3*02 lineages, subtype variations could be identified consistent with the detection of allelic subtype differences within the human DRB3*01, -DRB3*02 and -DRB3*03 lineages (Blasczyk et al. 1998; Kotsch et al. 1999; data not shown). The same feature was observed within the PatrDRB4*01 and DRB4*02 lineage as well as for the different GogoDRB5 alleles.

In order to get an impression of the degree of intron 1 variability, we estimated the variable positions for the separate aligned alleles of the chimpanzee, gorilla, bonobo, orangutan and rhesus macaque and compared them with our human data (Blasczyk et al. 1998; Kotsch et al. 1999) (Table 1). The determined GC content of intron 1 was estimated based on the consensus sequences of the different species (Table 2). No differences between the species were found.

Phylogenetic analysis

The phylogenetic tree inferred from the intron 1 sequence data mirrors a high sequence similarity for some lineages (Fig. 2). The alleles belonging to the PatrDRB1*02 and human DRB1*15 lineages for example constitute one cluster showing a high sequence stability of the introns. This finding is in concordance with the observation made for exon 2 sequences, indicating that these lineages have evolved in a trans-species manner. The same feature is observed for all

Table 1. Comparison of the intron 1 and exon 2 variability of the different species

Species	Intron 1 variability (%)	Exon 2 variability (%)	No. of analyzed alleles
Human	4.08	5.49	57 (data not shown)
Chimpanzee	4.78	6.27	19
Bonobo	3.54	4.07	3
Gorilla	4.81	6.27	9
Orangutan	5.93	5.19	4
Rhesus macaque	5.36	6.45	14

All pairs were counted from the consensus. The exon 2 variability is estimated based on the ARS and non-ARS codons. ARS, antigen recognition site

Table 2. Comparison of the GC content in intron 1 in the different species

Species	GC content (%)
Human	64.4
Chimpanzee	64.2
Bonobo	61.5
Gorilla	64.5
Orangutan	64.9
Rhesus macaque	65.3

All data were estimated based on the ACGT content of the consensus sequences (insertions and deletions were ignored).

DRB6 alleles, which form in the intron 1 tree as well in the exon 2 tree, a separated cluster (Fig. 3). The HLA-DRB6*0101 allele is very similar to the PatrDRB6*0109 and GogoDRB6*0102 alleles. The sequence homology between the human DRB1*0701 allele and the PatrDRB1*0702 in the second exon is also found in the intron sequences.

The most divergent branches separate mainly the MamuDRB1*03, -DRB1*04 and -DRB1*10 alleles from the other DRB loci. The MHC-DRB1*04 lineage is present in humans and in rhesus macaques but is missing in the chimpanzee, gorilla and orangutan. Although the exon sequences MamuDRB1*0404 and HLA-DRB1*0402 for example display a high degree of sequence similarity (Slierendregt et al. 1992), the intron 1 data suggest a closer relationship of MamuDRB1*04 alleles to other MamuDRB alleles than to the human DRB1*04 lineage. Interestingly, the MamuDRB1*0401 and MamuDRB1*0402 seem to have a different origin than the other MamuDRB1*04 alleles.

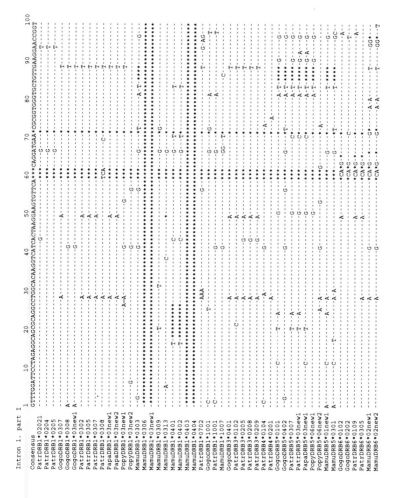

Fig. 1. Sequence alignment of the 3' end of HLA-MHC intron 1 of non-human primates. Nucleotide agreement with the consensus is indicated by a hyphen (-). Asterisks (*) are introduced to achieve maximum alignment. The numbers above the sequences refer to the nucleotide positions.

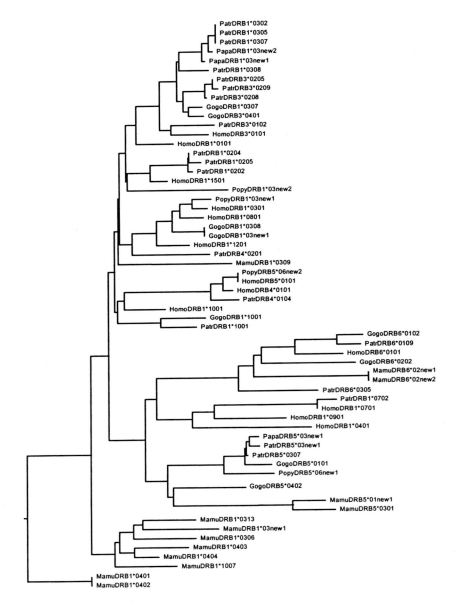

Fig. 2. Phylogenetic tree for the different DRB alleles based on intron 1 sequences. The tree was constructed by the neighbour-joining algorithm of Saitou and Nei (1987).

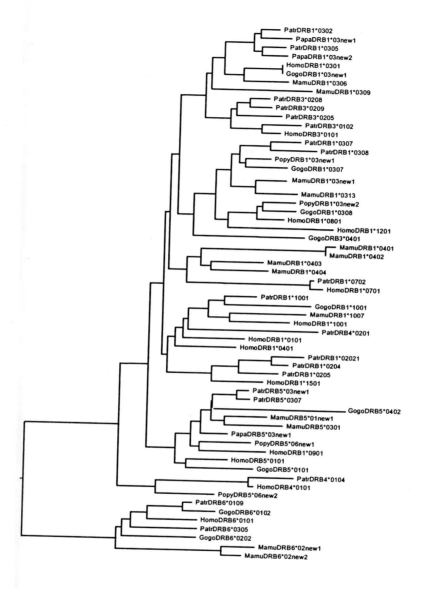

Fig. 3. Phylogenetic tree for the different DRB alleles based on exon 2 sequences. The tree was constructed by the neighbour-joining algorithm of Saitou and Nei (1987).

The next major branch in the intron tree splits off most of the non-human primates DRB5 alleles excluding the PopyDRB5*06new2 and the human DRB5*0101 allele which are most closely related to the PatrDRB4*0104 and the human DRB4*0101 allele. The next two branches include all DRB6 alleles and the alleles of the DR53 haplotype group. In the exon 2 tree, the HLA-DRB1*0901 clusters with several MHC-DRB5 genes from different non-human primates in contrast to the intron tree where this allele displays a great relationship to the alleles of the human DR53 haplotype group (DRB1*04 and DRB1*07).

The next two groups in the tree contain the DRB1*10 and the DRB4 lineage. The alleles of the DRB1*10 lineage of humans, chimpanzee, gorilla and rhesus macaque constitute one cluster in the exon 2 tree. The same feature is observed for the intron 1 tree with the exception of MamuDRB1*1007 which separates very early from the other MHC loci.

The following major branches contain the alleles of the DR52 and DR51 associated group. In one group mainly the DRB1*03 alleles of the gorilla, orangutan and the human DRB1*03, DRB1*08 and DRB1*12 were found and in the next group the DRB1*02 alleles are clustering. All alleles of the MHC-DRB3 lineage as well as the PatrDRB1*03, PapaDRB1*03 and the human DRB1*01 lineage were detected in the last cluster, indicating a common origin of these alleles.

Conclusion

In order to gain further knowledge of the origin of HLA-DRB alleles we present an analysis of the intron 1 variability of several non-human primate species. In the phylogenetic trees of the analyzed intron 1 and exon 2 data, the different sequences of the species tend to form separate subclusters within the lineages. This feature is especially observed for the MHC-DRB1*02, -DRB1*03, DRB1*07, -DRB1*10, DRB5 and DRB6 lineages, indicating that they were already present in an ancestral primate species. Therefore, the new DRB intron sequence data of non-human primates provide insight into the mechanisms by which polymorphism is generated at the MHC loci and are a helpful tool to estimate the age of the ancestral alleles that existed prior to speciation of higher primates.

References

Bergström TF, Josefsson A, Erlich HA, Gyllensten UB (1998) Recent origin of HLA-DRB1 alleles and implications for human evolution. Nat Genet 18:237-242

Blasczyk R, Kotsch K, Wehling J (1998) The nature of polymorphism of the HLA-DRB intron sequences is lineage specific. Tissue Antigens 52:19-26

Fan W, Kasahara M, Gutknecht J, Klein D, Mayer WE, Jonker M, Klein J (1989) Shared class II polymorphisms between humans and chimpanzees. Hum Immunol 26:107-121

Gyllensten UB, Erlich HA (1993) MHC class II haplotypes and linkage disequilibrium in primates. Hum Immunol 36:1-10

Jukes TH, Cantor CR (1969) Evolution of Protein Molecules. Academic Press, New York

Kenter M, Otting J, Anholts J, Jonker M, Schipper R, Bontrop RE (1992) MHC-DRB diversity of the chimpanzee (Pan troglodytes). Immunogenetics 37:1-11

Kenter M, Otting N, De Weers M, Anholts J, Reiter C, Jonker M, Bontrop RE (1993) MHC-DRB and –DQA1 nucleotide sequences of three lowland gorillas: implications for the evolution of primate MHC class II haplotypes. Hum Immunol 36:205-218

Klein J (1987) Origin of major histocompatibility complex polymorphism: the trans species hypothesis. Hum Immunol 19:155-162

Klein J, Bontrop RE, Dawkins RL, Erlich HA, Gyllensten UB, Heise ER, Jones PP, Wakeland EK, Watkins DI (1990) Nomenclature for the major histocompatibility complexes of different species: A proposal. Immunogenetics 31:217-219

Kotsch K, Wehling J, Blasczyk R (1999) Sequencing of HLA class II genes based on the conserved diversity of the non-coding regions: sequencing based typing of the HLA-DRB gene. Tissue Antigens 53:486-497

Kumar S, Tamura K, Nei M (1993) MEGA: Molecular evolution genetics analysis, version 1.0. Pennsylvania State University, University Park, PA

Kupfermann H, Mayer WE, O'hUigin C, Klein D, Klein J (1992) Shared polymorphism between gorilla and human major histocompatibility complex DRB loci. Hum Immunol 34:267-278

O'hUigin C (1997) Non-human primate MHC class II sequences: a compilation. In: Blancher A, Klein J, Socha WW (eds) Molecular Biology and Evolution of Blood Group and MHC Antigens in Primates. Springer, Berlin Heidelberg, pp 358-371

Saitou N, Nei M (1987) The neighbour-joining method: A new method for reconstructing phylogenetic trees. Mol Biol Evol 4:406-425

Slierendregt BL, van Noort JT, Bakas RM, Otting N, Jonker M, Bontrop RE (1992) Evolutionary stability of transspecies major histocompatibility complex class II DRB lineages in humans and rhesus monkeys. Hum Immunol 35:29-39

Slierendregt BL, Otting N, van Besouw N, Jonker M, Bontrop RE (1994) Expansion and contraction of rhesus macaque DRB regions by duplication and deletion. J Immunol 154:2298-2307

Evolution of Catarrhini *DPB1* exon 2 under intragenic recombination

Katsuko Hashiba[1], Shigeki Mitsunaga[2], Katsushi Tokunaga[3], and Yoko Satta[4]

[1]Saitama Prefectural University, 820 Sannomiya, Koshigaya, Saitama 343-8540, Japan
[2]Japanese Red Cross Central Blood Center, 4-1-31 Hiroo, Shibuya-Ku, Tokyo 150-0012, Japan
[3]Department of Human Genetics, School of International Health, Graduate School of Medicine, The University of Tokyo, Hongo 7-3-1, Bunkyo-Ku, Tokyo 113-0033, Japan
[4]Department of Biosystems Science, School of Advanced Sciences, Graduate University for Advanced Studies, Hayama, Kanagawa 240-0193, Japan

Summary. We examined 128 primate *Mhc* class II *DPB1* exon 2 sequences with special reference to the role of intragenic recombination in generating nucleotide diversity. The phylogenetic analysis shows that the beta-sheet coding region exhibits a trans-specific mode of evolution, while the alpha-helix coding region exhibits a monophyly of hominoids to OWMs (Old World Monkeys). Linkage among variable amino acid positions is tight in OWM sequences. In the hominoid alpha-helix coding region, however, amino acid sequence motifs are represented by fairly random combinations of several motifs in two subregions. The distribution of the number of nucleotide differences in all pairwise comparisons of *DPB1* alleles is multimodal in OWMs, whereas it is unimodal in hominoids. The same analysis shows that the alpha-helix region tends to be more unimodal than the beta-sheet region. It is concluded that intragenic recombination has played more important roles in *DPB1* diversity in hominoids than in OWMs.

Key words. *DPB1*, Intragenic recombination, Beta-sheet and alpha-helix regions, Hominoids, Catarrhini

Introduction

In primates, *Mhc* displays the highest level of polymorphism among the nuclear loci examined so far. Characteristic features of *Mhc* polymorphism include long persistence time and a large number of alleles, often resulting in a trans-specific mode of evolution (Klein 1980; Klein and Figueroa 1986; Klein et al. 1993). It is now firmly established that these unusual features of *Mhc* polymorphism are

386

attributable to balancing selection operating at the peptide binding region (PBR) of *Mhc* molecules (Nei and Hughes 1988, 1989; Takahata and Nei 1990).

As balancing selection becomes strong, the persistence time of alleles becomes long and the number of alleles becomes large (Takahata 1990). Selection intensity was estimated based upon the nonsynonymous substitution rate in the PBR coding region: the faster the rate, the stronger the selection (Takahata 1990, Satta et al. 1994). Among four polymorphic *HLA* class II loci (*DRB1*, *DQB1*, *DPB1*, and *DQA1*), the *DRB1* locus shows the fastest rate of PBR nonsynonymous substitutions and thereby the greatest selection intensity. This estimate of selection intensity is consistent with other observations that *DRB1* alleles show the longest persistence time and the largest number of alleles (Satta et al. 1994).

On the other hand, it appears that the *DPB1* locus was subjected to the weakest selection among the four loci. The trans-specific mode of polymorphism is evident at *DRB1*, *DQB1* and *DQA1* alleles, but not at *DPB1* alleles. For example, some *HLA-DRB1* alleles are more closely related to chimpanzee and OWM *DRB1* alleles (Bontrop et al. 1995). This situation is different at the *DPB1* locus. Phylogenetic study of Catarrhini *DPB1* exon 2 sequences revealed a monophyly of *HLA-DPB1* alleles relative to OWM alleles, although the trans-specific mode of polymorphism is evident within OWMs (Slierendregt et al. 1995). Nonetheless, the number of *DPB1* alleles is not particularly small and, in fact, second largest among the four loci. Linkage analyses of phylogenetically informative sites among *HLA-DPB1* alleles consistently suggested significant involvement of intragenic recombination in generating nucleotide diversity (Jakobsen and Easteal 1998, Takahata and Satta 1998). However, it has not been examined whether the significant role of intragenic recombination is restricted to *HLA-DPB1* alleles or extends to *DPB1* alleles in distantly related non-human primates.

Here, in order to address the role of intragenic recombination in generating nucleotide diversity at the Catarrhini *DPB1* locus, we make a detailed phylogenetic analysis of 125 published *DPB1* exon 2 sequences and those of three new alleles of *Macaca fascicularis* (*Mafa-DPB1*). We also examine the amino acid sequence motifs in exon 2 as well as the pattern of nucleotide differences within human, chimpanzee, and rhesus monkey *DPB1* alleles.

Materials and methods

Exon 2 sequences of *Mafa-DPB1* alleles were obtained from blood samples of cynomolgus monkeys that were taken from self-sustaining Malaysian and Indonesian colonies maintained at the Tsukuba Primate Center. Genomic DNA was extracted from the cells by a standard phenol-chloroform method. Exon 2 sequences of the *Mafa-DPB1* locus were amplified by PCR with a set of primers used to amplify *HLA-DPB1* exon 2. Subsequent cloning and sequencing were carried out as described in Mitsunaga et al. (1992). The nucleotide sequences of three new alleles thus obtained were deposited to DDBJ (accession numbers: D16608-D16610).

To perform phyrogenetic analysis, 124 Catarrhini *DPB1* exon 2 sequences were retrieved from DDBJ and the *HLA* database at www.anthonynolan.com/HIG: two from *Macaca fascicularis* (*Mafa*, Hashiba et al. 1993), 16 from *Macaca mulatta* (*Mamu*, Slieredregt et al. 1995), one from *Presbytis entellus* (*Pren*, Gyllensten et al. 1996), 64 from *Homo sapiens* (*HLA*, www.anthonynolan.com/HIG), 28 from *Pan troglodytes* (*Patr*, Otting et al. 1998), six from *Pan paniscus* (*Papa*, Otting et al. 1998), five from *Gorilla gorilla* (*Gogo*, Gyllensten et al. 1996), and two from *Pongo pygmaeus* (*Popy*, Otting et al. 1998). In addition, one cotton-top tamarin sequence (*Saoe-DPB1*0101*, Bidwell et al. 1994) was used as an outgroup sequence. The total number of exon 2 sequences used for these primates is therefore 128.

Results

Phylogenetic analysis of primate *DPB1* exon 2

The phylogenetic tree of the primate *DPB1* exon 2 sequences was constructed by the neighbor joining method (Saitou and Nei 1987) based on both synonymous and nonsynonymous nucleotide differences (Fig. 1). In agreement with previous results (Slieredregt et al. 1995), the hominoid sequences showed a monophyletic relationship to the OWM sequences, although the trans-specific mode of polymorphism was manifest within each of hominoids and OWMs. Bootstrap values within the OWM cluster ranged from 65 to 100%, whereas those within the hominoid cluster ranged from 4 to 98%. A high bootstrap value for a particular cluster indicates a relatively large number of informative sites supporting the cluster. Conversely, a low bootstrap value in the hominoid cluster implies that the sequences contain many phylogenetically incompatible sites or they do not contain a sufficiently large number of informative sites (Satta et al. 1999).

Incompatible sites result from either intragenic recombination or convergent evolution (Klein and O'hUigin 1993; Satta, Klein and Takahata 1999). In order to examine if hominoid *DPB1* genes are more prone to frequent intragenic recombination than OWM's, we carried out the following analyses. It was suggested that there might be a recombination hot spot between the beta-sheet and alpha-helix coding regions in exon 2 of hominoid *DRB1* genes, possibly mediated through the chi-like sequence located at the boundary (Gyllensten et al. 1991). Since a similar chi-like sequence is also found at the *HLA-DPB1* locus, we constructed phylogenetic trees of the beta-sheet and alpha-helix coding regions separately (Figs. 2 and 3). Although most bootstrap values were small, the phylogenetic relationships were clearly different between the two regions. In the beta-sheet coding region, the hominoid sequences intermingled with the OWM sequences (Fig. 2). In the alpha-helix coding region, however, the hominoid

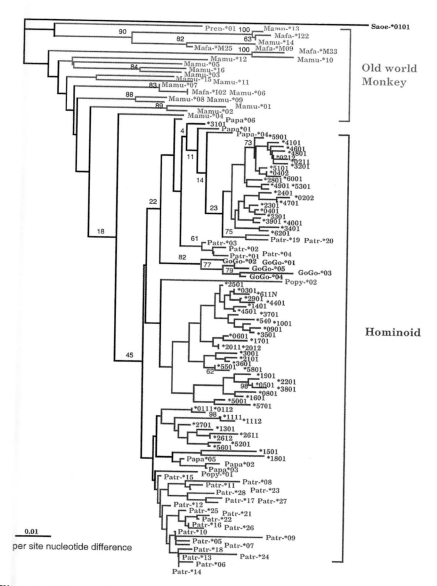

Fig. 1. Phylogenetic tree of exon 2 of *Mhc-DPB1*. The phylogenetic tree was constructed by the neighbor-joining method (Saito and Nei 1987). Bootstrap values were obtained based on 1000 replicates.

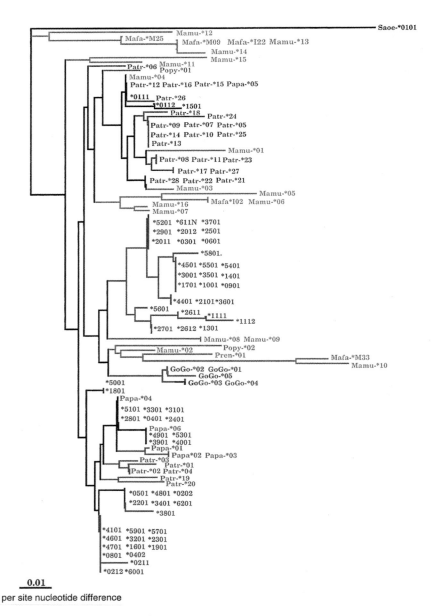

Fig. 2. Phylogenetic tree of exon 2 of *Mhc-DPB1* in the beta-sheet region. Bootstrap values not shown.

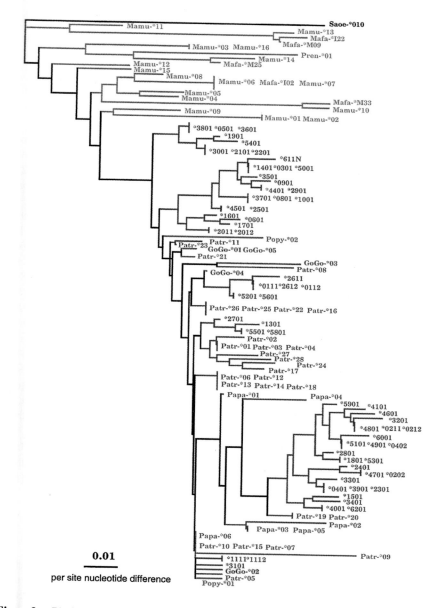

Fig. 3. Phylogenetic tree of exon 2 of *Mhc-DPB1* in the alpha-helix region. Bootstrap values not shown.

sequences tended to make a monophylectic cluster relative to the OWM sequences (Fig. 3). This suggests that molecular mechanisms responsible for generating the nucleotide diversity are different between the two regions.

Amino acid sequence motifs within *HLA-*, *Patr-*, and *Mamu-DPB1* alleles

Patchwork patterns of amino acid sequence motifs in the PBR have been reported for *HLA* class I sequences, providing evidence for intragenic recombination (Lawlor et al. 1990; Parham and Ohta 1996). We examined such patterns in *HLA-*, *Patr-* and *Mamu-DPB1* amino acid sequences. In the alpha-helix coding region of *HLA-DPB1*, we selected a pair of five amino acid positions (55, 56, 57, 65, and 69) and (76, 84, 85, 86 and 87). There are seven and four distinct motifs at these pairs of amino acid positions, respectively. If linkage is complete in the whole alpha-helix coding region, there cannot be more than seven amino acid motifs in this region. However, there were actually 20 combinations between the two subregion motifs, only a little smaller than the maximum number (7 x 4 = 28) of combinations. The proportion of the former to the latter was 20/28 = 71% (Fig. 4). This strongly suggests that the alpha-helix coding region at the *HLA-DPB1* locus underwent frequent intragenic recombination. The same analysis was made for the beta-sheet coding region, by defining two pairs of amino acid positions: (8, 9 and 11) and (35 and 36). There were four motifs in each of these subregions. The number of combined motifs in the beta-sheet was nine, the proportion being 9/16 = 56% (Fig. 4). Thus, within the beta-sheet coding region, intragenic recombination appears to be less frequent than in the alpha-helix coding region. As a result of the trans-specific polymorphism of hominoid *DPB1* alleles, these patterns also held true in the case of *Patr-DPB1* (Fig. 4).

However, *Mamu-DPB1* show much fewer combinations of motifs in the beta-sheet or alpha-helix coding region as well as in the whole exon 2. Most of these alleles are different from each other mainly by a few nucleotide substitutions (Fig. 4).

Mismatch distribution of *HLA-*, *Patr-* and *Mamu-DPB1* alleles

The mismatch distribution (the distribution of the number of nucleotide differences in all pairwise comparisons of alleles) is often used to infer the population demographic history, as previously done in human mitochondrial DNA sequences (Rienzo and Wilson 1991; Slatkin and Hudson 1991). If genes are sampled from a random mating population with a constant population size, the expected genealogy may be described by the coalescence process (Kingman 1982). The mismatch distribution is then expected to be multimodal in most of the cases. When the population has experienced a rapid expansion in the relatively recent past, genes tend to be equally closely related to each other so that the mismatch distribution becomes unimodal. However, since intragenic recombination shuffles nucleotide substitutions among alleles, they tend also to be equally closely related to each other and the mismatch distribution is expected to be unimodal (Satta 1997). To confirm this expectation, we carried out computer simulations. Results show that intragenic recombination indeed makes the mismatch distribution unimodal (Fig. 5). As the recombination rate increases, the range of the distribution becomes somewhat narrow and the average value of nucleotide differences is somewhat reduced (Fig. 5).

HLA-DPB1

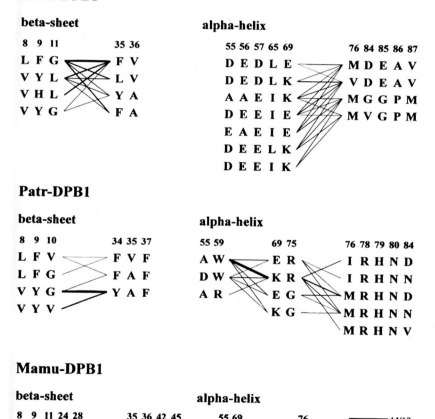

Fig. 4. Combination of motifs for *HLA-*, *Patr-* and *Mamu-DPB1*, and *HLA-* and *Mamu-DPB1* in the beta-sheet and alpha-helix regions. The thickness of the line indicates the number of combinations. The number above each amino acid code (= one letter) shows the position of the residue in the molecule.

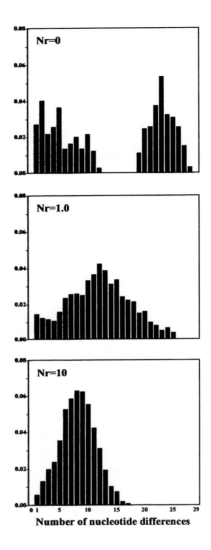

Fig. 5. Simulation results of the mismatch distribution with different recombination rates. Each figure shows the distribution of the nucleotide differences in pairs of alleles in a single replication. The parameters used in the simulation are $N = 100$, $Nu = 2.0$, and $Ns = 50$, where N is the number of individuals, u is the mutation rate per gene per generation and s is the selection intensity under symmetric balancing selection (Satta 1997). The simulation begins with a monomorphic population. After the attainment of equilibrium ($1,000N$ generations), the number of nucleotide differences in all pairs of segregating alleles was examined. The recombination rates (Nr) vary from 0 to 10.0.

We examined the mismatch distribution and the average number of nucleotide differences for the *HLA-*, *Patr-* and *Mamu-DPB1* sequences (Fig. 6). The *HLA* and *Patr* alpha-helix coding sequences show unimodal mismatch distributions, whereas both beta-sheet and alpha-helix coding regions of *Mamu-DPB1* sequences show multimodal mismatch distributions. The average number of nucleotide differences in the *HLA-DPB1* alpha-helix coding region (5.9 ± 2.9) is significantly smaller than that in the corresponding *Mamu-DPB1* region (11.2 ± 4.7), again consistently suggesting strong effects of intragenic recombination in *HLA-DPB1*. Thus, both mismatch distributions and average numbers of nucleotide sequences strongly suggest that intragenic recombination has occurred more frequently in *HLA-* and *Patr- DPB1* alpha-helix coding regions than in the *Mamu-DPB1* region.

Discussion

We examined primate *DPB1* exon 2 nucleotide or deduced amino acid sequences by the phylogenetic analysis, amino acid motifs, and the mismatch distribution. All these suggest a significant role of intragenic recombination in generating nucleotide diversity at the hominoid alpha-helix coding region, relative to that at the OWM counterpart. The role of intragenic recombination might be region-specific, as shown by the difference between beta-sheet and alpha-helix coding regions, or species-specific, as argued for the difference between hominoids and

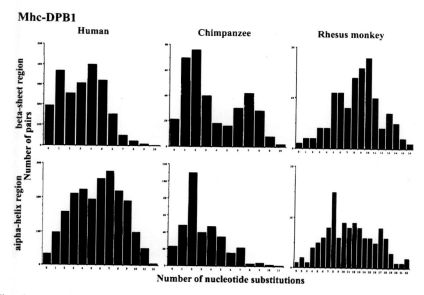

Fig. 6. Mismatch distribution in the sequences of *HLA-*, *Patr-* and *Mamu-DPB1* genes in the beta-sheet and alpha-helix regions.

OWMs. One possibility is that the recombination rate *per se* may differ from region to region as well as from species to species. This possibility seems most parsimonious, but it remains to be explained how the rate can differ in a species-specific and region-specific manner.

Alternatively, the recombination rate might have been kept constant in the whole of exon 2 as well as among closely related species, but balancing selection might have operated differently from region to region or from species to species. However, the computer simulation shows that the shape of mismatch distributions depends heavily on the assumed recombination rate. Even under strong selection, the unimodality is rarely observed without intragenic recombination. The observed multimodality seems to be rather strong evidence for intragenic recombination.

We therefore conclude that, in the hominoid but not in the OWM *DPB1*, the rate of intragenic recombination has been elevated particularly in the alpha-helix coding region. The elevation probably happened after the divergence of hominoids from OWMs, about 35 millions years ago (Martin 1993; Takahata 1999).

Acknowledgments

We thank Dr. Naoyuki Takahata for helpful discussions and comments.

References

Bidwell JL, Lu P, Wang Y, Zhou K, Clay TM, Bontrop RE (1994) DRB, DQA, DQB, and DPB nucleotide sequences of *Saguinus oedipus* B95-8. Eur J Immunogenet 21:67-77

Bontrop RE, Otting N, Slierendregt BL, Lanchbury JS (1995) Evolution of major histocompatibility complex polymorphism and T-cell receptor diversity in primates. Immunol Rev 143:33-62

Di Rienzo A, Wilson AC (1991) Branching pattern in the evolutionary tree for human mitochondrial DNA. Proc Natl Acad Sci USA 88:1597-1601

Gyllensten U, Sundvall M, Erlich HA (1991) Allelic diversity is generated by intra-exon exchange at class II Mhc DRB1 locus of primates. Proc Natl Acad Sci USA 88:3686-3690

Gyllensten U, Bergstrom T, Josefsson A, Sundvall M, Erlich HA (1996) Rapid allelic diversification and intensified selection at antigen recognition sites of the Mhc class II DPB1 locus during hominid evolution. Tissue Antigens 47:212-221

Hashiba K, Kuwata S, Tokunaga K, Juji T, Noguchi A (1993) Sequence analysis of DPB1-like genes in cynomolgus monkeys (*Macaca fascicularis*). Immunogenetics 38:462

Hughes AL, Nei M (1988) Pattern of nucleotide substitutions at major histocompatibility complex class I loci reveals overdominant selection. Nature 335:167-170

Hughes AL, Nei M (1989) Nucleotide substitution at major histocompatibility complex class II loci: Evidence for overdominant selection. Proc Natl Acad Sci USA 86:958-962

Jakobsen IB, Wilson SR, Easteal S (1998) Patterns of reticulate evolution for the classical class I and II HLA loci. Immunogenetics 48:312-323

Kingman JFC (1982) On the genealogy of large populations. J Appl Prob 19A:27-43

Klein J (1980) Generation of diversity at MHC loci: Implications for T cell receptor repertoires. In: Fougereau J, Dausset J (eds) Immunology 80. Progress in Immunology IV. Academic Press, London

Klein J, Figueroa F (1986) Evolution of the major histocompatibility complex. Crit Rev Immunol 6:295-386

Klein J, O'hUigin C (1995) Class II Mhc motifs in an evolutionary perspective. Immunol Rev 143:89-111

Klein J, Satta Y, O'hUigin C, Takahata N (1993) The molecular descent of the major histocompatibility complex. Annu Rev Immunol 11:269-295

Lawlor DA, Zemmour J, Ennis PD, Parham P (1990) Evolution of class-I MHC genes and proteins: from natural selection to thymic selection. Annu Rev Immunol 8:23-63

Martin RD (1993) Primate origins, plugging the gaps. Nature 363:223-234

Mitsunaga S, Kuwata S, Tokunaga K, Uchikawa C, Takahashi K, Akaza T, Mitomi Y, Juji T (1992) Family study on HLA-DPB1 polymorphism: linkage analysis with HLA-DR/DQ and two " new " alleles. Hum Immunol 34:203-211

Otting N, Doxiadis GGM, Versluis L, de Groot NG, Anholts J, Verduin W, Rozenmuller E, Claas F, Tilanus MGJ, Bontrop RE (1998) Characterization and distribution of Mhc-DPB1 alleles in chimpanzee and rhesus macaque populations. Hum Immunol 59:636-664

Parham P, Ohta T (1996) Population biology of antigen presentation by MHC class I molecules. Science 272:67-73

Saitou N, Nei M (1987) The neighbor-joining method: A new method for reconstructing phylogenetic trees. Mol Biol Evol 4:406-425

Satta Y (1997) Effects of intra-locus recombination on HLA polymorphism. Hereditas 127:105-112

Satta Y, O'hUigin C, Takahata N, Klein J (1994) Intensity of natural selection at the major histocompatibility complex loci. Proc Natl Acad Sci USA 91:7144-7188

Satta Y, Klein J, Takahata N (1999) DNA archives and our nearest relative: The trichotomy problem revisited. Mol Phyl Evol (in press)

Slatkin M, Hudson RR (1991) Pairwise comparisons of mitochondrial DNA sequences in stable and exponentially growing populations. Genetics 129:555-562

Slierendregt BL, Otting N, Kenter M, Bontrop RE (1995) Allelic diversity at the Mhc-DP locus in rhesus macaques (Macaca mulatta). Immunogenetics 41:29-37

Takahata N (1990) A simple genealogical structure of strongly balanced allelic lines and trans-species evolution of polymorphism. Proc Natl Acad Sci USA 87:2419-2423

Takahata N (1999) Molecular phylogeny and demographic history of humans. The Proceedings of the Dual Congress (in press)

Takahata N, Nei M (1990) Allelic genealogy under overdominant and frequency-dependent selection and polymorphism of major histocompatibility complex loci. Genetics 124:967-978

Takahata N, Satta Y (1998) Selection, convergence, and intragenic recombination in HLA diversity. Genetica 102/103:157-169

The effect of mutation, recombination and selection on HLA non-coding sequences

Diogo Meyer[1] and Rainer Blasczyk[2]

[1]Department of Integrative Biology, University of California, Berkeley, CA 94720-3140, USA
[2]Department of Transfusion Medicine, Hannover Medical School, Germany

Summary. Several lines of evidence implicate balancing selection in the maintenance of high levels of diversity at coding regions of HLA loci. In the present study we analyze polymorphism, phylogenetic relations, and evidence of recombination at introns adjacent to the polymorphic exons involved in peptide presentation. We confirm previous findings of lower nucleotide diversity at introns, relative to exons. Phylogenies of HLA-B non-coding sequences were more "starlike" and lineages were less differentiated than at HLA-A. HLA-B also presented higher overall levels of reticulation than HLA-A. Together, these results suggest that the processes shaping variation at coding regions are, to some extent, affecting the non-coding variation.

Key words. HLA, Introns, Recombination, Balancing selection

Introduction

Studies of polymorphism of the coding regions at HLA and MHC genes have offered key insights into the evolutionary patterns and processes that characterize these loci. Of particular importance are the long persistence times of allelic lineages (groups of closely related alleles) (Klein et al. 1998), the unusually high levels of polymorphism, an excess of non-synonymous substitutions in regions involved in peptide presentation (Hughes and Nei 1988) and an excess of alleles at intermediate frequencies, relative to neutral expectations (Hedrick and Thomson 1983). Taken together, these results support the hypothesis that variation at MHC loci is maintained by balancing selection (Takahata and Nei 1990; Edwards and Hedrick 1998; Hughes and Yeager 1998). The role of interallelic recombination or gene conversion in the creation of novel alleles has been demonstrated. Specific recombination events have been detected, and the intensity of recombination shown to vary across HLA loci (Parham et al. 1995; Jakobsen et al. 1998; Takahata and Satta 1998).

398

A series of recent studies have examined variation at non-coding regions of HLA. In an analysis of introns 1 and 2 of the class II DRB1 locus, Bergstrom and coworkers (1998) found remarkably little evidence of recombination at the introns, low levels of nucleotide diversity within lineages at introns, and relatively few lineages exhibiting long persistence times. Thus the intron variation did not exhibit the properties that characterize coding regions of HLA. These results were interpreted as evidence that alleles at DRB1 are in fact of fairly recent origin. Cereb and coworkers (1997) analyzed polymorphism at class I loci (A, B and C) and similarly found nucleotide diversity was reduced relative to coding regions involved in peptide presentation. However, these authors interpreted reduced intron diversity as an outcome of intra-locus homogenization, a result of the joint effect of recombination and drift acting on introns, leading them to present neutral levels of variation, lower than those of the adjacent coding regions (Cereb et al. 1997; Hughes and Yeager 1998).

Here we extend studies of non-coding regions by examining polymorphism of a larger number of introns at HLA-A and B and by employing an array of analytical methods. Our approach consists of examining indices of diversity, phylogenetic relatedness, differentiation among lineages and levels of compatibility. The results are discussed in the light of our knowledge on how selection and recombination affect polymorphism.

In the present study we measured nucleotide diversity at introns and the differentiation between allelic lineages (Takahata and Satta 1998). Silent variation in coding sequences was compared to non-coding variation, in order to assess the impact of linkage as an explanation of variation at non-coding sequences. Phylogenies from different introns were inferred and compared, in order to investigate whether the phylogenetic relations among introns are concordant. Compatibility analysis was employed to assess the degree to which introns have been subject to recombination. Recombination at coding (Parham et al. 1995) and non-coding regions (Gomez-Casado et al. 1999) has previously been addressed by directly examining alleles putatively involved. In the present study we complement this approach by examining levels of polymorphism and compatibility, which provide information on overall levels of recombination and are useful when multiple overlapping recombination events may have taken place.

Methods

Sequence generation

DNA samples

Genomic DNA was prepared by a standard salting out procedure. A set of 48 well defined lymphoblastoid B-cell lines mostly from the Ninth and Tenth International

Histocompatibility Workshops and 325 PCR-typed clinical samples representing all serological HLA-A and B antigens and most of their subtypes were analyzed.

Amplification and sequencing primers

For sequencing of the non-coding regions, haplotype-specific PCR-based template preparation was carried out by the use of group or allele-specific amplification primers. These primers were located in the 5' flanking region and the first through 4th exon. The resulting PCR products carried the intron of interest of a single allele flanked by exon sequences. The PCR primers and several nested primers were applied as sequencing primers.

Sequencing of the non-coding regions

Sequencing was performed by cycle sequencing dye terminator chemistry using AmpliTaq FS on Applied Biosystems Sequencers 373 and 377 as previously described (Kotsch et al. 1997). Sequences are currently being prepared for submission (Meyer and Blasczyk, in preparation).

Data analysis

Nucleotide diversity

This quantity (π) is estimated by the mean number of differences over all pairwise comparisons (Nei 1987). Nucleotide diversity within lineages (π_w), and between lineages (π_B) were computed. For exons, nucleotide diversity was estimated using synonymous substitution rates, computed by the method of Nei and Gojobori (1986) as implemented in the software DNAsp (Rozas and Rozas 1999).

Differentiation among lineages

Following Takahata and Satta (1998), differentiation among lineages was computed using $fst = (\pi - \pi_w)/\pi$. This value may range from 1 (when alleles are identical within lineages and distinct among lineages) to zero (when alleles from different lineages are as similar to each other as alleles from the same lineage).

Statistical test for differences in π among introns

Sites were randomly permuted among different introns. In this way the expected distribution of differences among introns attributable to sampling effects was generated. The observed difference in nucleotide diversity among introns was then compared to this distribution, in order to obtain a p-value. The analysis was performed using a program written by D.M.

Phylogenetic lineage assignment

Phylogenies were inferred using the Jukes-Cantor distance, with insertions/deletions ignored, and the neighbor-joining algorithm (Saitou and Nei 1987) as implemented in PAUP* (Swofford 1998). Both synonymous and non-synonymous substitutions were included. Clades with bootstrap support > 50% were used to define lineages.

Compatibility analyses

All pairs of parsimony informative sites were compared and scored as either compatible or incompatible, using the program *reticulate* (Jakobsen and Easteal 1996). When there were more than two nucleotides at a site the 'multiple nucleotides option' was used.

Results

Sequences from introns 1, 2 and 3 of HLA-A and B were analyzed (Table 1). Only the alleles for which introns and exons 2–7 are available were included in the nucleotide diversity analysis. All introns were used for phylogenetic and compatibility analyses.

Nucleotide diversity

There is heterogeneity in nucleotide diversity among introns. Intron 3 has the lowest nucleotide diversity at HLA-A and B (Table 2). The nucleotide diversity at intron 1 is significantly higher than that of intron 3, in accordance with the results of Cereb et al. (1997). Higher diversity at intron 2, relative to intron 3, was found at both loci, but the difference was not statistically significant ($p > 0.1$).

Table 1. Length of aligned sequences and number of alleles

Locus	Intron 1	Intron 2	Intron 3	#Alleles
A	130 bp	241 bp	601 bp	51
B	129 bp	254 bp	575 bp	46

Introns are generally less variable than the coding regions within which they are contained. With the exception of intron 1 at HLA-A, intron nucleotide diversity was much lower than silent nucleotide diversity of exons (Table 2). This is in contrast to the pattern normally seen in nuclear genes, where evolutionary rates at introns and synonymous rates at exons are similar (Hughes and Yeager 1997).

Nucleotide diversity of exons and introns are not correlated in a simple manner: HLA-B has higher synonymous nucleotide diversity than HLA-A at exons, yet the nucleotide diversity values of introns for HLA-B are lower than those of HLA-A.

Phylogenetic analyses

Phylogenies were inferred for each intron separately, at both HLA-A and B. Two aspects of the phylogenies were investigated: the extent of concordance among trees from different introns, and the general shape of the trees.

Concordance among phylogenies

We examined the agreement among phylogenies from different introns. At HLA-A, introns 1 and 3 define groups where alleles from the same serologically defined group are found (Fig. 1A and C). The only serologically defined lineage that was split into distinct groups was A2, which was subdivided into two clear subgroups at the three introns. In intron 2 of HLA-A (Fig. 1B) there are instances where individual alleles were placed in a group distinct from its lineage assignment (in intron 2, an A10 allele is placed within an A19 group and two A1 alleles are placed in a group with A9 alleles).

Although the groups of alleles are themselves fairly conserved at all introns, the way these groups are related does vary among introns. For example, although A19 alleles always group together, at intron 3 they are seen to be nearly identical to A2 and A10, while at introns 1 and 2 they are clearly distinct from these lineages. In summary, at HLA-A the different introns support similar groupings of alleles, but the relationship among these lineages differs among introns. Further investigation is required in order to assess whether these differences are significant, or are due to uncertainty in the placement of groups of alleles on the phylogeny.

At HLA-B a very different result was obtained. Alleles that are grouped by certain introns are spread throughout the phylogeny when others are analyzed. Fig.1D is a phylogeny of intron 2. Symbols next to allele names refer to the group that those alleles belong to in an intron 3 phylogeny. There are several cases where these symbols are spread throughout the tree. For example, the alleles B*0801, B*4101, B*4102 and B*4201 share identical intron 3 sequences (represented by filled squares) yet are divergent when intron 2 is analyzed (see Fig. 1D). This analysis shows that the level of discordance among topologies from different introns is greater at HLA-B than at A.

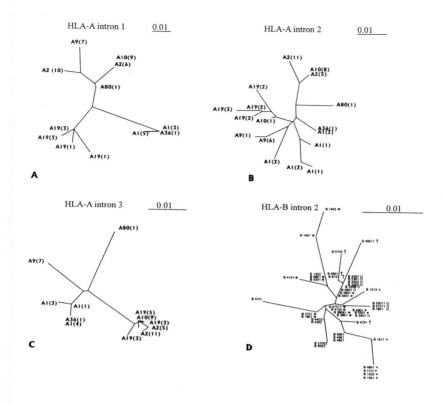

Fig. 1. Phylogenetic relations among HLA alleles. The bars provide the scale of the genetic distance represented on the tree. **A.** HLA-A intron 1. **B.** HLA-A intron 2. **C.** HLA-A intron 3. **D.** HLA-B intron 2. In the phylogenies for HLA-A the serological group to which alleles belong is indicated, and the number of alleles at each position on the tree is given in parentheses. For HLA-B each distinct symbol refers to a group of alleles that formed a single group when intron 3 was analyzed. Phylogenies were inferred using the Jukes-Cantor distance and the neighbor-joining method.

Table 2. Nucleotide diversity ($\pm$ s.d.) per 100 bases (π)

HLA-A

	π^a
Intron 1	4.114 (0.327)
Intron 2	2.421 (0.139)
Intron 3	1.687 (0.188)
Exon 2	3.656
Exon 3	2.846
Exons 4–7	5.095

HLA-B

	π
Intron 1	2.411 (0.166)
Intron 2	0.969 (0.253)
Intron 3	0.708 (0.113)
Exon 2	5.050
Exon 3	4.423
Exons 4–7	2.151

[a] Nucleotide diversity at exons refers to synonymous variation.

Shape of phylogenies

Tree shape provides information on the evolutionary processes acting at a locus. Trees can be broadly described as "starlike" or "structured". Starlike trees can be produced by several biological phenomena. It is of interest in the present context that such trees arise if the alleles experience extensive recombination. High recombination rates have the effect of making sites independent from one another. As a consequence, different sites may support different phylogenies. When these sites are analyzed in the context of a single data set the result is an unresolved (or starlike) tree.

Structured trees, on the other hand, present resolution, with sets of alleles being clearly grouped by shared mutations. This is the expected pattern under the process of diversification of alleles with low (or no) recombination.

At HLA-A, intron 2 displays the most starlike topology (Fig. 1B). For example, at intron 2 of HLA-A, alleles from the A1 lineage are spread out over a greater portion of the tree, whereas they are closely grouped for the other introns. Similarly, at intron 2 of HLA-B groups of alleles are not as clearly separated by long intermediate branches. Overall, we found that phylogenies for the HLA-B introns are more starlike than those of HLA-A.

Distribution of variation: diversity within and between lineages

The differentiation of alleles within and between lineages was quantified using *fst* (see "Differentiation among lineages" in Methods for details). High *fst* implies that most differences among alleles are due to inter-lineage comparisons, whereas low *fst* values are obtained when differences within and between lineages are similar. It is important to note that *fst* is a measure of differentiation among lineages that is scaled by the overall differentiation among alleles. Thus *fst* values for intron 3 are higher than those for introns 1 and 2, even though intron 3 has lower nucleotide diversity. This is because although the total differentiation among alleles at this intron is low, most of the differences are due to inter-lineage differentiation.

HLA-A

The serological classification of alleles was used to define lineages, and the fit of allelic variation to these was assessed using *fst*. The *fst* values for introns were high (> 0.74), implying that lineages defined by serology are well differentiated. Intron 3 has the highest *fst*, in accordance with our expectations from inspecting the phylogeny, which displays little differentiation within lineages. When the *fst* values for coding regions were computed, we found that the exons not involved in peptide presentation (4–7) show high *fst* values, a pattern similar to that of intron 3 (Table 3A), while exon 2 presented the poorest fit to the lineage assignments.

In order to assess whether our results were in anyway artifacts of the serological definition of HLA-A lineages, we also defined lineages based on groups supported by a phylogenetic analysis of exons 2 and 3. This resulted in groups of alleles that were virtually identical to those defined by serology. The differentiation among lineages was thus shown not to be an artifact of lineage definition.

HLA-B

Serological definitions of lineages at HLA-B are not as reliably established as those of HLA-A. We therefore assigned alleles to lineages based on a phylogeny inferred from exons 2 and 3 (see "Phylogenetic lineage assignment" in Methods). Our goal is to quantify the extent to which the genealogical relationships among alleles, reflected in coding sequences, is mirrored by variation in non-coding sequences.

Introns at HLA-B are not as strongly structured into groups that match exon 2 and 3 lineages, as was the case for HLA-A. This can be seen from the lower *fst* values at the HLA-B introns, relative to those of HLA-A (Table 3). This is in accordance with the starlike shape of the HLA-B allele genealogy.

Two main patterns emerge from the distribution of nucleotide diversity within and between lineages. Firstly, at intron 3 of HLA-A the amount of variation due to inter-lineage differences is greatest. Notice that the higher *fst* for intron 3 is another way of presenting the results obtained in the phylogenetic analysis, which showed that intron 3 variation was more "structured". Secondly, HLA-A intron variation is apportioned into lineages more clearly than variation at HLA-B.

The main caveat with this analysis is the sensitivity it presents to the definitions of lineages. The assignment of alleles to lineages was based on a phylogenetic analysis, and ambiguity arises due to differences in grouping according to inference method, and the bootstrap threshold used to define reliable groups. Thus the exact *fst* values presented for HLA-B are not as reliable as those of HLA-A, where lineages are not ambiguous. However, the overall lack of structure at HLA-B is a conclusion that can be reliably drawn from the present analysis.

Phylogenetic concordance among regions: compatibility analyses

Compatibility was studied using the method of Jakobsen and Easteal (1996). All pairs of parsimony informative sites in the data set are analyzed, and classified as either compatible or incompatible. Compatible pairs of sites are those at which the variants occurring at each one can be accommodated onto a phylogeny without invoking more than one change per site. In other words, the phylogenetic information provided by both sites does not present any disagreement. Incompatible sites, on the other hand, are those that require multiple changes along a phylogeny. Incompatibility may arise due to convergent mutations or recombination events.

At HLA-A, intron sequences are highly compatible to each other (see values in bold type, Table 4). We also found that exons 4–7 were highly compatible with introns. A similar hierarchy is seen at HLA-B: introns are more compatible to each other and to introns 4–7 than are exons 2 and 3.

However, HLA-A and B differ with respect to their overall values of compatibility. HLA-B presents lower compatibility than does HLA-A (compatibility values for the entire locus are 0.772 and 0.523, for HLA-A and B respectively).

The two loci thus share the property of having introns with higher compatibility, compared to exons 2 and 3 (indicating lower recombination and/or convergence at introns), yet differ in that HLA-B presents lower compatibility overall.

Table 3. Nucleotide diversity within and between lineages, per 100 bases, and differentiation between lineages for HLA-A and B

	HLA-A (7 lineages)[a]			HLA-B (22 lineages)[b]		
	π_B[c]	π_W[d]	*fst*	π_B	π_W	*fst*
Intron 1	5.1	1.1	0.74	2.57	1.4	0.43
Intron 2	2.9	0.7	0.72	1.02	0.6	0.39
Intron 3	2.1	0.1	0.94	0.75	0.3	0.58
Exon 2	3.9	2.5	0.32	5.41	1.0	0.80
Exon 3	3.3	0.7	0.75	4.51	3.5	0.21
Exons 4–7	6.2	0.3	0.94	2.26	0.9	0.58

[a] The seven lineages correspond to serological groups.
[b] The lineages were defined from the bootstrap consensus clade of a neighbor joining analysis of exons 2 and 3.
[c] π_B: between lineage nucleotide diversity.
[d] π_W: within lineage nucleotide diversity.

Table 4. Compatibility among regions of HLA-A and B

HLA-A		HLA-B		
Regions[a]	Compatibility[b]	Regions	Compatibility	
E2 vs E3	0.538	E2 vs E3	0.339	← most reticulation[c]
E2 vs E4–7	0.620	E2 vs I2	0.380	
E2 vs I3	0.664	E2 vs E4–7	0.383	
E3 vs E4–7	0.671	I1 vs E3	0.505	
I1 vs E3	0.675	I2 vs E3	0.540	
E3 vs I3	0.681	I1 vs E2	0.562	
I1 vs E2	0.682	E3 vs E4–7	0.562	
E2 vs I2	0.688	E2 vs I3	0.578	
I2 vs E3	0.738	I1 vs E4–7	0.625	
I2 vs E4–7	0.826	**I1 vs I2**	**0.655**	
I1 vs I2	**0.840**	**I1 vs I3**	**0.665**	
I1 vs I3	**0.859**	I2 vs E4–7	0.676	
I2 vs I3	**0.872**	E3 vs I3	0.797	
I1 vs E4–7	0.896	**I2 vs I3**	**0.823**	
I3 vs E4–7	0.930	I3 vs E4–7	0.926	← least reticulation

[a] Each line refers to a comparison between two regions. E: exons, and I: introns.
[b] Values are the proportion of parsimony informative sites that are compatible among two regions (Jakobsen and Easteal 1996).
[c] Compatibility values are ranked from lowest to highest, separately for each locus. Comparisons between two introns are in bold.

Discussion and conclusions

Nucleotide diversity at introns is not entirely explained by distance from selected exons

Intron 3 has the lowest nucleotide diversity at both HLA-A and B. Intron 3 is also the longest of the introns analyzed, making its sites on average further away from exons 2 and 3 than the sites of introns 1 and 2. Thus, low nucleotide diversity at intron 3 has previously been interpreted as an outcome of the lower linkage between this intron and exons 2 and 3, which are the targets of balancing selection. Increased diversity of introns 1 and 2 is expected to arise due to the greater linkage these display with respect to the selected exons (Cereb et al. 1997). However, nucleotide diversity at intron 2 was not significantly higher than that of intron 3. By inspecting the values for nucleotide diversity at the introns (Table 2), what we in fact find is that intron 2 diversity values are closer to those of intron 3, than to 1. These results suggest that linkage relations alone do not explain the observed levels of diversity.

We found that intron and exon nucleotide diversities are not coupled in a simple way: nucleotide diversity at HLA-B is higher than that of HLA-A at exons 2 and 3, but is lower at introns. Thus there is no single evolutionary factor which can fully account for both coding and non-coding levels of polymorphism.

There are differences in phylogenetic patterns between HLA-A and B

All three introns at HLA-A support well defined allelic lineages (groups of related alleles are separated be longer internal branches). The groups defined by the introns closely match those defined by serology or exon 2 and 3 sequence data. The results for HLA-B differ from those of A in two respects. Firstly, HLA-B alleles display a more starlike phylogeny. Secondly, there are several instances of discordance among introns, with respect to the phylogenetic placement of specific alleles. These results show that the lack of phylogenetic structure at HLA-B, which had previously been found to characterize coding regions (Jakobsen et al. 1998), also affects the adjacent non-coding regions.

There are differences in *fst* values between HLA-A and B

The *fst* values for HLA-B were lower than those of HLA-A. Lower values of *fst* may result when inter-lineage recombination occurs, as shown by Takahata and Satta (1998). Thus the lower *fst* found at HLA-B strengthen the preceding conclusions, implicating recombination as a force that has to some extent shaped HLA-B intron diversity.

Compatibility is highest outside exons 2 and 3

At both HLA-A and B, compatibility was highest among introns, and among introns and exons not involved in peptide presentation. Exons 2 and 3 were highly incompatible to each other, and to the adjacent non-coding regions. The compatibility analysis also shows that recombination does not extend in the 3' direction of the loci, since intron 3 and exons 4–7 are remarkably compatible to each other for both HLA-A and B. Analysis of introns 4, 5 and 6 will aid to better address this issue. From these results we conclude that incompatibility is not distributed randomly over the locus, and is in fact highly localized in the exons coding for regions involved in peptide presentation.

However, the localization of recombination in coding regions is not absolute. By comparing compatibility at HLA-A and B, we found that non-coding sequences of HLA-B display increased reticulation (reduced compatibility). For coding sequences HLA-B also displays reduced compatibility. Thus increased reticulation at HLA-B coding regions (Jakobsen et al. 1998) has left a mark on non-coding sequences, but has not been sufficiently intense as to erase the differences in compatibility between introns and exons.

It is interesting to note that the high levels of compatibility between intron 3 and exons 4–7 at HLA-B suggest a solution to the apparently intractable problem of obtaining an allelic genealogy for the alleles of HLA-B, since these regions appear to be evolving in the absence of reticulation. It will be interesting to study the lineages defined by these regions, and compare them to variation at exons 2 and 3 and the other introns.

Final conclusions

Variation of introns is the result of the combined effects of mutation, recombination, and selection. Selection on the peptide presenting exons is expected to increase variation at linked neutral regions. The lower nucleotide diversity values for intron 3 supports this scenario. However, intron 2 nucleotide diversity values were lower than expected. The effect of recombination at intron 2 may partly account for this result. Recombination has the effect of decreasing nucleotide diversity at the linked neutral region, and of reducing interlineage differentiation (Takahata and Satta 1998). We suggest that increased recombination at intron 2 may account for lower nucleotide diversity and lineage differentiation values observed at this region.

The compatibility analysis provides compelling evidence that the entire region of the HLA-A and B loci 3' of exon 3 is not experiencing recombination as are the other regions. This would have the effect of enhancing the differentiation among lineages in this region (as is seen by fst values for intron 3 and exons 4–7 at HLA-A).

Our results agree with earlier findings, showing that the evolution of HLA alleles is under different evolutionary forces in coding and non-coding regions (Cereb et al. 1997). However, there are differences between variation at the HLA-A and B non-coding regions which may be attributable to the joint effects of balancing selection (increasing nucleotide diversity among lineages) and recombination (reducing nucleotide diversity and fst). Thus, although introns are evolving in a more neutral manner, they nonetheless display patterns of variation that indicate they have been subject to evolutionary forces acting on their flanking sequences, the highly polymorphic exons.

Acknowledgements

We thank Glenys Thomson, Shannon McWeeney and Kristie Mather for helpful comments. D.M. is supported by a CAPES scholarship from the Brazilian government and NIH grant GM35326 to Glenys Thomson.

References

Bergstrom TF, Josefsson A, Erlich HA, Gyllensten U (1998) Recent origin of HLA-DRB1 alleles and implications for human evolution. Nat Genet 18:237-242

Cereb N, Hughes AL, Yang SY (1997) Locus-specific conservation of the HLA class I introns by intra-locus homogenization. Immunogenetics 47:30-36

Edwards SV, Hedrick PW (1998) Evolution and ecology of MHC molecules: From genomics to sexual selection. Trends Ecol Evol 13:305-311

Gomez-Casado E, Vargas-Alarcon G, Martinez-Laso J, Granados J, Varela P, Alegre R, Longas J, Gonzalez-Hevilla M, Martin-Villa JM, Arnaiz-Villena A (1999) Evolutionary relationships between HLA-B alleles as indicated by an analysis of intron sequences. Tissue Antigens 53:153-160

Hedrick P, Thomson G (1983) Evidence for balancing selection at HLA. Genetics 104:449-456

Hughes AL, Nei M (1988) Pattern of nucleotide substitution at major histocompatibility complex class I loci reveals overdominant selection. Nature 335:167-70

Hughes AL, Yeager M (1997) Comparative evolutionary rates of introns and exons in murine rodents. J Mol Evol 45:125-130

Hughes AL, Yeager M (1998) Natural selection at major histocompatibility complex loci of vertebrates. Annu Rev Genet 32:415-35

Jakobsen IB, Easteal S (1996) A program for calculating and displaying compatibility matrices as an aid in determining reticulate evolution in molecular sequences. Comput Appl Biosci 12:291-295

Jakobsen IB, Wilson SR, Easteal S (1998) Patterns of reticulate evolution for the classical class I and II HLA loci. Immunogenetics 48:312-323

Klein J, Sato A, Nagl S, O'hUigin C (1998) Molecular trans-species polymorphism. Annu Rev Ecol Syst 29:1-12

Kotsch K, Wehling J, Koehler S, Blasczyk R (1997) Sequencing of HLA class I genes based on the conserved diversity of the noncoding regions: Sequencing-based typing of the HLA-A gene. Tissue Antigens 50:178-191

Nei M (1987) Molecular Evolutionary Genetics. Columbia University Press, New York

Nei M, Gojobori T (1986) Simple methods for estimating the numbers of synonymous and nonsynonymous nucleotide substitutions. Mol Biol Evol 3:418-426

Parham P, Adams EJ, Arnett KL (1995) The origins of HLA-A,B,C polymorphism. Immunol Rev 143:141-180

Rozas J, Rozas R (1999) DNAsp version 3: an integrated program for molecular population genetics and molecular evolution analysis. Bioinformatics 15:174-175

Saitou N, Nei M (1987) The neighbor-joining method: a new method for reconstructing phylogenetic trees. Mol Biol Evol 4:406-425

Swofford D (1998) PAUP*. Phylogenetic Analysis Using Parsimony (*and Other Methods). Version 4. Sinauer, Sunderland, Massachusetts

Takahata N, Nei M (1990) Allelic genealogy under overdominant and frequency-dependent selection and polymorphism of major histocompatibility complex loci. Genetics 124:967-978

Takahata N, Satta Y (1998) Footprints of intragenic recombination at HLA loci. Immunogenetics 47:430-441

HLA-DQ haplotypes in 15 different populations

Ann B. Begovich[1], William Klitz[2], Lori L. Steiner[1], Sarah Grams[1], Vina Suraj-Baker[1,3], Jill Hollenbach[4], Elizabeth Trachtenberg[5], Leslie Louie[2], Peter A. Zimmerman[6], Adrian V.S. Hill[7], Mark Stoneking[8], Takehiko Sasazuki[9], Olga Rickards[10], Vincent P.K. Titanji[11], Vladimir I. Konenkov[12], and Marina L. Sartakova[12]

[1]Department of Human Genetics, Roche Molecular Systems, Alameda, CA, USA
[2]School of Public Health, University of California, Berkeley, CA, USA
[3]Department of Medicine, Tulane University School of Medicine, New Orleans, LA, USA
[4]Department of Integrative Biology, University of California, Berkeley, CA, USA
[5]Children's Hospital Oakland Research Institute, Oakland, CA, USA
[6]Laboratory of Parasitic Diseases, National Institutes of Health, Bethesda, MD, USA
[7]Institute of Molecular Medicine, University of Oxford, Oxford, UK
[8]Max-Planck-Institute for Evolutionary Anthropology, Leipzig, Germany
[9]Department of Genetics, Kyushu University, Fukuoka, Japan
[10]Dipartimento di Biologia, Universita di Roma "Tor Vergata", Rome, Italy
[11]University of Buea, Cameroon
[12]Institute of Clinical Immunology, Novosibirsk, Russia

Summary. In order to understand the forces governing the evolution of the DQ molecule, PCR-based methods have been used to type the DQA1 and DQB1 loci encoding this heterodimer on 2,807 chromosomes from 15 different populations including Africans, Asians, Amerindians and Caucasians. These ethnically diverse samples represent a variety of population substructures and include small, isolated populations as well as larger populations where admixture has occurred. Nine DQA1 alleles and 18 DQB1 alleles have been identified which make up 42 distinct DQ haplotypes. Some haplotypes are found in all ethnic groups while others are confined to a single ethnic group or population. Despite evidence of recombination between the DQA1 and DQB1 loci, there are no examples of a haplotype carrying a DQw1-associated alpha chain and a DQw2-, DQw3-, or DQw4-associated beta chain in *cis* (and vice versa). These data suggest that these haplotypes, which encode unstable heterodimers, are rapidly removed from the population through natural selection.

Key words. Haplotype, Heterodimer, HLA-DQA1, HLA-DQB1, Polymorphism

Introduction

The human class II molecule HLA-DQ is a glycoprotein consisting of an alpha and a beta chain, both encoded in the HLA-region, found on the cell surface of antigen presenting cells. This heterodimeric molecule, like the other class II molecules DR and DP, serves as a receptor for processed peptides, derived predominantly from membrane and extracellular proteins (e.g., bacterial peptides), which it presents to CD4[+] T cells initiating an immune response. Unlike the well-studied DR molecule, where only the beta chain is polymorphic, the genes encoding both the alpha and beta chains of the DQ molecule are highly polymorphic. There are 19 DQA1 and 39 DQB1 alleles which encode 16 and 33 unique polypeptide chains respectively (Bodmer et al. 1999).

In an individual heterozygous for both DQA1 and DQB1, each alpha chain should, in theory, be able to interact and pair with a beta chain encoded by the same chromosome (*cis*-heterodimer) or the other chromosome (*trans*-heterodimer) leading to four distinct DQ molecules, each capable of binding a unique array of peptides. In fact, both *cis*- and *trans*-heterodimers have been identified and observed to be functional (Giles et al. 1985; Nepom et al. 1987; Kwok et al. 1988; Lundin et al. 1990); although recent gene transfer experiments have shown that not all alpha and beta chains pair efficiently (Kwok et al. 1989, 1993). Specifically, the following heterodimers were unable to pair or be expressed on the cell surface: DQA1*0301-DQB1*0501, DQA1*0401-DQB1*0501, DQA1*0501-DQB*0501, DQA1*0101-DQB1*0201, and DQA1*0101-DQB1*0302. This led to the general understanding that DQβ chains that are members of the DQw1 specificity (DQB1*0501-0504, 0601-0615) do not form stable cell surface heterodimers with DQw2-, DQw3- or DQw4-associated DQα chains (DQA1*0201, 0301-0303, 0401, 0501-0505, and 0601). Similarly, heterodimers composed of DQw2-, DQw3- or DQw4-associated DQβ chains and DQw1-associated DQα chains are unstable.

The fact that particular DQ alpha and beta chains do not pair efficiently may explain the observation that in most populations studied to date one rarely, if ever, finds alpha and beta alleles encoding these unstable heterodimers on the same chromosome (i.e. a DQw1-associated beta chain is not found on the same haplotype as a DQw2-, DQw3-, or DQw4-associated alpha chain and vice versa) (Imanishi et al. 1992; Clayton et al. 1997; Gyllensten and Erlich 1989). Consequently, the haplotypes encoding these unstable heterodimers have been referred to as "forbidden". One reported exception to this general rule appears in a recent report on class II loci in seven indigenous Siberian populations (Grahovac et al. 1998). In this study, which included 343 individuals (686 chromosomes), 13 distinct "forbidden" DQ haplotypes were reported in 22 individuals. The authors attributed this unusual finding to a combination of recombination and genetic drift in these small, isolated populations.

In order to carefully ascertain the pattern of DQA1 and DQB1 haplotypes and their distribution in geographically diverse populations, high-resolution DNA-based methods have been used to characterize DQ allele and haplotype frequencies in 15 different populations representing 2,807 chromosomes from Africans, Amerindians, Asians, and Caucasians.

Materials and methods

Populations

The 15 populations included in this study and the number of chromosomes typed are listed in Table 1. Only one population, the Centre d'etude du polymorphisme humain (CEPH), is representative of Northern Europeans, which is the population that has been studied most extensively for HLA. The remaining 14 populations are from diverse geographical regions and include individuals of African, Asian and Amerindian origin. Some have evolved under virtual isolation while others have been subject to admixture. All populations, except for the CEPH, include unrelated individuals.

Table 1. Populations

Population	Geographic origin	Number of chromosomes tested (2N=2807)
Ugandans	Africa	94
Cameroon	Africa	344
Gambians	Africa	292
Ecuadorian Africans	South America	116
Cayapa (Ecuador)	South America	166
Mixe (Oxaca, Mexico)	North America	106
Mixtec (Oxaca, Mexico)	North America	126
Zapotec (Oxaca, Mexico)	North America	170
Indonesians	Asia	264
PNG Highlanders	Asia	156
PNG Lowlanders	Asia	96
Japanese	Asia	172
Tuvinians (Mongolia)	Asia	338
East Indians (Bombay, India)	Asia	118
CEPH (Northern European)	Europe/North America	249

Ugandans

The 47 samples from Uganda, which were collected at the Uganda Cancer Institute in Kampala as part of a study on Kaposi's sarcoma, were from individuals who identify themselves as Muganda, members of the Baganda ethnic group. They were provided by L. Louie (University of California, Berkeley, CA).

Cameroon

The 172 samples from Cameroon, which were collected as part of a study on the immune response to the filarial parasite *Onchocerca volvulus*, which causes river blindness, were provided by V. P. K. Titanji (University of Buea, Cameroon) and P. A. Zimmerman (NIAID, Bethesda, MD). They were obtained from individuals living in villages near Saa, a community 50 km northwest of Yaounde, the capital of Cameroon. The majority of this population are Bantu-speaking Manguissa and Etons, with a few immigrants from the Banileke, Ewondo, Ngwo, and Ngemba groups.

Gambians

The Gambia population, which consists of 146 samples, was the control population used in a study of HLA and susceptibility to malaria (Hill et al. 1991). The ethnic composition of this group is mixed; there are Mandinka, Jola, Wolof, Fula, and others from less common ethnic groups.

Cayapa & Ecuadorian Africans

The Cayapa (N=83) are an indigenous population from Ecuador of approximately 3,600 individuals. They define themselves as Chachi and speak a language classified as belonging to the Chibchan-Paezan branch of the Amerind family. They are unique in that they have undergone evolution in virtual isolation for 15,000-20,000 years. The Ecuadorian Africans (N=58) are ancestors of escaped slaves, arriving in Ecuador within the past 500 years. Despite their close proximity, there is little admixture between these two Ecuadorian populations. Samples from these two populations were provided by O. Richards (Universita di Roma "Tor Vergata", Rome, Italy) and E. Trachtenberg (CHORI, Oakland, CA).

Mixe, Mixtec, and Zaptotec

The Mixe (N=53), Mixtec (N=63) and Zapotec (N=85) are indigenous populations from Oaxaca, Mexico. Approximately 400,000 Zapotecs, 240,000 Mixtecs, and 100,000 Mixe live in Oaxaca today. The Zapotec and Miztec languages are part of the same language family and are thought to have diverged at least 3,500 years ago while the Mixe language belongs to a separate family. This linguistic data suggests that the Zapotec and Mixtec populations may be closer genetically than either is to the Mixe. Samples from these three populations, which were collected for a study on genetic diversity within Amerindian populations, were provided by J. Hollenbach (University of California, Berkeley, CA).

Indonesians and Papua New Guineans

The Indonesians (N=132), Papua New Guinean (PNG) highlanders (N=78) and PNG lowlanders (N=48) were collected for studies on mitochondrial diversity (Stoneking et al. 1990, Redd et al. 1995). The Indonesian samples were collected from two different groups of islands, the Nusa Tengaras and the Moluccas. The PNG-highlanders are an isolated population living in the mountains and are thought to be representative of the first people to colonize PNG. The lowlanders, on the other hand, live near the coast and have been subject to migration and gene flow from other Asian populations.

Japanese

The Japanese samples, which were collected from 86 medical students participating in a study on responsiveness to a hepatitis vaccine, were provided by T. Sasazuki (Kyushu University, Fukuoka, Japan).

Tuvinians

The Tuvinians are from a Central Asian republic of nearly 400,000 individuals located on a plateau wedged between northwestern Mongolia and southern Siberia. Samples from 169 individuals, who are thought to be the descendants of Genghis Khan, were provided by M. Sartakova and V. Konenkov (Institute of Clinical Immunology, Novosibirsk, Russia).

East Indians

The 59 East Indian samples, which were collected in Bombay, India as a control population for a study on Graves Disease, were provided by V. Suraj-Baker (Tulane University School of Medicine, New Orleans, LA).

Northern Europeans

The CEPH population (Dausset et al. 1990) consists of 37 of the original 39 CEPH families, all of Northern European descent (Begovich et al. 1992). In total, 478 samples were available for analysis providing information on 249 distinct haplotypes. For this analysis, the non-transmitted haplotype from each available grandparent was included; hence, the odd number of haplotypes.

HLA-typing methods

Nine of the populations were typed in-house using the reverse sequence-specific oligonucleotide (SSO) technology (Saiki et al. 1989; Begovich and Erlich 1995). DQA1 typing was performed using a modified version of the AmpliType PM + DQA1 PCR Amplification and Typing Kit (Perkin Elmer, Foster City). This modified assay distinguishes the following DQA1 alleles: 0101, 0102, 0103, 0201, 0300, 0401, 0501, and 0601. DQB1-typing was performed using the Dynal RELI™ SSO HLA-DQB1 test (Dynal, Norway). This assay distinguishes the following DQB1 alleles: 0501, 0502, 05031, 05032, 0504, 0601, 0602/06111, 0603/0614, 0604, 06051, 06052, 0606, 0607, 0608, 0609, 0610, 06112, 0612, 0613, 0615, 0201/0202, 0203, 0301, 0302/0307, 03032, 0304, 0305, 0306, 0308, 0401, and 0402. Note that two alleles separated by a slash (i.e. 0602/06111) indicates that these two alleles are not distinguished. Throughout this manuscript the first allele is used to designate both (i.e. 0602 indicates 0602/06111). The remaining six populations (Ugandans, Ecuadorian Africans, Cayapa, Mixe, Mixtec and Zapotec) were DQ-typed by our collaborators using standard SSO technology (Bugawan and Erlich 1991; Olerup et al. 1993; Carrington et al. 1992). Samples with unusual alleles or haplotypes were retyped in-house using the reverse SSO-technology.

Statistical analyses

Allele frequencies have been determined by gene counting. DQA1-DQB1 haplotypes, in all but the CEPH population, which consisted of family data, were estimated using the computer program HAUPT (Baur and Danilovs 1980).

Results

DQA1 allele frequencies

The DQA1 allele frequencies in all 15 populations are shown in Table 2. Nine DQA1 alleles were identified in these 2,807 chromosomes. The number of alleles per population varies from as few as four in the Cayapa and Mixe, both Amerindian populations, to as many as eight in the Cameroon, Japanese, Tuvinians, and East Indians. The DQA1*0300 and 0501 alleles were each found in all 15 populations while 0601 was only found in a subset of the Asian populations. The 0103 allele is absent in all four Amerindian populations studied here, while the 0201 allele, which is common in Caucasians and some African populations, is rare or absent in the Amerindian populations and relatively infrequent in the Japanese, Indonesians and the two PNG populations. The rare DQA1*0502 allele, which appears to be unique to Africans (Zimmerman et al.

1995), was only screened for in the Cameroon population. The DQA1*0502 allele, if present in the other populations, would have typed as DQA1*0501.

Table 2. DQA1 allele frequencies[a]

	0101	0102	0103	0201	0300	0401	0501	0502[b]	0601	#[c]
Ugandans	.191	**.394**	.053	.106	.053	.085	.117			7
Cameroon	.099	**.363**	.012	.029	.273	.081	.137	.006		8
Gambians	.110	.223	.031	.051	.120	.065	**.401**			7
Ecuadorian Af	.052	**.379**	.129	.103	.034	.095	.207			7
Cayapa		.006			**.620**	.211	.163			4
Mixe	.009				.292	.264	**.434**			4
Mixtec	.071	.024			**.349**	.222	.333			5
Zapotec	.047	.029		.023	**.382**	.212	.306			6
Indonesians	.216	**.492**	.023	.023	.053		.106		.087	7
PNG Highlanders	.320	**.359**	.077	.013	.160		.070			6
PNG Lowlanders	.104	**.542**	.042	.010	.115		.135		.052	7
Japanese	.093	.116	.233	.006	**.442**	.012	.070		.029	8
Tuvinians	.116	.180	.101	.089	.201	.044	**.260**		.009	8
East Indians	.178	.127	**.271**	.161	.068	.017	.161		.017	8
CEPH	.141	.205	.044	.141	.213	.036	**.221**			7

[a]The most frequent allele in each population is highlighted in bold.
[b]This rare allele, which appears to be unique to Africans, was only screened for in the Cameroon population. It would have typed as *0501 with the HLA-typing methods used in this study.
[c]Indicates the number of alleles found in each population.

DQA1*0102 is the most frequent allele in three of the four African populations as well as three of the Asian populations (Table 2). The most common allele in three of the Amerindian groups and the Japanese is 0300. The 0501 allele is the most frequent allele in the Gambians, Mixe, Tuvinians and CEPH, while DQA1*0103 is the most common allele in the East Indians.

DQB1 allele frequencies

The DQB1 allele frequencies in these 15 populations are shown in Table 3. Although the typing assay can detect 31 distinct DQB1 alleles, only 18 (16 known and two potential new alleles) were found in these populations. The number of alleles per population varies from as few as four in the Mixe to as many as 15 in the Tuvinians. DQB1*0301 was the only allele found in all 15 populations, while 0302 and 0402 were each found in 14 of the 15 populations. DQB1*0305 was found in a single individual from the CEPH population. DQB1*0401 was only

Table 3. DQB1 allele frequencies[a]

	0200	0301	0302	03032	0304	0305	0401	0402	0501
Ugandans	.181	.074	.011	.011				.085	**.255**
Cameroon	.145	.096	.006	.241				.038	.113
Gambians	.171	**.356**	.058					.051	.205
Ecuadorian Af	.164	.172	.009	.009				.086	.052
Cayapa		.163	**.452**	.211				.151	.006
Mixe		**.434**	.283					.274	.009
Mixtec	.008	**.325**	.309		.008			.254	.024
Zapotec	.023	.300	**.353**	.006	.006			.235	.035
Indonesians	.045	.174	.019	.008			.023		.178
PNG Highlanders	.013	.070		.006			.115	.038	
PNG Lowlanders	.010	.198	.031	.042			.010	.021	
Japanese	.012	.099	.081	.174			.163	.029	.052
Tuvinians	.130	**.266**	.044	.080	.003		.032	.047	.062
East Indians	.195	.102	.076	.034				.017	.076
CEPH	.181	**.185**	.145	.060		.004		.036	.096

	0502	05031	0601	0602	0603	0604	0609	New[b]	#[c]
Ugandans		.011		.191	.074	.053	.053		11
Cameroon				**.308**	.009	.009	.035		10
Gambians	.010	.003		.065	.027	.007	.044		11
Ecuadorian Af	.069			**.233**	.103		.103		10
Cayapa								.018	6
Mixe									4
Mixtec	.008	.048		.008			.008		10
Zapotec		.012				.006	.023		10
Indonesians	**.307**	.053	.167	.026					10
PNG Highlanders	.026	**.359**	.103	.269					9
PNG Lowlanders	**.312**	.062	.260	.052					10
Japanese	.023	.023	**.221**	.064	.012	.041	.006		14
Tuvinians	.030	.030	.047	.115	.053	.018	.041		15
East Indians	.017	.119	**.263**	.042	.034	.008	.008	.008	14
CEPH	.016	.044	.004	.157	.044	.024	.004		14

[a]The most frequent allele in each population is highlighted in bold.
[b]There are three samples in the Cayapa with unique probe hybridization patterns that suggest all three carry the same new allele. This allele appears to be different from the new allele found in the East Indians, which was observed only once.
[c]Indicates the number of DQB1 alleles found in each population.

seen in the Asian populations, while 0601 was found in Asians and Caucasians but not Africans or Amerindians.

The most frequent DQB1 allele in the Gambians, Mixe, Mixtec, Tuvinians, and the CEPH is 0301 (Table 3). DQB1*0302 is the most common allele in the Cayapa and the Zapotec, while 0502 is the most frequent allele in the Indonesians and PNG Lowlanders. In both the Japanese and the East Indians, 0601 is the most common allele, while 0602 is the most frequent allele in the Cameroon population and the Ecuadorian Africans. DQB1*05031 is the most frequent allele in the PNG highlanders, while in the Ugandans 0501 is the most common allele.

Two novel DQB1 alleles were also uncovered in this study. One, identified in a single East Indian individual, appears to be a variant of DQB1*0601 and was found on the same chromosome as DQA1*0103. The second new allele, found in three Cayapa, appears to be a variant of DQB1*0402. In all three individuals, it was found on the same chromosome as DQA1*0401. Both these alleles are currently being cloned for sequence analysis.

HLA-DQ haplotype frequencies

Assuming that alleles at the DQA1 and DQB1 loci are associating randomly, there are 9 (number of DQA1 alleles) x 18 (number of DQB1 alleles), or 162, possible DQ haplotypes in these 15 populations; only 42 distinct DQA1-DQB1 haplotypes were actually found (Table 4 and Fig. 1). The number of DQ haplotypes per population varies from as few as five in the Mixe to as many 23 in the Tuvinians. In general, the four Amerindian populations have the lowest number of DQ haplotypes; this is not unexpected as they have the lowest number of DQ alleles.

The most frequent haplotype in three of the African populations (the Ugandans, Cameroon, and Ecuadorian Africans) and the CEPH is DQA1*0102-DQB1*0602. In the fourth African population, the Gambians, the most common haplotype is 0501-0301. This haplotype is also the most frequent one in the Mixe, Mixtec and Tuvinians. The 0300-0302 haplotype is the most frequent haplotype in the Cayapa and Zapotec, while 0103-0601 is the most common haplotype in the Japanese and East Indians. In the Indonesians and PNG lowlanders the haplotype 0102-0502 is most common, while in the PNG highlanders the most common haplotype is 0101-05031.

While one haplotype, DQA1*0501-DQB1*0301, was found in all 15 populations, a number of others, including 0101-0501, 0102-0602, 0201-0200, and 0401-0402, were found in populations representing all the different ethnic groups (Table 4). There is also a subset of haplotypes that appears to be confined to particular ethnic groups. This seems to be especially true in the Asian populations where five unique haplotypes, 0101-0502, 0102-0601, 0103-05031, 0300-0401, and 0601-0301, were identified. Some of these haplotypes, such as 0102-0601 and 0300-0401, which are rare or absent in the other ethnic groups, are actually quite frequent in some of the Asian groups. For example, DQA1*0102-

Table 4. DQ haplotype frequencies[a]

DQA1	DQB1	UGA	CAM	GAM	EUC	CAY	MXE	MXT	ZAP
0101	-0501	.176	.096	.106	.052		.009	.024	.035
	-0502								
	-05031	.011		.003				.048	.012
	-0602		.003						
	-0603	.005							
0102	-0501	.079	.018	.099					
	-0502			.010	.069	.006		.008	
	-05031								
	-0601								
	-0602	**.181**	**.305**	.065	**.224**			.008	
	-0603	.038							
	-0604	.053	.006	.003					.006
	-0609	.043	.035	.044	.086			.008	.023
0103	-05031								
	-0601								
	-0602	.011			.009				
	-0603	.032	.009	.027	.103				
	-0604		.003	.003					
	-0609	.011			.017				
	-new								
0201	-0200	.106	.029	.051	.103				.024
	-0302								
	-03032								
0300	-0200	.032	.026	.062	.017			.008	
	-0301								
	-0302	.011	.006	.058	.009	**.410**	.283	.309	**.353**
	-03032	.011	.241		.009	.211			.006
	-0304								
	-0305								
	-0401								
	-0402						.009	.032	.023
0401	-0301		.044	.014	.009				
	-0302					.042			
	-0402	.085	.038	.051	.086	.151	.264	.222	.212
	-new					.018			
0501	-0200	.043	.090	.058	.043				
	-0301	.075	.046	**.342**	.164	.163	**.434**	**.325**	.300
	-0302								
	-03032								
	-0304							.008	.006
0502	-0301		.006						
0601	-0301								
	#[b]	18	17	16	15	7	5	11	11

Table 4. DQ haplotype frequencies[a] (continued)

DQA1	DQB1	IND	PNGH	PNGL	JAP	TUV	EIN	CEPH
0101	-0501	.140			.052	.062	.076	.096
	-0502	.023		.031	.017	.023		
	-05031	.053	**.320**	.062	.023	.030	.097	.044
	-0602							
	-0603							
0102	-0501	.037						
	-0502	**.284**	.026	**.271**	.006	.006	.017	.016
	-05031						.009	
	-0601	.146	.072	.219			.042	
	-0602	.026	.262	.051	.064	.115	.042	**.157**
	-0603							.004
	-0604				.041	.018	.009	.024
	-0609				.006	.041	.009	.004
0103	-05031		.038				.009	
	-0601	.021	.031	.031	**.221**	.047	**.220**	.004
	-0602		.008					
	-0603				.012	.053	.034	.040
	-0604							
	-0609							
	-new						.009	
0201	-0200	.023	.013	.010	.006	.068	.119	.084
	-0302						.009	
	-03032					.021	.030	.056
0300	-0200	.004				.008		
	-0301				.012	.060		.060
	-0302	.019		.031	.076	.044	.068	.145
	-03032	.008	.006	.042	.174	.050		.004
	-0304					.003		
	-0305							.004
	-0401	.023	.115	.010	.163	.032		
	-0402		.038	.021	.017	.003		
0401	-0301				.012			
	-0302							
	-0402				.012	.044	.017	.036
	-new							
0501	-0200	.019			.006	.052	.076	.096
	-0301	.087	.070	.135	.058	**.202**	.085	.124
	-0302				.006			
	-03032					.007		
	-0304							
0502	-0301							
0601	-0301	.087		.042	.017	.005	.017	
	#[b]	16	12	13	21	23	20	18

DQB1*0601 has a frequency of 0.219 in the PNG lowlanders and 0.146 in the Indonesians, while 0300-0401 has a frequency of 0.163 in the Japanese and 0.115 in the PNG highlanders (Table 4).

Twelve of the 42 DQ haplotypes were each observed in a single population (Table 4). Of these 12, four were found in two or more individuals. The DQA1*0401-DQB1*0302 and DQA1*0401-DQB1*"new" haplotypes, which are unique to the Cayapa, were found on seven and three independent chromsomes respectively. The 0501-03032 haplotype, which is unique to the Tuvinians, was found on two chromosomes. The fourth haplotype, 0502-0301, was found on two chromosomes in the Cameroon; however, the Cameroon population was the only one tested for DQA1*0502. Additional population studies have shown that this

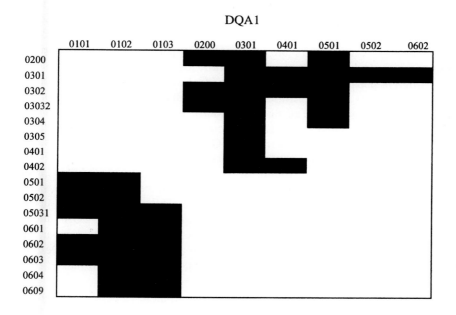

Fig. 1. DQ haplotypes found in the 15 populations. DQA1 alleles are listed along the top; DQB1 alleles are listed along the left side. A black box indicates that this DQA1-DQB1 haplotype was found in this population study. The two haplotypes with new DQB1 alleles are not included in this figure.

Table 4 footnotes

[a]The most frequent haplotype in each population is highlighted. The population names, which have been shortened, are listed in the same order as in Tables 1-3.
[b]Indicates the number of haplotypes found in each population.

haplotype, while rare, is found in African populations in Ecuador, Ghana and the United States (Zimmerman et al. 1995). The remaining eight haplotypes were each found on a single chromosome (Table 4); all eight samples were retyped and the haplotypes confirmed. A single unique DQ haplotype was found in the Cameroon (DQA1*0101-DQB1*0602), Ugandans (0101-0603), Japanese (0501-0302), CEPH (0301-0305), and Tuvinians (0300-0304). Three unique haplotypes were also identified in the East Indians (0102-05031, 0103-"new", and 0201-0302).

Discussion

This study of 2,807 chromosomes in 15 diverse populations has identified nine DQA1 and 18 DQB1 alleles which are arranged on 42 distinct haplotypes. Some of these 42 haplotypes are found in populations that span all the ethnic groups examined here, while others are confined to a particular ethnic group or population. Analysis of the composition of these haplotypes and their frequencies in the different populations (Fig. 1 and Table 4) shows that the *cis*-association of DQA1 and DQB1 alleles, while non-random due to patterns of linkage disequilibrium (Begovich et al., in preparation), is not absolute. A particular DQA1 allele can be found on the same chromosome with more than one DQB1 allele and vice versa. For example, DQB1*0602 is found on the same chromosome as the DQA1 alleles 0101, 0102 and 0103, although, in all 12 populations in which it is found, it is most frequently associated with DQA1*0102. Likewise, the DQB1*0301 allele can be found on the same chromosome with five different DQA1 alleles 0300, 0401, 0501, 0502, and 0601. Again, in most populations it is most frequently associated with one DQA1 allele, *0501. These data suggest that, although the DQA1 and DQB1 loci are separated by less than 50kb, inter-locus recombination can occur between them. Despite this fact, there are no examples of any "forbidden" haplotypes in these 2,807 chromosomes (Fig. 1 and Table 4).

The complete absence of "forbidden" haplotypes in these 15 populations agrees with the majority of the published data (Imanishi et al. 1992; Clayton et al. 1997; Gyllensten and Erlich 1989) and suggests one of two things. Either recombination does not occur between DQw1- and DQw2,3,4-associated haplotypes or, more likely, recombination does occur, although infrequently, and any resulting chromosomes carrying a "forbidden" haplotype are rapidly lost from the population via negative selection and/or genetic drift. Given the results from these 15 populations, which represent a large array of human ethnic groups and population structures from large groups subject to admixture to small groups evolving under virtual isolation, any reports of multiple "forbidden" haplotypes in a population (i.e. Grahovac et al. 1998) warrant further analysis. We conclude that the removal of haplotypes with DQA1 and DQB1 alleles that form unstable heterodimers through natural selection has been a consistent feature of DQ evolution throughout human history.

Acknowledgements

We are grateful to H.A. Erlich and S.J. Mack for valuable discussions and careful reading of this manuscript. This work was supported in part by grant AI29042 from the NIH.

References

Baur MP, Danilovs J (1980) Population genetic analysis of HLA-A, B, C, DR and other genetic markers. In: Terasaki PI (ed) Histocompatibility Testing 1980. UCLA Tissue Typing Laboratory, Los Angeles, pp 955-993

Begovich AB, McClure GR, Suraj VC, Helmuth RC, Fildes N, Bugawan TL, Erlich HA, Klitz W (1992) Polymorphism, recombination and linkage disequilibrium within the HLA class II region. J Immunol 148:249-258

Begovich AB, Erlich HA (1995) HLA typing for bone marrow transplantation, new polymerase chain reaction-based methods. JAMA 273:586-591

Bodmer JG, Marsh SGE, Albert ED, Bodmer WF, Bontrop RE, Dupont B, Erlich HA, Hansen JA, Mach B, Mayr WR, Parham P, Petersdorf EW, Sasazuki T, Schreuder G M Th, Strominger, JL, Svejgaard A, Terasaki PI (1999) Nomenclature for factors of the HLA system, 1998. Hum Immunol 60:361-395

Bugawan TL, Erlich HA (1991) Rapid typing of HLA-DQB1 DNA polymorphism using nonradioactive oligonucleotide probes and amplified DNA. Immunogenetics 33:163-170

Carrington M, Miller T, White M, Gerrard B, Stewart C, Dean M, Mann D (1997) Typing of HLA-DQA1 and DQB1 using DNA single-strand conformation polymorphism. Hum Immunol 33:208-212

Clayton J, Lonjou C, Whittle D (1997) Anthropology tables. In: Charron D (ed) HLA – Genetic Diversity of HLA, Functional and Medical Implication. EDK, Sevres, pp 665-775

Dausset J, Cann H, Cohen D, Lathrop M, Lalouel JM, White R (1990) Centre d'etude du polymorphisme humain (CEPH): collaborative genetic mapping of the human genome. Genomics 6:575-577

Giles R, DeMars R, Chang CC, Capra JD (1985) Allelic polymorphism and transassociation of molecules encoded by the HLA DQ subregion. Proc Natl Acad Sci USA 82:1776-1780

Grahovac B, Sukernik RI, O'hUigin C, Zaleska-Rutczynska Z, Blagitko N, Raldugina O, Kosutic T, Satta Y, Figueroa F, Takahata N, Klein J (1998) Polymorphism of the HLA class II loci in Siberian populations. Hum Genet 102:27-43

Gyllensten UB, Erlich HA (1989) Ancient roots for polymorphism at the HLA-DQα locus in primates. Proc Natl Acad Sci USA 86:9986-9990

Hill AVS, Allsopp CEM, Kwiatkowski D, Anstey NM, Twumasi P, Rowe PA, Bennett S, Brewster D, McMichael AJ, Greenwood BM (1991) Common West African HLA antigens are associated with protection from severe malaria. Nature 352:595-600

Imanishi T, Akaza T, Kimura A, Tokunaga K, Gojobori T (1992) Allele and haplotype frequencies for HLA and complement loci in various ethnic groups. In: Tsuji K, Aizawa M, Sasazuki T (eds) HLA 1991. Proceedings of the Eleventh International Histocompatibility Workshop and Conference. Oxford Science Publications, Oxford New York Tokyo, pp 1065-1220

Kwok WW, Schwarz D, Nepom BS, Hock RA, Thurtle PS, Nepom GT (1988) HLA-DQ molecules form α-β heterodimers of mixed allotype. J Immunol 141:3123-3127

Kwok WW, Thurtle P, Nepom GT (1989) A genetically controlled pairing anomaly between HLA-DQα and HLA-DQβ chains. J Immunol 143:3598-3560

Kwok, WW, Kovats S, Thurtle P, Nepom GT (1993) HLA DQ allelic polymorphisms constrain patterns of class II heterodimer formation. J Immunol 150:2263-2272

Lundin KE, Sollid LM, Qvigstad E, Markussen G, Gjertsen HA, Ek J, Thorsby E (1990) T lymphocyte recognition of a celiac disease-associated cis- or trans-encoded HLA-DQα/β heterodimer. J Immunol 145:136-139

Nepom BS, Schwarz D, Palmer JP, Nepom GT (1987) Transcomplementation of HLA genes in IDDM. HLA-DQ α- and β-chains produce hybrid molecules in DR3/4 heterozygotes. Diabetes 36:114-117

Olerup O, Aldener A, Fogdell A (1993) HLA-DQB1 and –DQA1 typing by PCR amplification with sequence-specific primers (PCR-SSP) in 2 hours. Tissue Antigens 41:119-143

Redd AJ, Takezaki N, Sherry ST, McGarvey ST, Sofro ASM, Stoneking M (1995) Evoluntionary history of the COII/tRNA Lys intergenic 9 base pair deletion in human mitochrondrial DNAs from the Pacific. Mol Biol Evol 12:604-615

Saiki RK, Walsh PS, Levenson CH, Erlich HA (1989) Genetic analysis of amplified DNA with immobilized sequence-specific oligonucleotide probes. Proc Natl Acad Sci USA 86:6230-6234

Stoneking M, Jorde LB, Bhatia K, Wilson AC (1990) Geographic variation in human mitochondrial DNA from Papua New Guinea. Genetics 124:717-733

Zimmerman PA, Phadke PM, Lee A, Elson LH, Aruajo EN, Guderian R, Nutman TB (1995) Migration of a novel DQA1* allele (DQA1*0502) from African origin to North and South America. Hum Immunol 42:233-240

HLA class I and class II loci in Pacific/Asian populations

Steven J. Mack[1,2], Teodorica L. Bugawan[1], Mark Stoneking[3], Muhammad Saha[4], Hans-Peter Beck[5], and Henry A. Erlich[1,2]

[1]Department of Human Genetics, Roche Molecular Systems, Alameda, CA, USA
[2]Children's Hospital Oakland Research Institute, Oakland, CA, USA
[3]Department of Anthropology, Pennsylvania State University, University Park, PA, USA
[4]Department of Paediatrics, National University of Singapore, Singapore
[5]Department of Medical Parasitology, Swiss Tropical Institute, Basel, Switzerland

Summary. PCR/SSOP typing was used to identify class I and II alleles in individuals from Malay, Filipino and Papua New Guinea (PNG) Highland populations, as well as two PNG Lowland and three Indonesian populations. Allele frequency distributions were analyzed via population genetic and phylogenetic methods. Three common Australian alleles (DRB1*0405, *0410 and *1407) were seen in the Highlands, and are rare or absent in the other populations. Phylogenetic analyses of DRB1, DPB1 and A allele frequency data indicate a close relationship between the Highlands and Australia. Three putatively new alleles in the HLA-B15 and B56 serogroups were identified in the Highlands (f = 0.01 - 0.05). Variation in homozygosity at the DRB1, DPB1 and A loci suggests that founder effects and positive selection have played significant roles in the history of these populations. A*2402 comprised 78% of the Highlands A-locus alleles, suggesting founder effects, or positive selection for this allele. However, the DRB1, DPB1 and HLA-B loci display no evidence of founder effects in this population, indicating strong directional selection for this allele.

Key words. Filipino, HLA-A, HLA-B, HLA-DRB, HLA-DPB, Indonesian, Malay, Molucca, New Allele, Nusa Tenggara, PCR, PNG, Selection

Introduction

As the most polymorphic coding sequence loci in the human genome (Marsh 1997), the class I and class II HLA loci represent highly informative and convenient markers for population genetics studies. Recently, simple PCR/SSO probe methods of HLA typing have been developed, capable of distinguishing the extensive allelic diversity at these loci (Bugawan et al. 1990; Bugawan and Erlich

427

1991; Erlich et al. 1991, 1993). Population genetics studies of the HLA class I and class II loci can reveal the operation of selective forces on HLA polymorphism, such as symmetric balancing selection (Klitz et al. 1986; Begovich et al. 1992) or positive directional selection for a specific allele (Bugawan et al. 1994), and phylogenetic analyses of the HLA allele and haplotype frequencies can illustrate the history of and relationships between modern human populations. Moreover, HLA typing of various populations can identify new (previously unreported) alleles, and sequence comparisons of newly generated alleles and their putative parental allele can reveal possible sites (specific polymorphic amino acid residues) for selection for new sequence variants (Titus-Trachtenberg et al. 1994). In addition, the identification of allele frequency distributions in different populations can be valuable in estimating the likelihood of obtaining an HLA matched donor for bone marrow transplantation in a given population.

Based on geographic, linguistic and cultural factors, the populations of the Pacific islands can be divided into the following five broad groups. Australia includes the Australian continent and the island of Tasmania. Austronesia, or island Southern Asia, includes Indonesia, Malaysia, Borneo, Singapore, and the Philippines. Melanesia includes New Guinea, Fiji and the intervening islands. Micronesia includes the islands north of Melanesia (with the exception of the Tuvaluan islands, which are Polynesian). Polynesia includes New Zealand, and all of the islands east of Melanesia and Micronesia, extending to Hawaii and Easter Island. Together, Melanesia, Micronesia and Polynesia are referred to as Oceania.

Previous genetic studies have suggested that Australian and PNG Highland populations may share an ancient common ancestry (Serjeantson 1989). These findings support the anthropological theory that peoples of the PNG Highlands and Australia were part of a single population which inhabited the Sahul landmass (connecting Australia, New Guinea and Tasmania) more than 9,000 years ago, but was split into three isolated populations on the islands known today as New Guinea, Australia and Tasmania subsequent to a dramatic rise in sea level that flooded Sahul. While this relationship was not supported by early studies of globin genes (Tsintsof et al. 1990), more recent studies of mtDNA support an ancestral connection between the PNG Highlands and Australia (Redd et al. 1995; van Holst Pellekaan et al. 1997).

Those Pacific populations other than Australians and PNG Highlanders are thought to have arrived more recently, and studies of mtDNA sequences indicate that island Southern Asian and Oceanian populations spread south from Taiwan and the Philippines subsequent to a migration from Asia, and are unrelated to Australian populations (Lum et al. 1994; Melton et al. 1995, 1998).

Here, we apply high resolution PCR/SSO typing of class I and class II HLA loci, coupled with population genetic analyses, to examine the evolutionary forces at work during the histories of Australian, island Southern Asian, and Melanesian (PNG Highland and PNG Lowland) populations. In particular, our analysis of allele frequencies for the HLA-A, B, DRB, and DPB loci for these populations serves to illustrate the utility of various population genetic methods.

Materials and methods

PCR amplification and typing

Genomic DNA was isolated from blood samples, PCR amplified, and typed for HLA-A, DRB1, and DPB1 (Bugawan et al. 1990, 1994, 1999; Erlich et al. 1991, 1993; Bugawan and Erlich 1991; Begovich et al. 1992; Steiner et al. 1996). HLA-B locus genotypes were determined using a new reverse line blot typing method (Bugawan et al., in preparation).

Population samples

Indonesian samples were obtained from two island chains of Eastern Indonesia, the Moluccas and the Nusa Tenggaras, as well as from Indonesian immigrants to Singapore. Malay samples were obtained from ethnic Malays living in Indonesia. Filipino samples were obtained from ethnic Filipinos living in Singapore. Papua New Guinea (PNG) samples were obtained from a Highland population and two Lowland populations. The PNG Lowland MV samples were collected during a public malaria vaccination program in the rural area of Wosera. While these samples are primarily derived from a PNG Lowland population, individuals from other populations may have been sampled as well.

Statistical analysis

The Ewens-Watterson homozygosity statistic (F) (Ewens 1972; Watterson 1978) was used to examine the selective forces influencing polymorphism at these HLA loci. The MONTECARLO program (Slatkin 1994, 1996) was used to perform the Ewens exact and homozygosity tests on the allele distributions for each population.

Phylogenetic analysis

The PHYLIP 3.5 package (Felsenstein 1995) was used to generate neighbor-joining trees/networks (Saitou and Nei 1987) from Nei's genetic distances calculated using the HLA allele frequency distributions presented here, as well as for the frequency distributions of several Australian populations (Gao et al. 1992, 1994; Lester et al. 1995, 1997) (A. Redd, unpublished data) at the A, DRB1 and DPB1 loci, and an additional PNG Highland population at the DRB1 locus. The number of Australian populations included in each tree varies by locus, depending on the availability of the allele frequency distributions.

Haplotype analysis

HLA-A-B haplotypes were interpreted by direct counting and were estimated using standard methods (Baur and Danilovs 1980; Long et al. 1995).

Results and discussion

HLA-A allele frequency distributions

The allele frequency distributions for HLA-A in the Malay, Nusa Tenggara, Molucca and PNG populations are shown in Table 1. A*1101 and *2402 are the most frequently observed alleles in all populations. In the PNG Highlands, A*2402 is present at the remarkably high frequency of 0.78, and the combined frequency of the A*2402 and A*1101 alleles is 0.89.

Table 1. HLA-A locus allele frequency distributions (from Bugawan et al. 1999, with permission)

A locus allele	Indonesian Molucca	Indonesian Nusa Tenggara	Malay
		0.0175	0.0070
0201/16		0.0088	0.0950
0203	0.0577	0.0439	0.0740
0206	0.0577	0.0175	0.0270
0301			0.0070
1101	0.3269	0.1404	0.2370
1102			0.0140
1104		0.0088	0.0070
2402	0.1538	0.4035	0.1820
2403			0.0140
2407	0.2308	0.1404	0.1420
2410		0.0088	0.0340
2601/2		0.0175	0.0140
2901/2			0.0070
3001			0.0070
3101		0.0175	
3201	0.0192		
3301/3			0.0740
3401	0.1538	0.1667	0.0270
3601		0.0088	
68011			0.0070
68012			0.0070
6802			0.0070
7401			0.0140
2n	52	114	148
h	0.80	0.77	0.88

Table 1. (continued)

A locus allele	PNG Lowland MV	PNG Lowland	PNG Highland
0101			
0201/16		0.0430	
0203			
0206			0.0050
0301			0.0050
1101	0.4370	0.2550	0.1030
1102			
1104			
2402	0.3920	0.4890	0.7830
2403			
2407		0.0320	
2410			
2601/2			
2901/2			
3001			
3101	0.0820	fv0.0110	
3201			
3301/3			
3401	0.0890	0.1700	0.1030
3601			
68011			
68012			
6802			
7401			
2n	158	94	188
h	0.68	0.68	0.37

The recently reported A*2407 allele, which was first identified in the Japanese (23) (f = 0.20) is common among Malays and Indonesians, but is rare in PNG populations. In addition to the novel probe reactivity patterns associated with *2407, two additional probe reactivity patterns, indicative of new A alleles *1104 and *2410 (Bugawan, et al. 1999), and are observed in the Nusa Tenggara and Malay populations.

Based on the average value (0.7) for the observed heterozygosity (h) calculated from the frequency distributions of these populations, the PNG Highlands show a reduced level of allelic diversity (h = 0.37). Given that the alleles at the HLA loci are, in most populations, subject to balancing selection (Klitz et al. 1986; Begovich et al. 1992), this unusual allele frequency distribution for the PNG Highlands may reflect founder effects and bottlenecks, genetic drift, or positive directional selection. In the case of founder effects, bottlenecks and genetic drift, neighboring class I loci would be expected to reflect allelic distribution patterns similar to those seen at the A-locus, whereas in the case of directional selection operating on the A-locus, other class I loci might be expected to display evidence of balancing selection. Alternatively, relatively recent directional selection operating on the A-locus might result in a "selective sweep" of closely linked loci, resulting in similarly high allele frequencies at these loci. To distinguish between these cases, we determined the allele frequency distributions for the HLA-B locus in the PNG Highland population.

HLA-B allele frequency distribution in the PNG Highlands

The HLA-B allele frequency distribution in the PNG Highland population is shown in Table 2. Unlike the HLA-A locus, for which only five alleles were observed, sixteen alleles were observed at the B-locus. The B*1506, *4001, *5601 and *5602 alleles account for 73% of the B-locus alleles in this population. Accordingly, the diversity (h) for the B-locus is much higher (0.91) than that observed at the A-locus in the PNG Highlands (0.37), and is also higher than the A-locus diversity of the other populations studied here. This observed diversity at the B-locus is difficult to reconcile with the hypothesis of a founder effect or population bottleneck proposed to account for the HLA-A data. Based on the relatively even frequencies of these four alleles, no evidence of strong directional selection is observed at the B-locus, suggesting that the high frequency of the A*2402 allele is the result of selective forces operating specifically at the HLA-A locus. To further examine the relationship between these HLA-A and B alleles, we determined the A-B haplotypes present in the PNG Highlands population (see below).

Three putatively new HLA-B alleles were detected via novel probe reactivity patterns, and were observed to occur at low frequencies in the PNG Highland

Table 2. HLA-B allele frequency distributions (from Bugawan et al. 1999, with permission)

B locus allele	Frequency
1301	0.0455
1502	0.0065
1506	0.1429
1521	0.0260
1525	0.0260
15N1[1]	0.0455
15N2[1]	0.0130
2704	0.0325
4001	0.1494
4002	0.0455
4701	0.0065
5501/02	0.0065
5601	0.2078
5602	0.2273
56N[1]	0.0130
2n	154
h	0.91

[1]15N1, 15N2, and 56N are putatively novel alleles which display probe reactivity patterns similar to alleles of the HLA-B15 and B56 serogroups.

population (f = 0.01 - 0.05). Denoted in Table 2 as B*15N1, *15N2 and *56N, the probe reactivity patterns for these alleles most closely resembled those for alleles in the B15 and B56 serogroups. The cloning and sequencing of these alleles is currently under way.

HLA-A-B haplotypes in the PNG Highlands

HLA-A-B haplotypes are presented in Table 3. The A*2402 allele is included in 12 of the 27 haplotypes identified in this population. In addition, four of these 12 haplotypes account for 66% of the haplotype diversity. While A*2402 comprises 78% of the HLA-A allelic diversity, 85% of the B*4001 and *5601 alleles, 92% of the *1506 alleles, and 100% of the *5602 alleles in this population are observed in a haplotype with the A*2402 allele. Additionally, the B*5602 allele is the most common B allele in this population. In comparison, the other B alleles observed on multiple haplotypes are more evenly distributed between A*2402 and other A allele haplotypes. The significance of the B*5602 allele's high frequency and its presence in a single A-B haplotype may become clearer after HLA-B locus high-resolution typing has been carried out on additional Pacific/Asian populations.

Table 3. HLA-A-B haplotype frequency distribution in the PNG Highlands (from Bugawan et al. 1999, with permission)

A locus allele	Associated B allele	Frequency
0206	-1525	0.006
0301	-0702	0.006
1101	-1301	0.009
	-1506	0.012
	-1521	0.012
	-15N1[1]	0.032
	-2704	0.006
	-4001	0.022
	-5501/02	0.006
2402	-1301	0.024
	-1502	0.006
	-1506	0.132
	-1521	0.015
	-1525	0.013
	-15N1[1]	0.013
	-4001	0.128
	-4002	0.032
	-4701	0.006
	-5601	0.176
	-5602	0.228
	-56N[1]	0.013
3401	-1301	0.013
	-1525	0.006
	-15N2[1]	0.013
	-2704	0.026
	-4002	0.013
	-5601	0.032

[1]15N1, 15N2, and 56N are putatively novel alleles which display probe reactivity patterns similar to alleles of the HLA-B15 and B56 serogroups.

While it remains possible that these four common A-B haplotypes are under positive selection, or are predominant in this population due to a selective sweep of HLA-B alleles on A*2402 haplotypes, the relatively equivalent frequencies of the four common B alleles, as well as the presence of other A*2402-B haplotypes, probably reflects balancing selection rather than directional selection operating on the B-locus. In addition, there are no sequence motifs common to these four B alleles which present an obvious candidate for positive selection. The number and relative diversity of A*2402 haplotypes also argues against the idea that the high frequency of the A*2402 allele in this population is due to a founder effect.

DRB1 allele frequency distributions

The DRB1 allele frequency distributions for the Filipino, Malay, Indonesian and PNG populations are shown in Table 4. The frequency of the DR2 serogroup is

Table 4. DRB1 allele frequency distributions

DRB1 allele	Filipino	Indonesian Nusa Tenggara	Indonesian Molucca	Indonesian Singapore
0101				0.0200
0103				
1501		0.0355	0.0625	0.1200
1502	0.4853	0.4793	0.4625	0.1700
1503				0.0300
1602		0.0414	0.0500	
03011	0.0294	0.0237		0.0300
0401				
0403/6	0.0294	0.0178		0.0300
0404				
0405	0.1029	0.0178	0.0250	0.0300
0407				
0408				
0410				
0411		0.0118		
1101	0.0294	0.0888	0.0750	0.0200
1104		0.0059		0.0100
1201	0.0147	0.0296	0.0375	0.0200
1202	0.1912	0.1538	0.1250	0.2600
1301		0.0059		0.0200
1302				0.0100
1401		0.0178		0.0300
1402				
1404		0.0296	0.0125	0.0200
1405		0.0059		
1407				
1408		0.0118	0.0375	
0701	0.0147		0.0750	0.0900
0802				0.0100
0803	0.0441	0.0118	0.0375	0.0300
0901	0.0441	0.0118		0.0200
1001	0.0147			0.0300
2n	68	169	80	100
h	0.72	0.74	0.76	0.88

strikingly high overall (f>0.5 in some populations); the DR2 allele, DRB1*1502, is common in all groups and reaches a frequency of 0.48 among Indonesians and Filipinos. DRB1*0405, a common allele in Japan and China (Imanishi et al. 1991) as well as in Australia (Gao et al. 1994; Lester et al. 1997), is frequent in the PNG Highlands and Filipinos but is rare among Indonesians, Malaysians, and the PNG. DRB1*1407, an allele reported in Australians (Gao et al. 1994), is common among PNG Highlanders, and rare among the PNG Lowlands. This allele differs from the more widely distributed DRB1*1401 allele by a Val to Gly change at position 86. DRB1*1408, a common Australian allele which differs from the putative parental *1401 allele by an Ala to Asp change at position 57, is found in the PNG Highlands, as well as in the PNG Lowlands and Indonesia. As was the case with the allele frequencies in the B-locus, the DRB1 allele frequency distribution and allelic diversity of the PNG Highlanders (0.85) are not easily reconciled with the hypothesis of a founder effect or population bottleneck proposed to account for the HLA-A data in this population.

Table 4. (continued)

DRB1 allele	Malay	PNG Lowland MV	PNG Lowland	PNG Highland
0101	0.0065			
0103				
1501	0.0974		0.0780	0.2820
1502	0.1364	0.1288	0.2470	0.1210
1503			0.2080	
1602	0.0584	0.2879		0.0290
03011	0.0455			
0401				0.0110
0403/6	0.0195	0.0152		
0404				
0405	0.0065		0.0390	0.1320
0407	0.0065			
0408				
0410				0.0290
0411	0.0065			
1101	0.0390	0.5682	0.1040	0.0060
1104	0.0065			
1201	0.0195		0.0780	0.0570
1202	0.3442		0.0390	0.0060
1301	0.0065			
1302	0.0130			
1401	0.0065		0.0390	0.0290
1402				
1404	0.0390			
1405				0.0060
1407			0.0130	0.1260
1408			0.0520	0.0920
0701	0.0714			0.0060
0802	0.0065			
0803	0.0260		0.0520	0.0630
0901	0.0325		0.0520	0.0060
1001	0.0065			
2n	154	132	98	175
h	0.85	0.58	0.87	0.86

DPB1 allele frequency distributions

The DPB1 allele frequency distributions for the Filipino, Malay, Indonesian and PNG populations are shown in Table 5. The DPB1*0501 allele, the most frequent DPB1 allele in Japan and China (Imanishi et al. 1992), is observed at very high frequencies in all these populations, but is observed at its lowest frequency in the PNG Highlands. Conversely, the DPB1*0201 reaches its highest frequency in the PNG Highlands, but is relatively rare among Indonesians and Malaysians. This allele is also common in Australia (A. Redd, unpublished Data). DPB1*0401, frequent in the PNG Lowlands and in some Australian groups (Gao et al. 1994; Lester et al. 1997), is the most common allele in the PNG Highlands.

Certain DPB1 alleles were seen at high frequencies in the Malay and

Table 5. DPB1 allele distributions

DPB1 allele	Filipino	Indonesian Nusa Tenggara	Indonesian Molucca
0101	0.3833	0.0291	0.3370
0201	0.0889	0.0291	0.0761
0202	0.0167		
0301	0.0500	0.0407	0.0326
0401	0.0667	0.1105	0.0870
0402		0.0756	0.0326
0501	0.2556	0.4360	0.2174
0601			
0901	0.0111		
1001	0.0111		
1301	0.0889	0.2151	0.0652
1401	0.0111	0.0407	0.0217
1501			
1601			
1701			
1901			
2001	0.0056		
2101			
2301			0.0217
2401			
2601			
2701			
2801			0.0109
2901			0.0109
3001	0.0111		
3101		0.0233	0.0870
3201			
3301			
3901			
4301			
2n	180	172	92
h	0.77	0.74	0.82

Indonesian populations, but not in the PNG populations. For example, DPB1*1301, a rare allele in most populations around the globe, is relatively frequent among Malays and the Nusa Tenggaras but is absent from PNG. DPB1*0101, rare among most Asian groups but frequent among Filipinos (Bugawan et al. 1994), appears at a high frequency in the Moluccas and is rare among Malays and in PNG.

In contrast to the low heterozygosity for HLA-A in the PNG Highlands, the heterozygosity for DPB1 (0.71) in this population is comparable to that observed for the other populations studied. The diversity values calculated from these DRB1 and DPB1 allele frequency distributions indicate a fairly high level of diversity for class II loci in all populations, with the possible exception of the PNG Lowland MV population.

Table 5. (continued)

DPB1 allele	Malay	PNG Lowland	PNG Highland
0101	0.0260	0.0729	0.0256
0201	0.0649	0.1354	0.3590
0202	0.0390		
0301	0.0584	0.0417	0.0385
0401	0.1883	0.3438	0.3718
0402	0.0390		
0501	0.2403	0.3229	0.1603
0601		0.0104	
0901	0.0130		
1001			
1301	0.1494		
1401	0.0390	0.0208	0.0449
1501	0.0130		
1601			
1701	0.0065		
1901	0.0065		
2001			
2101	0.0130		
2301			
2401	0.0065		
2601	0.0065		
2701		0.0104	
2801	0.0779		
2901		0.0312	
3001			
3101			
3201		0.0104	
3301	0.0065		
3901			
4301	0.0065		
2n	154	96	156
h	0.87	0.76	0.71

Taking an overall view of the allele frequency data for the HLA-A, DRB1 and DPB1 loci, there seem to be some general trends which serve to distinguish the island Southern Asian populations (Indonesian, Malay and Filipino) from the Melanesian populations (PNG) (see below).

Homozygosity

Selective pressures operating on a population can be inferred from the value of the homozygosity parameter (F) calculated from allele frequency distributions (Ewens 1979; Watterson 1978). In general, analysis of the Ewens-Watterson F-statistic for homozygosity can provide evidence of symmetric balancing selection or positive directional selection if the observed value of F differs significantly from estimated values of F calculated using a model of neutral evolution (i.e. genetic drift with no selection), for a sample of the same size, displaying the same number of alleles. Homozygosity values for the A, B, DRB1, and DPB1 loci in each of the populations reported here are shown in Table 6.

Table 6. Homozygosity (F) at the A, B, DRB1 and DPB1 loci

	Observed F	Neutral F	P values
A Locus			
Molucca	0.2145	0.3250	0.10
Nusa Tenggara	0.2335	0.2282	N.S.
Malay	0.1330	0.1355	N.S.
PNG Lowland MV	0.3593	0.5963	0.08
PNG Lowland	0.3366	0.4269	N.S.
PNG Highland	0.6387	0.5367	N.S.
B Locus			
PNG Highland	0.1467	0.1842	N.S.
DRB1 Locus			
Filipino	0.2898	0.2250	N.S.
Molucca	0.2522	0.2350	N.S.
Nusa Tanager	0.2682	0.2400	N.S.
Singapore	0.1280	0.1257	N.S.
Malay	0.1632	0.1306	N.S.
PNG Lowland MV	0.4225	0.5843	N.S.
PNG Lowland	0.1383	0.2252	0.05
PNG Highland	0.1474	0.1934	N.S.
DPB1 Locus			
Filipino	0.2358	0.2615	N.S.
Molucca	0.1893	0.2205	N.S.
Nusa Tenggara	0.2599	0.3373	N.S.
Malay	0.1352	0.1524	N.S.
PNG Lowland	0.2496	0.4269	N.S.
PNG Highland	0.2969	0.3074	N.S.

N.S.: P values were not significant.

Although an overall trend toward lower observed homozygosities is evident, only the value of F for the DRB1 locus in the PNG Lowland population deviates significantly from values expected for populations of the same size, with the same number of alleles operating under a model of neutral evolution. In addition, the homozygosities for the Molucca and PNG Lowland MV population at the A-Locus closely approach significantly low values.

The homozygosity values for the A, B and DPB1 loci are predominantly lower than expected; only the HLA-A allele frequency distribution of the PNG Highland population has a homozygosity which is higher than the expected, neutral value. These predominantly low homozygosities are consistent with the observations of Klitz et al. (1986) for the DRB1, DQA1 and DQB1 loci, which suggest that homozygosity values lower than the neutral F statistic reflect the operation of balancing selection on those populations. While their inferences of balancing selection did not apply to the DPB1 locus, balancing selection seems to be operating at this locus in the populations presented here.

Several homozygosity values for populations at the HLA-A and DRB1 loci are higher than the expected neutral value. Although none of these values are significantly higher than the expected neutral value, the most pronounced difference between the observed and expected values occurs at the HLA-A locus, for the PNG Highlands population. These high homozygosity values suggest that balancing selection cannot be the only selective force shaping HLA-A and DRB1 allele frequency distributions. In the case of the PNG Highlands, because the A*2402 allele has a frequency of 0.78, overdominance (heterozygote advantage) does not appear to play a major role in determining the allele distribution at the HLA-A locus.

In principle, a value of homozygosity higher than the expected neutral value could reflect founder effects and population bottlenecks, genetic drift, or positive directional selection for an allele. Populations in the PNG Highlands are very geographically isolated, and some reports have attributed the high homozygosity of HLA class I antigen frequency distributions (Serjeantson 1989), to genetic drift or founder effects. However, another study of HLA class I antigens in PNG populations reported altitudinal cline differences in HLA-A antigen frequencies and proposed that these reflected selection operating differentially along the altitude gradient (Smith et al. 1994). The homozygosity for the HLA-B locus in the PNG Highlands is lower than expected for neutral evolutionary conditions. While this difference is not statistically significant either, it does suggest the operation of balancing selection at this locus. Similarly, the DRB1 and DPB1 allele frequency distributions for the PNG Highlands do not suggest a major reduction in diversity due to drift or population bottlenecks in these populations. In addition, the degree of mitochondrial DNA diversity (h= 0.17-0.24) calculated for the PNG Highlands (Stoneking et al. 1990) does not indicate a recent bottleneck.

Overall, the data suggest that directional selection is operating on the A*2402 allele, and raises the possibility that A*2402 confers resistance to some infectious disease pathogen endemic in the PNG Highlands. While the HLA-A distribution could reflect historical founder effects or genetic drift, this scenario would require that the founding population had possessed numerous A*2402-B allele haplotypes, as no evidence of directional selection is observed at the B-locus. In theory, such a population could have been formed via the fusion of multiple, previously distinct populations, but such events are known to result in deviations from expected Hardy-Weinberg equilibrium proportions (i.e. the Wahlund effect), and no such deviations are observed in these populations.

Phylogenetics

The distribution of alleles in different populations can be used to construct a phylogenetic tree/network used to examine the historical relationships between groups. While the extensive polymorphism at the HLA loci makes them highly informative markers, this polymorphism may be subject to selection (see above), and phylogenetic trees based on population allele frequency distributions should be interpreted with some caution. Phylogenetic trees constructed using HLA-A, DRB1 and DPB allele frequencies are shown in Figures 1, 2, and 3.

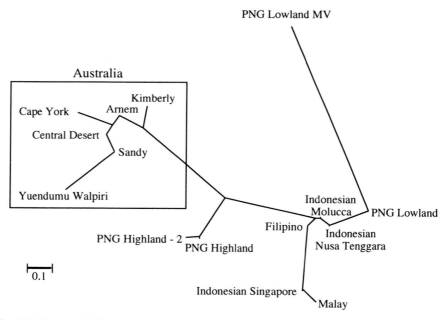

Fig. 1. Neighbor-joining tree constructed using the eight DRB1 allele frequencies reported here as well as DRB1 allele frequencies for six additional Australian and one additional PNG Highland population. Australian populations are indicated in the labeled box. Genetic distance scale is shown at the lower left.

The DRB1 tree in Figure 1 reflects the population relationships expected from previous linguistic and anthropological data, as well as other genetic studies. A notable feature of this tree is the separation between the PNG Highland and Lowland populations; as predicted by the Sahul landmass theory, the PNG Highlands populations are more closely related to Australian populations, while the PNG Lowland populations are much more closely related to island Southern Asian populations.

These population relationships are reflected in the trees in Figures 2 and 3, albeit to a lesser extent. The PNG Highlands population is closest to the Australian Arnem population in the DPB1 tree shown in Figure 2, but is quite distant from the Australian Sandy population also included in the tree analysis. This Australian population is closest to the Nusa Tenggara population.

Similarly, the Australian Yuendumu Walpiri population is closest to the Indonesian Nusa Tenggara population in the A-locus tree shown in Figure 3; while the PNG Highland population is close to this Australian population in this tree, the PNG Lowland population is actually closer. This close relationship between the PNG Lowlands and an Australian population is also reflected in the DPB1 tree in Figure 2; the PNG Lowland and Highland populations are approximately equidistant from the Australian Arnem population.

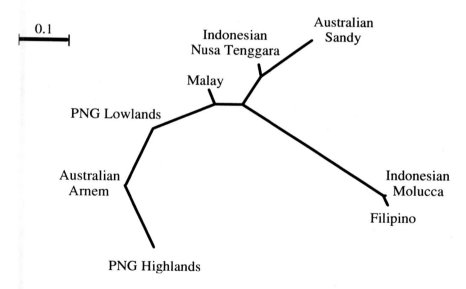

Fig. 2. Neighbor-joining tree constructed using the six DPB1 allele frequencies presented here, as well as the DPB1 allele frequencies for two Australian populations. A scale showing the genetic distance is shown at the upper left.

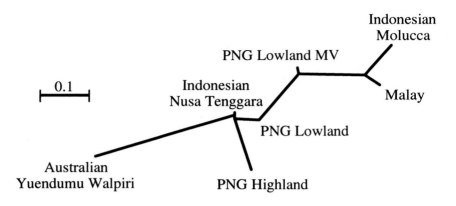

Fig. 3. Neighbor-joining tree constructed using the six HLA-A allele frequencies presented here, as well as the A-locus allele frequencies for an additional Australian population. A genetic distance scale is shown at the upper left.

The differences between these three trees are striking, and merit some discussion. At the HLA-A locus, the directional selection operating on the PNG Highlands, inferred from the high homozygosity values for this population, is presumably operating to a lesser extent on the PNG Lowland and Indonesian populations. These selective forces may have affected the HLA-A allele frequencies in these populations to a degree sufficient to perturb the relationship between Australian populations and PNG Highland populations observed in other studies (van Holst Pellekaan et al. 1997a, 1997b) and for other HLA loci.

The differences between relationships illustrated by the DRB1 and DPB1 trees are less easily understood. Although there does seem to be a uniform influence of balancing selection in operation at the DPB1 locus in each of the populations discussed here, the homozygosities at this locus indicates that these populations are evolving in a manner which is not significantly different from neutral expectations. The differences between these trees might reflect differences in the way the two loci evolve; the DRB1 locus, more polymorphic than all but the HLA-B locus, may be better suited for this type of population study that the DPB1 locus. This suitability may be reflected in the large overall branchlengths, and greater degree of resolution between populations observed for the DRB1 tree.

Conclusions

The distribution of HLA-A, B, DRB1, and DPB1 allele frequencies in these island Southern Asian and Melanesian populations suggests that balancing selection may be operating on them in general and that positive directional selection for A*2402 may account for the very high frequency of this allele in the PNG Highlands. The absence of a predominant HLA-B allele in the PNG Highlands population, as well as the diversity of DRB1 alleles, and A-B haplotypes, suggests that the high frequency of the A*2402 allele is not the result of a population bottleneck or founder effect.

Allele frequency distribution patterns and phylogenetic analyses of HLA-A, DRB1 and DPB1 allele frequency data indicate a close relationship between the PNG Highlands and Australian populations. This relationship is most clearly demonstrated at the DRB1 locus; in this case, the tree topology supports the theory that modern PNG Highlanders and Australians are descendants of a single population that inhabited the Sahul landmass which gave rise to New Guinea, Australia and Tasmania 9,000 years ago.

The variation in population relationships shown in phylogenies generated for different HLA loci suggests that the history of each locus varies with respect to historical selection pressures and founder effects; this may make some loci more suitable for generating phylogenies than others, or suggest multi-locus rather than single locus analyses.

References

Baur MP and Danilovs JA (1980) Population analysis of HLA-A, B, C, DR and other genetic markers. In: Terasaki PI (ed) Histocompatibility Testing 1980. UCLA Press, Los Angeles, CA

Begovich AB, McClure GR, Suraj VC, Helmuth RC, Fildes N, Bugawan TL, Erlich HA, Klitz W (1992) Polymorphism, recombination, and linkage disequilibrium within the HLA class II region. J Immunol 148:249-258

Bugawan TL, Apple R, Erlich HA (1994) A method for typing polymorphism at the HLA-A locus using PCR amplification and immobilized oligonucleotide probes. Tissue Antigens 44:137-147

Bugawan TL, Begovich AB, Erlich HA (1990) Rapid HLA-DPB typing using enzymatically amplified DNA and nonradioactive sequence-specific oligonucleotide probes. Immunogenetics 32:231-241

Bugawan TL, Chang J, Klitz W, Erlich HA (1994) PCR/Oligonucleotide probe typing of HLA Class II alleles in a Filipino population reveals an unusual distribution of HLA haplotypes. Am J Hum Genet 54:331-340

Bugawan TL, Erlich HA (1991) Rapid typing of HLA-DQB1 DNA polymorphism using nonradioactive oligonucleotide probes and amplified DNA. Immunogenetics 33:163-170

Bugawan TL, Mack SJ, Stoneking M, Saha N, Beck HP, Erlich HA (1999) HLA class I allele distributions in six Pacific/Asian populations: evidence of selection at the HLA-A locus. Tissue Antigens 53:311-319

Erlich HA, Bugawan TL, Begovich AB, et al. (1991) HLA-DR, DQ and DP typing using PCR amplification and immobilized probes. Eur J Immunogenet 18:33-55

Erlich HA, Bugawan T, Begovich A, Scharf S (1993) Analysis of HLA class II polymorphism using Polymerase Chain Reaction. Arch Pathol Lab Med 117:482-485

Ewens WJ (1972) The sampling theory of selectively neutral alleles. Theor Popul Biol 3:87-112

Felsenstein J (1995) PHYLIP (Phylogenetic Inference Package, version 3.5c), University of Washington, Seattle, WA

Gao X, Jakobsen IB, Serjeantson S (1994) Characterization of the HLA-A polymorphism by locus-specific polymerase chain reaction amplification and oligonucleotide hybridization. Hum Immunol 41:267-279

Gao X, Veale A, Serjeantson SW (1992) HLA class II diversity in Australian aborigines: unusual HLA-DRB1 alleles. Immunogenetics 36:333-337

Imanishi T, Tatsuya A, Kimura A, Tokunaga K, Gojobori T (1992) In: Tsuji K, Auzawa M, Sasazuki T (eds) HLA 1991. Vol 1. Oxford University Press, Oxford, pp 1065-1220

Klitz W, Thomson G, Baur MP (1986) Contrasting evolutionary histories among tightly linked HLA loci. Am J Hum Genet 39:340-349

Lester S, Cassidy S, Humphreys I, Bennett G, Hurley CK, Boettcher B, McCluskey J (1995) Evolution in HLA-DRB1 and major histocompatibility complex class II haplotypes of Australian aborigines. Definition of a new DRB1 allele and distribution of DRB1 gene frequencies. Hum Immunol 42:154-60

Lester S, Gao X, Varney M, Tilanus M, Boettcher B, McCluskey J (1997) Australian Aboriginal Normal. In: Terasaki PI, Gjertson DW (eds) HLA 1997. UCLA Tissue Typing Laboratory, Los Angeles, CA

Long JC, Williams RC, Urbanek M (1995) An E-M algorithm and testing strategy for multiple locus haplotypes. Am J Hum Genet: 56:799-810

Lum JK, Rickards O, Ching C, Cann RL (1994) Polynesian mitochondrial DNAs reveal three deep maternal lineage clusters. Hum Biol 66:567-590

Marsh SG (1997) Nomenclature for factors of the HLA system, update 1997 Eur J Immunogenet 24:239-40

Melton T, Peterson R, Redd AJ, Saha N, Sofro ASM, Martinson J, Stoneking M (1995) Polynesian genetic affinities with Southeast Asian populations as identified by mtDNA analysis. Am J Hum Genet 57:403-414

Melton T, Clifford S, Martinson J, Batzer M, Stoneking M (1998) Genetic evidence for the proto-Austronesian homeland in Asia: mtDNA and nuclear DNA variation in Taiwanese aboriginal tribes. Am J Hum Genet 63:1807-1823

Redd AJ, Takezaki N, Sherry ST, McGarvey ST, Sofro AS, Stoneking M (1995) Evolutionary history of the COII/tRNALys intergenic 9 base pair deletion in human mitochondrial DNAs from the Pacific. Mol Biol Evol 12:604-615

Saitou N, Nei M (1987) The neighbor-joining method: a new method for reconstructing phylogenetic trees. Mol Biol Evol 4:406-425

Serjeantson SW (1989) HLA genes and Antigens. In: Hill AVS, Serjeantson SW (eds) The Colonization of the Pacific: A Genetic Trail. Clarendon Press, New York, pp120-135

Slatkin M (1994) An exact test for neutrality based on the Ewens sampling distribution. Genet Res 64:71-74

445

Slatkin M (1996) A correction to the exact test based on the Ewens sampling distribution. Genet Res 68:259-260

Smith T, Bhatia K, Prasad M, Koki G, Alpers M (1994) Altitude, language, and class I HLA allele frequencies in Papua New Guinea. Am J Phys Anthropol 95:155-168

Steiner LL, McCurdy DK, Cavalli A, Moonsamy PV, Begovich AB (1996) Two new DPB1 alleles identified in a study of the genetics of susceptibility to pauciarticular juvenile rheumatoid arthritis. Tissue Antigens 49:262-266

Stoneking M, Jorde LB, Bhatia K, Wilson AC (1990) Geographic variation in human mitochondrial DNA from Papuan New Guinea. Genetics 124:717-733

Titus-Trachtenberg EA, Rickards O, De Stefano GF, Erlich HA (1994) Analysis of HLA Class II haplotypes in the Cayapa Indians of Ecuador: a novel DRB1 allele reveals evidence for convergent evolution and balancing selection at position 86. Am J Hum Genet 55:160-167

Tsintof AS, Hertzberg MS, Prior JF, Mickleson KNP, Trent RJ (1990) α-globin gene markers identify genetic differences between Australian Aborigines and Melanesians Am J Hum Genet 46:138-143

van Holst Pellekaan S, Frommer M, Sved JA, Boettcher B (1997) Mitochondrial D-loop diversity in Australian riverine and Australian desert Aborigines. Electrophoresis 18:1538-1543

van Holst Pellekaan S, Frommer M, Sved JA, Boettcher B (1997) Mitochondrial control-region sequence variation in Aboriginal Australians. Am J Hum Genet 62:435-449

Watterson GA (1978) The homozygosity test of neutrality. Genetics 88:405-417

HLA class I alleles in Australian aborigines and their peptide binding profiles

Xiaojiang Gao[1,2], Sue Lester [3], Anthony Veale[4], Barry Boettcher[5], Bart Currie[6], James McCluskey[3], and Gareth Chelvanayagam[2]

[1]IRSP, SAIC Frederick, NCI-FCRD, Frederick, MD, USA
[2]Human Genetics Group, John Curtin School of Medical Research, The Australian National University, Canberra, Australia
[3]Red Cross Blood Transfusion Service, Adelaide, Australia
[4]National Centre for Epidemiology and Population Health, The Australian National University, Canberra, Australia
[5]Department of Biological Sciences, The University of Newcastle, Newcastle, Australia
[6]Menzies School of Health Research, Darwin, Australia

Summary. The HLA system is under balancing selection. HLA alleles are maintained in populations by their divergent functions. The peptide binding profiles of HLA alleles may hold the key to a better understanding of HLA diversification and polymorphism maintenance. Long isolated Australian aboriginal populations provide a good model for these studies. The four HLA-A, seven B and five or six C alleles commonly detected in them represent the minimum class I repertoire carried by the founder group as well as the minimum class I polymorphism that needs to be maintained in these populations. All these alleles have a unique combination of the P2 and P9 anchor preferences indicating unique peptide binding profiles that have 'earned' their place in the minimum allele set. This study of Australian aborigines provides further insights into the mechanism of HLA evolution in general.

Key words. Major histocompatibility complex, Class I human leucocyte antigen gene, Human evolution, Peptide presentation, Australian aborigines

Introduction

Recent archeological findings indicate that Australia has been colonized for at least 65,000 years (Thorne et al. 1999). The origin of the first settlers remains a matter of debate. It is generally accepted, however, that Australian aborigines share a common ancestry with the descendants of the early settlers of New Guinea (Serjeantson 1989) as for most of its history, Australia and New Guinea were the

446

one landmass – the ancient continent of Sahul. Even since its separation from New Guinea about 7,000 years ago until European contact 200 years ago, Australia has been biologically relatively isolated despite some evidence of limited contact in the northern coastal areas with visitors from the neighboring Torres Strait Islands, New Guinea and Southeast Asia (White 1997).

DNA typing of HLA class II genes of Australian aborigines showed that this long period of isolation has resulted in a unique HLA profile. In particular the DRB1 gene has apparently adapted well to the local environment by generating and selecting several new alleles (Gao et al. 1992a; 1992b; Lester et al. 1995) despite the fact that the class II serology detected only a few DR antigens (Serjeantson 1989). Locally generated novel alleles account for up to 40% of the DRB1 gene frequency in the Australian aboriginal populations examined so far. Another prominent feature of their class II profile is that the distribution of alleles varies significantly between populations in different regions. Several alleles, including locally generated novel alleles, have highly confined population distributions. Very little is known about HLA class I genes in Australian aborigines. As for class II serology, early class I serology reported a rather restricted range of class I polymorphism in aborigines. Recently developed high resolution DNA typing technologies for class I genes provide an opportunity to examine the aboriginal class I polymorphism at allele level.

It is well established that HLA molecules function by directly binding and presenting antigenic peptides to T cells to initiate immune responses (Chen and Parham 1989; Bjorkman and Parham 1990). Class I molecules preferentially bind peptides of nine amino acid resides (Fremont et al. 1992). The backbone structures or side chains of these nine residues are physically bound or accommodated by correspondent pockets or binding environments. In all but a few alleles the pockets that bind the backbone structures of the second (P2) and the ninth (P9) residues of the peptide are highly selective and preferentially bind certain types of amino acids known as anchors (Barber et al. 1995). Other binding pockets or environments are more flexible and can accommodate a variety of different side chains. Therefore the peptide ligands presented by a given allele are largely determined by the P2 and P9 pockets (or B and F pockets as they have been classically named) which in turn are determined by the allelic variation in the peptide binding sites that form part of the pockets. Alleles sharing identical or similar P2 and P9 pockets present the peptide ligands with identical or similar P2 and P9 anchor residues though these peptides may differ in other residue positions. Alleles having strikingly different P2 or P9 or both pockets generally do not present identical peptides. Therefore, P2 and P9 pockets may be a useful marker for the functional similarity between different class I alleles. In recent years the anchor preferences of many class I alleles have been identified. Efforts have also been made to define each of the nine peptide-binding environments for different class I alleles (Chelvanayagam 1996).

It is believed that MHC polymorphism is generated and maintained by balancing selection or heterozygous advantage (Doherty and Zinkernagel 1975;

Hughes and Nei 1988). Many attempts have been made to reconstruct the evolutionary history of the HLA polymorphism by building phylogenetic trees that equally enlist all known HLA alleles. Although the phylogeny of the whole set of HLA alleles is useful in tracing the origins of HLA lineages, a better understanding of the diversification and maintenance of HLA polymorphism relies on the analysis of individual populations. Over 400 class I alleles have been officially recognized worldwide so far (WHO Nomenclature Committee for HLA Factors, May 1999. http://www.anthonynolan.com/HIG/nomenc.html). In any given population, however, only a small fraction of these alleles are present. As evolutionary forces operate on individual populations, only the alleles that are present in the population matter. The numbers, types and frequencies of the alleles in the populations are all relevant to the evolutionary process. Long isolated populations like the Australian aborigines may provide a good model for these studies because they have evolved in a relatively simplified genetic environment. In an open environment the frequent genetic exchange between different human groups will constantly break the genetic equilibrium, obscuring the source of alleles and haplotypes and complicating the pattern of polymorphism maintenance. Population studies of Australian aborigines may therefore shed more light on HLA evolution in general. In this study we examine the peptide binding profiles of class I alleles in Australian aboriginal populations to see whether the functional significance of peptide binding correlates with the maintenance of the HLA class I polymorphism.

Materials and methods

Study populations

Samples were collected from the field for previous studies. Community consent was obtained to use the samples for on-going HLA-related genetic studies. A total of 413 individuals from four Australian aboriginal populations were examined. The four populations were from the Cape York Peninsular in Northern Queensland (n=104), Groote Eylandt off the Northern Territory coast (n=75), the Kimberley region of Western Australia (n=41) and the Yuendumu community in the Central Desert (n=193). These regions cover a vast area of the continent with highly divergent geographic features. The Cape York population is known to have had an extensive genetic input from recent racial mixing (estimated at 20-30%), mainly of European origin (Gao et al. 1992b). The other three populations, however, have had relatively little Caucasoid admixture. The geographic isolation of the Yuendumu community in the Central Desert has effectively prevented genetic exchange with other Australian regions as well as the outside world. The three coastal regions, however, have had seasonal visitors from the neighboring Melanesia and Southeast Asia, which may have resulted in limited genetic input even before European contact (White 1997). All the samples except the Groote Eylandt series have been typed for HLA class II genes in previous studies (Gao et al. 1992a, 1992b; Lester et al. 1995).

HLA typing

Typing for HLA-A, B and C was carried out using high-resolution sequence specific oligonucleotide (SSO) hybridization. The typing method has previously been described in detail (Gao et al. 1994). Briefly, exons 2 and 3 of HLA-A, B and C genes were separately amplified from genomic DNA using locus-specific PCR primers and hybridized with a panel of SSO probes designed according to known class I sequence motifs. The washing procedure conditions were adjusted to allow discrimination of single-base mismatches. Alleles and genotypes were assigned to each sample by SSO hybridization patterns predicted from known sequences. Typing resolution is determined by the number of probes included in the protocol. In the present experiments allele level typing was achieved using 60 probes for each of the A and C loci and 100 probes for the B locus. Samples showing unusual SSO hybridization patterns indicating novel alleles were further analyzed by cloning and sequencing for the whole sequence of exons 2 and 3. In addition, at least two samples of each of the alleles commonly detected in Australian aboriginal populations were subject to sequencing analysis to confirm the SSO typing results.

Definition of peptide binding pockets of the class I molecule

The peptide binding pockets of class I molecules have previously been broadly defined by the set of MHC amino acids within a fixed neighborhood of a peptide residue (e.g. 2 Angstroms) (Chelvanayagam 1996). Using these environments as a starting point, MHC amino acids that are not in direct contact with side chain atoms of the respective P2 and P9 peptide anchor residues were excluded from the pocket definition. The polymorphic component of the P2 pocket comprises positions 9, 63, 66, 67, 70, 99 for HLA-A; 9, 24, 45, 63, 66, 67, 70, 99 for HLA-B; and 9, 24, 66, 99 for HLA-C. Likewise, the P9 pocket comprises positions 77, 80, 81, 95, 116 for HLA-A; 77, 80, 81, 95, 116, 143, 147 for HLA-B; and 77, 80, 95, 116, 143, 147 for HLA-C. These pockets must be regarded as an approximation as this process was only carried out on a sample of crystal structures with bound peptides, so that not all 20 amino acids were considered in the context of all class I allelic products. Nonetheless, the pockets correlate well with published pockets as well as known anchor motifs.

Construction of P2 and P9 distance trees

Distances were calculated separately between all pairs of possible P2 and P9 pocket types within a given locus to discriminate between the likely anchor preferences. The distance between a given pair of pockets was calculated by summing the cost of amino acid substitution for each of the amino acid positions in the pocket. The substitution cost takes into account the five physico-chemical properties that best represent amino acid substitution values (Argos 1987). The cost of substitution was determined by looking up a 20x20 matrix filled with the costs of all substitutions between any two amino acids. The cost matrix was constructed by representing each of the 20 amino acids as points in a 5 dimensional space and calculating the Euclidian distance between all pairs of points. Dendrograms were constructed using a group average clustering procedure.

Anchor preferences of HLA class I alleles

The anchor preferences for the P2 and P9 pockets were either taken from a roadmap of peptide binding specificities (Chelvanayagam 1996) or estimated on the basis of their similarity to other pockets. Although the latter estimates are unreliable they are at least indicative of the type of amino acids that can bind.

Results

All the HLA-A, B and C alleles commonly detected in the study populations are also present in other regional (Asia-Oceania) or world populations. Only two rare A alleles appear to have a local origin. These were A*2406 and 2413 (Gao et al. 1997a) sporadically detected in the Yuendumu population and in one case (A*2413) in Kimberley. The gene frequencies of the HLA-A, B and C alleles in the study populations are given in Table 1. Aboriginal communities have experienced varying degrees of racial admixture since the European contact about 200 years ago. The HLA repertoire of contemporary Australian aboriginal groups comprises alleles inherited from the ancestral founder groups, alleles from foreign genetic input and locally generated new alleles. The origin of most DNA-detected HLA alleles in this study can be explained by comparing their frequency distributions in local and world populations as well as by examining the linkage relationships of HLA class I and class II haplotypes. Caucasian contributions to the aboriginal gene pool can be estimated from the combined gene frequencies of those obvious Caucasian alleles indicated by 'others' in Table 1. They range from 0.12 to 0.20 in the Cape York population and from 0.01 to 0.06 in the other three populations. The actual Caucasoid contribution to the HLA frequencies could, however, be higher since several alleles including A*0201, A*2402, B*4001, B*4002, Cw*0303, Cw*0401 and Cw*0702 are shared by both Australian aborigines and Caucasoids.

The four Australian aboriginal populations share a common set of class I alleles including four A alleles: A*0201, 1101, 2402 and 3401; six B alleles: 1301, 1521,

Table 1. Gene frequencies of HLA-A, B, and C in Australian aboriginal populations

Allele	Cape York	Kimberley	Groote Eylandt	Yuendumu
HLA-A	n=103	n=36	n=75	n=191
*0201	.1748[1]	.1111	.1067	.1126
*1101	.1796	.0139	.2400	.0759
*2402	.2233	.0972	.2933	.2984
*2406	.0000	.0000	.0000	.0131
*2413	.0000	.0139	.0000	.0183
*3101	.0049	.0000	.0067	.0026
*3401	.2913	.6806	.3200	.4398
others[2]	.1261	.0139	.0334	.0131
HLA-B	n=100	n=38	n=75	n=193
*1301	.2700	.1316	.2333	.2435
*1521	.1350	.0263	.0600	.1244
*1525	.0000	.0000	.0800	.0570
*3901	.0350	.0132	.0000	.0207
*4001	.0800	.1842	.1733	.0492
*4002	.0800	.2763	.1800	.1891
*5601	.1650	.3553	.1800	.1140
*5602	.0300	.0132	.0267	.1891
others	.2050	.0000	.0669	.0130
HLA-Cw	n=89	n=28	n=73	n=192
*0102	.1854	.3750	.2671	.2474
*0303	.0281	.1250	.1507	.0547
*0401	.2753	.1786	.2534	.2682
*0403	.1573	.0357	.1233	.1510
*0702	.0955	.0000	.0000	.0677
*1502	.0674	.2857	.1370	.2031
others	.1910	.0000	.0685	.0079

[1] Gene frequencies were calculated by direct counting.
[2] Alleles from more recent population admixture.

Table 2. HLA class I haplotypes showing significant positive linkage disequilibrium in Australian aboriginal populations

Haplotype	Cape York	Kimberley	Groote Eylandt	Yuendumu
A*0201-B*1525	-	-	++	-
A*1101-B*4002	-	-	+	++
A*2402-B*1525	-	-	-	++
A*2402-B*4001	-	-	++	-
A*3401-B*56	-	++	++	+
A*1101-Cw*0401	-	++	-	-
A*1101-Cw*1502	-	-	+	++
A*2402-Cw*0303	-	-	++	-
A*3101-Cw*0403	-	-	-	++
B*1301-Cw*0401	++	++	++	++
B*1521-Cw*0403	++	++	++	++
B*4001-Cw*0303	++	++	++	++
B*4002-Cw*1502	++	++	++	++
B*56 -Cw*0102	++	++	++	++
B*5601-Cw*0702	-	-	-	++

++ haplotype frequency>0.02, χ^2>3.85 and relative delta value>0.6.

+ haplotype frequency>0.02, χ^2>3.85 and relative delta value<0.6.

- no significant linkage disequilibrium.

4001, 4002, 5601 and 5602; and five C alleles: Cw*0102, 0303, 0401, 0403 and 1502. In addition, a further three alleles were detected in some of the populations. These were A*2406 in Yuendumu, A*2413 in Yuendumu and Kimberley, and B*1525 in Groote Eylandt and Yuendumu. A*2406, A*2413 and B*1521 were all first discovered in Australian aborigines (Lienert et al. 1995; Gao et al. 1997) and B*1521 was later widely detected in other Oceanic and Southeast Asian populations (Gao et al. 1997b). Altogether six A alleles, seven B alleles and five C alleles were considered as aboriginal (being the alleles carried by the founders as well as locally generated new alleles). The origin of the remaining three alleles, A*3101, B*3901 and Cw*0702 can not readily be determined as they have confined distributions at lower frequencies in the study populations. Also they have been widely detected in other racial groups including Melanesians and Caucasoids. In the study populations, however, they were associated with unique aboriginal class I and class II haplotypes. These alleles could either have been introduced by early visitors from neighboring regions or more recently by Caucasian admixture and have since experienced haplotype recombination.

Linkage relationships between HLA-A, B and C loci were tested for linkage disequilibrium. Haplotypes having a significant delta value are shown in Table 2. The four study populations share very few common A-B and A-C haplotypes. The Cape York population had no significant A-B or A-C linkage relationships. The Kimberley population had only two significant haplotypes (one A-B and one A-C). Other haplotypes showing significant linkage relationships were observed in the Groote Eylandt and Yuendumu populations and generally did not overlap between the two populations despite the fact that they share most of the alleles involved in the haplotypes. In sharp contrast, strong linkage disequilibrium relationships were observed between B and C in all four Australian aboriginal populations. Six common B-C haplotypes (haplotype frequency>0.02) were found to have a significant delta value. Of these five haplotypes, B*1301-Cw*0401, B*1521-Cw*0403, B*4001-Cw*0303, B*4002-Cw*1502 and

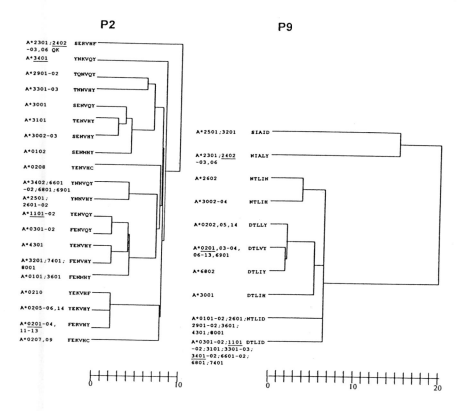

Fig. 1. P2 and P9 pockets of HLA-A alleles and distance trees. The aboriginal alleles are indicated with underlines. The P2 pocket definitions are positions 9, 63, 66, 67 and 99. The P9 definitions are positions 77, 80, 81, 95 and 116.

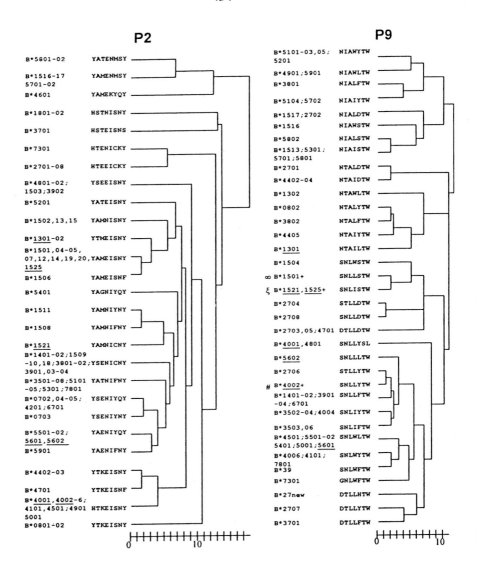

Fig. 2. P2 and P9 pockets of HLA-B alleles and distance trees. The aboriginal alleles are indicated with underlines. The P2 pocket definitions are positions 9, 24, 45, 63, 66, 67, 70 and 99. The P9 definitions are positions 77, 80, 81, 95, 116, 143 and 147. "This group also includes B*1503,05-12,14-15,18-19;3505;4003;4601. ⁵This group also includes B*1502,20;3501,07-08;4802. #This group also includes B*0702-05;0801;1509-10;41; 4201; 4005.

B*56(5601 or 5602)-Cw*0102 are shared by all four aboriginal groups. The sixth, B*5601-Cw*0702, was only detected in Yuendumu.

The peptide binding profiles of the class I alleles found in the study populations were evaluated by comparing the peptide binding residues of the nonamer bound by the class I molecule, particularly the polymorphic amino acid positions that are in direct contact with the P2 and P9 residues. The distance trees constructed from these variable positions took into account the impact of both the number and type of replaced amino acid residues in these positions on peptide binding.

In the HLA-A trees (Fig. 1) the four major aboriginal alleles, A*0201, 1101, 2402 and 3401 each represent one of the five major allele lineages or families that have been identified in world populations (Kato et al. 1989). If all the polymorphic peptide-binding sites are taken into account, the percentage of mismatched amino acids between these four A alleles ranges from 40-70%, indicating markedly different peptide binding environments. Indeed these alleles

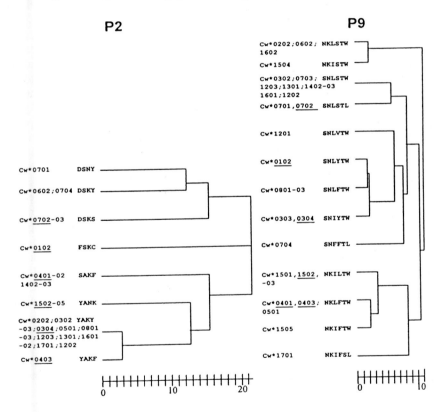

Fig. 3. P2 and P9 pockets of HLA-C alleles and distance trees. The aboriginal alleles are indicated with underlines. The P2 pocket definitions are positions 9, 24, 45, 66 and 99. The P9 definitions are positions 77, 80, 95, 116, 143 and 147.

are well separated in both the P2 and P9 trees except that A*1101 shares an identical P9 pocket with A*3401. The highly divergent peptide binding environments of the Australian aboriginal A alleles lead to different anchor preferences for both P2 and P9 pockets as shown in Table 3.

In the B distance trees (Fig. 2) not all the seven major aboriginal B alleles are as well separated as the A alleles. Three allele pairs, 1521/1525, 4001/4002 and 5601/5602 share identical P2 or P9 pockets. The percent of mismatched peptide binding residues between each of these pairs ranged from 3-11%, which is much lower than the percentages between the rest of the aboriginal B alleles that range from 30-50%. Nonetheless all six Australian aboriginal B alleles have different binding preferences for either P2 or P9 or both.

A closer look at the three pairs of B alleles that have a lower level of mismatched peptide binding residues revealed that, despite the smaller numbers of amino acid replacements the positions and types of the amino acids substituted have a significant impact on the peptide binding property. For example although the two B*56 subtypes, B*5601 and 5602, differ by only two amino acids in positions 95 (W>L) and 97 (T>R) both substitutions have a significant effect on peptide binding. Position 95 forms part of the P9 pocket. The substitution of tryptophan, the largest amino acid, with the much smaller leucine, significantly changes the shape and depth of the P9 pocket. Consequently the P9 binding preference changes from tiny alanine (B*5601) to the much larger phenylalanine (B*5602) as shown in Table 3. Furthermore, position 97, though not strictly part of the P2 or P9 pocket, has an indirect effect on five out of the nine peptide binding pockets including P9 if a major substitution occurs in this position (Chelvanayagam 1996). In the case of the two B56 alleles a large and charged arginine (B*5602) replaces the small but hydrophobic threonine (B*5601). This is likely to affect the overall peptide binding of the two alleles. It is a similar story for the other two pairs. B*4001 and 4002 have eight differences but only three of them (at positions 97, 143 and 147) have an effect on peptide binding. Two of the substitutions, 97 (R>S) and 147 (L>W) are major changes that lead to a change of the P9 preference from tryptophan (B*4001) to leucine (B*4002). B*1521 and 1525 differ by two substitutions in the P2 pocket, positions 63 (N>E) and 67 (C>S). These substitutions change the P2 preference from glutamic acid (B*1521) to a more flexible range of preferences including glutamine, proline, leucine and valine (Table 3). These substitutions are more conservative than the other allele pairs.

The major HLA-C alleles (including Cw*0702) are similar to HLA-A being well separated in both the P2 and P9 distance trees (Fig. 3). Only one pair of alleles, Cw*0401 and 0403 share an identical P9 pocket but the P2 pockets are highly divergent. Again each of the six C alleles have different anchor preferences for either P2 or P9 or both.

Table 3. Anchor preferences of HLA-A, B and C alleles detected in Australian aborigines

HLA-A	P2	P9	HLA-B	P2	P9	HLA-Cw	P2	P9
*0201	L	L/V	*1301	(Q)	(Y)	*0102	A/L/P	L
*1101	T/V	K	*1521	(E)	Y	*0303	A	L
*2402	Y	F/L	*1525	Q/P/L/V	Y	*0401	Y/F/P	L/F
*3401	(V)	(K)	*4001	E	(W)	*0403	A/S/T	L/F
			*4002	E	(L)	*0702	Y/R	Y
			*5601	P	A	*1502	V/L	(V)
			*5602	P	F			

P2 and P9 stand for second and ninth residues of the peptide accommodated respectively by the P2 and P9 pockets of the class I molecule. Anchor preferences are represented by the single-letter amino acid code. Parentheses indicate the anchor preferences predicted on the basis of similarity to other pockets.

Discussion

Australia had a relatively isolated biogeographical environment until European contact. Consequently the HLA alleles and haplotypes carried into Australia by the ancestral founder groups of Australian aborigines are readily identified in contemporary populations. Although contemporary Australian aborigines share a core set of HLA alleles with Melanesians, particularly non-Austronesian speaking Melanesians, it reflects their common ancestry rather than genetic exchange. Neither the class II data from our earlier studies nor the class I results from this study showed much evidence of any significant genetic input from neighboring Austronesian regions suggesting that the formation of the Torres Strait 7,000 years ago effectively blocked subsequent Austronesian migration to Australia. The HLA data are in agreement with some of the non-HLA genetic markers (Roberts-Thomson et al. 1999) and archeological and linguistic investigations (Bellwood 1989).

Our earlier studies showed marked variation between HLA class II profiles from different regions suggesting a lack of genetic exchange between these Australian populations. In contrast the class I polymorphisms are more evenly distributed. The four study populations share a virtually identical set of A, B and C alleles and B-C haplotypes at comparable frequencies. Such a homogenous HLA profile is usually indicative of frequent and extensive genetic exchanges between populations. In this case, however, the highly divergent A-B and A-C linkage relationships and markedly different class II polymorphisms do not support this interpretation. Rather, the relatively homogenous distribution of class

I polymorphisms suggests that the study populations descended from an ancestral founder group that had a limited number of alleles and haplotypes.

HLA genes diversify in response to the changing environment. Randomly occurring gene mutations that have a selective advantage are saved from extinction through genetic drift and will gradually reach an appreciable level in the population. Given enough time this process eventually leads to polymorphism turnover. As we previously reported, rapid polymorphism turnover is evident in class II HLA loci of Australian aborigines. In sharp contrast their class I genes are highly conserved. The lack of impetus to generate new alleles together with the relatively balanced distribution of the class I polymorphism suggests that a) the class I repertoire of the study populations may be in balance with environmental factors and b) the populations in different regions may have been subject to similar selective forces that operate on class I genes. The markedly different evolutionary patterns in class I and class II loci could be due to the functional differences of class I and class II molecules as they may deal with different environmental reagents. Interestingly, class I genes have an unusually active evolutionary history in another long isolated people, the South Amerindians in the tropical Amazon forests. Some of these populations have almost completely rebuilt their class I loci with locally generated new alleles (Belich et al. 1992; Watkins et al. 1992; Parham et al. 1997). The strikingly different evolutionary patterns of class I genes in these two long isolated peoples could be the result of a range of factors including the differences in their founder groups, colonization histories, density of inhabitance, population sizes, European encounters and biological environments.

Despite their long isolated history and severe population bottlenecks the aboriginal populations from different regions have retained the same set of class I alleles. These aboriginal alleles may represent not only the minimum class I repertoire carried into Australia by the ancient founder groups but also the minimum class I polymorphism that needs to be retained in the populations in order to maintain an appropriate level of heterozygosity. The minimum level of HLA polymorphism in a population involves both the number and type of alleles. In the study populations the minimum numbers of class I alleles are four A, six to seven B and five to six C alleles. Examination of HLA distributions in other world populations revealed that similar minimum numbers apply to most long isolated indigenous populations (Charron 1997). The type of alleles maintained in a population is just as important. Individual HLA alleles must be functionally significant to 'earn' their place in a population. Once an allele becomes functionally redundant it will be replaced by other more useful alleles. Therefore, a minimum HLA allele set is likely to comprise only the alleles with highly divergent functions. The peptide binding profiles of the class I alleles reported in this study may help to better understand the mechanism of polymorphism maintenance.

The four aboriginal A alleles, A*0201, 1101, 2402 and 3401 each represent one of the five major allele lineages or families in the A locus. Extensive sequence variation in the polymorphic peptide binding sites between these families (up to

70% mismatches) means their overall binding environments including P2 and P9 anchor preferences are highly divergent. Our previous investigations on HLA-A lineages (Chelvanayagam et al. 1996; Jakobsen et al. 1998) showed that, sequence phylogeny and unique surface motifs divide A lineages into three broad groups: the A2 group, the A9 (A23 and A24) group and the group comprising the remaining allele lineages (excluding A*8001). A2 and A9 have highly distinctive family structures that may translate into unique peptide binding properties that can not be replaced by any other lineages. Other lineages including A1/3/11, A10, A19, and A30, however, have been through a higher degree of inter-lineage sequence exchange, which may have obscured the functional boundaries between the lineages. Interestingly, the HLA distribution in world populations correlates well to this grouping. A2 and A9 are invariably present as two of the major A alleles in virtually all populations though the particular A2 and A9 subtypes may vary in different populations. One or more allele lineages from the third group also need to be present. Which particular 'third party' allele lineages are maintained in a given population may be decided by the local selective reagents or purely by chance depending on the allele lineages that happened to be carried by the ancestral founder group.

The HLA-B locus has a more active evolutionary history and a higher degree of diversity. Extensive inter-allelic sequence exchanges have obscured the ancient family structures. Sequence phylogeny can generally group the B alleles according to their serological specificities. In the Australian aboriginal populations six or seven B alleles from four lineages have been maintained. Again, all the B alleles have a unique combination of P2 and P9 anchor preferences, which may justify their inclusion in the minimum allele set. Three allele pairs, each from a same lineage (B15, B40 and B56) coexist in the minimum allele set despite the minor sequence variation in their peptide binding sites. The peptide binding profiles of these allele pairs especially the B40 and B56 pairs demonstrated that, depending on the position and type of the substituted amino acids, minor sequence variation including single replacements could make a big difference in peptide binding.

The six C alleles maintained in the study populations are also well separated in P2 and P9 distance trees and they all have a unique combination of the P2 and P9 anchor preferences. The more conserved HLA-C locus has a tight linkage relationship with the more active HLA-B. Therefore, the C alleles maintained in a population are not entirely dictated by their independent functional merits. Rather, their functional significance may be appreciated by their association with HLA-B or vise versa in a lesser degree. Indeed, the peptide binding profiles of the B and C alleles in each of the six most prominent B-C haplotypes detected in the study populations seem to complement rather than compete with each other, which may partly explain the strong linkage disequilibrium between B and C.

Peptide binding and presentation by HLA is a complex issue that is not fully understood. A single substitution in either the peptide or HLA molecule may affect several binding pockets. Some alleles may have secondary anchor pockets in addition to P2 and P9 (Ruppert et al. 1993; Chen et al. 1994). In some rare instances, alleles may even select other peptide residues instead of P2 and P9 as the principle anchors (DiBrino et al. 1994). Nonetheless given the canonical mode of peptide binding in which P2 and P9 always project into the class I cleft (Chelvanayagam et al. 1997) the P2 and P9 pockets and their anchor preferences are generally indicative of the overall peptide binding characteristics of class I alleles. In this study we demonstrated that the functional significance of HLA class I alleles predicated by their peptide binding profiles is highly relevant to the maintenance of HLA polymorphism in Australian aboriginal populations. The observations from this study provide further evidence in support of the hypothesis that HLA diversity is driven and maintained by balancing selection.

Acknowledgements

The authors would like to thank Ms Penny Main for her critical reading and assistance in the preparation of the manuscript. We also thank the Australian aboriginal communities involved in this study. Without their generous consent and participation this study would not have been possible.

References

Argos P (1987) A sensitive procedure to compare amino acid sequences. J. Mol Biol 193:385-396

Barber LD, Gillece-Castro B, Percival L, Li X, Clayberger C, Parham P (1995) Overlap in the repertoires of peptides bound in vivo by a group of related class I HLA-B allotypes. Curr Biol 5:179-190

Belich MP, Madrigal JA, Hildebrand WH, Zemmour J, Williams RC, Luz R, Petzl-Erler ML, Parham P (1992) Unusual HLA-B alleles in two tribes of Brazilian Indians. Nature 357:326-329

Bellwood P (1989) The colonisation of the Pacific: some current hypotheses. In Hill AVS. Serjeantson SW (eds) The colonisation of the Pacific: a genetic trail. Clarendon Press, Oxford, pp 1-59

Bjorkman PJ, Parham P (1990) Structure, function and diversity of class I major histocompatibility complex molecules. Annu Rev Biochem 59:253-288

Charron D (ed) (1997) Genetic diversity of HLA: Functional and Medical Implication. EDK Publishing, Paris

Chelvanayagam G (1996) A road map for HLA-A, HLA-B, and HLA-C peptide binding specificities. Immunogenetics 45:5-26

Chelvanayagam G, Jakobsen IB, Gao X, Easteal S (1996) Structural comparison of major histocompatibility complex class I molecules and homology modeling of five distinct human leukocyte antigen-A alleles. Protein Eng 9:1151-1164

Chelvanayagam G, Apostolopulos V, McKenzie IFC (1997) Milestones in the molecular structure of the major histocompatibility complex. Protein Eng 10:471-474

Chen W, Khilko S, Fecondo J, Margulies DH, McCluskey J (1994) Determinant selection of major histocompatibility complex class I-restricted antigenic peptides is explained by class I-peptide affinity and is strongly influenced by nondominant anchor residues. J Exp Med 180:1471-1483

Chen BP, Parham P (1989) Direct binding of influenza peptides to class I HLA molecules. Nature 337:743-745

DiBrino M, Parker KC, Shiloach J, Turner RV, Tsuchida T, Garfield M, Biddison WE, Coligan JE (1994) Endogenous peptides with distinct amino acid anchor residue motifs bind to HLA-A1 and HLA-B8. J Immunol 152:620-631

Doherty PC, Zinkernagel RM (1975) Enhanced immunological surveillance in mice heterozygous at the H-2 gene complex. Nature 256:50-52

Fremont DH, Matsumura M, Stura EA, Peterson PA, Wilson IA (1992) Crystal structures of two viral peptides in complex with murine MHC class I H-2Kb. Science 257:919-927

Gao X, Veal A, Serjeantson SW (1992a) AB1: a novel allele found in one third of an Australian population. Immunogenetics 36:64-66

Gao X, Veal A, Serjeantson SW (1992b) HLA class II diversity in Australian Aborigines: unusual HLA-DRB1 alleles. Immunogenetics 36:333-337

Gao X, Jakobsen I, Serjeantson SW (1994) Characterisation of the HLA-A polymorphism by locus-specific polymerase chain reaction amplification and oligonucleotide hybridisation. Hum Immunol 41:267-279

Gao X, Lester S, Matheson B, Boettcher J, McCluskey J (1997a) Three newly identified A*24 alleles: A*2406, A*2413 and A*2414. Tissue Antigens 50:192-196

Gao X, Lester S, Boettcher B, McCluskey J (1997b) Diversity of HLA genes in populations of Australia and the Pacific. In: Charron D (ed) Genetic diversity of HLA: Functional and Medical Implication. Vol.1. EDK Publishing, Paris, pp 298-306

Hughes AL, Nei M (1988) Pattern of nucleotide substitution at major histocompatibility complex class I loci reveals overdominant selection. Nature 335:167-170

Jakobsen IB, Gao X, Easteal S, Chelvanayagam G (1998) Correlating sequence variation with HLA-A allelic families: implications for T cell receptor binding specificities. Immunol Cell Biol 76:135-142

Kato K, Trapani JA, Allopenna J, Dupont B, Yang SY (1989) Molecular analysis of the serologically defined HLA-Aw19 antigens: A genetically distinct family of HLA-A antigens comprising A29, A31, A32, and Aw33, but probably not A30. J Immunol 143:3371-3378

Lienert K, McCluskey J, Bennett G, Fowler C, Russ G (1995) HLA class I variation in Australian aborigines: characterization of allele B*1521. Tissue Antigens 45:12-17

Lester S, Cassidy S, Humphreys I, Bennett G, Hurley CK, Boettcher B, McCluskey J (1995) Evolution in HLA-DRB1 and major histocompatibility complex class II haplotypes of Australian aborigines, Definition of a new DRB1 allele and distribution of DRB1 gene frequencies. Hum Immunol 42:154-160

Parham P, Arnett KL, Adams EJ, Little AM, Tees K, Barber LD, Marsh SG, Ohta T, Markow T, Petzl-Erler ML (1997) Episodic evolution and turnover of HLA-B in the indigenous human populations of the Americas. Tissue Antigens 50:219-232

Roberts-Thomson JM, Martinson JJ, Norwich JT, Harding RM, Clegg JB, Boettcher B (1999) An ancient common origin of aboriginal Australians and New Guinea highlanders is supported by α–globin haplotypes analysis. Am J Hum Genet 58:1017-1024

Ruppert J, Sidney J, Celis E, Kubo RT, Grey HM, Sette A (1993) Prominent role of secondary anchor residues in peptide binding to HLA-A2.1 molecules. Cell 74:929-937

Serjeantson SW (1989) HLA genes and antigens. In: Hill AVH, Serjeantson SW (eds) The Colonization of the Pacific: A Genetic Trail. Clarendo Press, Oxford, pp 121-173

Thorne A, Grün R, Mortimer G, Spooner NA, Simpson JJ, McCulloch M, Taylor L, Curnoe D (1999) Australia's oldest human remains: age of the Lake Mungo 3 skeleton. J Hum Evol 36:591-612

Watkins DI, McAdam SN, Liu X, Strang CR, Milford EL, Levine CG, Garber TL, Dogon AL, Lord CI, Ghim SH et al. (1992) New recombinant HLA-B alleles in a tribe of South American Amerindians indicate rapid evolution of MHC class I loci. Nature 357:329-333

White N (1997) Genes, languages and landscapes in Australia. In: McConvell P and Evans N (eds) Archaeology and linguistics: aboriginal Australia in global perspective. Oxford University Press, Oxford, pp 45-81

An evolutionary overview of the MHC-G polymorphism: clues to the unknown function(s)

Pablo Morales[1], Jorge Martinez-Laso[1], Maria Jose Castro, Eduardo Gomez-Casado, Miguel Alvarez, Ricardo Rojo, Javier Longas, Ernesto Lowy, Isabel Rubio, and Antonio Arnaiz-Villena

Department of Immunology and Molecular Biology, Hospital Universitario "12 de Octubre", Universidad Complutense de Madrid, Carretera de Andalucia s/n. 28041 Madrid, Spain

Summary. The functions of the major histocompatibility complex-G (MHC-G) molecule are still unknown. The idea that this molecule may be involved in preventing the rejection of fetuses is suggested by only indirect evidence. In the present paper, we review the structure, *in vitro* function and tissue distribution of MHC-G genes and proteins in different primate species. From available data, we conclude the following. First, the nomenclature of MHC-G alleles needs to be revised. Rhesus monkey A/G gene cannot be orthologous to MHC-G because of a lack of structural similarity. Cotton-top tamarin G molecules (which are also structurally similar to E molecules) cannot be orthologous to classical class I molecules. Second, selective pressure to maintain a low degree of polymorphism appears to operate only at the peptide-binding region (PBR) of the MHC-G molecule. Thus, this observation contradicts the idea that the MHC-G leader peptide is the only functional part of the molecule, and suggests that G proteins may have an antigen presenting function to clonotypic T cell receptors besides the ability to interact with NK receptors. Third, MHC-G may also have a function in the thymus, because it is expressed in the thymic epithelium. MHC-G may be important for creating an appropriate T-cell repertoire. Fourth, the presence of HLA-G proteins in tumours and the specific absence of soluble HLA-G mRNA isoforms in the tissues taken from patients with autoimmune diseases suggest that MHC-G may have a role in the immune response and inflammation control.

Key words. Natural killer cells, MHC-G, Evolution, Primates, Placenta

---

[1]The contributions of P. Morales and J. Martinez-Laso are equal and the order of authorship is arbitrary

Introduction

HLA-G was described in 1987 (Geraghty et al. 1987) and classified as a non-classical class I (or class Ib) gene. Although HLA-G shows a high degree of sequence similarity to classical class I (class Ia) genes and HLA-G shares most of the structural features with these genes, it has specific characteristics. First, the mature HLA-G mRNA lacks exon 7 (Ellis et al. 1990; Pook et al. 1991; Castro et al. 1998) and has a stop codon at exon 6; thus, the translation of most of exon 6 and full exons 7 and 8 is not done. Hence, the cytoplasmic tail of HLA-G is much shorter (it has only six amino acids) than the one found in classical class Ia molecules (Geraghty et al. 1987). Spontaneous endocytosis of HLA-G is severely reduced because of its short cytoplasmic tail, and the turnover of HLA-G/β2m/peptide complexes may be so slow that the presentation of exogenous peptides would not be very efficient. However, the long-lasting presence of the complexes on the surface of cytotrophoblast cells could make one of the proposed functions of HLA-G (that is, the inhibitory interaction with NK cells during pregnancy) more efficient (Davis et al. 1997). Second, the promoter region of HLA-G is different from those of other HLA class I genes. The interferon consensus sequence (ICS) is mostly deleted in the HLA-G gene (Flanagan et al. 1991) and HLA-G is not interferon-inducible; the palindromic sequence of the enhancer A region also has a five nucleotide deletion, and site α and enhancer B have nucleotide variations (van den Elsen et al. 1998) (Fig. 1). In addition, a putative locus control region (LCR) has been found at the extreme 5′end of the HLA-G gene. Consequently, these regulatory features of the HLA-G gene may be responsible for its unusual expression pattern and may be a determinant of its still unknown function(s). Third, the HLA-G gene shows six different transcriptional isoforms (all lacking exon 7) which encode both membrane-bound and soluble proteins (Ishitani and Geraghty 1992; Fujii et al. 1994). The membrane-bound HLA-G1 isoform contains three external domains (Fig. 2), whereas the G2, G3 and G4 isoforms are formed by splicing out exon 3 (coding for the α2 protein domain), exons 3 and 4 (coding for the α2 and α3 domains) and exon 4 (coding for the α3 domain), respectively. The soluble isoforms (G5 or G1s and G6 or G2s) retain a part of intron 4; the presence of a stop codon within this intron prevents the translation of the transmembrane and cytoplasmic domains. The outcome is the production of soluble proteins having three external domains (G5) or the α1 and α3 domains (G6). The other class I genes, including HLA-E and –F, do not appear to transcribe these short isoforms. Thus, this alternative splicing is HLA-G locus-specific (Le Bouteiller and Blaschitz 1999).

It has been established that cytotrophoblasts express the full-length HLA-G protein (Kovats et al.1990; Yelavarthi et al.1991; Chumbley et al.1993; McMaster et al.1995; Houlihan et al.1995; Jurisicova et al.1996; Rouas-Freiss et al.1997a) and secrete a truncated form of this molecule (Kovats et al.1990). Although HLA-G mRNA has been detected in other adult and fetal tissues by reverse-transcriptase PCR analyses (Onno et al. 1994), a subpopulation of thymic epithelial cells are the

CONSENSUS	ENHANCER A / R I -200 TGGGGATTCCCCA	ICS -178 AGTTTCACTTCT	SITE α -127 TGACGC	ENHANCER B -100 CATTGGGTGTC	CCAAT BOX -77 CCAAT	TATA BOX -48 TCTAAA
HLA-A2	-------------	------TT----	C-----	-----------	-----	------
HLA-B27	-------------	------------	------	-----------	-----	------
HLA-Cw1	--A------T---	------------	------	-----------	-----	------
HLA-E	*************	-------CG-TC	------	--A--------	-----	-A----
HLA-F	--A-A--------	------T---TC	-A----	-------C---	-----	-A----
HLA-G	--------*****	***C--T-C--C	CCGG--	----A----A-	-----	------

Fig. 1. Alignment of MHC class I promoter sequences. The ICS and the enhancer A regions show nucleotide deletions in HLA-G; the α site and enhancer B have nucleotide variations. A putative LCR (not shown) has been found in the 5' UT (untranslated) region of the HLA-G gene. Identity to the consensus sequence is indicated by a dash (-) and the deletions are denoted by asterisks (*).

only other cells known to express the HLA-G protein (Crisa et al. 1997).

Previous work has almost exclusively looked for the evidence that HLA-G is the key molecule in immunological interactions between the mother and the fetus during pregnancy without paying adequate attention to other possibilities.

Another interesting feature of the MHC-G gene concerns its evolution. The study of MHC-G polymorphism in different primate species may unravel the evolutionary pressures that act upon this gene and may provide clues to its biological function(s). In this chapter, we review currently available data on evolution and function of MHC-G.

MHC-G polymorphism

MHC class I genes are among the most polymorphic found in vertebrates. This is a very interesting feature from the evolutionary point of view because the study of MHC polymorphism is a useful tool with which to understand the origin of populations within species and to estimate different evolutionary pressures placed on the genome. The study of MHC-G polymorphism and characterization of this locus in different primate species have made it possible to obtain a considerable amount of valuable data that can be interpreted from both the evolutionary and the functional points of view.

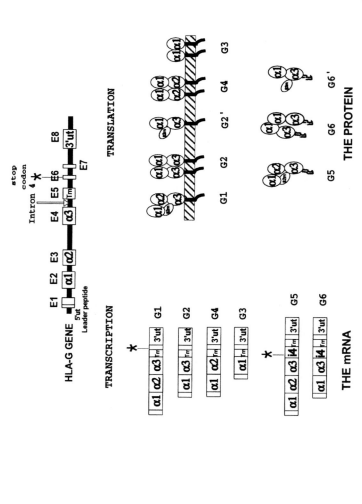

Fig. 2. The mRNA organization and proposed structure of the HLA-G isoforms. Asterisks indicate stop codons in the cytoplasmic domain (exon 6: E6) and in intron 4 (I4) in the soluble isoforms, G5 and G6. Exons 2, 3, and 4 (E2, E3 and E4) encode the α1, α2, and α3 domains, respectively. β2m indicates β2-microglobulin. Only protein isoforms G1, G2, G5 and G6 have been identified biochemically; the other protein isoforms remain theoretical. Modified from Fujii et al. (1994).

MHC-G polymorphism in humans

Initially, HLA-G was thought to be non-polymorphic (Geraghty et al. 1987; Kovats et al. 1990). However, subsequent studies have revealed the existence of HLA-G alleles (Ellis et al. 1990; Morales et al. 1993; Ober et al. 1996; Yamashita et al. 1996; Suarez et al. 1997; Kirzembaum et al. 1997). Fourteen HLA-G alleles have been described up until now at the DNA level (Fig. 3); four different HLA-G proteins, one of them a mutated protein (if it is translated at all), have been defined. The latter, presumably a null allele (*HLA-G*0105N*), has a deletion of cytosine at the third position of codon 129 or the first position of codon 130 that causes a frameshift mutation (Suarez et al.1997). Consequently, all amino acids encoded by the second half of exon 3 would be altered by this deletion and there would be a stop codon at the beginning of exon 4 (the α3 domain); this would give rise to a non-functional truncated protein (Suarez et al. 1997; Ober and Aldrich 1997). The HLA-G1 isoform was not detected in a first trimester placenta from a fetus homozygous for this allele, whereas low levels of HLA-G2 could be detected (Ober et al. 1998). The stop codon, TGA, may encode selenocysteine and thus a full α3 domain could be obtained; this reading-through mechanism has also been described for HLA-DRB6 (Moreno-Pelayo et al. 1999). In addition, and surprisingly, two adult women who are homozygous for the *HLA-G*0105N* allele have been identified (Ober et al. 1998; Castro et al., unpublished results). This suggests that HLA-G1 protein expression is not essential for fetal survival. HLA-G2 or the other isoforms lacking abnormal exon 3 may compensate for the absence of the G1 isoform (Fig. 3).

Regarding the other HLA-G alleles, non-synonymous substitutions are located in residues that, in the class Ia molecules, do not interact with peptides or the T cell receptor. In addition, there are only two non-synonymous substitutions that give rise to three different proteins (Fig. 3). The first one changes serine to threonine at codon 31 (allele *HLA-G*0103*) and the second one changes isoleucine to leucine at codon 110 (alleles *HLA-G*01041*, *-G*01042* and *-G*01043*). Both of them are conservative substitutions. Thus, these substitutions are unlikely to affect either the three-dimensional structure of the protein or the peptide binding region (Ober and Aldrich 1997), suggesting that HLA-G is probably under strong selective pressure against variability.

The HLA-*G*0101* lineage seems to be the oldest one. In the first place, HLA-G*01011 shows a significant linkage disequilibrium with HLA-A2 in all populations so far examined (Caucasoids, Orientals and Negroids) (Arnaiz-Villena et al. 1997a), while other alleles do not. This also argues against the existence of a recombination hot-spot suggested previously between HLA-A and –G loci (Shukla et al. 1991; Venditti and Chorney 1992). Secondly, the two non-synonymous substitutions found in the HLA-G alleles (codons 31 and 110) are conserved in all the MHC-G alleles described up until now (Castro et al. 1996; Boyson et al. 1996; Cadavid et al. 1997; Arnaiz-Villena et al. 1997b). Thirdly, the

	EXON 2				EXON 3			EXON 4		EXON 5	REFERENCES
	31	57	69	93	107	110	130	188	290	309	
	ACG	CCG	GCC	CAC	GGA	CTC	CTG	CAC	GGC	AGA	
HLA-G*01011	ACG	CCG	GCC	CAC	GGA	CTC	CTG	CAC	GGC	AGA	Geraghty et al 1987 Yamashita et al 1996 / Shukla et al 1990 Ober et al 1996 / Morales et al 1993
HLA-G*01012	---	--A	---	--T	---	---	---	---	--T	--G	Ellis et al 1990 Ober et al 1996 / Morales et al 1993 Castro et al 1998 / Yamashita et al 1997
HLA-G*01013	---	--A	---	--T	--T	---	---	---	···	···	Morales et al 1993 / Yamashita et al 1996 / Ober et al 1996
HLA-G*01014	---	---	--T	---	---	---	---	---	···	···	Ober et al 1996
HLA-G*01015	---	---	---	---	--T	---	---	---	···	···	Kirzenbaum et al 1997
HLA-G*01016	---	---	---	---	---	---	---	--T	···	···	Kirzenbaum et al 1997
HLA-G*01017	---	--A	---	--T	--T	---	---	---	···	···	Kirzenbaum et al 1997
HLA-G*01018	---	--A	---	---	---	---	---	---	···	···	Kirzenbaum et al 1997
HLA-G*0102	---	---	---	---	---	---	---	---	---	---	Pook et al 1992
HLA-G*0103	T--	---	---	---	---	---	---	---	---	---	Morales et al 1993
HLA-G*01041	---	--A	---	---	---	A--	---	---	···	···	Yamashita et al 1996 Ober et al 1996 / Kirzenbaum et al 1997
HLA-G*01042	---	--C	---	---	---	A--	---	--T	···	···	Kirzenbaum et al 1997
HLA-G*01043	---	---	---	---	---	A--	---	---	···	···	Kirzenbaum et al 1997
HLA-G*0105N	---	--A	---	--T	---	---	*--	---	···	···	Suarez et al 1997

Fig. 3. Polymorphic codons in HLA-G alleles and schematic representation of the HLA-G molecule showing non-synonymous variations. Sequences at codon 31 code for either a threonine (thr: ACG) or a serine (ser: TCG); sequences at codon 110 code for either a leucine (leu: CTC) or a isoleucine (ile: ATC); sequences at codon 130 code for either a leucine (leu: CTG) or a cysteine (cys: TGC) in the null allele HLA-G*0105N. Other nucleotide variations in the rest of the HLA-G alleles cause no amino acid change.

alleles of the *HLA-G**0101 lineage have accumulated more synonymous substitutions than the other HLA-G lineages (-*G**0103 and -*G**0104). Finally, the *HLA-G**0105N allele appears to be derived from the *HLA-G**01012 allele (Fig. 3).

The HLA-A19 cross-reactive group of antigens shows a random association with HLA-G alleles. This may be accounted for by one or both of the following possibilities. First, the HLA-A19 group includes alleles of different evolutionary origins. For example, there is evidence that the HLA-A30 allele belongs to an ancestral HLA-A group that is different from those of the other HLA-A19 subtypes (Kato et al. 1989; Lawlor et al. 1990; Firgaira et al. 1994). Second, serologically defined HLA-A alleles might be further divided into different subtypes at the molecular level, and each subtype might be associated with different HLA-G alleles.

It is notable that the BeWo cell line, an adherent human choriocarcinoma cell line used frequently to study HLA antigen expression (Trowsdale et al. 1980), is not homozygous for the HLA-G gene. Unlike the previous report (Ellis et al. 1990), this cell line carries both *HLA-G**01012 and -*G**01013 (Morales et al. 1993; Castro et al. 1996).

MHC-G polymorphism in non-human primates

The HLA-G orthologous gene has been identified in different primate species (Watkins et al. 1990a, 1991; Corell et al. 1994; Castro et al. 1996; Boyson et al. 1996; Cadavid et al. 1997; Arnaiz-Villena et al. 1997b). The polymorphism pattern found has different features depending on the species. The level of MHC-G polymorphism is low in *Pongidae* primates (chimpanzee, gorilla and orangutan) as in humans (Fig. 4), although it is somewhat higher in orangutan. The level of MHC-G polymorphism is higher in *Cercopithecinae* monkeys (cynomolgous, rhesus and green); however, all their alleles bear stop codons in exon 3 (coding for the $\alpha2$ domain). These mutations seem to have emerged in a common ancestor of *Cercopithecinae* after human and *Pongidae* lineages separated about 33 million years ago (Klein 1987). The presence of these stop codons suggests that MHC-G could be a pseudogene in *Cercopithecinae*. However, no premature stop codons have been found in the exons encoding the leader peptide or the $\alpha1$ domain. Thus, *Cercopithecinae* may need only the MHC-G isoforms lacking the $\alpha2$ domain [it has been found that the HLA-G2 isoform also inhibits the natural killer cell induced cytotoxicity (Rouas-Freiss et al. 1997b)]. Alternatively, as suggested for the HLA-DRB6 gene, stop codon reading-through mechanisms might be present and all isoforms might be synthesized (Arnaiz-Villena et al. 1999; Moreno-Pelayo et al. 1999).

Some data suggest that MHC-G could have been duplicated, at least, in the rhesus monkey (Castro et al. 1996; Boyson et al 1996). Thus, the duplication of this locus might have generated both a pseudogene and a functional gene, although no non-stop forms of MHC-G have been found in these species and all MHC-G alleles found in rhesus monkeys had pseudogene features (Boyson et al.

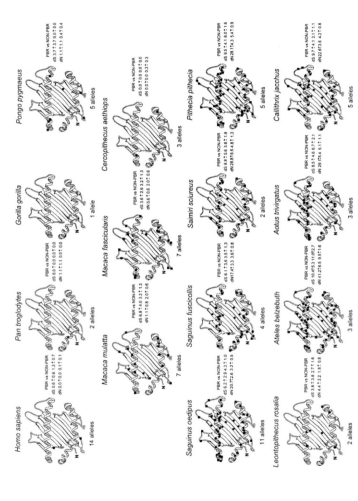

Fig. 4. MHC-G molecules in different primate species. Non-synonymous variations that occur at the PBR and non-PBR residues are indicated by black circles and black squares, respectively. The number of alleles found in each species and the mean number of synonymous (dS) and non-synonymous (dN) substitutions per 100 sites at the PBR and non-PBR sites are indicated under and at the right side of each molecule, respectively.

1996). The MHC-AG gene of the rhesus monkey has been proposed as a functional substitute for the MHC-G gene on the basis that its expression is restricted to placenta (Boyson et al 1997); it also has limited polymorphism and a truncated cytoplasmic tail. Furthermore, its mRNAs appears to undergo alternative splicing. Hence, this Mamu-AG gene might be a functional analogue of the human HLA-G gene, but not the orthologous one because it does not have the required similarity in its primary structure. This would be the first described MHC protein for which a postulated very specialised function (at the placenta) is taken over in another species by an altogether different molecule. Finally, it has been found that HLA-E preferentially binds a nonamer peptide derived from the leader sequence (Llano et al. 1998). Thus, it is possible that the function of HLA-G is to provide peptides to HLA-E molecules at the placental barrier. However, the high degree of conservation of the full MHC-G molecule (at least in the *Pongidae* and human) suggests the existence of additional function(s). Thus, the existence of potentially functional alternative G isoforms, the possible existence of stop codon reading-through mechanisms, and the function of the MHC-G leader peptide cast doubts about whether or not the *Cercopithecinae* MHC-G gene is a non-functional gene.

On the other hand, the New World monkeys (NWMs) seem to have MHC class I genes that are more closely related to HLA-G than to HLA-A or -B (Watkins et al. 1990a, 1990b; Cadavid et al. 1997). However, these NWM class I alleles seem to be related more closely to HLA-E than to -G in phylogenetic trees constructed with a larger number of class I alleles (Alvarez et al. 1997) (Fig. 5). This observation suggests that the MHC-E gene is more conserved than class Ia genes. Furthermore, MHC-G genes of NWMs seem to be much more polymorphic than those of the Old World monkeys (Fig. 4).

It has been proposed that different polymorphic patterns found in different primate species could be related with the sexual behaviour of each species. For instance, the cotton-top tamarins (*Saguinus oedipus*) breed within a long-lasting monogamous couple, and always give birth to monozygotic twins which then cling to the father. All the other non NWMs (including humans) are polygamous (with further within-species peculiarities) and the mothers are in contact with many different fetuses carrying different MHC antigens. *Cercopithecinae* mothers give birth to monkeys from the • or dominant male in more than 95% of the cases. It is likely that a mechanism which avoids high MHC-G allelism has been developed by the primate species that are most exposed to semiallogeneic fetuses (anthropoid monkeys and humans); this would protect the mother against frequent MHC incompatible trophoblast aggressions (as is seen in some pregnancy related tumors) and on the other hand it would provide the fetus with a simple non-polymorphic system that switches off maternal NK cells (Arnaiz-Villena et al. 1999). However, there is another theory that explains the observed differences in the G polymorphism pattern. The NWMs are able to utilize these G-like molecules as classical antigen presenting molecules. Analysis of the pattern of nucleotide substitutions in NWMs (Cadavid et al. 1997) revealed an elevated rate of

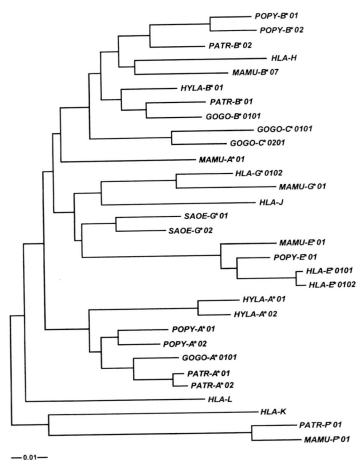

Fig. 5. Dendrogram depicting the evolutionary relationship of different MHC sequences. The tree was constructed using the sequences of exons 2 and 3. Note a close relationship between *Saoe-G* and *MHC-E*. Modified from Alvarez et al. (1997).

non-synonymous versus synonymous substitutions (dN>dS) in the peptide binding region (PBR), indicating that balancing selection is acting to diversify the PBR of these genes. However, in anthropoid monkeys and humans, the pattern of substitutions is inverted (dN<dS) in both PBR and non-PBR regions, indicating that monomorphism at the PBR is favored (see also Fig. 4). Several conclusions may be drawn from these findings: 1) MHC-G molecules may not have a normal (like a classical class I molecule) antigen presenting function in *Cercopithecinae*, *Pongidae* and humans. 2) MHC-G molecules present specialized peptides in

higher primates. Indeed, particular peptides have been eluted from HLA-G molecules (Rammensee 1995), which are mostly derived from intracellular proteins. 3) The selective pressure for maintaining low polymorphism throughout the MHC-G PBR in higher primates (Fig. 4) does not support the idea that only the leader peptide of MHC-G is functionally important (see below).

Intron sequences

Analysis of intronic sequences is a powerful tool for examining phylogenetic relationships of MHC genes (Satta et al. 1996). Since non-coding sequences are largely free from natural selection, they can provide a large number of nucleotide substitutions as well as insertions and deletions (indels) useful for phylogenetic comparison.

The MHC-G intron 2 sequence shows conserved motifs in most primate species studied so far, particularly a 23 base pair (bp) deletion between positions 161 and 183 that seems to be locus-specific (Castro et al. 1996). However, the cotton-top tamarin (*Saguinus oedipus*) MHC-G intron 2 does not have this deletion (Fig. 6). Moreover, its sequence presents other different and unique motifs like a 17 bp deletion and a 14 bp insertion at the end of this intron. The two most feasible explanations for this phenomenon are: (1) that the MHC-G-like sequences in the cotton-top tamarin did not give rise to the Old World Monkey and human MHC-G alleles (see above); and (2) that the indel motifs present in these species occurred after the separation of the New World/Old World monkey lineages about 38 million years ago.

HLA-G protein and its putative functions

The characteristic expression pattern has led to the belief that the HLA-G protein plays an important role in immunological interactions between the mother and the fetus during pregnancy. Consequently, most studies have focused on demonstrating an inhibitory effect of placental HLA-G proteins on maternal NK cells through their interaction with inhibitory NK receptors. Results have been accumulated by using cell lines that have not been well characterized and a small set of anti-HLA-G monoclonal antibodies. Hence, no sound evidence regarding the HLA-G function and its specific inhibitory receptors has been obtained. It is now established that the CD94/NKG2A ligand is HLA-E (Lee et al. 1998; Llano et al. 1998), but not HLA-G, as previously suggested. Interestingly, the complex comprising the HLA-G leader peptide nonamer bound to HLA-E constitutes the most stable ligand for CD94/NKG2A. Thus, it has been proposed that the role of HLA-G may be restricted to providing its nonamer peptide to the HLA-E molecule. This proposal is consistent with the observations that a fetus

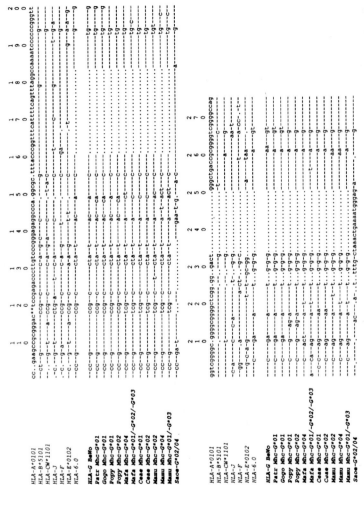

Fig. 6. MHC-G intron 2 nucleotide sequences compared to those of other MHC sequences. Cotton-top tamarins (*Saoe-G*02/04*) do not have the typical 23-bp deletion observed in the other species. Moreover, other differential and unique motifs such as a 17-bp deletion and a 14-bp insertion have been found at the end of this intron in tamarins. The identity between nucleotides is indicated by a dash (-) and the deletions are denoted by asterisks (*).

homozygous for the HLA-G null allele as well as *Cercopithecinae* monkeys with stop codons in their MHC-G gene can survive. It would be interesting to know whether or not the alternative G isoforms can play the same function as the G1 isoform.

Thus, the MHC-G protein may have two different NK-related functions. First, it may send inhibitory signals to NK cells through specific NK receptors [ILT2/LIR1 (Navarro et al. 1999); ILT4 (Colonna et al. 1998); p49 (Cantoni et al. 1998) or others] together with HLA-E. Second, it may provide its nonamer peptide to the HLA-E molecule. In the case of homozygosity for MHC-G null alleles, HLA-E with the -G leader peptide would totally assume this inhibitory function. Nevertheless, if the expression of HLA-G and -E is important for the placental tissue invasion of the maternal uterine decidua during embryo implantation, it would be interesting to test whether the MHC-G protein expression may help stop such invasion and avoid the presence of placental tumours.

Finally, there is a third putative function for the HLA-G molecule that has not been sufficiently studied. It has been demonstrated that HLA-G may present peptides. Peptide presentation by HLA-G expressed on trophoblasts might allow T cell-mediated surveillance of viral infection or malignant transformation in pregnancy (Diehl et al. 1996). This function could also be performed by the MHC-C protein (Schust et al. 1999). Other proposed G molecule functions are detailed in Le Bouteiller and Blaschitz (1999).

The question of whether HLA-G is a molecule that induces tolerance has also been addressed. Its expression under abnormal conditions might be used by cells to stop inflammation or autoimmune self-aggression. HLA-G expression in malignant cells and in inflammation is a new subject of investigation that is being actively pursued at present. HLA-G molecules are expressed in some human melanoma cell lines and biopsies; they may inhibit NK cells from attacking tumors (Paul et al. 1998). In addition, HLA-G transcripts (but not the soluble isoforms) have been observed in thyrocytes from patients with Graves' disease (Castro et al. 1998), raising the possibility that the soluble isoforms would be the most efficient tolerogenic ones. The question to be answered is whether the HLA-G expression in or around tumours (Pangault et al. 1999) or at the site of inflammation is a mechanism used by tumor cells and autoaggressive cells to escape from the immune surveillance system and to control inflammation, respectively. However, HLA-G expression may simply be a consequence of the stress to which these pathological cells are exposed.

Acknowledgements

This work was supported by the grants provided by the Ministerio de Educación (PM-57-95 and PM-96-21) and Comunidad de Madrid (06-70-97 and 8.3/14/98).

References

Alvarez M, Martinez-Laso J, Varela P, Diaz-Campos N, Gomez-Casado E, Vargas-Alarcon G, Garcia-Torre C, Arnaiz-Villena A (1997) High polymorphism of Mhc-E locus in non-human primates: alleles with identical exon 2 and 3 are found in two different species. Tissue Antigens 49:160-167

Arnaiz-Villena A, Chandanayingyong D, Chiewsilp P, Fan L, Fauchet R, Fumiaki N, Gómez-Casado E, Guo SS, Gyòdi E, Hammond M, Jersild C, Juji T, Lepoge V, Martinetti M, Martínez-Laso J, Mehra N, Pénzes M, Rajalingam R, Rajczy K, Tokunaga K (1997a) HLA-G polymorphism: 12th International Histocompatibility Workshop Study. In: Charron D et al. (eds) Genetic Diversity of HLA: Functional and Medical Implications. EDK, pp 155-159

Arnaiz-Villena A, Martinez-Laso J, Alvarez M, Castro MJ, Varela P, Gomez-Casado E, Suarez B, Recio MJ, Vargas-Alarcon G, Morales P (1997b) Primate Mhc-E and –G alleles. Immunogenetics 46:251-266

Arnaiz-Villena A, Morales P, Gomez-Casado E, Castro MJ, Varela P, Rojo R, Martinez-Laso J (1999) Evolution of Mhc-G in primates: a different kind of molecule for each group of species. J Reprod Immunol 43:111-125

Boyson JE, Iwanaga KK, Golos TG, Watkins DI (1996) Identification of the Rhesus monkey HLA-G ortholog. J Immunol 157:5428-5437

Boyson JE, Iwanaga KK, Golos TG, Watkins DI (1997) Identification of a novel MHC class I gene, Mamu-AG, expressed in the placenta of a primate with an inactivated G locus. J Immunol 159:3311-3321

Cadavid LF, Shufflebotham C, Ruiz FJ, Yeager M, Hughes AL, Watkins DI (1997) Evolutionary instability of the major histocompatibility complex class I loci in New World primates. Proc Natl Acad Sci USA 94:14536-14541

Cantoni C, Verdiani S, Falco M, Pessino A, Cilli M, Conte R, Pende D, Ponte M, Mikaelsson MS, Moretta L, Biassoni R (1998) p49, a putative HLA class I-specific inhibitory NK receptor belonging to the immunoglobulin superfamily. Eur J Immunol 28:1980-1990

Castro MJ, Morales P, Fernandez-Soria V, Suarez B, Recio MJ, Alvarez M, Martin-Villa M, Arnaiz-Villena A (1996) Allelic diversity at the primate MHC-G locus: exon 3 bears stop codons in all Cercophitecinae sequences. Immunogenetics 43:327-336

Castro MJ, Morales P, Catalfamo M, Fernandez-Soria V, Suarez B, Varela P, Perez-Blas M, Alvarez M, Jaraquemada D, Arnaiz-Villena A (1998) Lack of HLA-G soluble isoforms in Graves-Basedow thyrocytes and complete cDNA sequence of the HLA-G*01012 allele. Eur J Immunogenet 25:311-315

Chumbley G, King A, Holmes N, Loke YW (1993) In situ hybridization and Northern blot demonstration of HLA-G mRNA in human trophoblast populations by locus-specific oligonucleotide. Hum Immunol 37:17-22

Colonna M, Samaridis J, Cella M, Angman L, Allen RL, O´Callaghan CA, Dunbar R, Ogg GS, Cerundolo V, Rolink A (1998) Human myelomonocytic cells express an inhibitory receptor for classical and nonclassical MHC class I molecules. J Immunol 160:3096-3100

Corell A, Morales P, Martinez-Laso J, Martin-Villa JM, Varela P, Paz-Artal E, Allende LM, Rodríguez C, Arnaiz-Villena A (1994) New species-specific alleles at the primate MHC-G locus. Hum Immunol 41:52-55

Crisa L, McMaster MT, Ishii JK, Fisher SJ, Salomon DR (1997) Identification of a thymic epithelial cell subset sharing expression of the class Ib HLA-G molecule with fetal trophoblasts. J Exp Med 186:289-298

Davis DM, Reyburn HT, Pazmany L, Chiu I, Mandelboin O, Strominger JL (1997) Impaired spontaneous endocytosis of HLA-G. Eur J Immunol 27:2714-2719

Diehl M, Münz C, Keilholz W, Stevanovic S, Holmes N, Loke YW, Rammensee HG (1996) Nonclassical HLA-G molecules are classical peptide presenters. Curr Biol 6:305-314

Ellis SA, Palmer MS, McMichael AJ (1990) Human trophoblast and the choriocarcinoma cell line BeWo express a truncated HLA class I molecule. J Immunol 144:731-735

Firgaira FA, Male DA, Morley AA (1994) The ancestral HLA-A lineage split is delineated by an intron 3 insertion/deletion polymorphism. Immunogenetics 40: 445-448

Flanagan J, Murata M, Burke P, Shirayoshi Y, Appella E, Sharp P, Ozato K (1991) Negative regulation of the major histocompatibility complex class I promoter in embryonal carcinoma cells. Proc Natl Acad Sci USA 88:3145-3149

Fujii T, Ishitani A, Geraghty DE (1994) A soluble form of the HLA-G antigen is encoded by a messenger ribonucleic acid containing intron 4. J Immunol 153: 5516-5524

Geraghty DE, Koller BH, Orr HT (1987) A human major histocompatibility complex class I gene that encodes a protein with a shortened cytoplasmic segment. Proc Natl Acad Sci USA 84:9145-9149

Houlihan JM, Biro PA, Harper HM, Jenkinsori HJ, Holmes CH (1995) The human amnion is a site of MHC class Ib expression: evidence for the expression of HLA-E and HLA-G. J Immunol 154:5665-5674

Ishitani A, Geraghty DE (1992) Alternative splicing of HLA-G transcripts yields proteins with primary structures resembling both class I and class II antigens. Proc Natl Acad Sci USA 89:3947-3951

Jurisicova A, Casper RF, MacLusky NJ, Mills GB, Librach CL (1996) HLA-G expression during preimplantation human embryo development. Proc Natl Acad Sci USA 93:161-165

Kato K, Trapani JA, Allopenna J, Dupont B, Yang SY (1989) Molecular analysis of serologically defined HLA-A19 antigens. A genetically distinct family of HLA-A antigens comprising A29, A31, A32 and Aw33, but probably not A30. J Immunol 143:3371-3378

Kirzembaum M, Djouliain S, Djoulah S, Hors J, Legall I, de Oliveira EB, Prost S, Dausset J, Carosella E (1997) HLA-G gene polymorphism segregation within CEPH reference families. Hum Immunol 53:140-147

Klein J (1987) Origin of Major Histocompatibility Complex Polymorphism: the trans-species hypothesis. Hum Immunol 19:155-162

Kovats S, Main EK, Librach C, Stubblebine M, Fisher SJ, De Mars R (1990) A class I antigen, HLA-G, expressed in human trophoblasts. Science 248:220-223

Lawlor DA, Warren E, Ward FE, Parham P (1990) Comparison of class I MHC alleles in humans and apes. Immunol Rev 113:147-185

Le Bouteiller P, Blaschitz A (1999) The functionality of HLA-G is emerging. Immunol Rev 167:233-244

Lee N, Llano M, Carretero M, Ishitani A, Navarro F, Lopez-Botet M, Geraghty DE (1998) HLA-E is a major ligand for the natural killer inhibitory receptor CD94/NKG2A. Proc Natl Acad Sci USA 95:5199-5204

Llano M, Lee N, Navarro F, Garcia P, Albar JP, Geraghty E, Lopez-Botet M (1998) HLA-E bound peptides influence recognition by inhibitory and triggering CD94/NKG2 receptors: preferential response to an HLA-G derived nonamer. Eur J Immunol 28:2854-2863

McMaster MT, Librach CL, Zhou Y, Lim H, Janatpour MJ, DeMars R, Kovats S, Damsky G, Fisher SJ (1995) Human placental HLA-G expression is restricted to differentiated cytotrophoblasts. J Immunol 154:3771-3778

Morales P, Corell A, Martinez-Laso J, Martin-Villa JM, Varela P, Paz-Artal E, Allende M, Arnaiz-Villena A (1993) Three new HLA-G alleles and their linkage disequilibria with HLA-A. Immunogenetics 38:323-331

Moreno-Pelayo MA, Fernandez-Soria VM, Paz-Artal E, Ferre-Lopez S, Rosal M, Morales P, Varela P, Arnaiz-Villena A (1999) Complete cDNA sequences of DRB6 gene from humans and chimpanzees: a possible model of a stop codon reading-through mechanism in primates. Immunogenetics 49:843-850

Navarro F, Llano M, Bellon T, Colonna M, Geraghty DE, Lopez-Botet M (1999) The ILT2 (LIR1) and CD94/NKG2A NK cell receptors respectively recognize HLA-G1 and HLA-E molecules co-expressed on target cells. Eur J Immunol 29:277-283

Ober C, Aldrich CL (1997) HLA-G polymorphism: neutral evolution or novel function? J Reprod Immunol 36:1-21

Ober C, Rosinsky B, Grimsley C, van der Ven K, Robertson A, Runge A (1996) Population genetics studies of HLA-G: allele frequencies and linkage disequilibrium with HLA-A. J Reprod Immunol 31:111-123

Ober C, Aldrich C, Rosinsky A, Robertson A, Walker MA, Willadsen S, Verp MS, Geraghty DE, Hunt JS (1998) HLA-G1 protein expression is not essential for fetal survival. Placenta 19:127-132

Onno N, Guillaudeux T, Amiot L, Renard I, Drenov B, Hirel B, Girr M, Semana G, Le Bouteiller P, Fauchet R (1994) The HLA-G gene is expressed at a low mRNA level in different human cells and tissues. Hum Immunol 41:79-86

Pangault C, Amiot L, Caulet-Maugendre S, Brasseur F, Burtin F, Guilloux V, Drenou B, Fauchet R, Onno M (1999) HLA-G protein expression is not induced during malignant transformation. Tissue Antigens 53:335-346

Paul P, Rouas-Freiss N, Khalil-Daher I, Moreau P, Riteau B, Le Gal FA, Auril MF, Dausset J, Guillet JG, Carosella DE (1998) HLA-G expression in melanoma: a way for tumor cells to escape from immunosurveillance. Proc Natl Acad Sci USA 95:4510-4515

Pook MA, Woodcock V, Tassabehji M, Campbell RD, Summers CW, Taylor M, Strachan T (1991) Characterization of an expressible nonclassical class I HLA gene. Hum Immunol 32:102-109

Rammensee HG (1995) Chemistry of peptides associated with MHC class I and II molecules. Curr Opin Immunol 7:85-96

Rouas-Freiss N, Marchal RE, Kirszenbaum M, Dausset J, Carosella ED (1997b) The $\alpha 1$ domain of HLA-G1 and HLA-G2 inhibits cytotoxicity induced by natural killer cells: is HLA-G the public ligand for natural killer cell inhibitory receptors? Proc Natl Acad Sci USA 94:5249-5254

Rouas-Freiss N, Marchal-Bras, Goncalves R, Menier C, Dausset J, Carosella ED (1997a) Direct evidence to support the role of HLA-G in protecting the fetus from maternal uterine natural killer cytolysis. Proc Natl Acad Sci USA 94:11520-11525

Satta Y, Mayer WE, Klein J (1996) Evolutionary relationship of HLA-DRB genes inferred from intronic sequences. J Mol Evol 42:648-657

Schust DJ, Tortorella D, Ploegh HL (1999) HLA-G and HLA-C at the feto-maternal interface: lessons learned from pathogenic viruses. Sem Cancer Biol 9:37-46

Shukla H, Gillespie G, Shrivastava R, Collins F, Chorney M (1991) A class I jumping clone places the HLA-G gene approximately 100 Kilobases from HLA-H within the HLA-A subregion of the human MHC. Genomics 10:905-914

Suarez MB, Morales P, Castro MJ, Fernandez-Soria V, Varela P, Alvarez M, Martinez-Laso J, Arnaiz-Villena A (1997) A new HLA-G allele (HLA-G*0105N) and its distribution in the Spanish population. Immunogenetics 45:464-465

Trowsdale J, Travers P, Bodmer WF, Patillo RA (1980) Expression of HLA-A, -B and -C and β2m antigens in human choriocarcinoma cell lines. J Exp Med 152:11-17

Van den Elsen PJ, Peijnenburg A, van Eggermond MC, Gobin SJ (1998) Shared regulatory elements in the promoters of MHC class I and class II genes. Immunol Today 19:308-312

Venditti CP, Chorley MJ (1992) Class I gene contraction within the HLA-A subregion of the human MHC. Genomics 14:1003-1009

Watkins DI, Chen ZW, Hughes AL, Evans MG, Tedder TF, Letvin NL (1990a) Evolution of the MHC class I genes of a New World primate from ancestral homologues of human non-classical genes. Nature 346:60-63

Watkins DI, Letvin NI, Hughes AL, Tedder TF (1990b) Molecular cloning of cDNA that encode MHC class I molecules from a New World Primate (*Saguinus oedipus*): natural selection acts at points that may affect the peptide presentation to T cells. J Immunol 144:1136-1143

Watkins DI, Garber TL, Chen ZW, Toukatly G, Hughes AL, Letvin NL (1991) Unusual limited nucleotide sequence variation of the expressed major histocompatibility complex class I genes of a New World primate species (*Saguinus oedipus*). Immunogenetics 33:79-89

Yamashita T, Fujii T, Watanabe Y, Tokunaga K, Tadakoro K, Juji T, Taketani Y (1996) HLA-G gene polymorphism in a Japanese population. Immunogenetics 44:186-191

Yelavarthi KK, Fishback JL, Hunt JS (1991) Analysis of HLA-G mRNA in human placental and extraplacental membrane cells by in situ hybridization. J Immunol 146:2847-2854

MIC-A polymorphism and a *MIC-A-MIC-B* null haplotype with ∼100-kb deletion

Miki Komatsu-Wakui[1,2], Jun Ohashi[1], Katsushi Tokunaga[1], Yoshihide Ishikawa[2], Kouichi Kashiwase[2], Hitoshi Ando[3], Takashi Shiina[3], Daniel E. Geraghty[4], Hidetoshi Inoko[3], and Takeo Juji[2]

[1]Department of Human Genetics, Graduate School of Medicine, University of Tokyo, Tokyo, Japan
[2]Japanese Red Cross Central Blood Center, Tokyo, Japan
[3]Department of Genetic Information, Division of Molecular Life Science, Tokai University School of Medicine, Kanagawa, Japan
[4]Human Immunogenetics Program, Fred Hutchinson Cancer Research Center, Seattle, USA

Summary. The MHC class I chain-related gene (*MIC*) family has been identified as a new gene family in the *HLA-B* region. *MIC-A* shows a high degree of polymorphism, and more than 30 alleles have been reported. In the present study, we analyzed the polymorphisms of exons 2, 3, and 4, which encode α1, α2, and α3 domains, using PCR-single-strand conformation polymorphism methods and nucleotide sequencing in the Japanese population. Eight *MIC-A* alleles were observed, in which one allele, tentatively named *MIC-AMW*, was new. A strong linkage disequilibrium was observed between the *MIC-A* and *HLA-B* loci, with the exception that *B*3901* showed association with multiple *MIC-A* alleles. Interestingly, a *MIC-A-MIC-B* null haplotype, which is strongly associated with *B*4801*, was also identified. In this haplotype, a large-scale deletion (of approximately 100 kb) including the entire *MIC-A* gene was indicated and the *MIC-B* gene (*MIC-B0107N*) possessed a stop codon.

Furthermore, in order to examine whether balancing selection is operating at the *MIC-A* locus, we performed a computer simulation analysis, which assumed two linked loci (*MIC-A* and *HLA-B*) each with infinite alleles. The high degree of polymorphism at the *MIC-A* locus could not be explained solely by the hitch-hiking effect of the *HLA-B* gene, even if tight linkage was assumed between these loci. We, therefore, conclude that balancing selection is likely to operate at the *MIC-A* locus.

Key words. *MIC-A* polymorphism, *HLA* association, Deletion, Null haplotype, Balancing selection

480

Introduction

The MHC class I chain-related gene (*MIC*, alternatively termed *PERB 11*) family was identified during the search of the *HLA-B* region for other expressed genes (Bahram et al. 1994; Leelayuwat et al. 1994). Two of the five members, *MIC-A* and *MIC-B*, are located 46.4 kb and 141.2 kb centromeric from *HLA-B*, respectively (Mizuki et al. 1997; Shiina et al. 1998). Although the gene organization of *MIC-A* is similar to those of typical class I genes, MIC-A shares only 18, 25, and 30% similarity in the $\alpha 1$, $\alpha 2$, and $\alpha 3$ extracellular domains, respectively, to other class I genes. Although the function of the MIC-A molecule is still unknown, one of the remarkable features of this gene is a high degree of genetic polymorphism. More than 30 *MIC-A* alleles have been identified, and of 27 nucleotide variations in the coding sequence of *MIC-A*, 22 are nonsynonymous.

We have previously performed *MIC-A* allele typing using PCR-SSCP and PCR-sequence-specific primer (SSP) methods on a small number of *HLA-B27* positive Japanese patients with spondylarthropathies and healthy individuals (Tsuchiya et al. 1998). We have studied *MIC-A* allele frequencies in the Japanese population and found several common alleles. In addition, we identified a *MIC-A-MIC-B* "null" haplotype with a large-scale deletion (Komatsu-Wakui et al. 1999).

We found six *MIC-A* alleles with frequencies of more than 10% in Japanese. Also, the heterozygosity was 84%. These observations and the existence of many alleles strongly suggest that balancing selection is operating at the *MIC-A* locus. However, it is known that the gene, which is subjected to balancing selection, enhances the degree of polymorphism at the linked locus. Thus, the polymorphism of the *MIC-A* gene could have been maintained by the hitch-hiking effect of the *HLA-B* gene. In order to clarify this question, we performed a computer simulation analysis.

Materials and methods

Analysis of *MIC-A* polymorphism

Genomic DNAs from 114 unrelated healthy Japanese were examined, for which sequence-level typing of *HLA* genes has been performed by Tokunaga et al. (1997). Genomic DNAs from 10 HTCLs with previously reported *MIC-A* alleles (Fodil et al. 1996) were used as the standards. For the analysis of polymorphic exons 2, 3, and 4, specific PCR primers were designed, and then the amplified DNA fragments were analyzed by the PCR-SSCP method. The PCR products were directly sequenced by a cycle-sequencing method. For the analysis of a new allele, cloned PCR fragments were sequenced. *MIC-A* allele frequencies and haplotype frequencies of *MIC-A, HLA-A, -B,* and *-DRB1* loci were estimated using a maximum likelihood method (Imanishi et al. 1992) assuming the Hardy-

Weinberg law. Significance of two-locus associations was assessed by chi-square values calculated from 2×2 tables.

Analysis of *MIC-A-MIC-B* null haplotype

To confirm a large-scale deletion including the entire *MIC-A* gene, conventional-PCR, long-PCR, and nucleotide sequencing were performed. A total of 21 pairs of primers amplifying 200-250 bp fragments were designed at 2 kb-20 kb intervals between *MIC-B* and *HLA-B* loci based on the genome sequence of this region (Mizuki et al. 1997; Shiina et al. 1998). A long-PCR encompassing the presumed deletion was performed, and then the amplified fragments were directly sequenced. Exon 3 of *MIC-B*, possessing a stop codon, was sequenced directly after specific PCR amplification (Ando et al. 1997).

Mean number of nucleotide changes

The mean numbers of nucleotide changes based on 17 *MIC-A* alleles were calculated using the MEGA software (Kumar et al. 1993).

Simulation analysis

It is assumed that symmetric balancing selection operates at two linked loci (*MIC-A* and *HLA-B*) each with infinite alleles. The fitnesses of homozygote and heterozygote were assumed to be 1 and 1-*s*, respectively, where *s* is the homozygote disadvantage. In the simulation, several parameter sets of the selection intensity and the mutation rate at the *MIC-A* locus were examined.

Results

MIC-A allele frequencies

Eight different *MIC-A* alleles were identified from 114 healthy Japanese individuals, of which *MIC-A*008* was noted at the highest frequency of 25.2 %, followed by *MIC-A*009* (18.4 %), *002* (12.5 %), *010* (12.5 %), *004* (11.1 %), and *012* (10.9 %). A new allele tentatively named *MIC-AMW* was observed at a frequency of 1.8 %. Heterozygosity at the *MIC-A* locus was calculated to be 0.84, which is comparable to the value of *HLA-A* (Tokunaga et al. 1997). Of interest, "blank" alleles were estimated to be 6.7 % by a maximum likelihood method (Imanishi et al. 1992). In fact, this observation was found to be caused by a large-scale *MIC-A* gene deletion as described below.

Association of *MIC-A* with *HLA-B*

As expected from the close chromosomal location of *MIC-A* and *HLA-B*, a strong linkage disequilibrium was observed between these loci. Each *MIC-A* allele was often associated with multiple *HLA-B* subtypes. Furthermore, each *MIC-A* allele appeared to be associated with alleles in certain serological groups of *HLA-B*. In addition, *B*4801* was in strong linkage disequilibrium with *MIC-A* deletion. By contrast, *B*3901* showed association with multiple *MIC-A* alleles. Each common *HLA* haplotype described before (Tokunaga et al. 1997) carried specific *MIC-A* alleles.

Analysis of the large-scale deletion including the entire *MIC-A* gene

In order to perform further analysis of the possibility of the *MIC-A* deletion, six samples of *B48* homozygotes were collected from healthy donors. Twenty-one primer pairs were designed at 2-20 kb intervals between the *MIC-B* and *HLA-B* loci, in which each primer pair gave a 200-250 bp amplified product. Positive (undeleted) controls gave successful amplification with all primer pairs. On the other hand, in most of the other samples homozygous for *B*4801*, no amplification was observed in the middle of the two loci (of approximately 100 kb), but in the flanking region, amplification was clearly observed.

We then tried to obtain a PCR product encompassing the deleted region using different primer pairs. Since the *MIC-A* and *MIC-B* genes, together with their upstream regions, are highly homologous (Shiina et al. 1998), non-specific products were often obtained. A 4.1-kb fragment was chosen as a candidate encompassing the deletion. The 0.2-kb segment at the centromeric side of the 4.1-kb fragment had the same nucleotide sequence as the upstream segment of *MIC-B*, while the sequence of the 0.7-kb segment at the telomeric side corresponded to that of the upstream segment of *MIC-A*. However, it was not possible to judge whether the sequence of the area including the $(GA)_{21}$ repeat (ca. 3.2 kb), which is localized in between the above two segments, belongs to the upstream region of *MIC-A* or *MIC-B*. A large-scale deletion of some 100 kb in size, including the entire *MIC-A* gene, may have occurred somewhere in the central 3.2-kb segment of the 4.1-kb fragment.

The *MIC-A-MIC-B* null haplotype associated with *HLA-B*4801*

The *B48* homozygotes were typed for the *HLA-A, -B, -DRB1, MIC-A,* and *MIC-B* genes. We confirmed by direct sequencing that most of the *B*4801* homozygous samples possessed the *MIC-B* null allele, *MIC-B0107N* (Ando et al. 1997). On the other hand, these samples possessed a variety of alleles at the *HLA-A* and *-DRB1* loci. Consequently, *B*4801*, *MIC-A* deletion, and *MIC-B0107N* are considered to form a conservative haplotype with a frequency of 3.8 % in Japanese.

Synonymous and nonsynonymous changes

Because we do not know selection operating sites in the *MIC-A* gene, mean numbers of nucleotide changes were calculated for the α1, α2, and α3 domains. In the α2 domain, the mean number of nonsynonymous changes was larger than that of synonymous changes. Therefore, although the polymorphic residues of *MIC-A* are quite different from those of *HLA-A*, *-B*, and *-C*, the selection operating sites of *MIC-A* may be located in the α2 domain.

Balancing selection at the *MIC-A* locus

Using a simulation analysis, we further evaluated the hitch-hiking effect of the *HLA-B* gene on the polymorphism of *MIC-A*, because *HLA-B* is known to be subjected to balancing selection (Hughes and Nei 1988). In the simulation, it was assumed that the selection coefficient of the homozygote disadvantage at the *HLA-B* locus was several percents (Satta et al. 1994). When we assumed that balancing selection has not operated at the *MIC-A* locus (i.e. $s=0$), the number of polymorphic residues and the level of heterozygosity obtained from the simulation were below actual observations. If weak balancing selection at the *MIC-A* locus was supposed, the simulation results agreed well with the actual observations.

Discussion

In order to identify the polymorphism of exons 2-4 of the *MIC-A* gene and to investigate the allele frequencies in the Japanese population, the PCR-SSCP method was applied. Based on the SSCP and sequencing results, seven out of the sixteen alleles previously reported (Fodil et al. 1996) were observed in the Japanese population, and the estimated heterozygosity (0.84) was as high as that of *HLA-A*. In addition, the new allele tentatively named *MIC-AMW*, which showed association with *B*1518*, was observed.

A strong linkage disequilibrium was observed between *MIC-A* and *HLA-B* loci. Although most *MIC-A* alleles were associated with multiple *HLA-B* alleles, each *MIC-A* was often associated with alleles belonging to certain *HLA-B* serological group(s). Fodil et al. (1996) reported that non-synonymous substitutions occur more frequently than synonymous ones in the *MIC-A* gene. These findings suggest that *MIC-A* polymorphism is very old, comparable to the ages of the major lineages of *HLA-B* alleles, and that it has been maintained to date by some kinds of natural selection. Our simulation approach also supported the balancing selection operating at the *MIC-A* locus.

The frequencies of multi-locus haplotypes were estimated based on sequence-level genotyping results. The common *HLA* haplotypes in Japanese (Tokunaga et al. 1997) carried specific *MIC-A* alleles. On the other hand, Japanese *B*3901* showed association with multiple *MIC-A* alleles (*MIC-A*002*, **008*, and **010*) as determined from the population study and additional experiments. Recently, MIC-A was reported to confer specificity for recognition by $V_\delta 1\gamma\delta$ T cells (Groh et al. 1996, 1998). In view of these characteristics, *MIC-A* is an attractive candidate gene that could be associated with multi-factorial disorders and graft-versus-host disease.

The features of *MIC-A* polymorphism, such as high heterozygosity, the existence of more than 30 alleles, and a high rate of nonsynonymous substitution, suggest that the polymorphism has been maintained by balancing selection for a long period of evolutionary time. If the *MIC-A* molecule is subjected to positive selection, in other words, if this molecule is biologically important, individuals with deficiency of MIC-A expression are expected to be at a disadvantage. An unexpected finding in this study was that the *MIC-A-MIC-B* "null" haplotype occurs with an appreciable frequency in the population. Furthermore, individuals homozygous for the null haplotype were blood donors and apparently healthy. Thus, detailed studies should be carried out to evaluate the biological significance of this naturally occurring 'double knock-out' of the *MIC-A* and *MIC-B* genes.

References

Ando H, Mizuki N, Ota M, Yamazaki M, Ohno S, Goto K, Miyata Y, Wakisaka K, Bahram S, Inoko H (1997) Allelic variants of the human MHC class I chain-related B gene (*MICB*). Immunogenetics 46:449-508

Bahram S, Bresnahan M, Geraghty DE, Spies T (1994) A second lineage of mammalian major histocompatibility complex class I genes. Proc Natl Acad Sci USA 91:6259-6263

Fodil N, Laloux L, Wanner V, Pellet P, Hauptmann G, Mizuki N, Inoko H, Spies T, Theodorou I, Bahram S (1996) Allelic repertoire of the human MHC class I *MICA* gene. Immunogenetics 44:351-357

Groh V, Bahram S, Bauer S, Herman A, Beauchamp M, Spies T (1996) Cell stress-regulated human major histocompatibility complex class I gene expressed in gastrointestinal epithelium. Proc Natl Acad Sci USA 93:12445-12450

Groh V, Steinle A, Bauer S, Spies T (1998) Recognition of stress-induced MHC molecules by intestinal epithelial gammadelta T cells. Science 279:1737-1740

Hughes AL, Nei M (1988) Pattern of nucleotide substitution at major histocompatibility complex class I loci reveals overdominant selection. Nature 335:167-170

Imanishi T, Akaza T, Kimura A, Tokunaga K, Gojobori T (1992) Estimation of allele and haplotype frequencies for *HLA* and complement loci. In: Tsuji K, Aizawa M, Sasazuki T (eds) *HLA* 1991: Proceedings of the Eleventh International Histocompatibility Workshop and Conference. Vol. 1, Oxford University Press, Oxford, pp 76-79

Komatsu-Wakui M, Tokunaga K, Ishikawa Y, Kashiwase K, Moriyama S, Tsuchiya N, Ando H, Shiina T, Geraghty DE, Inoko H, Juji T (1999) *MICA* Polymorphism in Japanese and a *MICA-MICB* null haplotype. Immunogenetics 49:620-628

Kumar S, Tamura K, Nei M (1993) MEGA: molecular evolutionary genetic analysis. Version 1. Pennsylvania State University, University Park

Leelayuwat C, Townend DC, Degli-Esposti MA, Abraham LJ, Dawkins RL (1994) A new polymorphic and multicopy MHC gene family related to nonmammalian class I. Immunogenetics 40:339-351

Mizuki N, Ando H, Kimura M, Ohno S, Miyata S, Yamazaki M, Tashiro H, Watanabe K, Ono A, Taguchi S, Sugawara C, Fukuzumi Y, Okumura K, Goto K, Ishihara M, Nakamura S, Yonemoto J, Kikuti YY, Shiina T, Chen L, Ando A, Ikemura T, Inoko H (1997) Nucleotide sequence analysis of the *HLA* class I region spanning the 273-kb segment around the *HLA-B* and *-C* genes. Genomics 42:55-66

Satta Y, O'hUigin C, Takahata N, Klein J (1994) Intensity of natural selection at the major histocompatibility complex loci in primates. Proc Natl Acad Sci USA 91:7184-7188

Shiina T, Tamiya G, Oka A, Yamagata T, Yamagata N, Kikkawa E, Goto K, Mizuki N, Watanabe K, Fukuzumi Y, Taguchi S, Sugawara C, Ono A, Chen L, Yamazaki Y, Tashiro H, Ando A, Ikemura T, Kimura M, Inoko H (1998) Nucleotide sequencing analysis of the 146-kilobase segment around the *IκBL* and *MICA* genes at the centromeric end of the *HLA* class I region. Genomics 47:372-382

Tokunaga K, Ishikawa Y, Ogawa A, Wang H, Mitsunaga S, Moriyama S, Lin L, Bannai M, Watanabe Y, Kashiwase K, Tanaka H, Akaza T, Tadokoro K, Juji T (1997) Sequence-based association analysis of *HLA* class I and II alleles in Japanese supports conservation of common haplotypes. Immunogenetics 46:199-205

Tsuchiya N, Shiota M, Moriyama S, Ogawa A, Komatsu-Wakui M, Mitsui H, Geraghty DE, Tokunaga K (1998) *MICA* allele typing of *HLA-B27* positive Japanese patients with seronegative spondylarthropathies and healthy individuals. Arthritis Rheum 41:68-73

Repertoire forecast of MHC class I binding peptides with peptide libraries

Keiko Udaka[1], Karl-Heinz Wiesmüller[2], and Günther Jung[3]

[1]Department of Biophysics, Kyoto University, Sakyo, Kyoto 606-8502, Japan
[2]Evotec BioSystems GmbH, Grandweg 64, D-22529 Hamburg, Germany
[3]Institut für Organische Chemie, Eberhard-Karls-Universität Tübingen, Auf der Morgenstelle 18, D-72076 Tübingen, Germany

Summary. Understanding the repertoire of MHC binding peptides is a pivotal step towards looking into the genetic forces that keep the MHC molecules polymorphic and polygenic. We have attempted to forecast the repertoire of MHC class I binding peptides by means of positional scanning using peptide libraries. Position independent assessment of the binding was useful for roughly estimating the binding capacity of peptides, but the prediction was clearly limited. Namely, for approximately 80% of randomly chosen peptides, MHC binding capacity could be predicted within an order of magnitude in Kd values. When pairwise comparisons were made between different MHC class I molecules, the degree of repertoire overlap was 2 to 5%. The results support the leading hypothesis that the necessity to increase repertoire size acts as a driving force for diversification of peptide binding specificities in the MHC class I molecules.

Key words. MHC class I, Peptide, Libraries, Repertoire, Prediction

Introduction

Identification of anchor amino acids in MHC binding peptides has paved the way to predicting T cell epitope peptides as well as selecting promising peptides for vaccine development (Falk et al. 1991; Rammensee et al. 1995). However, the information about anchors is usually not sufficient to make accurate predictions. Peptides bearing both of two MHC class I binding major anchor amino acids can vary in their binding capacity by as much as 10,000 fold (Jameson and Bevan 1992, and our unpublished observations). We and others have developed a peptide library-based method to analyze the fine specificity of peptide binding by MHC class I molecules position by position (Udaka et al. 1995a, b; Pridzun et al. 1996; Stryhn et al. 1996; Udaka 1996). The method was successfully applied to specificity analyses of TAP (transporter associated with antigen processing) transporters and MHC class II molecules (Uebel et al. 1995; Fleckenstein et al. 1996; Hemmer et al. 1997). The analysis of MHC class I molecules underscored the importance of including not only the anchor amino acids but also all the amino acids within peptides in prediction of peptide binding. The profiles of amino acid preferences so obtained have improved the prediction of MHC binding peptides and helped identify T cell epitope peptides (Udaka et al. 1995a, 1996; Brock et al. 1996; Fleckenstein et al. 1996; Gundlach et al. 1996; Ohtsuka et al. 1998). On the other hand, it became obvious that there is a clear limit to understanding the peptide binding characteristics by position independent scanning (Udaka et al. 1995a). Here we summarize our recent experiences with peptide libraries on the capacity and the limit of the library scanning in predicting MHC class I binding peptides. In addition, we attempted to forecast the repertoire overlaps between different MHC class I molecules.

Synthetic peptide libraries

Synthetic peptide libraries of nine amino acids were designed according to the positional scanning formula shown in Figure 1. A position of the interest was fixed with one of 19 proteinogenic amino acids but cysteine. The other positions were synthesized with an equimolar mixture of 19 amino acids to allow semiquantitative assessment. Cysteine was omitted to avoid intra- and/or inter-chain disulfide formation. Fractional occupancies by individual amino acids were examined by pool sequencing and mass spectrometry, and deviations from the equimolarity were ~3% at most (Udaka, 1996; Wiesmüller, 1996). The use of an amino acid mixture instead of a single amino acid was aimed at randomizing the interactive effects that could occur between amino acids on the peptide or with the MHC molecules and at isolating the impact of a given amino acid at the position of interest.

$$^{N-} \text{X X X D X X X X X} ^{-C}$$
$$\text{- - - A - - - - -}$$
$$\text{- - - E - - - - -}$$

Fig. 1. Random peptide libraries used for positional scanning of MHC class I molecules. 9-mer random peptide libraries were synthesized. X indicates the position where a mixture of 19 amino acids but cysteine was used for coupling instead of a single amino acid. A position to be probed was fixed with one of 19 amino acids. The entire library consisted of 19 x 9 = 171 sublibraries, each bearing a single amino acid at one of 9 amino acid positions and a reference library, X_9 which had mixtures at all positions.

Specificity analysis of three MHC class I molecules

Three mouse MHC class I molecules, K^b, D^b and L^d, were analyzed by positional scanning (manuscript in preparation). Binding of library peptides was measured by MHC stabilization assay using TAP deficient RMA-S cells or its transfectants. By comparing the titration curve of X_9 consisting of ~3 x 10^{11} distinct peptides with known viral epitope peptides, it was estimated that peptides having comparable binding affinities to epitope peptides were present at a frequency of approximately one in a couple of hundred random 9-mer peptides.

As shown in Figures 2a, b, c, the profiles of amino acid preferences were characteristic for individual class I molecules and they were consistent with the major anchor motifs described by Falk et al. (1991) (Y, F at P5 and L, M at the C-terminus for K^b; N at P5 and M, I at P9 for D^b; P at P2 and F, L, M at P9 for L^d). These anchor positions were most selective for a few amino acids, and other amino acids exhibited large negative contributions to the binding compared with the reference library, X_9. Auxiliary anchors could be defined as positions of intermediate selectivity. Positions shown to be more exposed to solvent by crystal analysis are often known to serve as TCR contact sites and are relatively permissive to various amino acids. Most importantly, library scanning yielded a quantitative measure of amino acid preferences at every position of the 9-mer peptides. This has prompted us and others to investigate how the binding capacity of each amino acid measured in isolation from its sequence context is reflected in the binding of the whole peptide (Stryn et al. 1996; and our manuscript in preparation).

A scoring method to estimate the MHC binding capacity of peptides

Peptides bind to MHC class I molecules in an extended β conformation where individual amino acid side chains are accommodated in fairly distinct subsites (pockets) (Saper et al. 1991; Madden 1995). If individual side chains were to make relatively independent engagement with the corresponding pockets, the binding energy would be supplied additively by individual amino acids. This is, however, a quite wild assumption. Structure of proteins is realized by an interactive network of amino acids. Although interfaces of specific protein-protein interactions are often likened to a lock and key, the interaction is usually not that solid. Even a small change of the amino acids can lead to readjustment of the structure so that the protein settles into a conformation that requires the lowest energy. Such flexible accommodation of amino acids could be seen in the crystal structure of MHC-peptide interactions (Madden et al. 1993; Fremont et al. 1995). Nevertheless, if additive energies supplied from the individual amino acids were of a substantial fraction of the whole binding energy, some correlation should be seen between the occurrence of positively contributing amino acids and the binding of the whole peptide. To estimate such an additive fraction of the binding energies, we investigated the correlation between library binding data and binding of the whole peptides.

Library binding data are given in log relative Kd values as shown in the ordinate of Figure 2. Gibbs free energy of association, ΔG is defined as $\Delta G = RT\ln Kd$, where R is Boltzmann's gas constant and T is absolute temperature. $\Delta\Delta G$ of sublibraries in comparison with X_9 is expected to be relative to the difference of log Kd values. If an independent, therefore, additive contribution of each amino acid was assumed, the binding energy of the whole peptide would be relative to the sum of log Kd values from all the amino acids on peptide. For this reason, we used this sum as a score to estimate the MHC binding capacity of the peptide. "+/-" signs were reversed so that positive scores indicate high binders. Cysteine was scored as neutral zero because it was excluded from the synthetic libraries.

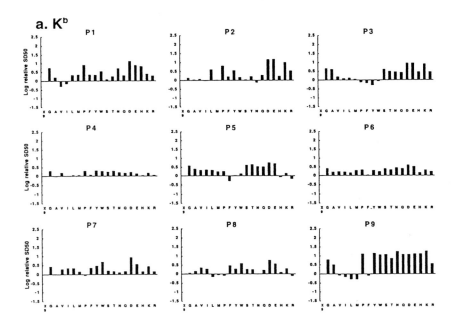

Fig. 2. Profiles of positional scanning using peptide libraries. Three MHC class I molecules, a. K^b, b. D^b and c. L^d were analyzed. Amino acid positions scanned were indicated as P1–P9. The binding capacity of the sublibraries bearing one of 19 amino acids was measured by stabilization assay and expressed as log Kd (M) values that gave a half maximal stabilization. Amino acids used to fix the position of sublibraries were indicated in single letter codes.

Fig. 2. (continued)

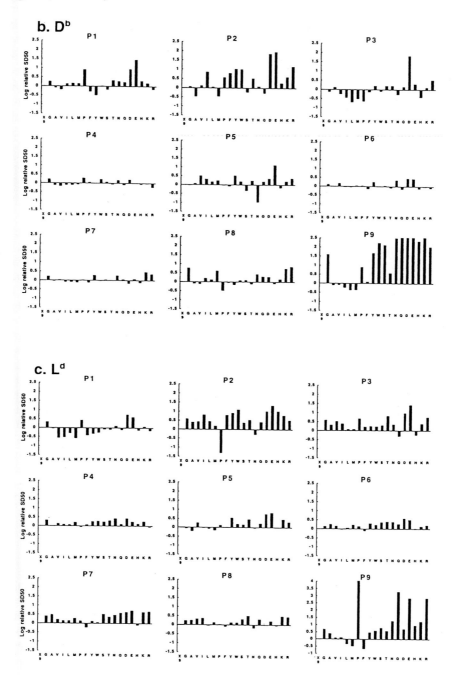

When a given protein sequence was divided into all possible 9-mer peptides and their MHC binding scores were calculated, the values were distributed over a range between ~ -8 and ~ +1. T cell epitope peptides harbored a number of amino acids that have shown positive contributions to MHC binding and tended to have high scores (Table 1). Most of the known epitope peptides (24/27) were scored with the values higher than 2SD (two times standard deviation) from the mean score. This indicated that an additive component of MHC-peptide interaction was, in fact, substantial and that the library scanning data might be useful in estimating the MHC binding capacity. Therefore, we next synthesized a number of peptides chosen from known protein sequences and compared their binding scores with experimental binding data.

Correlation between predicted MHC binding scores and actual binding

About 30 to 40 peptides for each of K^b, D^b and L^d were synthesized and their MHC binding was measured by MHC stabilization assay (manuscript in preparation). Stryn et al. (1996) have made a similar analysis previously by using a competition assay. In their assay, only the anchor bearing peptides were analyzed. We included a number of peptides that were not conforming to the major anchor motifs described by Falk et al. (1991). As shown in Figure 3, some correlation was seen between the predicted scores and actual binding of the peptides. However, individual peptides deviated substantially from the linear correlation, indicating limitations to the prediction methods. When a Gaussian distribution of the data points was assumed along the linear correlation, approximately 80% of the new-comer peptides were expected to fall within a ten-fold order of magnitude in Kd values from the linear correlation. Besides minor reasons, these deviations are likely to be caused mainly because position independent scoring ignored the interactive nature of amino acids in the MHC peptide complexes. We have reported one of such interactive cases where a K^b binding OVA peptide bears a pair of amino acids that have acted cooperatively in the sequence context of the peptide, leading to an unexpectedly high binding (Udaka et al. 1995a). Flexible nature of interface amino acids has also been reported in various proteins and protein-protein interfaces (Carter et al. 1984; Graf et al. 1988; Gregoret and Sauer 1993; Blaber et al. 1995).

Prediction of MHC binding peptides in proteins

Positional scanning-based scoring thus proved to have principal limitations to predicting the MHC binding capacities of peptides. However, it still offers more detailed information than from the major anchor motifs alone. Moreover, library scanning can be applied to all the 9-mer binding MHC class I molecules with the same set of the libraries. This is a clear advantage over the specificity analysis method using single amino acid substituted epitope peptides reported by Parker et

Table 1. MHC binding scores predicted for known T cell epitope peptides

a. K^b 8-mer

Sequence	Source	Score
SSIEFARL	HSV gpB	1.72
IIYRFLLI	Rota VP7	1.25
HIYEFPQL	endogenous	1.21
KSPWFTTL	MuLCp15E	1.08
RGYVYQGL	VSV NP	0.99
SIINFEKL	OVA	0.76
FEQNTAQP	MUT1	-2.80

+ 2SD (K^b8-mer) (0.46)

b. K^b 9-mer

Sequence	Source	Score
VGPVFPPGM	Rota VP6	0.37
FAPGNYPAL	Sendai NP	-1.17
YSGYIFRDL	Rota VP3	-2.31

+ 2SD (K^b9-mer) (-1.24)

c. D^b

Sequence	Source	Score
FQPQNGQFI	LCMV NP	2.24
RAHYNIVTF	HPV16E7	1.86
ASNENMETM	Flu NP	1.75
ASNENMEDA	Flu NP	1.10
IQVGNTRTI	Yersinia YOP	1.07
QGINNLDNL	SV40 T	0.97

+ 2SD (D^b) (0.92)

d. L^d

Sequence	Source	Score
YPNVNIHNF	endogenous	2.04
YPHFMPTNL	MCMV pp89	1.88
YPALGLHEF	Measles NP	1.72
QPQRGRENF	endogenous	1.65
TPHPARIGL	E. coli β-gal	1.58
APQPGMENF	endogenous	1.56
RPQASGVYM	LCMV NP	1.55
LPYLGWLVF	Tum⁻ P815	1.24
APTAGAFFF	JHMV NP	1.22
SPGRSFSYF	Measles HA	0.77
DPVLDRLYL	Measles HA	-0.98

+ 2SD (L^d) (-0.02)

MHC binding scores were calculated for known T cell epitope peptides.
Dotted lines indicate the location of 2SD values that were tentatively set as a threshold for binder peptides.
Peptide sequences were obtained from Rammensee et al. (1995).

al. (1994), because the latter requires a set of custom made variant peptides for every different MHC class I molecule. Since library scanning yields a semiquantitative measure of peptide binding, it may be used for automated scoring of MHC binding peptide screening from protein sequences in a database. Therefore, we have created such a scoring program and scanned protein sequences (manuscript in preparation).

Table 2 lists the peptides receiving the scores higher than 2SD away from the mean score. The known T cell epitope peptides were among the top scored peptides in most of the proteins analyzed. However, a number of peptides that had only one or no major anchor amino acid were scored high. We synthesized some

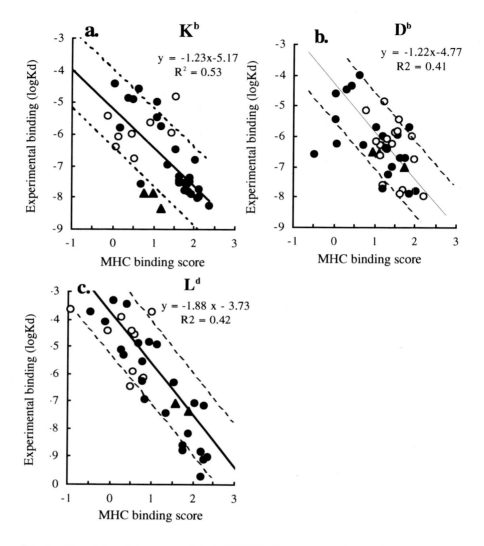

Fig. 3. Correlation between predicted MHC binding scores and experimental binding. MHC binding capacities of randomly chosen peptides from known protein sequences and a few artificially designed peptides were predicted with position independent scoring. Their actual binding was measured experimentally by MHC stabilization assay. Solid lines indicate the correlation deduced by the least square method. Broken lines indicate the range where binding is expected for 80% of the new-comer peptides assuming the Gaussian distribution of the data points. Closed symbols are peptides bearing two major anchor amino acids. Open symbols are peptides bearing only one or no major anchor amino acid. Triangles indicate natural T cell epitope peptides while other peptides are shown in circles.

Table 2. Prediction of MHC class I binding peptides in protein sequences

a. K^b 9-mer binding scores

Sendai virus NP

position	sequence	score
158	QIYGYPACL	-0.34
257	GNYIRDAGL	-0.53
229	VSLMVETLV	-0.67
452	VSGGHFVTL	-0.86
170	IVQVWIVLV	-0.91
26	VIPGQRSTV	-0.94
458	VTLHGAERL	-1.02
35	SVFVLGPSV	-1.04
173	VWIVLVKAI	-1.07
169	IIVQVWIVL	-1.14
323	**FAPGNYPAL**	**-1.17**
392	RLRHHLANL	-1.19
221	VMRSQQSLV	-1.26

Rotavirus VP6

position	sequence	score
357	**VGPVFPPGM**	**0.37**
388	VASIRSMLI	-0.39
169	SQPMHDNLM	-0.49
292	FQLMRPPNM	-0.64
191	AGFDYSCAI	-0.77
245	TTWFFNPVI	-0.78
269	IINTYQARF	-0.79
57	RNWTFDFGL	-0.83
244	ATTWFFNPV	-0.86
197	CAINAPANI	-0.87

b. D^b binding scores

SV 40 T antigen

21	SAWGNIPLM	1.56
284	VLLLLGMYL	1.19
222	**CKGVNKEYL**	**1.12**
671	QAPQSSQSV	1.10
462	VAIDQFLVV	1.05
232	YSALTRDPF	0.97
488	**QGINNLDNL**	**0.97**
148	AVFSNRTLA	0.92
445	KALNVNLPL	0.77

b. D^b binding scores (continued)

LCMV-NP

position	sequence	score
395	**FQPQNGQFI**	**2.24**
182	MAKQSQTPL	1.49
35	TSLLNGLDF	1.15
245	AAVKAGAAL	1.02
251	AALLHGGNM	0.98
443	GALPQGMVL	0.84
265	IKPSNSEDL	0.72

Flu-NP APR

295	YSLVGIDPF	1.99
269	VAHKSCLPA	1.93
54	RLIQNSLTI	1.86
365	**ASNENMETM**	**1.75**
457	FQGRGVFEL	1.21
163	CSLMQGSTL	1.13
261	SALILRGSV	0.98

c. L^d binding scores

LCMV NP

117	**RPQASGVYM**	**1.55**
514	TPHCALMDC	1.47
445	LPQGMVLSC	1.06
174	MPSLTMACM	0.79
147	RPQQGASGV	0.57
543	LPHDLIFRG	0.54
35	TSLLNGLDF	-0.06

MCMV pp89

167	**YPHFMPTNL**	**1.88**
297	VCQMMCNNM	0.94
1	EPAAPSCNM	0.53
4	APSCNMIMI	0.43
214	LPGEFKSEM	0.33
235	PPCYTKPFL	0.25
128	LCQLANDIF	-0.03

Peptides bearing major anchors are underlined.

T cell epitopes are in bold.

of these peptides and measured MHC binding. Somewhat to our surprise, as shown in Figure 2 with open symbols, those peptides with incomplete anchors did not act much differently from the anchor-bearing peptides. Peptides with no anchor at all seemed to have a little higher incidence of lower-than-expected binding. Although the number of peptides tested was still small, this suggested that other peptides not conforming to the major anchor motifs could exist at the level of MHC binding. Among known T cell epitopes, the incidence of peptides not bearing anchors is obviously less frequent. This most likely reflects some selective biases during antigen processing (Udono and Srivastava 1993; Schirmbeck and Reimann 1994; Niedermann et al. 1995; Bieland et al. 1996; Niedermann et al. 1996; Ossendorp et al. 1996) and/or peptide transport (Schirmbeck and Reimann 1994; Uebel et al. 1995). Coevolution of peptide selectivities between MHC class I molecules and TAP transporters has also been suggested (Heemels et al. 1993; Momburg et al. 1994; Neefjes et al. 1995; van Endert et al. 1995).

Repertoire forecast of MHC binding peptides by pair wise comparisons of MHC class I molecules

MHC class I and class II genes exhibit an extensive polymorphism within a species throughout the evolutionary history of jawed vertebrates (Klein and Figueroa 1986; Klein et al. 1993a; Okamura et al. 1997) with only a few exceptions (O'Brien et al. 1985; Klein and Figueroa 1986). So far, no evidence for hypermutation at the MHC loci has been found (Satta et al. 1993). Genetic analyses revealed that such variations have been accumulated over an evolutionary scale of time as a result of point mutations or some gene conversion-like segmental exchanges between homologous genes (Nathenson et al. 1986; Pease et al. 1991b; Klein et al. 1993b; Parham et al. 1995). Expression of MHC class I and class II genes is also polygenic. In mice and humans, two to three MHC class I and class II molecules are expressed codominantly. Amino acid differences between alleles and homologous genes are mainly located in the peptide binding region.

Genetic analyses suggest some evolutionary forces that favor the maintenance of genetic variations for the sake of peptide presentation (Hughes and Nei 1988, 1989; Klein et al. 1993b; Bergström and Gyllenstein 1995). A major advantage having polymorphic and polygenic MHC molecules is most likely a better protection against pathogens. Harboring a variety of MHC molecules either on an individual basis (heterozygote advantage) (Hughes and Nei 1988, 1989; Takahata et al. 1992) or in a species' population (frequency dependent selection) (Hill et al. 1992; Wills and Green 1995) may have endowed better survival. This leading hypothesis is, however, valid only if different MHC molecules present a diverse set of peptides. The more diverse, the stronger is the impact placed on the fate of the allele or the gene. On the other hand, if the T cell repertoire to be deleted due to self reactivity occupies a substantial fraction, deleting the repertoire by a variety of MHC-peptide complexes may have left too small a functional repertoire, leading to a less favorable outcome. In an attempt to gain insight into the impact of MHC diversity on the TCR repertoire, we wished to look into the repertoire of

MHC class I binding peptides and see the degree of repertoire overlaps between different MHC class I molecules.

Previous attempts to estimate the peptide repertoire overlap were either made with much longer peptides than optimal MHC class I binding peptides (Carreno et al. 1990) or with synthetic peptide libraries (Schumacher et al. 1992; Abastado et al. 1993; Rohren et al. 1993). Although these were elegant experiments, the latter, unfortunately, did not identify or quantitate the responsive peptide species. As described above, the positional scanning with the libraries does have a limited capacity to predict MHC binding peptides. Nevertheless, for ~80% of the arbitrarily chosen peptides, MHC binding capacity can be predicted within a ten fold order of magnitude in Kd values (manuscript in preparation). If we analyze a sufficiently large number of peptides, we may be able to catch a glimpse of a peptide repertoire and see the degree of repertoire overlaps between different MHC class I molecules.

As a trial, we calculated the MHC binding scores for 6018 peptides arbitrarily chosen from five protein sequences that have been shown to yield at least one T cell epitope and compared their scores in pairwise comparisons among K^b, D^b and L^d. D^b and L^d are allelic products and K^b is encoded in a different locus. In Figure 4, only the regions with peptides scored near or above the lines indicating 2SD higher than the mean score are shown. 2SD lines were adopted as a tentative threshold indicating high potentials of being presented naturally by the MHC class I molecules to a degree sufficient to be recognized by TCRs. As mentioned above, most of the known T cell epitope peptides are scored above this threshold.

Interestingly, most of the highly scored peptides are expected to be presented almost exclusively by either one of the MHC class I molecules, and peptides that may be presented by both of the MHC class I molecules are rare. When comparing K^b and D^b binding peptides, 94 (70%) peptides are scored higher than the 2SD threshold only with K^b and 33 (24.6%) with D^b. As few as 7 peptides (5.2%) have a potential to be presented by both K^b and D^b. In comparison between K^b and L^d, 83 (72%) are scored as >2SD for K^b binding and 27 (23.5%) for L^d. Only 5 peptides (4.3%) could be presented by both K^b and L^d. Between allelic products of D^b and L^d, 41 (49%) are scored as >2SD for D^b, 41 (49%) for L^d. Only 2 peptides (2.4%) are scored high for both D^b and L^d. Summing up, the forecast suggested a fairly small repertoire overlap between two different MHC class I molecules, allelic or non-allelic. A recent crisscross binding experiment with a limited number of synthetic peptides also suggested this degree of repertoire overlap not far from the actual binding data (manuscript in preparation). These results indicate that individuals who have two different MHC class I molecules would be able to present nearly twice as big a repertoire of peptides compared with those with only one type of MHC class I molecules. This observation is consistent with the view that the driving force leading to polymorphism has been in a direction to expand the repertoire of peptide presentation.

Three MHC class I molecules analyzed here are thought to have diverged from a primordial mouse MHC class I molecule a long time ago and have kept

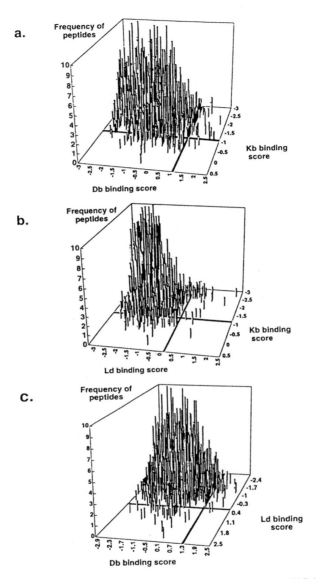

Fig. 4. Two dimensional display of frequency distribution of MHC binding scores for peptides present in the model proteins in pairwise comparisons between, **a.** K^b and D^b, **b.** K^b and L^d, and **c.** D^b and L^d. Solid lines indicate a tentative threshold of 2SD away from the mean score, suggesting the high potential of MHC binding.

diversifying further until now (Pease et al. 1991a). Their pairwise homologies are around ~70% on both amino acid and nucleotide levels in the $\alpha 1$ and $\alpha 2$ peptide binding domains. Repertoire diversification to this extent seems to have had positive advantages during evolution. This also suggests that the blind spot of TCR repertoire caused by deletion of self reactive TCRs may not have been so big a problem compared with the advantage of repertoire widening by gaining a new MHC class I molecule.

Genetic analyses indicate that MHC class I genes have diverged by gradual accumulation of small changes either of point mutations or segmental exchanges due to some gene conversion-like events. At each single step of MHC evolution, changes in the repertoire of peptides must have been small. In fact, as suggested by pool sequencing, a group of recently arisen allelic variants of HLA class I molecules seem to share a largely overlapping repertoire of peptides, with each member bearing some distinct peptides even in the cases of a single amino acid substitution (Barber et al. 1995; Sudo et al. 1995; Barber et al. 1997). It would be interesting to analyze the peptide repertoire of these variants more accurately and follow up the fate of these alleles in the future.

Acknowledgement

We thank Drs. Yoko Satta and Naoyuki Takahata for stimulating discussions. Encouragement by Profs. Hideo Yamagishi and Nobuhiro Go is gratefully acknowledged. This work was supported by PRESTO, JST, and by grants from The Ministry of Education and Kato Memorial Bioscience Foundation.

References

Abastado J-P, Casrouge A, Kourilsky P (1993) Differential role of conserved and polymorphic residues of the binding groove of MHC class I molecules in the selection of peptides. J Immunol 151:3569-3575

Barber LD, Gillece-Castro B, Percival L, Li X, Clayberger C, Parham P (1995) Overlap in the repertoires of peptides bound in vivo by a group of related class I HLA-B allotypes. Curr Biol 5:179-190

Barber LD, Percival L, Arnett KL, Gumperz JE, Chen L, Parham P (1997) Polymorphism in the a1 helix of the HLA-B heavy chain can have an overriding influence on peptide-binding specificity. J Immunol 158:1660-1669

Bergström T, Gyllenstein U (1995) Evolution of Mhc class II polymorphism : The rise and fall of class II gene function in primates. Immunol Rev 143:13-31

Bieland TJF, Tan MCAA, Monnee-van Muijen M, Koning F, Kruisbeek AM, van Bleek GM (1996) Isolation of an immunodominant viral peptide that is endogenously bound to the stress protein GP96/GRP94. Proc Natl Acad Sci USA 93:6135-6139

Blaber M, Baase WA, Gassner N, Matthews BW (1995) Alanine scanning mutagenesis of the α-helix 115-123 of phage T4 lysozyme. J Mol Biol 246:317-330

Brock R, Wiesmüller K-H, Jung G, Walden P (1996) Molecular basis for the recognition of two structurally different major histocompatibility complex/peptide complexes by a single T-cell receptor. Proc Natl Acad Sci USA 93:13108-13113

Carreno BM, Anderson RW, Coligan JE, Biddison WE (1990) HLA-B37 and HLA-A2.1 molecules bind largely nonoverlapping sets of peptides. Proc Natl Acad Sci USA 87:3420-3424

Carter PJ, Winter G, Wilkinson AJ, Fersht AR (1984) The use of double mutants to detect structural changes in the active site of the Tyrosyl-tRNA synthetase (Bacillus stearothermophilus). Cell 38:835-840

Falk K, Rötzschke O, Stevanovic S, Jung G, Rammensee H-G (1991) Allele-specific motifs revealed by sequencing of self-peptides eluted from MHC molecules. Nature 351:290-296

Fleckenstein B, Kalbacher H, Muller CP, Stoll D, Halder T, Jung G, Wiesmüller K-H (1996) New ligands binding to the human leukocyte antigen class II molecule DRB0101 based on the activity pattern of an undecapeptide library. Eur J Immunol 240:71-77

Fremont DH, Stura EA, Matsumura M, Peterson PA, Wilson IA (1995) Crystal sturucture of an H-2Kb-ovalbumin peptide complex reveals the interplay of primary and secondary anchor positions in the major histocompatibility complex binding groove. Proc Natl Acad Sci USA 92:2479-2483

Graf L, Jancso A, Szilagyi L, Hegyi G, Pinter K, Naray-Szabo G, Hepp J, Medzihradszky K, Rutter WJ (1988) Electrostatic complementarity within the substrate-binding pocket of trypsin. Proc Natl Acad Sci USA 85:4961-4965

Gregoret LM, Sauer RT (1993) Additivity of mutant effects assessed by binomial mutagenesis. Proc Natl Acad Sci USA 90:4246-4250

Gundlach BR, Wiesmüller K-H, Junt T, Kienle S, Jung G, Walden P (1996) Specificity and degeneracy of minor histocompatibility antigen-specific MHC-restricted CTL. J Immunol 156:3645-3651

Heemels MT, Schumacher TN, Wonigeit K, Ploegh HL (1993) Peptide translocation by variants of the transporter associated with antigen processing. Science 262:2059-2063

Hemmer B, Fleckenstein BT, Vergelli M, Jung G, McFarland H, Martin R, Wiesmüller K-H (1997) Identification of high potency microbial and self ligands for a human autoreactive class II-restricted T cell clone. J Exp Med 185:1651-1659

Hill AVS, Elvin J, Willis AC, Aidoo M, Allsopp CEM, Gotch FM, Gao XM, Takiguchi M, Greenwood BM, Townsend ARM, McMichael AJ, Whittle HC (1992) Molecular analysis of the association of HLA-B53 and resistance to severe malaria. Nature 360:434-439

Hughes AL, Nei M (1988) Pattern of nucleotide substitution at major histocompatibility complex class I loci reveals overdominant selection. Nature 335:167-170

Hughes AL, Nei M (1989) Nucleotide substitution at major histocompatibility complex class II loci : Evidence for overdominant selection. Proc Natl Acad Sci USA 86:958-962

Jameson SC, Bevan MJ (1992) Dissection of major histocompatibility complex (MHC) and T cell receptor contact residues and an assessment of the predictive power of MHC-binding motifs. Eur J Immunol 22:2663-2667

Klein D, Ono H, O'hUigin C, Vincek V, Goldschmidt T, Klein J (1993) Extensive MHC variability in cichlid fishes of Lake Malawi. Nature 364:330-334

Klein J, Figueroa F (1986) Evolution of the major histocompatibility complex. CRC Crit Rev Immunol 6:295-386

Klein J, Satta Y, O'hUigin C, Takahata N (1993b) The molecular descent of the major histocompatibility complex. Annu Rev Immunol 11:269-295

Madden DR (1995) The three dimensional structure of peptide-MHC complexes. Annu Rev Immunol 13:587-622

Madden DR, Garboczi DN, Wiley DC (1993) The atomic identity of peptide-MHC complexes: A comparison of the comformations of five viral peptides presented by HLA-A2. Cell 75:693-708

Momburg F, Roelse J, Howard JC, Butcher GW, Hämmerling GJ, Neefjes JJ (1994) Selectivity of MHC-encoded peptide transporters from human, mouse and rat. Nature 367:648-651

Nathenson SG, Geliebter J, Pfaffenbach GM, Zeff RA (1986) Murine major histocompatibility complex class-I mutants: Molecular analysis and structure-function implications. Ann Rev Immunol 4:471-502

Neefjes J, Gottfried E, Roelse J, Gromme M, Obst R, Hämmerling GJ, Momburg F (1995) Analysis of the fine specificity of rat, mouse and human TAP peptide transporters. Eur J Immunol 25:1133-1136

Niedermann G, Butz S, Ihlenfeld H-G, Grimm R, Lucchiari M, Hoschötzky H, Jung G, Maier B, Eichmann K (1995) Contribution of proteasome-mediated proteolysis to the hierarchy of epitopes presented by major histocompatibility complex class I molecules. Immunity 2:289-299

Niedermann G, King G, Butz S, Birsner U, Grimm R, Shabanowitz J, Hunt DF, Eichmann K (1996) The proteolytic fragments generated by vertebrate proteasomes : Structural relationships to major histocompatibility complex class I binding peptides. Proc Natl Acad Sci USA 93:8572-8577

O'Brien SJ, Roelke ME, Marker L, Newman A, Winkler CA, Meltzer D, Colly L, Evermann JF, Bush M, Wildt DE (1985) Getetic basis for species vulmerability in the Cheetah. Science 227:1428-1434

Ohtsuka Y, Udaka K, Yamashiro Y, Yagita H, Okumura K (1998) Dystrophin acts as a transplantation rejection antigen in Dystrophin-deficient mice : Implication for gene therapy. J Immunol 160:4635-4640

Okamura K, Ototake M, Nakanishi T, Kurosawa Y, Hashimoto K (1997) The most primitive vertebrates with jaws possess highly polymorphic MHC class I genes comparable to those of humans. Immunity 7:777-790

Ossendorp F, Eggers M, Neisig A, Ruppert T, Groettrup M, Sijts A, Mengede E, Kloetzel P-M, Neefjes J, Koszinowski U, Melief C (1996) A single residue exchange within a viral CTL epitope alters proteasome-mediated degradation resulting in lack of antigen presentation. Immunity 5:115-124

Parham P, Adams EJ, Arnett KL (1995) The origins of HLA-A, B, C polymorphism. Immunol Rev 143:141-180

Parker KC, Bednarek MA, Coligan JE (1994) Scheme for ranking potential HLA-A2 binding peptides based on independent binding of individual peptide side-chains. J Immunol 152:163-175

Pease LR, Horton RM, Pullen JK, Cai ZC (1991) Structure and diversity of class I antigen presenting molecules in the mouse. Crit Rev Immunol 11:1-32

Pridzun L, Wiesmüller K-H, Kienle S, Jung G, Walden P (1996) Amino acid preferences in the octapeptide subunit of the major histocompatibility complex class I heterotrimer H-2Ld. Eur J Immunol 236:249-253

Rammensee H-G, Friede T, Stevanovic S (1995) MHC ligands and peptide motifs : first listing. Immunogenetics 41:178-228

Rohren EM, Pease LR, Ploegh HL, Schumacher TNM (1993) Polymorphism in pockets of major histocompatibility complex class I molecules influence peptide preference. J Exp Med 177:1713-1721

Satta Y, O'hUigin C, Takahata N, Klein J (1993) The synonymous substitution rate of the major histocompatibility complex loci in primates. Proc Natl Acad Sci USA 90:7480-7484

Schirmbeck R, Reimann J (1994) Peptide transporter-independent, stress protein-mediated endosomal processing of endogenous protein antigens for major histocompatibility complex class I presentation. Eur J Immunol 24:1478-1486

Schumacher TNM, Bleek GMv, Heemels M-T, Deres K, Li KW, Imarai M, Vernie LN, Nathenson SG, Ploegh HL (1992) Synthetic peptide libraries in the determination of T cell epitopes and peptide binding specificity of class I molecules. Eur J Immunol 22:1405-1412

Stryhn A, Pedersen LO, Romme T, Holm CB, Holm A, Buus S (1996) Peptide binding specificity of major histocompatibility complex class I resolved into an array of apparently independent subspecificities : quantitation by peptide libraries and improved prediction of binding. Eur J Immunol 26:1911-1918

Sudo T, Kamikawaji N, Kimura A, Date Y, Savoie CJ, Nakashima H, Furuichi E, Kuhara S, Sasazuki T (1995) Differences in MHC class I self peptide repertoires among HLA-A2 subtypes. J Immunol 155:4749-4756

Takahata N, Satta Y, Klein J (1992) Polymorphism and balancing selection at major histocompatibility complex loci. Genetics 130:925-938

Udaka K (1996) Decrypting class I MHC-bound peptides with peptide libraries. Trends Biochem Sci 21:7-11

Udaka K, Wiesmüller K-H, Kienle S, Jung G, Walden P (1995a) Decrypting the structure of MHC-I restricted CTL epitopes with complex peptide libraries. J Exp Med 181:2097-2108

Udaka K, Wiesmüller K-H, Kienle S, Jung G, Walden P (1995b) Tolerance to amino acid variations in peptides binding to the MHC class I proetin H-2Kb. J Biol Chem 270:24130-24134

Udaka K, Wiesmüller K-H, Kienle S, Jung G, Walden P (1996) Self MHC-restricted peptides recognized by an alloreactive T lymphocyte clone. J Immunol 157:670-678

Udono H, Srivastava PK (1993) Heat shock protein 70-associated peptides elicit specific cancer immunity. J Exp Med 178:1391-1396

Uebel S, Meyer TH, Kraas W, Kienle S, Jung G, Wiesmüller K-H, Tampe R (1995) Requirements for peptide binding to the human transporter associated with antigen processing revealed by peptide scans and complex peptide libraries. J Biol Chem 270:18512-18516

van Endert PM, Riganelli D, Greco G, Fleischhauer K, Sidney J, Sette A, Bach J-F (1995) The peptide-binding motif for the human transporter associated with antigen processing. J Exp Med 182:1883-1895

Wiesmüller KH, Fiertag S, Fleckenstein B, Kienle S, Stoll D, Herrmann M, Jung G (1996) Peptide and cyclopeptide libraries. In: Jung G (ed) Combinatorial peptide and nonpeptide libraries. VCH, Weinheim, pp 203-246

Wills C, Green DR (1995) A genetic herd-immunity model for the maintenance of MHC polymorphism. Immunol Rev 143:263-292

Sequence conditions for gene conversion of mouse MHC genes

Jan Böhme[1] and Kari Högstrand[2]

[1]Center of Biotechnology, NOVUM, University College of South Stockholm/
Karolinska Institute, S-141 57, Huddinge, Sweden
[2]Department of Immunology, Stockholm University, S-106 91 Stockholm, Sweden

Summary. We have developed a semi-nested PCR assay which measures gene conversion in mouse MHC class genes. Detectable frequencies occur in DNA from sperm, but not in somatic DNA, such as liver or fibroblast DNA. The frequency of gene conversion events in MHC class II genes varies strongly from one MHC haplotype to another, with the highest detected frequencies as high as 1/40 000 for an individual heterozygous for both donor and acceptor sequences. Deletions or insertions in one gene relative to the other seem to lower the efficiency of gene conversion considerably. However, not all the variation in gene conversion frequency can be readily ascribed to the MHC genes themselves, implying that variation in the non-MHC background genes also might be of importance. Stretches within MHC genes amenable to gene conversion are located in CpG clusters, whereas MHC genes not involved in gene conversion have background CpG levels. This feature extends to some, but not all, of the non-MHC genes that have been reported to undergo gene conversion in mammals. DNA damage, either chemical or radiation-induced, increases the frequency of gene conversion of MHC class II genes in cultured cells of the fibroblastoid lineage. The effect of chemical DNA damage seems roughly dose-dependent, whereas irradiation has maximal effect at low doses.

Key words. MHC, Evolution, Gene conversion, CpG islands, DNA repair

Introduction

De novo mutations of MHC genes often have a particular appearance. Rather than consisting of a single point mutation, they are characterised by a whole stretch of mutated nucleotides. This stretch of nucleotides can often be found, at the same position within the gene, in another MHC gene. This "template" gene can either be another allele of the same MHC locus, or belong to a different MHC locus (see, for example, Pease et al. 1983; Nathenson et al. 1986; Gorski and Mach 1986).

This type of mutation, where a stretch of nucleotides are mutated seemingly to the model of another gene, occurs also during the affinity mutation of human (Kawamura et al. 1992; Silvain et al. 1993) and mouse (Baltimore 1981; Clarke et al. 1981; Ollo and Rougeon 1983; Maizels 1989) Ig genes, and during the rearrangement of Ig genes at least in fowl (Thompson and Neiman 1987; Reynaud et al. 1987) and in the rabbit (Becker and Knight 1990).

This type of mutation is highly reminiscent of a phenomenon called gene conversion. Gene conversion was studied in fungi as the non-Mendelian 3:1 segregation in a heterozygous diploid strain (Holliday 1964). The original definition of gene conversion was thus "the transfer of genetic material from one gene to another without the donor gene being changed". This type of strict definition is only possible in fungi, where the four products of an individual meiosis are observable. When the term "gene conversion" started to be used also for the type of mutations described above in vertebrates, (Baltimore 1981) the definition broadened to "a segmental mutation with a recognisable template outside the mutated gene".

Gene conversion has also been implicated to explain the existence of vast exact homologies, including silent substitutions, between certain genes, so called concerted evolution (Edelman and Gally 1968; Slightom et al. 1980; Liebhaber et al. 1981). It sounds paradoxical at first that the phenomenon of gene conversion is evoked both to explain pronounced diversity and extreme similarity. However, the transfer of fragments from a donor to an acceptor can lead to both diversification and homogenisation, depending on at least two parameters. The first is the length of the transferred fragment. The longer the fragment, the greater the probability for homogenisation (McKee 1996). The second is the degree of selection for a specific sequence, and the consequent amount of diversity available before gene conversion took place. The more original diversity, the greater the probability for a gene conversion event to produce even more diversity. This is illustrated both in variable exons of MHC genes (see Fig. 1) and in immunoglobulin genes (Miyata et al. 1980; Kawamura et al. 1992), where certain constant parts display the extreme similarity characteristic of concerted evolution, in striking contrast to neighbouring stretches of segmentally transferred diversity.

The notion of gene conversion as a generator of diversity proved to be controversial. It could be argued that the patchwise appearance of mutations was due to selective constraints, rather than to specific mechanisms for transfer of genetic information (Klein 1984). With the advent of the PCR technique (Saiki et al. 1985) a direct assay for *in vivo* gene conversion of unmanipulated genes became conceivable, and the amplification of DNA from individual sperm (Li et al. 1988) showed that such an assay would actually be technically feasible. In order to resolve whether gene conversion actually served as a mechanism for the generation of diversity in MHC genes, we proceeded to develop an assay for detection of an analogue of the spontaneous mutation *bm12* (McIntyre and Seidman 1984; Denaro et al. 1984) in mouse sperm. The outcome proved that such an assay was, indeed, technically feasible, and that a mutation of the *bm 12* type occurred once in about

500 000 mouse sperm (Högstrand and Böhme 1994). This frequency, albeit low, was high enough to be compatible with the frequency of spontaneous MHC class II mutations, and thus provided a clear indication that gene conversion not only existed as a theoretically possible means of creating new MHC diversity, but that it actually performed this task in real life as well. Subsequently, we were able to determine that this real-life event occured during the mitotic phase of the development of the male germline, because it was completed before the spermatogenetic cells entered meiosis (Högstrand and Böhme 1997).

Frequency of MHC class II gene conversion depends both on sequence of involved genes and on genetic background

Our first studies of gene conversion in the MHC (Högstrand and Böhme 1994, and 1997) measured interchromosomal gene conversion. A technique measuring individual mutation events in a pool of more than hundred thousand sperm needs to be rigorously controlled to be credible. The kind of control we employed involved measuring interchromosomal gene conversion in an F1 animal, using a mixture of DNA from the two parental strains as a control for all kinds of in vitro artefacts, in particular "jumping PCR", (Pääbo et al. 1989) as such DNA would be chemically indistinguishable from the F1 DNA. When we found out that we only obtained detectable signals from germline F1 DNA, whereas somatic DNA,

```
Abk gene  g ttc cag ccc ttc tgc tac ttc acc aac ggg acg cag cgc ata cgg ctt gtg atc aga tac atc
Abd gene  - --- a-- gg- gag --- --- a- --- --- --- --- --- --- --- c --- -c --- --- --- ---
Ebk gene  c -gt a-a t-t gag --t c-t --- ta- --- --- --- --- --- g-g --- --- g-- g-a --- --- t--
Ebd gene  c g-t aca t-t gag --t c-t --- ta- --- --- --- --- a- dtg --- t-- c-- gag --- t- ---

Abk gene  tac aac cgg gag gag tac gtg cgc ttc gac agc gac gtg ggc gag tac cgc gcg gtg acc
Abd gene  --- --- --- --- --- --- --- --- a- --- --- --- --- --- --- --- --- --- --- ---
Ebk gene  --- --- -t- --- --- a-- c- --- --- --- --- --- --- --- -t- --- --- --- ---
Ebd gene  --- --- --- --- a-- c- --- --- --- --- --- --- --- --- --- --- --- --- -a

Abk gene  gag ctg ggg cgg cca gac gcc gag tac tgg aat ... ... aag cag tac ctg gag cga acg
Abd gene  --- --- --- --- --- --- --- --- --- c agc cag cc- g-- at- --- --- --- ---
Ebk gene  --- --- --- --- --- --- --- a-- --- --c agc cag cc- g-- -t- --- --- -a- -a-
Ebd gene  --- --- --- --- --- --- --- a-- --- --- agc cag cc- g-- at- --- --- gat g -g
```

Fig. 1. Sequence of the two *Ab* and the two *Eb* alleles that have been analysed for gene conversion. Sequence shown is the part of the second (first domain) exon that is amplified in our PCR assay. The *Abk* sequence, and all sequences identical to it, are given no background. Nucleotides specific to *Abd*, or common to *Abd* and either or both of *Ebk* and *Abd*, are in open boxes, nucleotides specific for *Ebk* or common to *Ebk* and *Ebd*, are in light-shaded boxes and *Ebd*-specific nucleotides are in dark-shaded boxes. The bar underneath part of the sequence denotes a 65-bp nucleotide stretch, where there is sequence identity between an *Ab* and an *Eb* allele, although the *Ab* and *Eb* loci have been separated for at least 70 million years, which is thus highly reminiscent of concerted evolution. Note the gap of six nucleotides in the 3' part of the *Abk* gene.

such as liver DNA, had as low a level of detected PCR events as mixed parental DNA (Högstrand and Böhme 1994), DNA from somatic tissue became useful as a control for PCR artefacts. This enabled us to extend our studies to study intrachromosomal gene conversion in sperm, i. e. gene conversion between two genes which exist on the same chromosome, where it is not possible to control for PCR artefacts by mixing parental strains.

To compare the results in this study with the results obtained previously MHC (Högstrand and Böhme 1994, and 1997), we analysed intrachromosomal gene conversion in the same F1 (Balb/c × C3H/HeJ) strain as before, as well as in the reciprocal F1(C3H/HeJ × Balb/c) strain, and both inbred parental strains. Figure 2 shows that, interestingly enough, the gene conversion frequency in genes of the different H2 haplotypes differed by more than one order of magnitude (Högstrand and Böhme 1998). The gene conversion frequencies for the k haplotype, where a $bm12$ corresponding segment of Ebk is transferred to the Abk acceptor gene were $1.2 - 1.4 \times 10^{-6}$ in both F1 crosses, as well as in the inbred C3H/HeJ mice. However, the gene conversion frequencies for the d haplotype on the other hand, where a corresponding segment of Ebd was transferred to the acceptor gene Abd, were $2.6 - 2.8 \times 10^{-5}$ for both of the F1 crosses, i.e. more than an order of magnitude higher. In the inbred Balb/c mice the frequency was higher still, over 9.6×10^{-5}. Because the gene conversion frequencies for both the k and d haplotypes were obtained in cells from the same F1 animals, the difference in

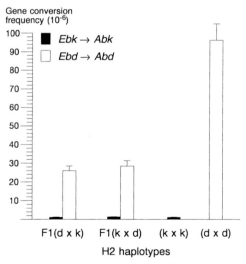

Fig. 2. Intrachromosomal gene conversion between Ab and Eb genes from two different haplotypes. F1 mice used were F1(Balb/c × C3H/HeJ) and the reciprocal crossing F1(C3H/HeJ × Balb/c) for the detection of $Ebk \rightarrow Abk$ or $Ebd \rightarrow Abd$ cis gene conversion. The parental strains C3H/HeJ (haplotype k) and Balb/c (haplotype d) were used to detect the sum of inter- and intrachromosomal gene conversion in the two haplotypes, respectively.

frequency between k→k and d→d conversions must depend entirely on factors peculiar to the two different haplotypes.

Because both reciprocal F1 crosses have approximately the same gene conversion frequencies when assayed for either the *k*- or the *d*-haplotype conversions, it does not appear to matter whether the chromosome analysed has maternal or paternal origin. Instead, one conspicuous difference between the analysed sequences of the *k* and *d* haplotypes is the 6 bp deletion in the *Abk* gene shown in Figure 1, situated immediately 5' to the donor primer sequence. This creates a break of homology between the *Abk* gene and the other assayed genes, which can be expected to make the recognition of the homologous donor gene less efficient. The fact that the frequency of interchromosomal gene conversion between *Ebd* and *Abk* (Högstrand and Böhme 1994, 1997), where this break of homology also exists, is $1.5 - 2 \times 10^{-6}$, i.e. more in agreement with the lower *k*-haplotype frequency than to the higher *d*-haplotype frequency, argues in favour of this interpretation.

In the F1 crosses, which are MHC heterozygous for both the *k* and the *d* haplotypes, each haplotype is only present on one chromosome, and only one type of detectable gene conversion event per cell is possible. However, each inbred strain possesses two chromosomes bearing the same haplotype. Therefore, in those mice we should detect the sum of both intrachromosomal and interchromosomal gene conversion, making a total of four possible events per cell. As can be observed from Figure 2, the gene conversion frequency of the homozygous Balb/c stain was 3-4 times higher than that of the heterozygous F1 mice. This finding indicates that both intra- and interchromosomal gene conversion in the *d* haplotype have approximately equal frequencies. However, in the *k* haplotype the frequency of gene conversion in homozygous C3H/HeJ mice and heterozygous F1 mice was almost identical. One possible interpretation of this finding is that the frequency of gene conversion does not only depend on the sequence of the donor and acceptor genes involved, but also on the existence of background genes that could facilitate or obstruct gene conversion. In that case, our results would indicate either that dominant genes facilitating gene conversion exist in the Balb/c strain, or that recessive genes obstructing gene conversion exist in the C3H/HeJ strain.

Strain differences in the frequency of gene conversion have been noted previously for the H2 mutants, where most mutations were found to occur in chromosomes originating from C57BL/6 mice, which are of *b*-haplotype (Nairn and Yamaga 1980). We have not yet been able to analyse gene conversion of the MHC in C57BL/6 mice due to specificity problems with PCR primers for the *b* haplotype. However, C57BL/6 mice do not *per se* appear to be associated with high gene conversion frequencies. *LacZ* transgenic mice were reported to have decreased their gene conversion frequency from 0.8% in CF1 mice to 0.001%, or 10^{-5}, similar to our frequency in the *d* haplotype, in C57BL/6J mice (Hanneman et al. 1997). This dramatic drop in gene conversion frequency was not further commented upon by the authors, but it indicates at least that the C57BL/6 mice strain alone was not responsible for the apparent preference of gene conversion in the b-haplotype of the H2 mutants.

Apart from the differences in gene conversion frequencies between the haplotypes, there is also a difference in the length pattern of the segments transferred in the *k*- and *d*-haplotype. Because only one breakpoint in the conversion by definition is included in our amplifications, we can only asses minimal transfer length of the fragments by sequencing the positive PCR products. All sequenced fragments transferred in the "high-frequency" *d*-haplotype, are similar and short in length with a minimal transfer of 6-35 nucleotides, depending on where in the area of sequence identity between the donor and acceptor genes the actual 5' breakpoint appears. However, whenever the "low-frequency" *k*-haplotype was used as an acceptor gene, a greater variability in the length of the transferred segments sequenced was evident. The smallest fragment had a minimal length of 27 nucleotides, whereas the largest fragment was at least 104 nucleotides long. Thus, it is clear that the average transferred fragment is considerably longer in the *k* haplotype. Given the much lower overall frequency of gene conversion in that haplotype, it is, on the other hand, not obvious that the transfer of such long fragments is *more* common in the k haplotype. It might well be the case that the homology break just 5' to the donor primer efficiently blocks all transfer of very short fragments, whereas the transfer of long fragments is about equally efficient in the two haplotypes.

The deletion in the *Abk* gene is a rare one in mouse MHC class II genes. Thus, the average frequency of gene conversion of a fragment corresponding to the bm12 mutation, irrespective of haplotype might be expected to be closer to the *d* value than to the *k* value. However, only one MHC class II mutant was found, (in the *Abb* gene), when more than 100,000 mice were tested with reciprocal skin grafting. Although it is difficult to estimate the true frequency from only one detection, the value seems to between our high and low frequency values for the single event we detect. One might speculate from this whether the stretch transferred in the *bm12* mutation is, in fact, the only stretch transferred at a comparable frequency in class II genes. Indeed, the equally polymorphic region, at the 5' end of the first domain exon, in the Abk gene that we chose as site for our acceptor primers in our original assay (Denaro et al. 1984) is not transferred at frequencies detected by our assay system. If it would, we would have detected PCR events, with our "acceptor" sequences as real-life donors, when we excluded unequal crossing-over as the mechanism behind our signals by using a "donor" primer located more than 100 bp into the downstream intron, finding no PCR signals (Denaro et al. 1984). The gene by far most affected among the more than 100.000 mice screened for H2 mutations was the *Kb* gene, with an overall mutation frequency of approximately 2×10^{-4} per locus and gamete, (reviewed in Nairn and Yamaga 1980) or about one order of magnitude higher than our high figures. However, it is clear that several different stretches can be transferred to that gene, so the frequency for an individual event in the Kb gene might be similar to what we detect in out "high converting" *d* haplotype.

Gene conversion in the MHC can be induced by DNA damage

Many questions remain regarding the mechanism of gene conversion. DNA damage has been shown to induce gene conversion in lower eukaryotes like fungi (Schiestl and Wintersberger 1983, 1992), as well as in mammalians (Hanneman et

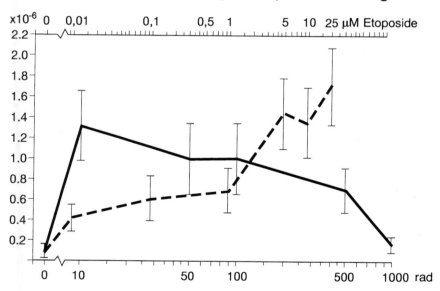

Fig. 3. Gene conversion frequencies in a non-germline testis cell line with low intrinsic gene conversion treated with DNA-damaging agents at varying intensity or concentration. Solid line and abscissa: Gene conversion after treatment with γ-irradiation. Dotted line and abscissa: Gene conversion after treatment with etoposide. Measure points at "0" correspond to untreated cells.

al. 1997; Wang et al. 1988; Murti et al. 1994; Egorov and Egorov 1998). Could gene conversion of the MHC be induced by DNA damage as well? Earlier studies had shown that paternal irradiation or mutagen treatment had no apparent effect on the gene conversion frequency in the series of H2 mutants examined (Egorov and Egorov 1998; Nairn and Yamaga 1980). In order to investigate this matter further, we made a cell line derived from non-germline mouse testis cells. This cell line was found to have an intrinsic low background level of gene conversion between *Ebd* and *Abk* of about 1×10^{-7}. Figure 3 shows that after treatment with DNA interfering agents such as γ-irradiation and etoposide this gene conversion frequency was found to increase up to 15-fold (Högstrand and Böhme 1999). Etoposide treatment resulted in an almost dose dependent induction of gene conversion, whereas irradiation gave a higher response when low doses of γ-rays were used. High and low doses of radiation have been shown to induce separate sets of DNA polymerases in the resulting DNA repair process (Mirzayans et al. 1992; Mirzayans et al.1996) which might explain the discrepancy from the more expected dose-dependent induction. DNA damage is also one of the major factors involved in cell transformation, so another explanation could be that low doses of irradiation result in transformation of some cells, and these cells might be both more gene conversion prone and more fast growing, whereas higher doses of irradiation impairs so many vital cellular functions so that the cells can not divide properly. One mechanism responsible for such a transformation might be gene conversion.

Nevertheless, DNA damage induces gene conversion of at least MHC class II genes, thus it is highly probable that the DNA repair system plays a major role in the event. The fact that gene conversion can be induced by DNA damage from a frequency of essentially zero in non-germline cells indicates that the specific gene conversion promoting factor present in the germline mitoses either is a DNA damaging activity or a constitutively high expression of a repair activity. This makes an interesting parallel to the situation in yeast, where DNA damage alone could compensate for a deficient endonuclease, and induce mating-type conversions in a mutant yeast strain (Kostriken et al. 1983). Therefore, it is highly probable that the first step in the gene conversion event is some kind of DNA damage. DNA damage can be assumed to be ubiquitous in the genome, yet all genes are not affected by gene conversion. Is it then possible to speculate what makes MHC genes such a favoured target for gene conversion?

Gene conversion prone regions of MHC genes are CpG rich

A prerequisite for gene conversion is that there are genes with a high degree of similarity in at least parts of the sequence. It also seems as if gene conversion is favoured if the genes are located in the same genetic region. However, there are many homologous members of gene families that do not undergo gene conversion, whereas others do. In particular, some MHC genes have frequently been reported to take part in gene conversion events, either as donors or as acceptors. It is worth noting that also essentially invariant MHC genes, such as *Q4* or *Q10*, have been remarked to take part in gene conversion events as donors, even though they never have been reported as acceptors. On the other hand, other invariant MHC genes, such as Ea/DRA or DOB, never seem to take part in such events, and neither does the polymorphic Aa/DQA gene.

An interesting sequence trait worth investigating is the so called CpG island. In normal vertebrate DNA, the occurrence of a C 5′ to a G, a so called CpG dinucleotide, is suppressed, such that CpG dinucleotides occur at about 25% of the frequency expected from the frequency of C and G nucleotides, so called GC content (Bird 1987; Cardon et al. 1994). Because the overall GC content of vertebrate DNA is around 40%, the overall frequency of CpG dinucleotides in the vertebrate genome is around one percent. However, this is not true for all stretches of DNA. There are stretches of DNA that both have a higher GC content, over 60%, and the expected number of CpG dinucleotides for this frequency, i.e. above ten percent, or more than an order of magnitude higher than in the rest of the genome (Bird 1987). Many CpG islands are associated with promoter regions of housekeeping genes, where they are thought to facilitate the general accessibility of the promoter DNA. However, there are some CpG islands that are associated neither with housekeeping genes or with promoter regions. As was first noted more than ten years ago, (Jaulin et al. 1985; Rask et al. 1985) a number of those are located in polymorphic exons of MHC genes. Figure 4 shows the CpG profile of a couple of different mouse MHC genes. As can be seen, the two polymorphic exons of a class I gene such as *Kb*, and the polymorphic exon of a class II gene such as *Ab*, display a clearly elevated frequency of CpG dinucleotides. Not only polymorphic MHC genes display this trait, as shown by the *Q4* gene (Högstrand and Böhme 1999). On the other hand, it is clear this is not an obligatory trait for MHC genes, as it is absent from both the *Aa* and the *Ea* gene. Table 1 summarises the GC content, CpG content, conclusion of status as a CpG island, and report of gene conversion as donor or acceptor of a number of mouse MHC genes.

As can be seen, all genes that are suspected to have taken part in a gene conversion event possess a typical or very likely CpG island, whereas the Aa and Ea genes, as well as the human DNA, DOB, DPA, DQA1, DQA2 and DQB2 genes, that have never been reported to undergo gene conversion, do not display such a sequence feature. The CpG island seems to be correlated with the propensity for gene conversion itself, rather than with polymorphism, because, on one hand, a number of nonclassical class I genes of the Q series display CpG islands in the absence of notable polymorphism (Högstrand and Böhme 1999), and on the other, the polymorphic Aa gene, which has not been associated with gene conversion, has background CpG values.

Are CpG clusters present also in genes outside the MHC that have been implicated in gene conversion? This is sometimes, but not always, the case. CpG clusters are present in the regions of the Ig that have been reported to take part in

512

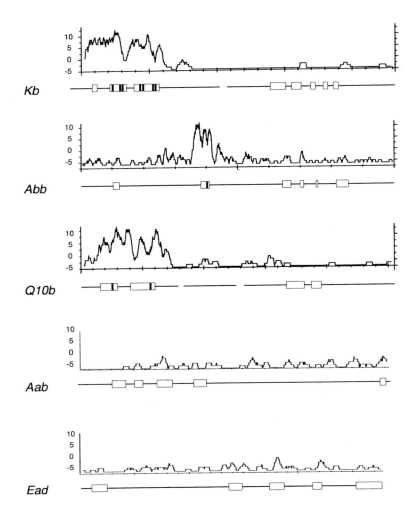

Fig. 4. Pattern of CpG dinucleotides in the mouse MHC genes *Kb*, *Abb*, *Q10b*, *Aab*, and *Ead*. The numbers of observed CpG minus the expected global pattern over a sliding window of 100 bp with a shift increment of 3 bp are shown. Exons are indicated as open boxes and introns as solid lines. A gap in the solid line denotes a gap in the known intron sequence. Recorded gene conversion events are marked with vertical lines on the exon boxes.

gene conversions. Furthermore, they are present in the two expressed α-globin genes. However, they are not in γ-globin genes, serum amyloid proteins or mouse urinary protein, which have been equally implicated in gene conversion. The significance of this is uncertain. Outside the MHC, Ig genes are the only genes

Table 1. CpG island status and implication in gene conversion for human and mouse MHC genes.

Gene	G+C content [a]	CpG content [a]	Bona fide CpG island [b]	Acceptor [c]	Donor [c]
Kb	67%	10%	++	+	+
Kk	69%	11%	++	+	+
Db	68%	10%	++	+	+
Abb	72%	13%	++	+	-
Ebb	63%	9%	+	-	+
Ead	48%	2%	-	-	-
Aab [d]	46%	2%	-	-	-
Q2k	65%	8%	+	-	+
Q4b	67%	10%	++	-	+
Q8b	66%	9%	+	-	+
Q10b	67%	9%	+	-	+
K1	64%	8%	+	-	+
DPB	66%	8%	+	+	+
DQB1	69%	11%	++	+	+
DRB1	64%	8%	+	+	+
DRA	49%	2%	-	-	-
DNA [e]	59%	4%	-	-	-
DOB	50%	1%	-	-	-
DPA	47%	3%	-	-	-
DQA1 [f]	51%	0,3%	-	-	-
DQA2 [g]	51%	0,7%	-	-	-
DQB2 [g]	42%	1%	-	-	-

[a] Within a stretch with elevated G+C or CpG figures,
or within a corresponding stretch if no elevated G+C or CpG present.
[b] Where ++ signifies G+C ≥ 60% and CpG ≥ 10%,
+ signifies G+C ≥ 60% and CpG ≥ 8%, and - signifies G+C < 60% and CpG < 4%.
[c] As implicated in the literature according to the corresponding reference(s)
[d] GenBank accession number AF027865
[e] GenBank accession number Z81310
[f] GenBank accession number Z84489
[g] GenBank accession number Z84490

where gene conversion has been implicated as a mechanism for generation of additional diversity. They possess elevated CpG, whereas the vast majority of genes displaying concerted evolution, normally also thought to be caused by gene conversion, do not. In the light of this, we feel that it at least has to be considered whether concerted evolution might be caused by mechanisms other than the gene conversion we describe in this article.

Concluding remarks

It is no longer controversial whether MHC genes really do undergo gene conversion. Still, we know very little about how this phenomenon comes about. It occurs during mitosis, but only in very special cells. Sequence traits, such as degree of homology and CpG content of the involved genes seem to of vast importance for its efficiency, but also non-MHC genes seem to play a role. As it can be enhanced *in vitro* by DNA-damaging agents, it is tempting to suggest that some kind of DNA nicking activity, perhaps with a high processivity for CpG rich DNA stretches, is needed for gene conversion to take place in vivo, and that this activity is only present in certain cells, such as early spermatogenetic cells and, most likely, early oogenetic cells and B lymphocytes at certain stages of development. Because it is enhanced, and thus seems conditioned, by DNA damage, the DNA repair system is most likely the next step in the process. Our inverted dose-response of radiation can be interpreted as if some repair polymerases are more efficient than others at promoting gene conversion, or alternatively, as if fast-growing or transformed cells have a higher intrinsic capacity for gene conversion.

A number of these issues are possible to address in the near future. The importance of the sequence homology and the frequency of CpG dinucleotides for the frequency of gene conversion can be studied in transgenic mice. The contribution of background non-MHC genes to variations in the frequency of gene conversion can be addressed by analysing MHC-congenic mouse strains, and such genes can be mapped by the use of mouse microsatellite markers. The frequency of MHC gene conversion events in B cells can be studied directly. B cells and spermatogonia can be analysed for the expression of homologues to yeast proteins implicated in fungal gene conversion. The frequency of gene conversion events can be compared in transformed and nontransformed cells of the same lineage. It is our firm conviction that considerable progress in the field of mammalian gene conversion is possible within a foreseeable amount of time.

Acknowledgements

This work was supported by grant B96-13X-10835-03 from the Swedish Medical Research Council and grant B-AA/BU 2069-300 from the Swedish Natural Science Research Council.

References

Baltimore D (1981) Gene Conversion: Some implications for immunoglobulin genes. Cell 24:592-594

Becker RS, Knight KL (1990) Somatic diversification of immunoglobulin heavy chain VDJ genes: Evidence for somatic gene conversion in rabbits. Cell 63:987-997

Bird AP (1987) CpG islands as gene markers in the vertebrate nucleus. Trend Genet 3:342-347

Cardon LR, Burge C, Clayton DA, Karlin S (1994) Pervasive CpG suppression in animal mitochondrial genomes. Proc Natl Acad Sci USA 91:3799-3803

Clarke SH, Claflin JL, Rudikoff S (1982) Polymorphisms in immunoglobulin heavy chains suggesting gene conversion. Proc Natl Acad Sci USA 79:3280-3284

Denaro M, Hammerling U, Rask L, Peterson PA (1984) The E β^b gene may have acted as the donor gene in a gene conversion-like event generating the Aβ^{bm12} mutant. EMBO J 3:2029-2032

Edelman GM, Gally JA (1968) Antibody structure, diversity, and specificity. Brookhaven Symp Biol 21:328-344

Egorov IK, Egorov OS (1998) Detection of new MHC mutations in mice by skingrafting, tumor transplantation and monoclonal antibodies: A comparison. Genetics 118:287-298

Gorski J, Mach B (1986) Polymorphism of human Ia antigens: gene conversion between two DR β loci results in a new HLA-D/DR specificity. Nature 322:67-70

Hanneman WH, Schimenti KJ, Schimenti JC (1997) Molecular analysis of gene conversion in spermatids from transgenic mice. Gene 200:185-192

Högstrand K, Böhme J (1994) A determination of the frequency of gene conversion in unmanipulated mouse sperm. Proc Natl Acad Sci USA 91:9921-9925

Högstrand K, Böhme J (1997) Gene conversion of major histocompatibility complex genes in the mouse spermatogenesis is a premeiotic event. Mol Biol Cell 8:2511-2517

Högstrand K, Böhme J (1998) Intrachromosomal gene conversion frequency in the H2 differs between haplotypes. Immunogenetics 48:47-55

Högstrand K, Böhme J (1999) DNA damage caused by etoposide and γ-irradiation induces gene conversion of the MHC in a mouse non-germline testis cell line. Mutation Res 423:155-169

Högstrand K, Böhme J (1999) Gene conversion of major histocompatibility complex genes is associated to CpG rich regions. Immunogenetics 49:446-455

Holliday R (1964) A mechanism for gene conversion in fungi. Genet Res 5:282-304

Jaulin C, Perrin A, Abastado J, Dumas B, Papamatheakis J, Kourilsky P (1985) Polymorphism in mouse and human class I H-2 and HLA genes is not the result of random independent point mutations. Immunogenetics 22:453-470

Kawamura S, Saitou N, Ueda S (1992) Concerted evolution of the primate immunoglobulin α-gene through gene conversion. J Biol Chem 267:7359-7367

Klein J (1984) Gene conversion in MHC genes. Transplantation 38:327-329

Kostriken R, Strathern JN, Klar AJS, Hicks JB, Heffron F (1983) A site-specific endonuclease essential for mating-type switching in Saccharomyces cerevisiae. Cell 35:167-174

Li H, Gyllensten UB, Cui X, Saiki RK, Erlich HA, Arnheim N (1988) Amplification and analysis of DNA sequences in single human sperm and diploid cells. Nature 335:414-417

Liebhaber SA, Goossens M, Kan YW (1981) Homology and concerted evolution at the α1 and α2 loci of human α-globin. Nature 290:26-29

Maizels N (1989) Might gene conversion be the mechanism of somatic hypermutation of mammalian immunoglobulin genes? Trends Genet 5:4-8

McIntyre KR, Seidman JG (1984) Nucleotide sequence of mutant I-A β^{bm12} gene is evidence for genetic exchange between mouse immune response genes. Nature 308:551-553

McKee BD (1996) Meiotic recombination: a mechanism for tracking and eliminating mutations? Bioessays 18:411-419

Mirzayans R, Andrais B, Paterson MC (1992) Synergistic effects of aphidicolin and 1-β-D-arabinofuranosylcytesine on the repair of γ ray induced DNA damage in normal human fibroblasts. Int J Radiat Biol 62:417-425

Mirzayans R, En's L, Detroit K, Barley RDC, Paterson MC (1996) Faulty DNA polymerase δ/ε-mediated excision repair in response to γ radiation or ultraviolet light in p53-deficient fibroblast strains from affected members of a cancer-prone family with Li-Fraumeni syndrome. Carcinogenesis 17 :691-698

Miyata T, Yasunga T, Yamawaki-Kataoka Y, Obata M, Honjo T (1980) Nucleotide sequence divergence of mouse immunoglobulin γ1 and γ2b chain genes and the hypothesis of intervening sequence-mediated domain transfer. Proc Natl Acad Sci USA 77:2143-2147

Murti JR, Schimenti KJ, Schimenti JC (1994) A recombination-based transgenic mouse system for genotoxicity testing. Mutation Res 307:583-595

Nairn R, Yamaga K (1980) Biochemistry of the gene products from murine MHC mutants. Annu Rev Genet 14:241-27

Nathenson SG, Geliebter J, Pfaffenbach GM, ZeffRA (1986) Murine major histocompatibility complex class-I mutants: molecular analysis and structure-function implications. Annu Rev Immunol 4:471-502

Ollo R, Rougeon F (1983) Gene conversion and polymorphism: generation of mouse immunoglobulin γ2a chain alleles by differential gene conversion by γ2b chain gene. Cell 32:515-523

Pääbo S, Higuchi RG, Wilson AC (1989) Ancient DNA and the polymerase chain reaction. The emerging field of molecular archaeology. J Biol Chem 264:9709-9712

Pease LR, Schulze DH, Pfaffenbach GM, Nathenson SG (1983) Spontaneous H-2 mutants provide evidence that a copy mechanism analogous to gene conversion generates polymorphism in the major histocompatibility complex. Proc Natl Acad Sci USA 80:242-246

Rask L, Gustafsson K, Larhammar D, Ronne H, Peterson PA (1985) Generation of class II antigen polymorphism. Immunol Rev 84:123-143

Reynaud C, Anquez V, Grimal H, Weill J (1987) A hyperconversion mechanism generates the chicken light chain preimmune repertoire. Cell 48:379-388

Saiki RK, Scharf SJ, Faloona F, Mullis KB, Horn GT, Erlich HA, Arnheim N (1985) Enzymatic amplification of beta-globin sequences and restriction site analysis for diagnosis of sickle cell anemia. Science 230:1350-1354

Schiestl R, Wintersberger U (1983) Induction of mating type interconversion in a heterothallic strain of Saccharomyces cerevisiae by DNA damaging agents. Mol Gen Genet 191:59-65

Schiestl RH, Wintersberger U (1992) DNA damage induced mating type switching in Saccharomyces cerevisiae. Mutation Res 284:111-123

Schimenti KJ, Hanneman WH, Schimenti JC (1997) Evidence for cyclophosphamide-induced gene conversion and mutation in mouse germ cells. Toxicol Appl Pharmacol 147:343-350

Silvain C, Aucouturier P, Leduc I, Mihaesco E, Preud'Homme J, Cogné M (1993) A human myeloma IgA with a hybrid heavy chain resulting from putative somatic gene conversion. Eur J Immunol 23:364-368

Slightom JL, Blechl AE, Smithies O (1980) Human fetal $^G\gamma$- and $^A\gamma$- globin genes: complete nucleotide sequences suggest that DNA can be exchanged between these duplicated genes. Cell 21:627-638

Thompson CB, Neiman PE (1987) Somatic diversification of the chicken immunoglobulin light chain gene is limited to the rearranged variable gene segment. Cell 48:369-378

Wang Y, Maher VM, Liskay RM, McCormick JJ (1988) Carcinogens can induce homologous recombination between duplicated chromosomal sequences in mouse L cells. Mol Cell Biol 8 :196-202

Mhc class II genes of Darwin's Finches: divergence by point mutations and reciprocal recombination

Akie Sato[1], Felipe Figueroa[1], Werner E. Mayer[1], Peter R. Grant[2], B. Rosemary Grant[2], and Jan Klein[1]

[1]Max-Planck-Institut für Biologie, Abteilung Immungenetik, D-72076 Tübingen, Germany
[2]Department of Ecology and Evolutionary Biology, Princeton University, Princeton, NJ 08544-1003, USA

Summary. The 15 species of Darwin's Finches are one of the classical examples of adaptive radiation — the divergence of a single ancestral species into multiple forms adapted to various ecological niches. As such they provide an opportunity to study the evolution of the major histocompatibility complex (*Mhc*) genes in the period of adaptive radiation. In the present study two families of the Ground Finch *Geospiza scandens* from the Daphne Major Island in the Galápagos Archipelago were sampled and exon 2 of their *Mhc* class II *B* genes was sequenced. The sequences indicate that unlike the domestic fowl with its compact, gene-poor *Mhc*, the *Mhc* of the Darwin's Finches contains more than half a dozen of class II *B* loci alone. Both the exon 2 and intron 2 sequences suggest that the genes at the different loci are closely related and that some of them may have diverged from one another during the adaptive radiation of the Darwin's Finches on the Galápagos Archipelago. The intron 2 sequences seem to have arisen by multiplication of a hexamer TCCCAG, ultimately reaching the length of ~ 2 kb. The repeats are still clearly recognizable in the introns of the genes at different loci. Sequence motifs in exon 2 of the class II *B* genes specify the polymorphic residues in the putative peptide-binding region (PBR) of the β2 domain. The evolution of the motifs can be explained by the sequential accumulation of point mutations and their incorporation (or fixation) by balancing selection. The divergence of the individual class II *B* genes is best explained by this process combined with occasional reciprocal recombination that shuffled the motifs.

Key words. Galápagos Islands, Adaptive radiation, Trans-species polymorphism, Bird *Mhc*, Multigene family

Introduction

Darwin's Finches are a group of 15 species of fringillid birds endemic to the Galápagos Archipelago (13 species) and the Cocos Island (one species) in the Pacific Ocean (Lack 1947; Grant 1986). Mitochondrial sequence analysis indicates that the ancestors of the Darwin's Finches were related to the extant representatives of the genus *Tiaris*, the grassquit of Central and South America (Sato et al. 1999; A. Sato, H. Tichy, C. O'hUigin, F. Figueroa, P. R. Grant, B. R. Grant, and J. Klein, in preparation). Taxonomically, therefore, they belong to the family Fringillidae, subfamily Emberizinae, and tribe Thraupini (Sibley and Ahlquist 1990). A flock or flocks of these ancestors no less than 40 individuals in size (Vincek et al. 1997) reached the Galápagos Archipelago less than two million years (my) ago (A. Sato et al., in preparation) and its descendants began to radiate by adapting to the different ecological niches available on the Archipelago. Three or four major branches of the initial adaptive radiation have survived to this day (Sato et al. 1999; Petren et al. 1999a). The earliest to diverge from the ancestral stock were the warbler-like finches, now represented only by two surviving species of the Warbler Finch, *Certhidia olivacea (Ceol;* the symbols in the parentheses are abbreviations consisting of the first two letters of the genus and first two letters of the species names; see Klein et al. 1990), distinguishable by microsatellite DNA length variation (Petren et al. 1999a). The next branch to diverge is now also represented by a single species, the Vegetarian Finch, *Platyspiza crassirostris (Plcr)*. In contrast to these two branches, which may be the remnants of ancient radiations, the remaining two branches, the Tree Finches and the Ground Finches, appear to comprise an ongoing radiation. In the Tree Finch, but in particular in the Ground Finch group, the species are differentiated morphologically by their adaptations to the different types of food they predominantly eat, and to a lesser degree behaviorally (interspecies hybridization has been documented for both groups; see Lack 1947; Bowman 1961; Grant 1986, 1993, 1994; Grant and Grant 1994), but not biochemically (Yang and Patton 1981) or molecularly (Sato et al. 1999, but see Petren et al. 1999a). The mtDNA cytochrome *b* and control region analysis has revealed the presence of several haplotypes which are intermingled among the various species of the Tree and Ground Finches (Sato et al. 1999). All this suggests that the Tree Finches and the Ground Finches are in the midst of separate adaptive radiations, in which the polymorphisms have not yet been sorted out among the various morphologically identifiable species. The Tree Finch assemblage comprises five such species in two currently recognized genera (Grant 1986): the Woodpecker Finch, *Cactospiza pallida, (Capa)*; the Mangrove Finch, *Cactospiza heliobates (Cahe)*; the Large Tree Finch, *Camarhynchus psittacula (Caps)*; the Medium Tree Finch, *Camarhynchus pauper (Capu)*; and the Small Tree Finch, *Camarhynchus parvulus (Capr)*. They are all insectivorous, but the individual species differ in the manner in which they catch their prey. Correspondingly, they differ in body and beak size (Lack 1947; Grant 1986). The Ground Finches are currently differentiated into six species of a single genus: the

Large Ground Finch, *Geospiza magnirostris (Gema)*; the Medium Ground Finch, *Geospiza fortis (Gefo)*; the Small Ground Finch, *Geospiza fuliginosa (Gefu)*; the Large Cactus Finch, *Geospiza conirostris (Geco)*; the Small Cactus Finch, *Geospiza scandens (Gesc)*; and the Sharp-beaked Ground Finch, *Geospiza difficilis (Gedi)*. They are all characterized by heavy, finch-like beaks which vary in size and shape according to the size and type of seeds they feed on. Related to the Tree Finches is the Cocos Finch, *Pinaroloxias inornata (Piin)*, which feeds on insects on the ground and in the trees and has a correspondingly slender, slightly decurved beak. The mtDNA data (Sato et al. 1999) suggest that the ancestors of the Cocos Finch may have flown to the Cocos Island from the Galápagos Archipelago approximately 0.5 my ago.

There are two good reasons to study the *Mhc* genes of Darwin's Finches. One reason is that the knowledge of the Darwin's Finch phylogeny can be used to make inferences about the evolution of the *Mhc* polymorphism in an adaptively radiating group of taxa. The other reason is that the knowledge of the extent and nature of the *Mhc* polymorphism can be applied to make inferences about the size, composition, and origin of the group's founding population (Vincek et al. 1997). In an earlier study (Vincek et al. 1997), we identified four groups of class II *Mhc* genes in the Galápagos finches and deduced from the intragroup polymorphism that the alleles and allelic lineages now detected in the finch populations were founded by at least 40 birds. Although the phylogenetic analysis suggested that each of the groups represented at least one locus, the uncertainty about the genetic relationships among the sequences limited their usefulness in terms of drawing conclusions about the founding population. For this reason we extended the analysis in the present study. Instead of picking up sequences randomly from unrelated individuals, we focused on two bird families (parents and offspring) and obtained a large number of sequences from each individual to better assess the presence of different loci and alleles at the individual loci.

Material and methods

DNA isolation

Blood samples were collected from the wing vein on filter papers. A small piece of the filter ($\sim$ 5 mm^2) was soaked in 200 µl of phosphate buffer saline and genomic DNA was extracted by using the QIAamp Blood Kit (Qiagen, Hilden, Germany).

Polymerase chain reaction (PCR) amplification

Two microliters of genomic DNA were added to 50 µl of PCR buffer, 0.2 mM of each of the four deoxyribonucleotides, 0.2 µM of each of the sense and antisense primers, 2.5 units of *Taq* DNA polymerase (Amersham Pharmacia Biotech,

Freiburg, Germany), and 0.4 units of *Pfu* DNA polymerase (Stratagene, Heidelberg, Germany). The DNA was amplified in the PTC-200 Thermal Controller (Biozym, Oldendorf, Germany) in 30 cycles, each cycle consisting of 15 sec denaturation at 94 °C, 15 sec annealing at the annealing temperature determined for each primer combination, and 1 min extension at 72 °C. A hot-start PCR was carried out by adding $MgCl_2$ as Hotwax Mg^{2+} beads (Invitrogen, NV Leek, The Netherlands). The primers used were as follows (Fig. 1): HOPE1, 5'-GAAAGCTCGAGTGTCACTTCACGAACGGC-3' (sense); HOPE2, 5'-GGGTG-ACAATCCGGTAGTTGTGCCGGCAG-3' (antisense; see Vincek et al. 1997); HOPE3, 5'-TGTCACTTCACGAACGGCACGGAGAAG-3' (sense); F2, 5'-GG-GGGTTCACCCCTWTGGGGAGATGAAT-3' (sense); F3, 5'-GGGTTCACCCC-TWTGGGGAGAGGGTG-3' (sense); F32, 5'-GGAAATCTACAACCGGCTGAT-GCATGTGA-3' (sense); F4, 5'-GTCCATCTACAACCGGGAGCAGTTCATAA-3' (sense); H4, 5'-GTTTGGATTGGGAATGGTTTGGGACGAAC-3' (antisense); and FA, 5'-ACCTCACCTGGATCGGGGCAGGGTAGAA-3' (anti-sense).

Cloning of PCR products

Forty microliters of PCR product were purified with the aid of the QIAquick Gel Extraction Kit (Qiagen). The eluted DNA was blunt-ended, phosphorylated, and ligated to *Sma* I-digested pUC18 plasmid vector with the help of the SureClone Ligation Kit (Amersham Pharmacia Biotech) and used to transform *E. coli* XL-1 blue competent bacteria.

DNA sequencing

Double-stranded DNA was prepared with the help of the QIAGEN Plasmid Kit and 0.5 µg of the DNA was then used in the dideoxy chain-terminating cycle

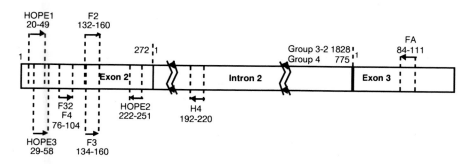

Fig. 1. Positions of PCR primers used in the present study. The sites are numbered from the 5' to 3' end in each segment (exon 2, intron 2, exon 3) separately; arrows indicate sense and antisense orientation.

sequencing reaction. The Thermo Sequenase Fluorescent Labelled Primer Cycle Sequencing Kit was used with Thermo Sequenase DNA polymerase to carry out the reaction, which was then processed by the LI-COR Long ReadIR DNA sequencer (MWG Biotech, Ebersberg, Germany). To confirm the sequences, 5-10 µg of DNA were used for the dideoxy chain-termination sequencing reaction carried out with the AutoRead Sequencing Kit (Amersham Pharmacia Biotech), which contained 5' fluorescent labeled sequencing primers and T7 DNA polymerase. The reactions were processed by the Automated Laser Fluorescent (A.L.F.) Sequencer (Amersham Pharmacia Biotech). The sequences are available in the GenBank database under the accession codes AF163618-AF163660, AF164158-AF164168, and AF165155-AF165-157.

Sequence analysis

Sequences were aligned by the ClustalW multiple sequence alignment software (Thompson et al. 1994). The alignment was improved by visual inspection with the help of the SeqPup software, version 0.6f (Gilbert 1996). Evolutionary relationships were evaluated with the help of the MEGA program package (Kumar et al. 1993) using two-parameter distances (Kimura 1980) and the neighbor-joining algorithm for phylogenetic tree construction (Saitou and Nei 1987).

Results

Family study

Blood samples were obtained from two families of the Ground Finch *Geospiza scandens*, each consisting of both parents and four offspring. The parentage in each family was initially identified by observation and subsequently confirmed by analysis of microsatellite DNA markers (Petren et al. 1999b). The DNA isolated from these samples was subjected to PCR by using primer pairs HOPE1-HOPE2 and HOPE3-HOPE2, designed to amplify the bulk (~ 170 bp) of exon 2 of the *Mhc* class II β chain-encoding genes (Fig. 1). The amplified products were then cloned and sequenced. Altogether, 12 to 16 clones per individual were sequenced from four independent PCRs. The alignment of the unique nucleotide sequences, together with those obtained in an earlier study (Vincek et al. 1997) is presented in Figure 2; the corresponding deduced protein sequences are displayed in Figure 3; and a phylogenetic tree based on this collection of nucleotide sequences is shown in Figure 4.

The study identifies 43 new exon 2 class II *B* sequences among the total of 148 sequences obtained. By phylogenetic analysis (Fig. 4), the new sequences are assigned to two of the four previously identified groups (Vincek et al. 1997),

Fig. 2. Exon 2 sequence alignment of Darwin's Finch class II *B* genes. The highlighted sequences are new and are accessible in the GenBank database under accession codes following the species abbreviations, the remaining sequences are from Vincek et al. (1997). Species abbreviations are explained in the **Introduction**. The sites are numbered from the first (5') nucleotide of exon 2. The simple majority consensus sequence is given at the top; identity with the consensus is indicated by a dash (-), absence of sequence information by a dot (.), and insertions/deletions (indels), introduced to optimize the alignment, by asterisks (*).

```
                   57         67         77         87         97         107
     CONSENSUS ─ GTGAGGTTCG TGGAGAGGTA CATCTACAAC CGGCAGCCCT ACGCGGTGTT CGACAGCGAC
  Gefo-Z74449   -----A-A-- -CC-----C  ---------- ---G---AG- T-ATAA---- ----------
  Gema-Z74430   -----A-A-- -CC-----C  ---------- ---G---AG- T-ATAA---- ---------T
  Geco-Z74421   -----A-A-- -CC-----C  ---------- ---G---AG- T-ATAA----
  Gefo-Z74460   -----A-A-- -CC-----C  -G-------- ---G---AG- T-ATAA----
  Gefo-Z74448   -----A-A-- -CC-----C  ---------- ---G---AG- T-ATAA----
  Gema-Z74431   -----A-A-- -CC-----C  ---------- ---G---AG- T-ATAA----
  Gefo-Z74444   -----A-A-- -CC-----C  ---------- ---G---AG- T-ATAA----
  Geco-Z74422   -----A-A-- -CC-----C  ---------- ---G---AG- T-ATAA----
  Gesc-AF163636 -----A-A-- ----------            ------AG- T-CT-A----
  Gesc-AF163618 ---A-A-- ----------              ------AG- T-CT-A----
  Gesc-AF163619 ---A-A-- --C-------             ------AG- T-CT-A----
  Gesc-AF163620 ---A-A-- ----------             ------AG- T-CT-A----
  Gesc-AF163623 ---A-A-- ----------             ------AG- T-CT-A----
  Gesc-AF163621 ---A-A-- ----------             ------AG- T-CT-A----
  Gesc-AF163622 ---A-A-- ----------             ------AG- T-CT-A----
  Gesc-AF163624 ---A-A-- ----------             ------AG- T-CT-A----
  Gesc-AF163625 -----A-A-- ----------            ------AG- T-CT-A----
  Gesc-AF163659 -----C--- ----------            ------AG- ---T-A----
  Gesc-AF163660 -----C--- ----------            ------AG- ---T-A----
  Gema-Z74432   ---------- --C-------            ------AG- ---T-A----
  Gefo-Z74455   ------A-- --C-C---G- A-------          AGC T-A--A----
  Gefo-Z74454   ---------- --C-C---G- A-------          AGC T-A--A----
  Gefo-Z74461   ----------  --C----G- A---------  --T-ATGC -T-T-A----
  Ceol-Z74466   -----A-A-- --C-C---G- A---------  --T-ATGC -T-T-A----
  Ceol-Z74468   ------A-- --C-C---G- A---------   --T-ATGC -T-T-A----
  Gefo-Z74443   ------A-- --C-C---G- G---------              ---T------
  Ceol-Z74467   ------A-- --C-C---G- A---------              ---T------
  Plcr-Z74426   -----A-A-- --C-C---G- A-------              ---T------
  Gefu-Z74441   ------A-- --C-C---G- A-------               ---T------
  Gefo-Z74458   ------A-- --C-C---G- A-------               ---T------
  Ceol-Z74459   ---------- --C-C---G- A-------              ---T------
  Gefo-Z74452   ---------- --C-C---G- A-------              ---T------
  Gefu-Z74439   ---------- ---C----G- A-------              ---T------
  Gefo-Z74446   ---------- ---C----G- A-------              ---T------
  Gefo-Z74447   ---------- --C-C---G- A-------              ---T------
  Gefo-Z74445   ---------- --C-C---G- A-------              ---T------
  Gesc-AF163654 ------C--- --------C- A-------               ---G ---T------
  Gesc-AF163634 ------C--- --------C- A-------               ---G ---T------
  Gesc-AF163653 ------C--- --------C- A-------               ---G ---T------
  Gesc-AF163635 ------C--- --------C- A-------               ---G ---T------
  Gesc-AF163657 ------C--- --------C- A-------               ---G ---T------
  Gefo-Z74450   ---------- --------C- A-------               ---G ---T------
  Plcr-Z74424   ---------- --------C- A-------               ---G ---T------
  Gesc-AF163637 ---------- --------C- A-------               ---G ---T------
  Plcr-Z74428   -----A-A-- ----------              ---G ---T------
  Gema-Z74462   ---------- ----------               ---G -T-T------
  Gema-Z74470   ---------- --C-------              ---G ---T------
  Gesc-Z74415   -----A-A-- --------C-              -T---G ---T------
  Gesc-Z74414   ---------- --------                -T---G ---T------
  Gesc-Z74412   ---------- --------                -T---G ---T------
  Gesc-AF163639 ---------- --------                -T---G ---T------
  Gesc-AF163658 ------C--- ----------
  Gesc-AF163640 ---------- ----------
  Gesc-AF163641 ---------- ----------
  Gesc-AF163648 ---------- ----------
  Gesc-AF163649 ---------- ----------
  Geco-Z74420   ------C--- ----------
  Gema-Z74433   ------C--- ----------
  Geco-Z74464   ------C--- ----------
  Gefo-Z74451   ------C--- ----------
  Gefo-Z74453   ------C--- --------C-
  Gefu-Z74438   ---------- --C-------
  Gesc-AF163632 -----A-A-- --C-------
  Gesc-AF163633 ----A-A-- --C-------
  Gefo-Z74459   -----A-A-- --C-------
  Gesc-Z74411   ---------- A------C-
  Gesc-Z74417   ---------- A------CG
  Gesc-Z74416   ---------- A------C-
  Capa-Z74436   ---------- -------C-
  Geco-Z74419   ---------- ----------
  Plcr-Z74425   ---------- ----------
  Ceol-Z74418   ---------- ----------
  Gesc-AF163644 ---------- ----------
  Gesc-AF163645 ---------- ----------
  Gesc-AF163646 ---------- ----------
  Gesc-AF163647 ---------- ----------
  Gesc-AF163638 ---------- ----------
  Gesc-AF163651 ------C--- --------C- A-------
  Gesc-AF163655 ------C--- --------C- A-------
  Gesc-AF163652 ------C--- --------C- A-------
  Gesc-AF163656 ------C--- ----------
  Gefu-Z74440   ------C--- --------C- A-------
  Gefu-Z74442   ------C--- --------C
  Gefo-Z74456   ------C--- ----------
  Gema-Z74435   ------C--- ----------
  Gema-Z74465   ------C--- C---------
  Geco-Z74423   ------C--- --------C-
  Gema-Z74434   ------C--- ----------
  Plcr-Z74427   ------C--- ----------
  Gesc-AF163650 ------C--- --------C- A-------
  Gesc-AF163642 ---------- ----------
  Gesc-AF163627 ----A-A-- ----------
  Gesc-AF163626 ----A-A-- ----------
  Gefo-Z74457   ----A-A-- ----------
  Gefo-Z74463   ----A-A-- ----------
  Gesc-AF163643 ---------- --C-------
  Gesc-AF163629 -----A-A-- --C-------
  Gesc-AF163630 -----A-A-- --C-------
  Gesc-AF163628 -----A-A-- --C-------
  Gesc-AF163631 -----A-A-- --C-------
  Gema-Z74429   -----A-A-- --C-------
  Capa-Z74437   -----A-A-- --C-------
  Gesc-Z74413   -----A-A-- --C-------
  Geco-AF164158 ---------- ----------
  Plcr-AF164159 ---------- ----------
  Ceol-AF164160 ---------- ----------
  Gefo-AF164161 ------C--- ----------          ---C------
  Ceol-AF164162 -----A-A-- --C-C---G- A---------  ---T-ATGC -T-T-A----
  Ceol-AF164163 ---------- --C-C---G- A---------       --TA TT--------
```

Fig. 2 (continued)

```
                117       127       137       147       157       167       177       187       197       207       217
CONSENSUS ▬▬ GTGGGGCACT ACGTGGGGTT CACCCCCTTT GGGGAGATGA ATGCCAAGCA CTGGAACAGC GACCCGGCAA TACTGGAGAA CAGACCGGACT GCGGTGGACT GGTA
Gefo-Z74449  .......... TT........ ........-A- ..-C--A-T TG......T .........-GT- ...-T--T-T .CA----G- .........-A-- ..........
Gema-Z74430  .......... TT........ ......-G--A- ..-C--A-T TG......-G .........-GT- ...-T--T-T .CA----G- .........-A-- ..........
Geco-Z74421  .......... TT........ ........-A- ..-C--A-T TG......-G .........-GT- ...-G--T-T .CA----G- .........-A-- ..........
Gefo-Z74460  .......... TT........ ........-A- ..-C--A-T TG......-G .........-GT- ...-G--T-T .CA----G- .........-A- CG-- ......
Gefo-Z74448  .......... TT........ ........-A- ..-C--A-T TG......-G .........-GT- ...-G--T-T .CA----G- .........-A-- ..........
Gema-Z74431  .......... TT........ ........-A- ..-C--A-T TG......-G .........-GT- ...-GGC-T-T .CA----G- .........-A-- ..........
Gefu-Z74444  .......... TT---.--C-- ........-A- ..-C--A-T TG......-G .........-GT- ...-G--T-T .CA----G- .........-A-- ..........
Geco-Z74422  .......... TT........ ........-A- ..-C--A-T TG......-G .........---GT- ...-G--T-T .CA----G- .........-A-- .....A C---
Gesc-AF163636 .......... .......... .......... .......-G .......... .......-A-T GGA---G- -GC-- .....A C---
Gesc-AF163618 .......... .......... .......... .......-G .......... .......-A-T GGA---G- -GC-- .....A C--
Gesc-AF163619 .......... .......... .......... .......-G .......... .......-A-T GGA---G- -GC-- .....A C--
Gesc-AF163620 .......... .......... .......... .......-G .......... .......-A-T GGA---G- -GC-- .....A C--
Gesc-AF163623 .......... .......... .......... .......-G .......... A----A-T GGA---G- -GC-- .....A C-GT
Gesc-AF163621 .......... .......... .......... .......-G .......... A----A-T GGA---G- -GC-- .....A C-GT
Gesc-AF163622 .......... .......... .......... ......-T- .......... A-----C- .......... .......... TT----A C--
Gesc-AF163624 .......... .......... .......... ......-T- .......... A-----C- .......... .......... TT----A C--
Gesc-AF163625 .......... .......... .......... .......-G .......... A-----C- .......... .......... .....A C--
Gesc-AF163659 .......... .......... .......... .......... A-------- AG-----C- G---A--T- -TG---G-- .....A C-CT
Gesc-AF163660 .......... ....----- .....-A--AAC -G--CG--- .......... AG-----C- G---A--T- -TG---G-- .....A C-C
Gema-Z74432  .......... TT-A----- ....-A- --G-G TG----- .......... -T-------C- .......... ----T-- CA----
Gefo-Z74455  .......... .......... ....-A- --G-G TG----- .......... ----AGT --GA---- ..........
Gefo-Z74454  .......... .......... ....-A- --G-G TG----- ......-T -------T- A------- ..........
Gefo-Z74461  .......... .......... ....-A- --G-G TG----- .......... ----A-T --GA----- -T-- -A-- CA----
Ceol-Z74466  .......... .......... ....-A- --G-G TG----- .......... --------G- .......... ..........
Ceol-Z74468  .......... TT........ ....-A- .......... ------G -T------ -A-T GGA---- G-- CA----
Gefu-Z74443  .......... .......... ....-A- --G-G TG----- .......... -----C- .......... ..........
Ceol-Z74467  ----G--.. .......... ....-A- --G-G TG----- .......... -A-T -CA---- -A---- ..........
Plcr-Z74426  .......... .......... ....-A- --G-G TG------G .......... -A-T GGA----- ----T- ----TT----
Gefu-Z74441  .......... .......... ....-A- .......... -G .......-**- *------C- .........-T- ..........
Gefo-Z74458  .......... .......... .......... .......... .........-A-T -CA---- ..........
Ceol-Z74469  .......... .......... ....-A- --G-G TG----- .......... -A-T -CA---- ..........
Gefo-Z74452  .......... .......... ....-A- --G-G TG-----T- .......-T -T- A------ ..........
Gefu-Z74439  ---AC-- .......... ....-A- --G-G TG-----T- .......-T -T- A------ ----CA----
Gefo-Z74446  .......... .......... ....-A- --G-G TG-----T- .......-T -A-T -GA----T- -A---- ..........
Gefo-Z74447  .......... .......... .......... .......... -CG A----A-T GGA----G- ..........
Gefo-Z74445  .......... .......... .......... .......-G -CG A----A-T GGA----G- ..........
Gesc-AF163654 .......... .......... ....--G-G TG------ .......... ------C- .........-G-- -A----A C-GT
Gesc-AF163634 .......... .......... ....--G-C TG------ .......... -T- A------ .........-A ----
Gesc-AF163653 .......... .......... ....--G-G TG------ .......... -T- A------ .....A C--
Gesc-AF163635 .......... .......... ....--G-G TG------ .......... -T- A------ ----T------ ..........
Gesc-AF163657 .......... .......... ....--G-G TG------ .......... -T- A------ ----T------ ..........
Gefo-Z74450  ---AC-- .......... ....--G-C TG------ -CG A----A-T GGA----G- ..........
Plcr-Z74424  .......... TT........ ....--G-C TG------ -CG A----A-T GGA----G- ..........
Gesc-AF163637 .......... -AC-- .......... .......-G .......... ------C- .........-- TT----A C--
Plcr-Z74428  .......... -AC-- .......... ----G-T------ .......... -A-T -GA---- -T- -A-- ..........
Gema-Z74462  ----T-- .......... .......... .......-G .......... -A-T -GA---T- -A-- ..........
Gema-Z74470  .......... .......... .......... .......-G .......... -A-T -GA---- -GT- ..........
Gesc-Z74415  .......... .......... .......... .......... A----A-T GGA----G- -GT- ..........
Gesc-Z74414  .......... .......... ....-A- .......... A----A-T -GA---- -G-- ..........
Gesc-Z74412  .......... .......... ....--G-G TG------ .......... -T- A------ .........-T- ..........
Gesc-AF163639 .......... .......... ....--G-G TG------ .......-T -T- A------ .........-T- ..........
Gesc-AF163658 .......... .......... ....--G-G TG------ .......-T -T- A------ ..........
Gesc-AF163640 .......... .......... ....--G-G TG------ .......-T -T- A------ .........-G-- -A----A C-GT
Gesc-AF163641 .......... .......... ....--G-G TG------T- .......-T -T- A------ .........-G-- -A----A C-GT
Gesc-AF163648 .......... .......... ....--G-G TG------ .......-T -T- A------ .........-G-- -A----A C-GT
Gesc-AF163649 .......... .......... ....--G-G TG------ .......-T -T- A------ .........-T- ..........
Geco-Z74420  ---AC-- .......... ....--G-G TG------ .......-T -T- A------ .........-T- ..........
Gema-Z74433  ---AC-- ---A-- .......... ....--G-G TG------ .......-T -T- A------ .........-T- ..........
Geco-Z74464  .......... .......... ....--G-G TG------ .......-T -T- A------ .........-T- ..........
Gefo-Z74451  .......... .......... ....--G-G TG------ .......-T -T- A------ .........-T- ..........
Gefo-Z74453  .......... .......... ....--G-G TG-----T- .......-T -A-T GGA----G- ---T-- CA----
Gefu-Z74438  .......... TT........ ....-A- --G-G TG------ .......... ------C- .........-T- .....A C---
Gesc-AF163632 .......... .......... ....--G-C TG------ .......... ------C- .........-G-- -A----A C-GT
Gesc-AF163633 .......... .......... ....--G TG------G -T------ ------C- .........-T- .....A C---
Gefo-Z74459  .......... .......... ....--G TG------G -T------ -A-T -GA----G- ..........
Gesc-Z74411  .......... .......... .......... -CG A----A-T -GA----G- ..........
Gesc-Z74417  .......... .......... .......... -CG A----A-T -GA----G- ..........
Gesc-Z74416  .......... .......... .......-G A-----A-T -GA---G- ---G-- ..........
Capa-Z74436  .......... .......... .......-G -CG A----A-T -GA----G- ----TT----
Geco-Z74419  .......... .......... .......-T- A-----A-T -GA---G- ---G-- -A----
Plcr-Z74425  .......... .......-C- .......-G -T------ ------C- .........-T- ..........
Ceol-Z74418  .......... .......... .......-G -T------ ------C- .........-T- ..........
Gesc-AF163644 .......... .......... .......-T- A-----A-T -GA----G-- CA----A C-GT
Gesc-AF163645 .......... .......... .......-T- A-----A-T -GA----G-- ---G--G- CA----A C-GT
Gesc-AF163646 .......... .......... .......-G A-----A-T -GA---- ----T------A C--
Gesc-AF163647 .......... .......... .......-G -A-T GGA----T- -A---- .....A C--
Gesc-AF163638 .......... .......... .......-G ------C- .........--- TT----A C--
Gesc-AF163651 .......... .......... .......-G ------C- .........--- TT----A C--
Gesc-AF163655 .......... .......... .......-G ------C- .........--- TT----A C--
Gesc-AF163652 .......... .......... .......-G ------C- .........--- TT----A C---
Gesc-AF163656 .......... .......... .......-G ------C- .........--- TT----A C--
Gefu-Z74440  .......... .......-A- .......-AG ----**- *G----A-C -CA---G- ----G- ..........
Gefu-Z74442  .......... .......-A- .......-AG ----**- *G----A-C -CA---G- ----G- ..........
Gefo-Z74456  .......... TT........ .......-A- .......-T- ------C- .........-T- ..........
Gema-Z74435  .......... TT........ .......-A- .......-G -T------ -A-T -CA---T- -A---- ----TT----
Gema-Z74465  .......... .......... .......-A- .......-G -T------ -A-T -CA---- ..........
Geco-Z74423  .......... .......... .......... -A-T -GA---- .........-T- ..........
Gema-Z74414  .......... .......-T- .......-T- ---GC- -A-T -GA---- ..........
Plcr-Z74427  .......... .......... .......-T- A-----C- .........-- .....A C---
Gesc-AF163650 .......... .......... .......-T- A-----C- .........-- .....A C---
Gesc-AF163642 .......... .......... .......-G A-----C- .........-- .....A C---
Gesc-AF163627 .......... .......... .......-G A-----C- .........-- .....A C---
Gesc-AF163626 .......... .......... .......-G A-----A-T GGA---- ..........
Gefo-Z74457  .......... .......-G G-----T- ------T- A------ ..........
Gesc-AF163643 .......... .......... .......-T- A-----C- .........-- .....A C--
Gesc-AF163629 .......... .......... .......-T- A-----C- .........-- ..........
Gesc-AF163630 .......... .......... .......-G A-----C- .........-- ----TT----A C--
Gesc-AF163628 .......... .......... .......-G A-----C- .........-- ----TT----A C--
Gesc-AF163631 .......... .......... .......... -A-T GGA---G- -GC---- ..........
Gema-Z74429  .......... .......... .......-T- A-----C- .........-T- ..........
Capa-Z74437  .......... .......... .......-T- ------C- .........-- -A----
Gesc-Z74413  .......... .......... .......-G --GA- ----C- .........-T- ..........
Geco-AF164158 .......... TT........ .......-A- ..-C--A-T TG------G T--G--GT- ---T-T -CA---G- ------A- .......-T
Plcr-AF164159 .......... .......... .......-A- .......-T- ------C- .........-T- .....A C--
Ceol-AF164160 .......... .......... .......-T- ------C- .........-T- .....A C--
Gefo-AF164161 .......... .......... ....--G-G TG------ .......... -T- A------ .........-G- .....A C--
Ceol-AF164162 .......... .......... ....-A- --G-G TG------ .......... ------G- ------G-- CA----A C-CT
Ceol-AF164163 .......... .......... ....-A- --G-G TG------ A-----A-T -GA------ ..........
```

Fig. 3.

```
                 19        29        39        49        59        69
CONSENSUS ==>  VRYVERYIYN RQPYAVFDSD VGHYVGFTPF GEMNAKRWNS DPAILENRRT AVDW
Gefo-Z74449    ----Q-S--- -EQFIM---- ---F-----Y -QKL--L--V -SVFM-D--N ----
Gema-Z74430    ----Q-S--- -EQFIM---- ---F----RY -QKL-----V -SVFM-D--N ----
Geco-Z74421    ----Q-S--- -EQFIM---- ---F-----Y -QKL-----V -AVFM-D--N ----
Gefo-Z74460    ----Q-SV-- -EQFIM---- ---F-----Y -QKL-----V -AVFM-D--N ----
Gefo-Z74448    ----Q-S--- -EQFIM---- ---F-----Y -QKL-----V -AVFM-D--N R---
Gema-Z74431    ----Q-S--- -EQFIM---- ---F-----Y -QKL-----V -GVFM-D--N ----
Gefo-Z74444    ----Q-S--- -EQFIM---- ---F-----Y -QKL-----V -SVFM-D--N ----
Geco-Z74422    ----Q-S--- -EQFIM---- ---F-----Y -QKL-----V -AVFM-D--N ----
Gesc-AF163636  --F------- --QFLM---- ---------- ---------- --EWM-DA-- ---T
Gesc-AF163618  ---------- --QFLM---- ---------- ---------- --EWM-DA-- ---T
Gesc-AF163619  ----Q----- --QFLM---- ---------- ---------- --EWM-DA-- ---T
Gesc-AF163620  ---------- --QFLM---- ---------- ---------- --EWM-DA-- ---T
Gesc-AF163623  ---------- --QFLM---- ---------- ---------- --ELM-YK-- ---R
Gesc-AF163621  ---------- --QFLM---- ---------- ---------- N-EWM-DA-- ---T
Gesc-AF163622  ---------- --QFLM---- ---------- ---------- N-EWM-DA-- ---T
Gesc-AF163624  ---------- --QFLM---- ---------- ------Y--- N--------- ----
Gesc-AF163625  ---------- --QFLM---- ---------- ------Y--- N--------- L--T
Gesc-AF163659  --L------- --Q-VM---- ---------- ---------- ---------- ---T
Gesc-AF163660  --L------- --Q-VM---- ---------- --KQ-RH--- S--R-KYM-A ---T
Gema-Z74432    --F-Q----- --QLTM---- ---FE----Y --RV--H--- V--------S Q---
Gefo-Z74455    ----H-E--- --QLTM---- ---------Y --RV--H--- --ELM----- ----
Gefo-Z74454    --F---Q--- --QLTM---- ---------Y --RV--H--- ---K------ ----
Gefo-Z74461    --F-D-E--- -LMHVM---- ---------Y --RV--Y--- --ELM----A Q---
Ceol-Z74466    ----H-E--- -LMHVM---- ---------Y --RV--H--- ------D--- ----
Ceol-Z74468    ----H-E--- -LMHVM---- ---------Y --RV--Y--- ------D--- ----
Gefo-Z74443    ----H-E--- ----V----- ---F-----Y -------L-- --EWM----A Q---
Ceol-Z74467    ----H-E--- ----V----- ---------Y --RV--H--- --EFM--K-- ----
Plcr-Z74426    ----H-E--- ---------- --D------- --RV--Y--- N-ELM-YK-- ----
Gefu-Z74441    ----H-E--- ---------- ---------Y ---------* N--------- L---
Gefo-Z74458    ----H-E--- ---------- ---------- ------H--- -------Y-- ----
Ceol-Z74469    --F-H-E--- ---------- ---------Y --RV--H--- N-ELM--â-- ----
Gefo-Z74452    --F-D-E--- ---------- ---------Y --RV--Y--- ---K------ ----
Gefu-Z74439    --F-D-E--- ---------- ---T-----Y --RV--Y--- ---K------ ----
Gefo-Z74446    --F-D-E--- ---------- ---------Y --RV--Y--- --ELM-Y--- Q---
Gefo-Z74447    --F-D-E--- ---------- ---------Y ------H--- --ELM-YK-- ----
Gefo-Z74445    --F-D-E--- ---------- ---------- --------T N-EWM-D--- ----
Gesc-AF163654  --L---Q--- ---------- ---------┐ --RV--H--- --------A E--T
Gesc-AF163634  --F---Q--- ---DV----- ---------- --RL--H--- ---K------ ---R
Gesc-AF163653  --L---Q--- ---DV----- ---------- --RL--H--- ---K------ ---T
Gesc-AF163635  --F---Q--- ---DV----- ---------- --RL--H--- ---K------ ---T
Gesc-AF163657  --L------- ---DV----- ---------- --RV--H--- ---K------ S---
Gefo-Z74450    --F---Q--- ---DV----- --T------- --RL--H--- ---K------ ----
Plcr-Z74424    --F---Q--- ---DV----- ---F-----Y --RL--H--T N-EWM-D--- ----
Gesc-AF163637  --F------- ---DV----- ---------- ---------- ---------- L--T
Plcr-Z74428    ---------- ---DV----- --T------- -------L-- N-ELM----- ----
Gema-Z74462    --F-D----- ---DV----- ---Y------ ---------- --ELM-YK-- ----
Gema-Z74470    --F-D----- ---DV----- ---------- ---------- --ELM-YK-- ----
Gesc-Z74415    ------H--- -L-DV----- ---------- ------H--- --EWM-DV-- ----
Gesc-Z74414    --F------- -L-DV----- ---------- ---------- N-EWM-DV-- ----
Gesc-Z74412    --F------- -L-DV----- ---------Y --RL--H--- Y-ELM-D--- ----
Gesc-AF163639  --F------- ---------- ---------- --RV--H--- ---K------ S---
Gesc-AF163658  --L------- ---------- ---------- --RV--H--- ---K------ S---
Gesc-AF163640  --F------- ---------- ---------- --RL--H--- ---K------ ---R
Gesc-AF163641  --F------- ---------- ---------- --RV--Y--- ---K------ ---R
Gesc-AF163648  --F------- ---------- ---------- --RV--H--- --------A E--T
Gesc-AF163649  --F------- ---------- ---------- --RV--H--- --------A E--T
Geco-Z74420    --L------- ---------- --T------- --RV--H--- ---K------ S---
Gema-Z74433    --L------- ---------- --T-E----- --RV--H--- ---K------ S---
Geco-Z74464    --L------- ---------- ---------- --RV--H--- ---K------ S---
Gefo-Z74451    --L------- ---------- ---------- --RV--H--R Y--K------ S---
Gefo-Z74453    --L---H--- ---------- ---------Y --RV--Y--- ---K--D--- ----
Gefu-Z74438    --F-Q----- ---------- ---F-----Y --RV--Y--- --EWM-D--S Q---
Gesc-AF163632  ----Q----- ---------- ---------- --RL--H--- ------Y--- ---R
Gesc-AF163633  ----Q----- ---------- ---------- --RV--H--- --------A E--T
Gefo-Z74459    ----Q----- ---------- ---------- ---V---L-- ------Y--- ----
Gesc-Z74411    --FE--H--- ---------- ---------- --------T N-ELM-D--- ----
Gesc-Z74417    --FE--R--- ---------- ---------- --------T N-ELM-D--- ----
Gesc-Z74416    --FE--H--- ---------- ---------- ---------- N-ELM-D--- ----
Capa-Z74436    --F---H--- ---------- ---------- --------T N-ELM-D--- L---
Geco-Z74419    --F------- ---------- ---------- ------Y--- N-ELM-D--A Q---
Plcr-Z74425    --F------- ---------- ---------S -------L-- -----D--A E---
```

```
Ceol-Z74418     --F------- ---------- --------Y ------Y--- ------Y--- ----
Gesc-AF163644   --F------- ---------- --------- ------Y--- N-ELM-D--A Q--T
Gesc-AF163645   --F------- ---------- --------- ------Y--- N-ELM-D--A Q--T
Gesc-AF163646   --F------- ---------- --------- --------- --ELM-D--- S--T
Gesc-AF163647   --F------- ---------- --------- --------- --ELM-YK-- ---R
Gesc-AF163638   --F------- ---------- --------- --------- --------- L--T
Gesc-AF163651   --L---Q--- ---------- --------- --------- --------- L--T
Gesc-AF163655   --L---Q--- ---------- --------- --------- --------- L--T
Gesc-AF163652   --L---Q--- ---------- --------- --------- --------- L--T
Gesc-AF163656   --L------- ---------- --------- --------- --------- L--T
Gefu-Z74440     --L---Q--- ---------- --------Y ---------* N--------- L---
Getu-Z74442     --L---S--- ---------- --------Y ------S--* S-ELM-D--A ----
Gefo-Z74456     --L------- ---------- ---F----- ------H--- --------- ----
Gema-Z74435     --L------- ---------- ---F-----Y ------H--- ------Y--- ----
Gema-Z74465     --LA------ ---------- --------Y -------L-- ------YK-- L---
Geco-Z74423     --L---H--- ---------- --------- ------H--- --EFM----- ----
Gema-Z74434     --L------- ---------- --------L- ------H--- --ELM----- S---
Plcr-Z74427     --L------- ---------- --------Y ------Y--A --ELM----- ----
Gesc-AF163650   --L---Q--- ---------- --------- ------Y--- N--------- ---T
Gesc-AF163642   --F------- ---------- --------- ------Y--- N--------- ---T
Gesc-AF163627   ---------- ---------- --------- ------Y--- N--------- ---T
Gesc-AF163626   ---------- ---------- --------- --------- --------- ---T
Gefo-Z74457     ---------- ---------- --------- --------- N-EWM----- ----
Gefo-Z74463     ---------- ---------- --------Y ---G--Y--- ---K------ ----
Gesc-AF163643   --F-Q----- ---------- --------- ------Y--- N--------- ---T
Gesc-AF163629   ----Q----- ---------- --------- ------Y--- N--------- ---T
Gesc-AF163630   ----Q----- ---------- --------- ------Y--- N--------- L--T
Gesc-AF163628   ----Q----- ---------- --------- --------- --------- L--T
Gesc-AF163631   ----Q----- ---------- --------- --------- --EWM-DA-- ---T
Gema-Z74429     ----Q----- ---------- --------Y ------Y--- ------Y--- ----
Capa-Z74437     ----Q----- ---------- --------Y ------Y--- ------Y--- ----
Gesc-Z74413     ----Q----- ---------- --------- .--------D --------A E---
Geco-AF164158   .......... ......---- ---F-----Y -QKL----SV -AVFM-D--N ----
Plcr-AF164159   --F------- ---------- --------S -------L-- ------D--A E--R
Ceol-AF164160   --F------- ---------- --------Y ------Y--- ------Y--- ---R
Gefo-AF164161   --L-----H- ---------- --------- --RV--H--- ---K------ S---
Ceol-AF164162   ----H-E--- -LMHVM---- --------Y --RV--H--- ------D--A Q--T
Ceol-AF164163   --F-H-E--- ---I------ --------Y --RV--H--- N-ELM----- ----
```

Fig. 3. Amino acid sequences translated from the nucleotide sequences in Figure 2. The amino acid residues are given in the international single-letter code. For other explanations see Figure 2.

namely group 1 and group 2. (In the previous study, no group 3 or 4 sequences were found in *G. scandens*; see Vincek et al. 1997.) The number of different sequences per individual ranged from four to ten. Assuming complete heterozygosity, a minimum of five class II *B* loci must therefore be postulated for this species of Darwin's Finches. Since the phylogenetic analysis assigns several of the sequences from one individual to a single group, it must be concluded that at least some of the previously defined four groups comprise more than one locus.

In spite of a considerable effort to sample exhaustively all the class II *B* genes of each individual, the segregation pattern among the offspring clearly shows that this goal was not achieved. Not only could some of the sequences found in the parents not be ascertained among the progeny, but the reverse was also true. For example, in family 1 only six of the 17 genes obtained from the two parents were found among the four offspring, while nine sequences present in the offspring were not detected in the parents. We attribute this lack of full "penetrance" to the stochasticity of the PCR when dealing with multiple closely related genes, and to the well known phenomenon of preferential amplification of some sequences.

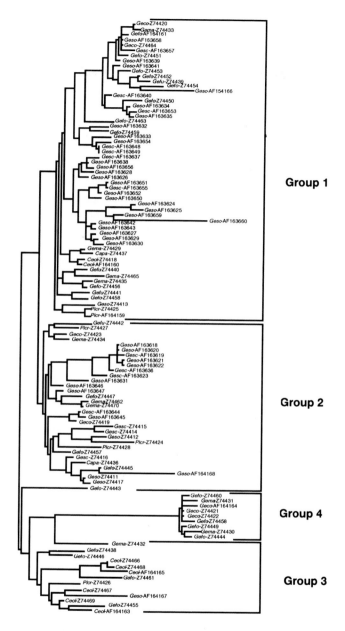

Fig. 4. Phylogenetic tree of the Darwin's Finch exon 2 class II *B* sequences from Figure 2. The tree was obtained by using the neighbor-joining algorithm of Saitou and Nei (1987), the MEGA program (Kumar et al. 1993), and the Kimura two-parameter (1980) distances. Species abbreviations are explained in the **Introduction**. The sequences are listed by their accession codes.

Not enough material was available to carry out other tests that could have resolved the ambiguities in the segregation pattern. However, virtually all the new sequences displayed in Figure 2 were obtained more than once and most duplicates were obtained from independent PCRs so that the correctness of the sequences is not to be doubted.

Intron sequences

Exon 2 codes for the part of the class II B genes that specifies the PBR of the β chain. This part is subject to balancing selection (Hughes and Yeager 1998) which influences the sequence identity of the individual genes and may obscure the genetic relationships among the genes. Convergent evolution, for example, may cause exon 2 sequences to appear more related to one another than is actually the case (see the chapter by K. Kriener et al., this volume). In this respect, intron sequences are generally better indicators of *Mhc* gene identities than the PBR-encoding exon sequences. Hence, in an attempt to differentiate loci from alleles, parts that included intron 2 of the Darwin's Finch class II B genes were amplified. Three such attempts were made. In the first, the primer pair HOPE1-FA was used to amplify the bulk of exon 2 and the entire intron 2 from *Certhidia olivacea*, a product ~ 2.2 kb in length (Fig. 1). In the second, primers specific for the previously identified exon 2 groups of sequences (F3 for groups 1 and 3-1, F2 for group 2, F32 for group 3-2, and F4 for group 4) were used in combination with the exon 3 primer FA to amplify the 3' part of exon 2 and the entire intron 2 from *Geospiza scandens* (group 1, 2, and 3-1), *Certhidia olivacea* (group 3-2), and *Geospiza conirostris* (group 4). In the third attempt, the primer pair HOPE3-H4 was used to amplify the bulk of exon 2 and ~ 190 bp of intron 2 (the H4 primer was based on a relatively conserved segment of intron 2 as determined from the sequences in the other two approaches). The complete and partial intron 2 sequences are displayed in Figures 5 and 6, respectively (without the exon 2 sequences, which — where informative — are included in Figure 2).

The entire intron 2 sequence consists of a highly degenerate repeat or repeats, in which the most frequently reiterated motif is the hexamer TCCCAG (Fig. 5). However, longer, less frequently repeated sequence stretches are also recognizable. The intron 2 sequences of the four groups of genes can be aligned, but only by postulating numerous indels. These range in length from single nucleotide deletions or insertions to stretches involving > 650 bp. The two most conserved parts of the intron are the first ~ 70 bp at its 5' end and ~ 70 bp at its 3' end. The intron sequences do not appear to be group-specific. Both the indels and the single nucleotide substitutions are scattered among the groups. Similarly, no clear intra-group subdivision of sequences has as yet emerged from the limited number of sequences available (Fig. 6).

Fig. 5.

```
                          1              11          21          31          41          51          61          71          81          91
CONSENSUS ==>             GTGAGCGCGG GGCAGAGCGT GTGCCCTCGG GCCCTGCCCT GCCAGTGACC ATCCCCAGTCA ATCCCCAGTCC ATCCCCAGTCC ATCCCCAGTCC ATTCCCTGAC
Group3-2  Ceol-AF164165   ---------- --*------- --*------- ----*----- ---------- ---------- ---------- ---------- *-----*-**  **********
Group1    Gesc-AF164166   ---------- --*------- --*------- ---------- ---------- ---------- ---------- ---------- *--*------  **********
Group2    Gesc-AF164168   ---------- ----*----- ---*------ ---------- ---------- ---------- -- C------ -- C----*** **********  **********
Group3-1  Gesc-AF164167   ---------- ---------- ---C------ --C-T----- ---------- ---------- ---------- -----A----  **********  *******T
Group4    Geco-AF164164   ---------- ---------- --C-T----- --A------- ---A------ C-GGAGC-TC TCAAA-C-TC C-TGGGAATG CC-AGAG--  C-CAG-CCT-

                          101            111         121         131         141         151         161         171         181         191
CONSENSUS ==>             CCCTCCCAGT CCATTCCCAG TCCCTTCCAG TCCACTCCAG TCCATTCCAA ATCCATCCCA GTCTCTCCCA CTCCCTCCCA GTCCCTCTCA GTCCCATCCCA
Group3-2  Ceol-AF164165   ---------- ---------- ---------- ---------- ---------- ---------- ---------- ----T----- --G-------  ----------
Group1    Gesc-AF164166   ---------- ---------- ---------- ---------- ---------- ---------- ---------- --T------- ---G------  ----------
Group2    Gesc-AF164168   ---T------ ----G----- ---C------ ---------- ---------- ---------- ---------- ---------- ----------  --C-------
Group3-1  Gesc-AF164167   ---------- ---------- ---------- ---------- ---------- ---------- ---------- ---------- ----------  ----------
Group4    Geco-AF164164   GTTGTG-TCA T-CACGGAGC CT-G---CT C---G-DATC C---CCAGTC C---CAGT-T *C-C-CAGT* **********  **********  **********

                          201            211         221         231         241         251         261         271         281         291
CONSENSUS ==>             AACCCTCCAG TCCATGCCTA GGGCCTTCCA GTCCATTCCC AGTTCGTCCC AAACCATTCC CAATCCAAAC CAAGTCCCTC CCAGTCCCTC CCAGTCCATT
Group3-2  Ceol-AF164165   -G-------- ---------- ---------- ---------- ---T------ ---------- ---------- ---------- --G-------  ----------
Group1    Gesc-AF164166   -G-------- ---------- ---------- ---------- ---T------ ---------- ---------- ---------- --*------*  ----------
Group2    Gesc-AF164168   ---------- ---------- ---------- ---------- *----*---- ---------- ---------- ---------- ----------  ----------
Group3-1  Gesc-AF164167   ---------- ---------- ---------- ---------- ---------- ---------- ---------- ---------- ----*-----  **********
Group4    Geco-AF164164   ********** ********** ********** ********** ********** ********** ********** ********** **********  **********

                          301            311         321         331         341         351         361         371         381         391
CONSENSUS ==>             CCCAGTCATC ATCCCAGTCC ATTCCCAGTC CATTCCCAGT CATTCCCAGT TCACTCCCCA ACTCCCACCC TCTCTCCCAG TACCGAGTGC CCCTCCCATC
Group3-2  Ceol-AF164165   -T----**-- -T-------- ---------- ---------- *--------- ---------- ---------- ---------- --C-------  ----------
Group1    Gesc-AF164166   -*-------T --*------- ********** ********** *----*---- ---------- ---------- ---------- ---T------  ----------
Group2    Gesc-AF164168   ------**-- --**------ -********** ********** **-------- ---------- ---------- ---------- ----------  ----------
Group3-1  Gesc-AF164167   ********** ********** ********** ********** ********** ---------- -T-------- ---------- ----------  **********
Group4    Geco-AF164164   ********** ********** ********** ********** ********** ********** **********  -T-------- **********  **********

                          401            411         421         431         441         451         461         471         481         491
CONSENSUS ==>             TTTCTCAGTG CCACTCCAGT CCTTCCCAGT TCCTCCCAGT CCCACCCAA CCCCCTCCAA CTCCATTCCC AGTCCATTCC CAGTTCATTC CCAGTCCCTC
Group3-2  Ceol-AF164165   ---------- ---------- ---------- ---------- ---T------ ---------- ---------- ---------- ----------  ----------
Group1    Gesc-AF164166   ---------- ---------- ---------- ---------- ---------- ---------- ---------- ----*----- ----------  --*-------
Group2    Gesc-AF164168   ---------- ---------- ---------- --T------- ---*------ ---------- ---------- ---------- ----------  --**------
Group3-1  Gesc-AF164167   ---------- ---------- ---------- --A------- -*-*------ ---------- ---------- ---------- ----------  --*****---
Group4    Geco-AF164164   ********** ********** ********** ********** ********** ********** ********** ********** **********  **********
```

Fig. 5. (continued)

```
              501        511        521        531        541        551        561        571        581        591
CONSENSUS ==> CCACTCCATC CCAGTCCATT CCCAGTTCAC TCTCAAACCT ACACCCACTC TCCCAGTACC CGATCCCTCC CACCTCCCTC AGTGCCACCC CAGTTCCGCC
Group3-2 Ceol-AF164165 ********** ********** ********** ********** ********** ********** ********** ********** ********** ********--
Group1    Gesc-AF164166 ********** ********** ********** ********** ********** ********** ********** ********** ********** *******---
Group2    Gesc-AF164168 ********** ********** ********** ********** ********** ********** ********** ********** ********** *******---
Group3-1  Gesc-AF164167 ---------- ---------- ---------- ---------- ---------- ---------- ---------- ---------- ---------- --*-*T----
Group4    Geco-AF164164 ********** ********** ********** ********** ********** ********** ********** ********** ********** **********

              601        611        621        631        641        651        661        671        681        691
CONSENSUS ==> TTGTGCCATC CCAGTCCATT CCCACTCACA CCCAGCCCGT TCCCAGTCCA TTCCCCATCC CTCCCAGTTC ATTCCCAGTT CCTCCCAGTC CCTCCCAAGTG
Group3-2 Ceol-AF164165 ---------- ---------- ---------- ---------- ---------- ---------- ---------- ---------- ---------- ----------
Group1    Gesc-AF164166 ---------- TT-------- ---------- TT------A- ---------- ---------- --CT----- ---------- --A-****** 
Group2    Gesc-AF164168 ****T-AC-- T---AT-TCC A--CA*-T-T ---TA-C* ********** ****----- --CT-C**** -C-T-T---G --A-****** 
Group3-1  Gesc-AF164167 CCAGT---T -CAGT-GC- ----G--G-T ---T-*-C ---T---T ----AG-- A------- ----T--- ---------- **********
Group4    Geco-AF164164 ********** ********** ********** ********** ********** ********** ********** ********** ********** **********

              701        711        721        731        741        751        761        771        781        791
CONSENSUS ==> TATTCCCAGG CCAATCACAGT CCAATCCCAGT CCTCCCAGT CCATTCCCAG CCAATCCCAG ACAATTCTCA GTCCATTCCC AGCAACTCCC AGTCCATTCCC CAGTCCATTT
Group3-2 Ceol-AF164165 ---------- ------*--- ---------- ---------- ---------- ---------- -T-------- ---------- ---------- ----------
Group1    Gesc-AF164166 ---------- -------*-- ---------- ---------- ---------- ---------- ---------- ---------- --C------- ----------
Group2    Gesc-AF164168 ****---- ---T----C -C--C--- ---------- ---------- ---------- ---------- ---------- ---------- ----------
Group3-1  Gesc-AF164167 ----- -T---*-- ---------- ---------- ---------- ---T----- ---------- ---------- ---------- ----------
Group4    Geco-AF164164 ********** ********** ********** ********** ********** ********** ********** ********** ********** **********

              801        811        821        831        841        851        861        871        881        891
CONSENSUS ==> CCATCCCAAT CCATTCCCAG TCCATTCCCAG TTCCTCCCAG AGCCCCCCCG CGCGTCCATT TCAGCGACGC GTTGAAGCTG ACGCTTCTCA GCCAATCAGA
Group3-2 Ceol-AF164165 -------G-- ---*------ ---------- ---------- ---------- ---------- ---------- --*--?*-- ---------- ----------
Group1    Gesc-AF164166 ---------- ----*----- ---------- ---------- ---------- ---------- ---------- --G--?*-- ---------- ----------
Group2    Gesc-AF164168 ---------- --*------- ---------- ---------- ---------- ---------- ---------- --*------- ---------- ----------
Group3-1  Gesc-AF164167 ---------- ---*------ ---------- ---------- C--------- ---------- ---------- --A------- --GA-A*-- C--C--G---
Group4    Geco-AF164164 ********** ********** ********** ********** ***-?---? ---------- ---------- ---------- ---------- ----------

              901        911        921        931        941        951        961        971        981        991
CONSENSUS ==> GCGCGTCTCT CTGATGACTC ATCAGTTGCC AGGCAGACCC GCAGCGGCCGA GTGCCCCGCG CTGGACTCGG CCTCATTCCC AGTTCCTCCC AGGCCACCA
Group3-2 Ceol-AF164165 ---------- ---------- ---------- ---------- ---------- ---------- ---------- ---------- ---C------ ----------
Group1    Gesc-AF164166 ---------- ---------- ---------- ---------- ---------- ---------- ---------- ---------- ---------- ----------
Group2    Gesc-AF164168 ---------- ---------- ---------- ---------- ---------- ---------- ---------- ---------- ---------- ----------
Group3-1  Gesc-AF164167 ---------- ---------- ---------- ---------- ---------- ---------- ---------- ---------- ----***-- -G--G-G-
Group4    Geco-AF164164 T--------- ---T------ ---------- ---------- ---------- ---------- ---------- ---------- ---------- GCTT-ATT-C
```

Fig. 5. (continued)

```
                              1001       1011       1021       1031       1041       1051       1061       1071       1081       1091
CONSENSUS ==>                 AAATCCCTCC CAGCGCCACC AAAATCCCTC CCAGCGCCAC CAAAATCCCA CCCAGTGCCA CCAAAATCCC TCCCAGCGC ACCAAAAATCC CTCCCAGCGC
Group3-2  Ceol-AF164165       ---------- ---------- ---------- ---------- ---------- ---------- ---------- --------- ----------- ----------
Group1    Gesc-AF164166       -----?---- ------C--- ------AT-- -*-*-T---- ------T-*- T-*------- ---------- --------- ----------- -----***
Group2    Gesc-AF164168       ---------- ------C--- ---------- -*-*-T---- ------C--C ---------- ---------- --------- ----------- ----------
Group3-1  Gesc-AF164167       ---------- ---------- ----AT---- -*-*-AT--- -------T-- ---------- C--------- ----?---- ----------- ----------
Group4    Geco-AF164164       C-G-T----- --T------- --G------- ---------- ---------- ---------- ---------- --------- ---------AT-* -----*****
                              ********** ********** ********** ********** ********** ********** ********** ********* *********** **********

                              1101       1111       1121       1131       1141       1151       1161       1171       1181       1191
CONSENSUS ==>                 CCCCAAAACA ACTCCCAGCG CCACCAAAAT CCCTCCCAGC GCCACCAAAA TCACTCCCAG CGGCACCAAA ATCCCTCCCA GTGCCACCAA AATCCCTCCC
Group3-2  Ceol-AF164165       --------TC C--------- ---------- ---------- ---------- CA-------- ---------- ----AT-*-- ---------- ----------
Group1    Gesc-AF164166       ********** ********** ********** ********** ********** ********** ********** ********** ********** **********
Group2    Gesc-AF164168       ---------- ---------- --AT-*--T- ------C--- ---*-CAT-- T--------- ---------- ---------- ---------- ----------
Group3-1  Gesc-AF164167       --------T- ---T------ ---------- -----C---- ----C----- --C------- -CAA------ -C-------- ----AT---- ----------
Group4    Geco-AF164164       ********** ********** ********** ********** ********** ********** ********** ********* ********** **********

                              1201       1211       1221       1231       1241       1251       1261       1271       1281       1291
CONSENSUS ==>                 AGGCGCCACCA AAATCCCTCC CAGTGCCACC AAAATCCCTC CCAGTGCCAC CAAAATCCCT CCCAGCGGCCA TCCCAGTCCC TCCAAGCCCA TCTCCAATTC
Group3-2  Ceol-AF164165       ***T-CAT --------- ---------- ------C--- ---*------ ---------- ***** ---------- ----------- ----------
Group1    Gesc-AF164166       ********** ********** ********** ********** ********** ********** ********** ********** ********** **********
Group2    Gesc-AF164168       ---------- --------- --C------- ------AT-- ---*------ ---T------ ---CAA---- ----------- -----A---- ----------
Group3-1  Gesc-AF164167       --T------- --------- ---------- ---------- ---------- ---------- ---------- ----------- ---------- ----------
Group4    Geco-AF164164       ********** ********* ********** ********** ********** ********** ********** ********** ********** **********

                              1301       1311       1321       1331       1341       1351       1361       1371       1381       1391
CONSENSUS ==>                 CTCCCAGTC CCTCCCAGTC CATTTCCAGT CTTTTCCCAG TTCATTCCCA GTTCACTCTC AGATCTCCAC CCTCTCTCCC AGGCGCCCCC ATTCCCTCCC
Group3-2  Ceol-AF164165       --------- ---------- ---------- ---------- ---------- ---------- ---------- -------C-- ---------- ----------
Group1    Gesc-AF164166       --------- ---------- --G------- ---G------ ---------- ---------- ---------- ----A----- ----T---G- ----------
Group2    Gesc-AF164168       --------- ---------- --G------- ---------- ---------- ---------- --C------- ---------- ---T------ ----------
Group3-1  Gesc-AF164167       --------- ---A------ ---------- ---------- ---------- ---------- ---------- ---------- ---------- ----------
Group4    Geco-AF164164       --T-A---- ---A------ ---------- ---TT----- ---T------ ---------- ---------- G--C------ --T-T-T--- ----------

                              1401       1411       1421       1431       1441       1451       1461       1471       1481       1491
CONSENSUS ==>                 ATCTCTCTCA GTGCCAACCC AATGAACCCC AGTCCCTCCC AGTCCCATCCC AGTTCATTCC AGTTCATTCC CAGTCCCTCC CAGTCCACAC CCAGCCCCTC
Group3-2  Ceol-AF164165       ---------- ---------- ---------- ---------- ---------- ---------- ---------- CAGTCCATCC ---*---*-- ---T-A----
Group1    Gesc-AF164166       ---------- ---------- ---------- ---------- ---------- ---------- ---------- ---------- ------*--- -------T--
Group2    Gesc-AF164168       ---------- ---------- ---A------ ---------- ---A------ ---------- ---------- ------C--- ---T--TT-- -------T--
Group3-1  Gesc-AF164167       ---------- ---------- ---------- ---A------ ---C-*---- ---G---C-- ---------- ********** ********** **********
Group4    Geco-AF164164       ---------- ------G--- C-A-*---- ---------- ---G---C-- ---C------ ---------- --T--A -A-T--T*-- -T---T----
```

Fig. 5. (continued)

```
                         1501       1511       1521       1531       1541       1551       1561       1571       1581       1591
CONSENSUS ==>      CAAGTCAATT CCAAGCCTAC CCAGACCCTC CAAGTCCAAT TCCCAGTTCA TTCCCAGTTC AGTCCCAGAC CATCCACCCT CTCTCCCAGT GCTCCCCATG
Group3-2 Ceol-AF164165  -C---TC--- --C--T--** ********** ********** ---------- ---------- ---------- --------A- ---------- -------C
Group1   Gesc-AF164166  ---A----- --C--T--** ********** ---------- ---------- ---------- -----A--GT --*------- -------*- ----------
Group2   Gesc-AF164168  -C---C--C --GT--C-- ---------- ---------- ----CT--- ---------- --C-?-T-G- --C-?-T-G- --A----? --C----C
Group3-1 Gesc-AF164167  ********** ---------* ********** ********** ********** ********** ********** ********** ********** **********
Group4   Geco-AF164164  ********** --------** --------*C- --------*C- --C------- -C------** ********** ********** ********** **********

                         1601       1611       1621       1631       1641       1651       1661       1671       1681       1691
CONSENSUS ==>      CCCTCCCACC TCTCTCAGTG CCACCCAAGC TCCCTCCCAA GTCTACACCC AGTGCCTCCC AGTACA TCCC AGACCATTCC CAGTCCCTCT CCATCTCATG
Group3-2 Ceol-AF164165  ---------- ---------- ---------- ---------- ---------- --------*C ---------- ----T---- ---------- ----------
Group1   Gesc-AF164166  ---------- ---------- ---------- ----C----- ---------- ---T----- ---------- ---------- ---------- ----------
Group2   Gesc-AF164168  T-------AA ---------** --------*- ---------- ---------- ---------- ---------- ---------- ---------- ----------
Group3-1 Gesc-AF164167  ********** ********** ********** ********** ********** ********** ********** ********** ********** **********
Group4   Geco-AF164164  ********** ********** ********** ********** ********** ********** ********** ********** ********** **********

                         1701       1711       1721       1731       1741       1751       1761       1771       1781       1791
CONSENSUS ==>      CGAGTCCCTC CCAGTCCATC CCCAGTCCCT CCCAGTCCCT CCCAGTCCCT TCCTAGTCCA TTCCCAGTCC ATCCCAGTCC CTCCCAGTCC ATTCCCAGTC
Group3-2 Ceol-AF164165  T--------- ---------* ---------- ---------- ---------- ---------- ---------- -C-------- ---------- ----------
Group1   Gesc-AF164166  -?-------- ---------C- ---------- ---------- ---------- ---------- ---------- ---------- ---------- ----------
Group2   Gesc-AF164168  ********** ********** ********** ********** ********** ********** ********** ********** ********** **********
Group3-1 Gesc-AF164167  ********** ********** ********** ********** ********** ********** ********** ********** ********** **********
Group4   Geco-AF164164  ********** ********** ********** ********** ********** ********** ********** ********** ********** **********

                         1801       1811       1821       1831       1841       1851       1861       1871       1881       1891
CONSENSUS ==>      CCACACAGTC AATCCCAGTC CCTCCCAATC CATTCCCAGT CCACTCCCAG TCCATCCCAA GCCCCTCCCA GTCCATCCCA AGTCCCTCCC AGTCCATTCC
Group3-2 Ceol-AF164165  --T-C----- CC------- -C*------- --C*------ ---T----- --------*- G-T-*----- ----------- --------* --------*-
Group1   Gesc-AF164166  --T-T---- ---------- ---------- ----T--- --T-T---- ---------- T-AT----- ---------- ------*** ----------
Group2   Gesc-AF164168  ********** ********** ********** ***----**TA GGG-CA--CA GT-CAT--C- T-------- ---------- ---------- --*------
Group3-1 Gesc-AF164167  ********** ********** ********** ***CT--CAG T-C------ -TC-A-T-C A----T--C A---T--C -C--C--- --C-C-*--
Group4   Geco-AF164164  ********** ********** ********** ******GAC -TC-A-T-TC -TTCC-ATTG C-----C-- A--CCT--T- CACTT--TG --G-CACT-

                         1901       1911       1921       1931       1941       1951       1961       1971       1981       1991
CONSENSUS ==>      CAGTCCCTC CCAGTCCCTC CCAGTCCCTC CCAGCCCCAG CCCAGCCGTC CCCTCTCTCT CCCTCTCTCT CCCAGTCGCC CCCAGCTGAT CCCGCTCTC
Group3-2 Ceol-AF164165  -------*- ---------- ---------- ---------- ---------- ---------- ---------- ---------- ---------- ----------
Group1   Gesc-AF164166  -------*- ------A-- ---------- ---------- ------A-- ---------- ---------- ---------- ---------- ----------
Group2   Gesc-AF164168  ------A-- ---------- ---------- ---------- ---------- ---------- ---------- ---------- ---------- ----------
Group3-1 Gesc-AF164167  ------A-- ---------- ---------- ---------- ---------- ---------- ---------- ---------- ---------- ----------
Group4   Geco-AF164164  ACC-----C- -AGTC-AT-- ---------- G-T-----A-C T-C------ ---------- ---------- ---------- ---------- ----------
```

```
                        2001        2011
  CONSENSUS ==>         TGTCTCTCTC  TCCCCCAG
Group3-2  Ceol-AF164165 ----------  --------
Group1    Gesc-AF164166 -C--------  --------
Group2    Gesc-AF164168 ----------  --------
Group3-1  Gesc-AF164167 **--------  --------
Group4    Geco-AF164164 -****-----  --------
```

Fig. 5. Full intron 2 sequence alignments of Dawin's Finch class II *B* genes representing five phylogenetically defined clades (groups I, II, III-1, III-2, and IV; see Vincek et al. 1997). In view of the repetitive nature of the sequences, no attempts were made to resolve the ambiguities. Species abbreviations are explained in the **Introduction**; the sequences are listed under their accession codes. The sites are numbered starting from the 5' end of the introns. The simple majority consensus sequence is given at the top; identity with the consensus is indicated by a dash (-), and indels by asterisks (*). The presumed core repeat is highlighted.

Discussion

Number of loci

Until a few years ago, the only avian *Mhc* studied by molecular methods was the *B* complex of the domestic fowl, the "chicken" (Plachy et al. 1992). Most recently, the *B* complex has become the target of a sequencing effort which has revealed its organization and has confirmed some of its unusual features (Kaufman et al. 1999) The fowl *Mhc* is located on one of the minichromosomes and consists of two parts, separated by the nuclear organizer region (NOR). As the region is the site of frequent recombinations (Miller et al. 1994, 1996), the two parts, referred to as the *B* and the *Rfp-Y* complexes (Briles et al. 1993), segregate independently of each other in genetic crosses as if they were physically unconnected. The *B* complex, as currently defined, occupies a chromosomal segment of < 40 kb and contains only two class I *A* and two class II *B* loci, of which one in each pair is more strongly expressed than the other (Kaufman et al. 1999). A single class II *A* locus is at a distance of ~ 5 cM from the *B* complex. Similarly, the *Rfp-Y* complex contains two class I *A* and two class II *B* loci which, however, because of their low polymorphism, low sequence divergence of alleles, and low expression levels are believed to be of the nonclassical type (Zoorob et al. 1993). The relative compactness and simplicity of the fowl *Mhc* contrasts with the baroqueness of the organization encountered in most other *Mhc*s studied thus far, in particular the human *HLA* and the mouse *H2* complexes, each of which occupies a chromosomal segment of > 4 Mb and contains dozens of class I and class II, classical and nonclassical loci (Klein 1986; Trowsdale 1995). The question therefore arises: Is the fowl *Mhc* typical for birds or is it an isolated case of an accidental drastic reduction in the number of class I and class II loci and of compacting them into a short region?

The data described in the present study indicate that *Geospiza scandens* of the Darwin's Finches probably possesses more than half a dozen of the class II *B* loci

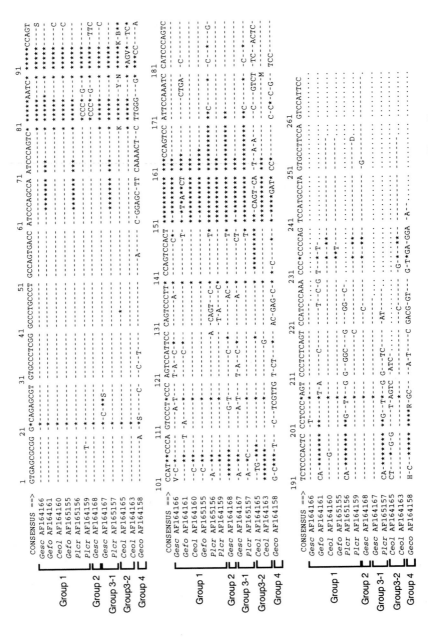

Fig. 6. Partial intron 2 sequences of Darwin's Finch class II *B* genes. For explanations see Figure 5.

alone. This conclusion derives from two observations. First, at least ten genes could be demonstrated in some of the individuals, necessitating the postulate of at least five loci. But in reality, the number of loci must be higher, because genes were found in the progeny that were not detected (for reasons mentioned earlier), but must have been present in the parents. Second, the genes detected are all members of only two of the four groups of loci identified in an earlier study (Vincek et al. 1997) in an assortment of other Darwin's Finch species. The number of class II *B* loci in the Darwin's Finches is, therefore, probably comparable to that found in humans and mice. An indication of a large number of class II *B* loci in song birds has also been reported by Edwards and colleagues (1999). Likewise, in other vertebrate classes, such as the teleost fishes (e.g., Málaga-Trillo et al. 1998), the presence of a number of class II *B* loci in each *Mhc* appears to be the rule. The case of the domestic fowl might, therefore, be unusual, indeed, limited perhaps to this particular species or to a group of birds within the gallinacean family. According to the "birth and death" (Nei and Hughes 1992; Nei et al. 1997) and the "accordion" (Klein et al. 1993) models of *Mhc* evolution, the complex undergoes periodic contractions and expansions, changing the numbers of resident class I and class II loci. In most species, the *Mhc* is seen in the expanded or expanding phase, but in the "chicken" it may have been caught in the contracted phase.

Contraction and expansion is one of the mechanisms of concerted evolution leading to a higher resemblance of genes within, as compared to that between, groups of taxa. Indeed, in a phylogenetic analysis, the Darwin's Finch class II *B* genes cluster together, separate from other avian *Mhc* genes (not shown; but see also Edwards et al. 1999), as would be expected of genes that have arisen from a different last common ancestor than other bird taxa.

Intron evolution

The relatively recent origin of the various Darwin's Finch class II *B* genes from a common ancestor is indicated not only by the exon 2 sequences, but even more strongly by the intron 2 sequences and organization. A striking feature of the intron 2 sequences in Darwin's Finches is their repetitive nature. The core of the repeat, the TCCCAG hexamer, or a variant thereof, appears to be derived from a longer, not easily recognizable repeat (perhaps CCCTCCCAG). This repeat is found in the introns of genes representing all four groups of class II B loci, with approximately the same degree of degeneration. The observation that large parts of intron 2 from different groups of loci can be aligned reasonably well supports the relatively recent derivation of the genes from a common ancestor. Presumably the repeats were in place in the last common ancestor before the four groups began to diverge from one another. In fact, intron 2 of the class II *B* genes from the Red-winged Blackbird, *Agelaius phoeniceus* (family Fringillidae, tribe Icterini; Sibley and Monroe 1990), has a similar structure to intron 2 of the Darwin's Finch genes

(Edwards et al. 1995). At the 5' end of the intron, ~ 50 bp which are relatively free of repeats, show high sequence similarity between the Blackbird and the Darwin's Finch genes. This segment is then followed in both by the highly repetitive part which extends all the way to the short conserved segment flanking exon 3. The principal element of the repetitive part in the Blackbird is, just like in Darwin's Finches, the hexamer TCCCAG. Darwin's Finch genes have, however, the repetitive segment interrupted in the middle by a block of ~ 80-130 bp, which is relatively free of repeats. This block is present in intron 2 of the genes in all four groups, but it is absent in the Blackbird genes.

The origin of the TCCCAG motif is obscure. The motif is also present in exon 3 of the Darwin's Finch genes, ~ 40 bp downstream of the 5' splice site. Another possible source could have been the 3' splice site of exon 2 together with the 5' splice site of exon 3. A hypothetical scenario for the origin of the intron 2 repetitive segment is depicted diagrammatically in Figure 7. Although the exon 2 of Darwin's Finch genes now ends with GAG and exon 3 begins with TGC, in other *Mhc* genes, CAG and TCC nucleotides are present at these sites, respectively (e.g., Ono et al. 1993), as in the TCCCAG motif. Degenerate multiplication of this motif may have generated the different variants now seen in the repetitive part of the intron. The process need not have started in the way shown — from a state in which exons 2 and 3 were originally contiguous. Interestingly, however, the origin of a new intron within exon 3 of class II *B* genes documented by Figueroa et al. (1995) for perch-like fishes could be explained in a similar fashion. The new intron arose apparently by multiplication of a hexamer which is present in the spliced transcript at the site interrupted by the intron in the genomic DNA.

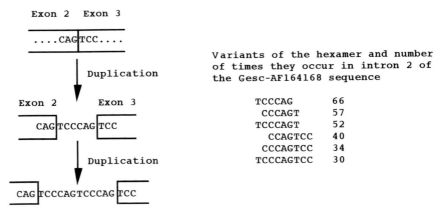

Fig. 7. Hypothetical origin of the repetitive segment in intron 2 of the Darwin's Finch class II *B* genes. The basic hexameric repeat is postulated to have arisen in an ancestor sequence from the last three nucleotide pairs of exon 2 and the first three nucleotide pairs of exon 3. Degenerate multiplication of the hexamer then produced the different variants of the repeat now found in the intron at high copy numbers.

The evolution of the intron subsequent to the initial split into the four groups apparently consisted of the degeneration of the individual repeats and the variation in the number of repeats through duplications and deletions. Both processes seem to have affected certain parts of the intron more than others, the former the ends of the original long repeat, the latter the middle part of the intron. If the postulate of the relative recency of the gene divergence can be upheld, it would be tempting to speculate that some of the expansions occurred concurrently with the adaptive radiation of the Darwin's Finch ancestors after their arrival on the Galápagos Archipelago. It may be possible to test this hypothesis by examining the homologous class II *B* genes of the nearest living relative of the Darwin's Finches from Central and South America.

Exon 2 evolution

Like class II *B* exon 2 sequences in many other species, those aligned in Figure 2 reveal the existence of characteristic motifs — short stretches of sequence differentiating individual genes. Motifs are apparent in both the nucleotide (Fig. 2) and amino acid (Fig. 3) alignments. Assuming that the three-dimensional structure of the avian class II molecules is similar to that of the human molecules (Brown et al. 1993), it can be deduced that the motifs in the Darwin's Finch class II β chains encompass amino acid residues involved in binding self and foreign peptides — the PBR. However, the alignments in Figures 2 and 3 also reveal something that similar alignments from other species usually do not. Perhaps because of the close evolutionary relationship among the Darwin's Finch genes, the alignments provide a glimpse into the evolutionary mechanisms involved in the generation of the motifs and divergence of the genes. The information gleaned from the analysis of the Darwin's Finch sequences is summarized diagrammatically in Figures 8 and 9. It suggests the involvement of two main processes in sequence divergence — point mutations and recombination, both presumably coupled to balancing selection. The dominant role of point mutations in the assembly of the motifs is indicated by the fact that the protein motifs can be arranged into networks in which most neighboring motifs can be derived from each other by a single nucleotide substitution at the DNA level (Fig. 8). Where two quite different motifs occur at identical positions, their distribution suggests that they may have had a longer time to diverge than the more closely related motifs. Considering that the sampling of the sequences has not been exhaustive and that some variants may have been lost by random genetic drift and by negative selection, the set suggests an evolutionary pathway in which the various intermediates are generated by point mutations exclusively. The involvement of balancing selection in this process is suggested by the observation that many of the observed substitutions at the DNA level are of the nonsynonymous type.

Examples of putative intra-exon 2 recombinants are given in Figure 9, together with their presumed parental types. The recombinants can be explained by postulating that *reciprocal* crossing over occurred between the exchanged parts. Where the recombinants apparently participated in reciprocal crossing over again, the resulting sequences appear as if they were generated by double crossing over or

Motif 31-34

Ⓐ QFLM —¹— QFIM —²— QLTM —²— QYVM —²— MHVM

Ⓑ PDVV —¹— PYVV —¹— PYAV

Motif 51-55

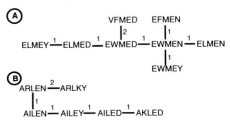

MNAKH —²— RVAKH —¹— RVAKY

MVAKR —¹— MNAKR   RLAKH —²— KLAKR —¹— KLAKL

MNAKY —¹— MGAKY   KQARH

Motif 61-65

Ⓐ
                    VFMED   EFMEN
ELMEY —¹— ELMED —¹— EWMED —¹— EWMEN —¹— ELMEN
                    EWMEY

Ⓑ ARLEN —²— ARLKY

AILEN —¹— AILEY —¹— AILED —¹— AKLED

Fig. 8. Proposed pathway for the evolution of amino acid motifs characterizing the sequences in Figure 3. The numbers on connecting lines indicate the number of replacements necessary to convert one motif in the other. "Motif number" refers to the position of the sequence in Figure 3. Amino acid residues are given in the international single-letter code.

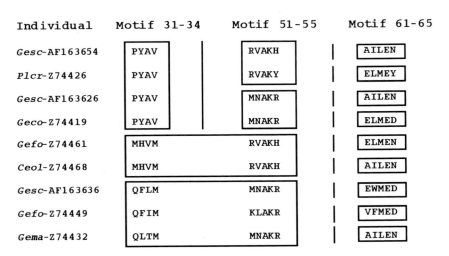

Individual	Motif 31-34	Motif 51-55	Motif 61-65
Gesc-AF163654	PYAV	RVAKH	AILEN
Plcr-Z74426	PYAV	RVAKY	ELMEY
Gesc-AF163626	PYAV	MNAKR	AILEN
Geco-Z74419	PYAV	MNAKR	ELMED
Gefo-Z74461	MHVM	RVAKH	ELMEN
Ceol-Z74468	MHVM	RVAKH	AILEN
Gesc-AF163636	QFLM	MNAKR	EWMED
Gefo-Z74449	QFIM	KLAKR	VFMED
Gema-Z74432	QLTM	MNAKR	AILEN

Fig. 9. Examples of putative recombinations found among the sequences in Figure 3. Only the amino acid motifs (boxed) characterizing the sequences are listed in the international single-letter code. Position of recombinations are indicated by vertical lines. Species abbreviations are explained in the **Introduction**.

gene conversion. In fact, once closely related sequences become sequentially involved in more than one recombination, it is impossible to differentiate between reciprocal and nonreciprocal exchanges. The shuffled motifs may then give the appearance that gene conversion is responsible for the shuffling, while in reality the arrangements can be equally well explained by classical, reciprocal recombination. The presence of obvious reciprocal recombinants in a sequence set such as that in Figure 2 makes the postulate of nonreciprocal recombination unnecessary.

Acknowledgments

We thank Ms. Sabine Rosner for technical and Ms. Jane Kraushaar for editorial assistance; the Galápagos Park Services and the Charles Darwin Research Station at Galápagos, and the Smithsonian Tropical Research Institute for administrative and logistical support; as well as the National Science Foundation for financial support to B.R.G. and P.R.G.

References

Bowman RI (1961) Morphological Differentiation and Adaptation in the Galápagos Finches. University of California Press, Berkeley

Briles WE, Goto RM, Auffray C, Miller MM (1993) A polymorphic system related to but genetically independent of the chicken major histocompatibility complex. Immunogenetics 37:408-414

Brown JH, Jardetzky TS, Gorga JC, Stern LJ, Urban RG, Strominger JL, Wiley DC (1993) Three-dimensional structure of the human class II histocompatibility antigen HLA-DR1. Nature 364:33-39

Edwards SV, Wakeland EK, Potts WK (1995) Contrasting histories of avian and mammalian *Mhc* genes revealed by class II B sequences from songbirds. Proc Natl Acad Sci USA 92:12200-12204

Edwards SV, Hess CM, Gasper J, Garrigan D (1999) Toward an evolutionary genomics of the avian *Mhc*. Immunol Rev 167:119-132

Figueroa F, Ono H, Tichy H, O'hUigin C, Klein J (1995) Evidence for insertion of a new intron into an *Mhc* gene of perch-like fish. Proc R Soc Lond B Biol Sci 259:325-330

Gilbert DG (1996) SeqPup version 0.6f: a biosequence editor and analysis application. http://iubio.bio.indiana.edu/soft/molbio

Grant PR (1986) Ecology and Evolution of Darwin's Finches. Princeton University Press, Princeton, N.J

Grant PR (1993) Hybridization of Darwin's finches on Isla Daphne Major, Galápagos. Philos Trans R Soc Lond B Biol Sci 340:127-139

Grant PR (1994) Population variation and hybridization: comparison of finches from two archipelagos. Evol Ecol 8:598-617

Grant PR, Grant BR (1994) Phenotype and genetic effects of hybridization in Darwin's finches. Evolution 48:297-316

Hughes AL, Yeager M (1998) Natural selection at major histocompability complex loci of vertebrates. Annu Rev Genet 32:415-435

Kaufman J, Jacob J, Shaw I, Walker B, Milne S, Beck S, Salomonsen J (1999) Gene organisation determines evolution of function in the chicken MHC. Immunol Rev 167:101-117

Kimura M (1980) A simple method for estimating evolutionary rates of base substitutions through comparative studies of nucleotide sequences. J Mol Evol 16:111-120

Klein J (1986) Natural History of the Major Histocompatibility Complex. John Wiley, New York

Klein J, Bontrop RE, Dawkins RL, Erlich HA, Gyllensten UB, Heise ER, Jones PP, Parham P, Wakeland EK, Watkins DI (1990) Nomenclature for the major histocompatibility complexes of different species: a proposal. Immunogenetics 31:217-219

Klein J, Ono H, Klein D, O'hUigin C (1993) The accordion model of *Mhc* evolution. In: Gergely J, Petranyi G (eds) Progress in Immunology. Springer-Verlag, Heidelberg, pp 137-143

Kumar S, Tamura K, Nei M (1993) MEGA: Molecular Evolutionary Genetic Analysis, version 1.01. The Pennsylvania State University, University Park, PA 16802

Lack D (1947) Darwin's Finches. Cambridge University Press, Cambridge

Málaga-Trillo E, Zaleska-Rutczynska Z, McAndrew B, Vincek V, Figueroa F, Sültmann H, Klein J (1998) Linkage relationships and haplotype polymorphism among cichlid *Mhc* class II *B* loci. Genetics 149:1527-1537

Miller MM, Goto RM, Bernot A, Zoorob R, Auffray C, Bumstead N, Briles WE (1994) Two *Mhc* class I and two *Mhc* class II genes map to the chicken *Rfp-Y* system outside the *B* complex. Proc Natl Acad Sci USA 91:4397-4401

Miller MM, Goto RM, Taylor RLJ, Zoorob R, Auffray C, Briles RW, Briles WE, Bloom SE (1996) Assignment of *Rfp-Y* to the chicken major histocompatibility complex/*NOR* microchromosome and evidence of high-frequency recombination associated with the nucleolar organizer region. Proc Natl Acad Sci USA 93:3958-3962

Nei M, Hughes A (1992) Balanced polymorphism and evolution by the birth-and-death process in the MHC loci. In: Tsuji K, Aizawa M, Sasazuki T (eds) HLA 1991, Proceedings of the Eleventh International Histocompatibility Workshop and Conference. Vol 2. Oxford University Press, Oxford, pp 27-38

Nei M, Gu X, Sitnikova T (1997) Evolution by the birth-and-death process in multigene families of the vertebrate immune system. Proc Natl Acad Sci USA 94:7799-7806

Ono H, O'hUigin C, Vincek V, Klein J (1993) Exon-intron organization of fish major histocompatibility complex class II *B* genes. Immunogenetics 38:223-234

Petren K, Grant BR, Grant PR (1999a) A phylogeny of Darwin's finches based on microsatellite DNA variation. Proc R Soc Lond B Biol Sci 266:321-329

Petren K, Grant BR, Grant PR (1999b) Extrapair paternity in the cactus finch, *Geospiza scandens*. Auk 116:252-256

Plachy J, Pink JRL, Hala K (1992) Biology of the chicken MHC (B complex). Crit Rev Immunol 12:47-79

Saitou N, Nei M (1987) The neighbor-joining method: A new method for reconstructing phylogenetic trees. Mol Biol Evol 4:406-425

Sato A, O'hUigin C, Figueroa F, Grant PR, Grant BR, Tichy H, Klein J (1999) Phylogeny of Darwin's finches as revealed by mtDNA sequences. Proc Natl Acad Sci USA 96:5101-5106

Sibley CG, Ahlquist JE (1990) Phylogeny and Classification of Birds: A Study in Molecular Evolution. Yale University Press, New Haven, CT

Sibley CG, Monroe BL (1990) Distribution and Taxonomy of Birds of the World. Yale University Press, New Haven, CT

Thompson JD, Higgins DG, Gibson TJ (1994) CLUSTAL W: improving the sensitivity of progressive multiple sequence alignment through sequence weighting, positions-specific gap penalties and weight matrix choice. Nucl Acids Res 22:4673-4680

Trowsdale J (1995) "Both man & bird & beast": comparative organization of MHC genes. Immunogenetics 41:1-17

Vincek V, O'hUigin C, Satta Y, Takahata N, Boag PT, Grant PR, Grant BR, Klein J (1997) How large was the founding population of Darwin's finches? Proc R Soc Lond 264:111-118

Yang SY, Patton JL (1981) Genetic variability and differentiation in Galápagos finches. Auk 98:230-242

Zoorob R, Bernot G, Renoir DM, Choukri F, Auffray C (1993) Chicken major histocompatibility complex class II genes: analysis of interallelic and inter-locus sequence variance. Eur J Immunol 23:1139-1145

Contrasting patterns of MHC and microsatellite diversity in social and solitary tuco-tucos (Rodentia: Ctenomyidae)

Tina M. Hambuch and Eileen A. Lacey

Museum of Vertebrate Zoology and Department of Integrative Biology, University of California, Berkeley, CA 94720, USA

Summary. MHC diversity in two species of tuco-tuco (Rodentia: Ctenomyidae) is compared and evaluated in light of behavioral and demographic differences between the study species. MHC diversity is also compared to diversity at nine microsatellite loci to explore interspecific patterns of genetic variation at functional and non-functional nuclear markers. While it is impossible to tease apart the relative effects of selection, history, current demography, and social structure, it is clear that these factors have differentially affected MHC and microsatellite loci in tuco-tucos. Microsatellite polymorphism in *Ctenomys sociabilis* is consistently less than that in *Ctenomys haigi* although microsatellite diversity in both species is consistent with Hardy-Weinberg expectations. In contrast, MHC polymorphism is greater in *C. sociabilis*. Heterozygosity levels in both species appear to be reduced relative to Hardy-Weinberg expectations although this deviation is significant only for *C. haigi*. These data suggest that the factors shaping genetic diversity in tuco-tucos differ markedly between microsatellite and MHC loci.

Key words. MHC, Tuco-tucos, *Ctenomys*, Microsatellites, Population genetics

Introduction

The major histocompatibility complex

The major histocompatibility complex (MHC) is present in and performs immune functions in all vertebrates. The complex is divided into three classes. Classes I and II encode for cell surface proteins that bind and present small peptides to T-cells, thus initiating the adaptive immune response. The general organization of this cluster of genes is well conserved among most vertebrates and, in particular, among most terrestrial taxa (Trowsdale 1995). Somewhat paradoxically, within species the degree of polymorphism and even the number of loci comprising the

MHC can vary markedly. In most of the vertebrates studied to date, this gene complex is the most polymorphic known (Figueroa et al. 1990; She et al. 1990). Mammalian populations typically harbor 6 to 12 alleles per locus, with greater than 200 alleles reported for some loci.

This diversity is one of the most intriguing aspects of MHC genes. Balancing selection, which maintains high levels of variation that would otherwise be lost due to drift, is thought to be responsible for the levels of polymorphism and heterozygosity commonly observed at MHC loci (Hedrick and Thomson 1983; Hughes et al. 1994). Heterozygote advantage and frequency-dependence are two forms of balancing selection that may underlie this variation and there has been considerable debate regarding which of these is primarily responsible for maintaining MHC diversity. Resolving this debate is difficult because although these forms of selection operate in different ways, expectations yield indistinguishable results and are not mutually exclusive. Interpreting data from natural populations is particularly challenging because the relative intensities of these forces may vary temporally, as well as among habitats or taxa.

Regardless of the specific form(s) of balancing selection involved, the agent driving selection on MHC loci is believed to be pathogens. This hypothesis has immense logical appeal because different MHC alleles are known to recognize and to bind foreign peptides. Several diseases have been identified in which particular alleles preferentially bind specific disease strains. Examples include at least one strain of malaria (Hill et al. 1994), Marek's disease in chickens (Briles et al. 1983), and Epstein Barr virus in humans (Campos-Lima et al. 1993). Klein and O'hUigin (1994) have suggested that the majority of MHC diversity is due to selection driven by long-term co-evolution between hosts and pathogens, such that one would expect to see a suite of specific allele-disease associations. Other factors thought to help maintain the diversity of at MHC loci are maternal-fetal interactions (Doherty and Zinkernagel 1975) and inbreeding avoidance (Penn and Potts 1999; Potts et al. 1990, 1994). In short, MHC diversity may be the result of multiple forces including pathogen exposure and assortative mating.

Few studies of MHC sequence variation have been conducted for natural populations and those data that are available yield variable impressions of MHC diversity. The majority of studies of natural populations have revealed the expected high levels of polymorphism and heterozygosity (e.g., Edwards 1995; Nadeau et al. 1981; Nevo and Beiles 1991). These data cover a broad range of taxa including fish (Graser et al. 1996), birds (Edwards 1995), and mammals (Nadeau et al. 1988; Nevo and Beiles 1991), suggesting the general conclusion that the extreme MHC diversity evident in humans and mice occurs in most vertebrates. In the few species for which low levels of MHC diversity have been detected, this absence of variability has typically been attributed to extreme demographic conditions. Most studies of natural populations that have reported reduced levels of MHC diversity have also documented low levels of diversity at other loci (Ellegren et al. 1993; Slade 1992), suggesting that demography and population history have constrained all forms of genetic variability in these animals. Thus understanding patterns of MHC diversity in natural populations requires an understanding of the overall genetic structure of the animals in question.

Microsatellite evolution

Microsatellites provide a valuable tool for characterizing the genetic structures of natural populations. Microsatellites are non-coding regions that consist of a 2-5 basepair sequence of DNA that is repeated in series a variable number of times. Variability in the number of repeat units that are strung together is generally attributed to slippage during DNA replication (Levinson and Gutman 1987). Mutation rates for microsatellites are variable and depend on several factors such as the length of the repeat unit (e.g., di- vs. tri-nucleotide repeat) and the total length of the repeated segment of DNA (Amos et al. 1996; Jin et al. 1996; Primmer et al. 1996). As a result, caution must be exercised when selecting microsatellite loci for interspecific comparisons of diversity. In particular, care should be taken to select loci that do not vary in repeat motif and that are not subject to ascertainment bias due to their method of isolation (Hutter et al. 1998; Pepin et al. 1995; van Treuren et al. 1997).

The typically high mutation rates of microsatellites make them valuable markers for studies of parentage, kinship, and population structure. Because microsatellites are non-coding, they are also generally considered to be selectively neutral. Although there are a few cases in which the neutrality of microsatellites has been questioned (Thornton 1997; Wierdl et al. 1997), these tend to be unusual cases in which either microsatellites are either "hitch-hiking" with nearby functional genes or specific microsatellite alleles have become so long that they interfere with the activity of neighboring functional loci. In general, however, microsatellites appear to conform to expectations for neutral loci. As a result, variation at these loci should be independent of selection but should reflect demographic parameters such as effective population size as well as neutral evolutionary processes such as genetic drift. Thus analyses of microsatellite loci provide a potentially valuable complement to studies of functional loci such as those found in the MHC.

Population biology, genetics and behavior of tuco-tucos

Studies of tuco-tucos (Rodentia: Ctenomyidae) provide an ideal opportunity to explore relationships between demography, behavior, and genetic variation. Like other subterranean rodents, tuco-tucos exhibit a number of ecological and demographic attributes that lend themselves to studies of population genetic structure. In particular, these animals tend to be characterized by low vagility, disjunct population distributions, and solitary lifestyles (Lacey et al., in press; Nevo 1979; Patton 1990), all of which are expected to affect genetic variation in predictable ways. Molecular studies of subterranean rodents, however, have revealed marked interspecific differences in genetic diversity that have generated considerable debate regarding the selective and other forces shaping genetic variation in these animals (e.g., Nevo et al. 1990; Patton and Yang 1977; Steinberg and Patton, in press). Efforts to interpret patterns of genetic variation have been hampered by the lack of detailed field studies of subterranean rodents (Smith and Patton 1999); comparative demographic and genetic analyses of these animals promise to yield important new insights into the processes underlying genetic diversification in small mammals.

The subterranean genus *Ctenomys* first appears in the fossil record approximately 1.8 million years ago, after which it radiated rapidly throughout sub-Amazonian South America (Lessa and Cook 1998; Ruedas et al. 1993). Although the 38 to 56 species of *Ctenomys* currently recognized (Reig et al. 1990; Woods 1991) are difficult to distinguish morphologically, marked interspecific differences in karyotype are evident, with diploid numbers ranging from 10 to 70 (Reig et al. 1990). Commonly known as tuco-tucos, these animals occur in habitats ranging from coastal deserts to montane meadows, with populations found at elevations ranging from sea level to over 4000 meters (Ruedas et al. 1993). This ecological variation suggests that tuco-tucos are subject to marked interspecific differences in demography and population structure that are likely to influence patterns of genetic diversity.

This study makes use of genetic samples and associated demographic data from two species of tuco-tuco: *Ctenomys sociabilis* and *Ctenomys haigi*, both occur in the Limay Valley, at the border of Neuquen and Rio Negro Provinces, Argentina. *C. haigi*, which occupies the eastern Limay Valley, is solitary; each adult inhabits its own burrow system (Lacey et al. 1998). *C. sociabilis* occurs in a similar habitat, yet is social; burrow systems are inhabited by multiple adult females, their dependent young, and in some cases, an adult male (Lacey et al. 1997). Burrow sharing arises due to natal philopatry by females; although male *C. sociabilis* do not remain in their natal burrow, they frequently disperse within the local population (e.g., from one burrow system to the next), suggesting that, within a population, animals of both sexes are closely related. In contrast, in *C. haigi*, neither sex is philopatric and adult males and females are rarely found within 0.5 kilometers of their natal burrow (Lacey, unpubl. data).

The close physical proximity of congeners with such different spatial and social arrangements provides an unusual opportunity to study relationships between social behavior, demography, and population genetics. Population genetics theory predicts that for neutral markers, *C. haigi* should be more evenly partitioned within and among populations than in *C. sociabilis*, in which variation should be evident primarily among populations. Tests of these predictions using microsatellite data are currently in progress and are providing one of the first detailed analyses of population genetic structure in tuco-tucos.

At the same time, microsatellite data provide an essential backdrop for studies of MHC variation in these species. The latter analyses are particularly intriguing given that increased exposure to pathogens has long been proposed as a fundamental cost of sociality (e.g., Brown and Brown 1986; Hoogland and Sherman 1976; May and Anderson 1979). Although the movement of tuco-tucos is restricted by the Limay River, other species occur on both sides of the river and individuals of some taxa cross the river on a daily basis. This suggests that the same pathogens are present on either side of the river. Because adult *C. sociabilis* share burrows, however, pathogen exposure is expected to be more intense in this species than in *C. haigi*. Given that variability at MHC loci is thought to be driven by pathogen exposure, the selective pressures operating on MHC loci in the study species should differ, yielding different patterns of MHC diversity.

Methods

Non-destructive tissue samples were collected from 35 adults per species during 1996-1998. All conspecifics were from the same local population; the populations sampled are the subject of ongoing behavioral and demographic studies. The individuals chosen for genetic analysis were resident throughout each population. In general, the same animals were used for both MHC and microsatellite analyses. For all samples, genomic DNA was isolated by proteinase K digestion followed by phenol-chloroform extraction and ethanol precipitation.

MHC analyses

Exon 2 of the MHC DQ Beta locus was chosen for analysis because it is known to code for the antigen binding region of an MHC molecule and is known to contain the hypervariable regions of MHC sequences that are thought to be most subject to selection. Tuco-tuco DNA was amplified using primers GH28 and GH29 (Scharf et al. 1988), which recognize the DQ Beta exon 2 locus in humans and other taxa. PCR products for tuco-tuco species were approximately 230 bp in length. PCR products were cleaned (Qiaquick PCR purification kit, Qiagen) and then screened using Denaturing Gradient Gel Electrophoresis (DGGE) (Lessa 1992). Although this procedure allowed us to distinguish heterozygotes from homozygotes, it did not allow us to characterize the underlying sequence differences between alleles. Consequently, cleaned PCR products were also cloned (TA cloning kits, Invitrogen, or Perfectly Blunt cloning kits, Novagen) and sequenced on an ABI 377 automated sequencer using both vector-specific primers and primers GH28 and GH29. Initially, 20 clones per individual were sequenced for three individuals per species. Because no more than two alleles per individual were detected and because the number of alleles revealed by cloning and by DGGE was always the same, the number of clones sequenced per individual was reduced to 2 for homozygotes and 5 for heterozygotes. Together, DGGE and sequencing of cloned

PCR products were used to characterize variation in a total of 30 *C. haigi* and 33 *C. sociabilis*.

Microsatellite analyses

Allelic variation was quantified at nine microsatellite loci isolated from *C. haigi*. The genotypes of 35 adults per species were characterized. The microsatellite loci used were isolated and amplified as described by Lacey et al. (in press); only loci exhibiting no interspecific differences in repeat motif were included in this analysis. Because the loci analyzed were not selected on the basis of allele size or polymorphism in *C. haigi*, interspecific comparisons of microsatellite data should not have been subject to ascertainment bias (Hutter et al. 1998). To quantify microsatellite variation, one primer per locus was radioactively end-labeled with ^{32}P prior to amplification (Richardson 1965). Amplification products were electrophoresed on a 6% polyacrylamide gel and visualized via autoradiography. Radioactively labeled m13amp18 sequence was used as a size standard on all runs. A detailed description of the amplification and screening procedures is given in Lacey et al. (in press).

Data analysis

All MHC DQ Beta exon-2-like alleles identified were subjected to GENBANK BLAST sequence similarity searches. In all cases, the BLAST search results confirmed that the sequences obtained from the study animals were most similar to the MHC DQ Beta exon 2 locus. Occasionally, an individual identified as homozygous using DGGE yielded two slightly different allele sequences. In all cases, sequencing of additional clones failed to reveal more than a single allelic variant and thus the initial sequence discrepancies for these animals were attributed to copy errors that arose during either the cloning or sequencing processes. In the event that some sequence variants were incorrectly identified as copy errors, the resulting data should underestimate the levels of allelic variation in each study population.

Results

MHC variation

The 230 bp of DNA amplified and sequenced for tuco-tucos corresponds to amino acids 11-85 of the typically 90 amino acid protein produced by the DQ Beta exon 2. Thus, the putative first and last hypervariable regions of this exon are missing from our amplification products. Analysis of our partial sequences revealed a total of 7 alleles in *C. sociabilis* and 5 alleles in *C. haigi*. Within each species, allelic variation consisted of 1-2 bp sequence differences that were restricted to a hypervariable motif resembling hypervariable region 2 of exon 2 of the DQ Beta locus.

No alleles were shared between species. Allele frequencies for each study population are reported in Table 1. In both populations, a single allele predominated, although the observed frequencies of alleles were more equitably distributed in *C. haigi*. Observed heterozygosity for *C. haigi* was not consistent with Hardy-Weinberg expectations due to an apparent deficiency of heterozygotes (Table 2). Although observed heterozygosity for *C. sociabilis* was not significantly different from Hardy-Weinberg expectations, a tendency toward heterozygote deficiency was also evident in this population (Table 2).

Table 1. Polymorphism and observed allele frequencies at the MHC DQ Beta exon-2-like locus in tuco-tucos

C. sociabilis		C. haigi	
Allele	Frequency	Allele	Frequency
Soc1	0.75	Hai1	0.03
Soc2	0.03	Hai2	0.60
Soc3	0.05	Hai3	0.13
Soc4	0.05	Hai4	0.22
Soc5	0.03	Hai5	0.02
Soc6	0.03		
Soc7	0.06		

Data are based on analyses of 30 *C. haigi* and 33 *C. sociabilis*.

Table 2. Observed and expected heterozygosities at the MHC DQ Beta exon-2-like locus in tuco-tucos

	Observed	Expected	χ^2 (P)
C. sociabilis	0.25	0.43	2.33 (0.13)
C. haigi	0.27	0.55	4.38 (0.04)

Data are based on analyses of 30 *C. haigi* and 33 *C. sociabilis*.
Expected values were generated according to Hardy-Weinberg.

Microsatellite variation

Striking interspecific differences in microsatellite variation were revealed by our analyses. Polymorphism in *C. sociabilis* was clearly reduced relative to that in *C.*

haigi at all loci surveyed (Table 3). For both species, observed heterozygosities at all variable loci were consistent with Hardy-Weinberg expectations (Table 4).

Table 3. Observed polymorphism at nine microsatellite loci in *C. sociabilis* and *C. haigi*

	Number of alleles	
Locus	*C. sociabilis*	*C. haigi*
Hai 2	1	8
Hai 3	1	9
Hai 4	3	13
Hai 5	1	6
Hai 7	4	9
Hai 8	1	10
Hai 9	1	2
Hai 11	2	10
Hai 12	1	7

Data are based on analyses of 35 adults per species.
Detailed descriptions of these loci are provided in Lacey et al. (in press).

Table 4. Observed and expected heterozygosities at nine microsatellite loci in *C. sociabilis* and *C. haigi*

	C. sociabilis			*C. haigi*		
Locus	Observed	Expected	χ^2	Observed	Expected	χ^2
Hai 2	0.00	0.00	---	0.84	0.85	0.54
Hai 3	0.00	0.00	---	0.87	0.90	1.40
Hai 4	0.62	0.61	0.00	0.87	0.84	0.60
Hai 5	0.00	0.00	---	0.46	0.48	0.01
Hai 7	0.67	0.68	0.43	0.73	0.73	0.00
Hai 8	0.00	0.00	---	0.66	0.63	0.10
Hai 9	0.00	0.00	---	0.50	0.53	0.05
Hai 11	0.16	0.17	0.00	0.94	0.88	1.16
Hai 12	0.00	0.00		0.84	0.77	1.44

Data are from analyses of 35 adults per species.
Chi-square values are given for all variable loci; in all cases, p > 0.05.
Detailed descriptions of these loci are provided in Lacey et al. (in press).
Expected values were generated according to Hardy-Weinberg.

Discussion

Our data indicate that levels of within-population microsatellite diversity differed markedly between species, with diversity being considerably reduced in *C. sociabilis*. In contrast, both polymorphism and heterozygosity at the DQ Beta-like locus were similar across species. Although microsatellite data were consistent with Hardy-Weinberg expectations, data from the DQ Beta exon-2-like locus revealed a significant departure from expected values in *C. haigi* and a similar but non-significant trend in *C. sociabilis*. Somewhat surprisingly, this departure consisted of a deficiency, rather than an excess, of heterozygotes. These results suggest that different forces are contributing to the maintenance of diversity at microsatellite loci and presumptive MHC loci.

One possible explanation for this difference in conformity to Hardy-Weinberg expectations is the inclusion of molecular artifacts in our data set. In particular, amplification of the DQ Beta-like locus may have been subject to primer bias (i.e., differential amplification of alleles from the same locus). The MHC primers used in this study were designed to amplify only approximately 85% of exon 2 of the DQ Beta locus in humans. Assuming that we amplified a homologous region, allelic variation in the remaining portion of the exon may have been missed. Further, because one of the primers used (GH28) appears to sit in a hypervariable region, it is possible that our amplifications detected only a subset of the DQ Beta-like alleles present in tuco-tucos. In this case, polymorphism and possibly heterozygosity at the DQ Beta exon-2-like locus may have been underestimated for our study populations. We are currently working to address this issue by developing primers that amplify the complete DQ Beta exon-2-like region present in *C. sociabilis* and *C. haigi*. By sequencing the missing portions of this locus, we will quantify exactly the diversity present at this locus.

Assuming that complete sequence data for the DQ Beta-like locus do not yield substantially different levels of polymorphism or heterozygosity, a second possible explanation for the departure from Hardy-Weinberg evident in our MHC data is the occurrence of demographic phenomena such as inbreeding or assortative mating. Both processes can lead to decreased heterozygosity within populations, thus potentially generating deviations from Hardy-Weinberg expectations. If, however, such demographic parameters underlie the heterozygote deficiency detected at the DQ Beta exon-2-like locus, then similar departures from Hardy-Weinberg should be evident in our microsatellite data. This was not the case (Table 4) and thus we suggest that inbreeding, assortative mating, and other demographic processes cannot explain the difference in conformity to Hardy-Weinberg expectations evident between microsatellite and presumptive MHC loci.

A third possible explanation for this difference is that selection is acting differentially on the microsatellite and MHC loci present in tuco-tucos. This interpretation follows logically from the assumed properties of the loci being compared. Indeed, microsatellite and MHC data were chosen for comparison because they are expected to provide different perspectives on the forces maintaining genetic variation within natural populations. Although more rigorous, quantitative analyses of the strength of selection on MHC versus microsatellite loci have yet to be conducted for tuco-tucos, we expect that differences in selection have contributed substantially to the patterns of genetic variability documented in this study.

If variability at the DQ Beta exon-2-like locus is the result of selection then our data suggest that the magnitude of selection may differ between the study species. Effective population size is believed to be smaller in *C. sociabilis*, suggesting that if selective pressures on the study species were equal, fewer DQ Beta-like alleles should be present in this species than in *C. haigi*. Allelic diversity in both species, however, was roughly equal (Table 1), implying that selection on the former must have been stronger or in some other way different from that experienced by *C. haigi*. This suggestion is intriguing given the behavioral differences between the study species (Lacey et al. 1997, 1998) and the frequently cited cost of increased pathogen exposure in group-living species (e.g., Alexander 1974; Hoogland and Sherman 1976).

Clearly, much remains to be learned regarding patterns and processes of genetic diversity in tuco-tucos. Data acquired to date suggest that further genetic studies of these animals are warranted and will yield important new insights into the factors shaping variation at functional and non-functional nuclear loci. Combined with detailed field data on population structure, future genetic analyses of *Ctenomys* should significantly improve our understanding of relationships between behavior, demography, and gene dynamics. Studies of MHC variation in tuco-tucos should prove particularly informative regarding the different selective and demographic forces shaping genetic variation in natural populations.

Conclusions and future work

This report characterizes diversity at an MHC DQ Beta exon-2-like locus amplified from two species tuco-tucos. Data on diversity at nine microsatellite loci are also presented for each species. This represents the first analysis of either microsatellite or MHC variability in tuco-tucos. Our data reveal striking interspecific differences in patterns of diversity at microsatellite loci, although patterns of diversity at the DQ Beta exon-2-like locus are similar in both taxa. Differences in conformity to Hardy-Weinberg expectations evident between microsatellite and presumptive MHC markers suggest that different forces are acting to maintain diversity at these loci. Differential selection on microsatellite and MHC loci provides an intriguing potential explanation for this pattern. This possibility will be evaluated as part of future studies of genetic variation in tuco-tucos.

Acknowledgments

We thank Jim Patton, Kristie Mather, Diogo Meyer, Shannon McWeeney, and Luis Fernando Garcia for their careful reading of previous drafts of the manuscript. Funding for this project was provided by an NSF Doctoral Dissertation Improvement Grant and an NIH Graduate Group in Genetics Training Grant, both awarded to TMH.

References

Alexander RD (1974) The evolution of social behaviour. Annu Rev Ecol Syst 5:325-383

Amos W, Sawcer SJ, Feakes RW, Rubinsztein DC (1996) Microsatellites show mutational bias and heterozygote instability. Nat Genet 13:390-391

Briles WE, Briles RW, Stone HA (1987) Resistance to a malignant lymphoma in chickens is mapped to a subregion of the major histocompatibilty complex. Science 219:977-979

Brown CR, Brown MB (1986) Ectoparasites as a cost of coloniality in Cliff Swallows (*Hirundo pyrrhonota*). Ecology 67:97-107

Campos-Lima P-O, Givioli, Zhang Q-J, Wallace LE, Dolcetti R, Rowe M, Rickin AB, Masucci MG (1993) HLA-A11 Epitope loss isolates of Epstein-Barr virus from a highly A11+ population. Science 260:98-100

Doherty PC, Zinkernagel RM (1975) Enhanced immunological surveillance in mice heterozygous at the H-2 gene complex. Nature 256:50-52

Edwards S (1995) Contrasting histories of avian and mammalian Mhc genes revealed by class II B sequences from songbirds. Proc Natl Acad Sci USA 92:12200-12204

Ellegren H, Hartman G, Johansson M, Andersson L (1993) Major histocompatibility complex monomorphism and low levels of DNA fingerprinting variability in a reintroduced and rapidly expanding population of beavers. Proc Natl Acad Sci USA 90:8150-8153

Figueroa F, Gutknecht J, Tichy H, Klein J (1990) Class II Mhc genes in rodent evolution. Immunol Rev 113:27-46

Graser R, O'uHuigin C, Vincek V, Meyer A, Klein J (1996) Trans-species polymorphism of Class II Mhc loci in danio fishes. Immunogenetics 44:36-48

Hamilton WD, Zuk M (1982) Heritable true fitness and bright birds: A role for parasites? Science 218:384-387

Hedrick PW, Thomson G (1983) Evidence for balancing selection at HLA. Genetics 104:449-456

Hill AVS, Yates SNR, Allsopp CEM, Gupta S, Gilbert SC, Lalvani A, Aidoo M, Davenport M, Plebanski M (1994) Human leukocyte antigens and natural selection by malaria. Philos Trans R Soc Lond B Biol Sci 346:379-385

Hoogland JL, Sherman PW (1976) Advantages and disadvantages of bank swallow (*Riparia riparia*) coloniality. Ecol Monographs 46:33-58

Hughes AL, Hughes MK, Howell CY, Nei M (1994) Natural selection at the class II major histocompatibility complex loci of mammals. Philos Trans R Soc Lond B Biol Sci 346:359-366

Hutter CM, Schug MD, Aquadro CF (1998) Microsatellite variation in *Drosophila melanogaster* and *Drosophila simulans*: A reciprocal test of the ascertainment bias hypothesis. Mol Biol Evol 15:1620-1636

Jin L, Macuabas C, Hallmayer J, Kimura A, Mignot E (1996) Mutation rate varies among alleles at a microsatellite locus: Phylogenetic evidence. Proc Natl Acad Sci USA 93:15285-15288

Klein J, O'uHuigin C (1994) Mhc polymorphism and parasites. Philos Trans R Soc Lond B Biol Sci 346:351-358

Lacey EA (1999) Spatial and social systems of subterranean rodents. In: Lacey EA, Patton JL, Cameron GN (eds) Life Underground: The Biology of Subterranean Rodents. University of Chicago Press, Chicago, IL:in press

Lacey EA, Braude SH, Wieczorek JR (1997) Burrow sharing by colonial tuco-tucos (*Ctenomys sociabilis*). J Mammal 78:556-562

Lacey EA, Maldonado JE, Clabaugh JP, Matocq MD (1999) Interspecific variation in microsatellites isolated from tuco-tucos (Rodentia: Ctenomyidae). Mol Ecol:in press

Lessa EP (1992) Rapid surveying of DNA sequence variation in natural populations. Mol Biol Evol 9:323-330

Lessa EP, Cook JA (1998) Molecular phylogenetics of tuco-tucos. Mol Phyl Evol 9:88-99

Levinson G, Gutman G (1987) Slipped-strand mis-pairing: a major mechanism for DNA sequence evolution. Mol Biol Evol 6:198-212

May RM, Anderson RM (1979) Population biology of infectious diseases, pt. 2. Nature 280:455-461

Nadeau JH, Wakeland EK, Gotze D, Klein J (1981) The population genetics of the H-2 polymorphism in European and North African populations of the house mouse (*Mus musculus L.*) Genet Res 37:17-31

Nadeau JH, Britton-Davidian J, Bonhomme F, Thaler L (1988) H-2 polymorphisms are more uniformly distributed than allozyme polymorphisms in natural poplulations of house mice. Genetics 118:131-140

Nevo E (1979) Adaptive convergence and divergence of subterranean mammals. Annu Rev Ecol Syst 10:269-308

Nevo E, Beiles A (1991) Selection for class II Mhc heterozygosity by parasites in subterranean mole rats. Experientia 48:513-515

Patton JL (1990) Geomyid evolution: the historical, selective, and random basis for divergence patterns within and among species. In: Nevo E, Reig OA (eds) Evolution of Subterranean Mammals at the Organismal and Molecular Levels. Wiley-Liss, New York, pp 49-70

Patton JL, Yang SY (1977) Genetic variation in *Thomomys bottae* pocket gophers: macrogeographic patterns. Evolution 31:697-720

Pearson P (1985) Los tucos-tucos (Genero *Ctenomys*) de los parques nacionales Lanin y Nahuel Huapi, Argentina. Historia Natural 5:337-343

Penn DJ, Potts WK (1999) The evolution of mating preferences and major histocompatibility complex genes. Amer Naturalist 153:145-164

Pepin L, Amigues Y, Lepingle A, Berthier J-L, Bensaid A, Vaiman D (1995) Sequence conservation of microsatellites between *Bos taurus* (cattle), *Capra hircus* (goat), and related species. Examples of use in parentage testing and phylogeny analysis. Heredity 74:53-61

Potts WK, Wakeland EK (1990) Evolution of diversity at the major histocompatibility complex. Trends Ecol Evol 5:181-186

Potts WK, Manning CJ, Wakeland EK (1994) The role of infectious disease, inbreeding and mating preferences in maintaining MHC genetic diversity: an experimental test. Philos Trans R Soc Lond B Biol Sci 346:351-357

Primmer CR, Ellegren H, Saino N, Moller AP (1996) Microsatellites show mutational bias and heterozygote instability. Nat Genet 13:390-393

Reig OA, Busch C, Ortells MO, Contreras JR (1990) An overview of evolution, systematics, population biology, cytogenetics, molecular biology and speciation in *Ctenomys*. In: Nevo E, Reig OA (eds) Evolution of Subterranean Mammals at the Organismal and Molecular Levels. Wiley-Liss, New York, pp 71-96

Richardson CC (1965) Phosphorylation of nucleic acid by an enzyme from T4 bacteriophage-infected *Escherichia coli*. Proc Natl Acad Sci USA 54:158-165

Ruedas LA, Cook JA, Yates TL (1993) Conservative genome size and rapid chromosomal evolution in the South American tuco-tucos (Rodentia: Ctenomyidae). Genome 36:449-458

Scharf SJ, Long CM, Erlich HA (1988) Sequence analysis of the HLA-DRB and HLA DQB loci from three *Pemphigus vulgaris* patients. Hum Immunol 22:61-69

She JX, Boehme S, Wang TW, Bonhomme F, Wakeland EK (1990) The generation of Mhc class II gene polymorphism in the genus *Mus*. Biol J Linnean Soc 41:141-161

Slade R (1992) Limited Mhc polymorphism in the southern elephant seal: implications for Mhc evolution and marine mammal population biology. Proc Royal Soc Lond B Biol Sci 249:163-171

Smith MF, Patton JL (1999) Phylogenetic relationships and the radiation of sigmodontine rodents in South America: evidence from cytochrome b. J Mammal Evol 6:89-129

Steinberg EK, Patton JL (1999) Genetic structure and the geography of speciation in subterranean rodents: opportunities and constraints on evolutionary diversification. In: Lacey EA, Patton JL, Cameron GN (eds) Life Underground: The Biology of Subterranean Rodents. University of Chicago Press, Chicago, IL:in press

Thornton CA, Wymer JP, Simmons Z, McClain C, Moxley III RT (1997) Expansion of the myotonic dystrophy CTG repeat reduces expression of the flanking DMAHP gene. Nat Genet 16:407-409

Trowsdale J (1995) "Both man and bird and beast": Comparative organization of MHC genes. Immunogenetics 41:1-17

van Treuren R, Kuittinen H, Karkkainen K, Baena-Gonzalez E, Savolainen O (1997) Evolution of microsatellites in *Arabis petraea* and *Arabis lyrata*, outcrossing relatives of *Arabidopsis thaliana*. Mol Biol Evol 14:220-229

Wierdl M, Greene CN, Datta A, Jinks-Robertson S, Petes TD (1996) Destabilization of simple repetitive DNA sequences by transcription in yeast. Genetics 143:713-721

Woods CA (1993) Suborder Hystricognathi. In: Wilson DE, Reeder DM (eds) Mammal Species of the World: A Taxonomic and Geographic Reference. 2nd ed, Smithsonian Institution Press, Washington D.C., pp 771-806

Author index

Key word index